PROGRESS IN COLLOID & POLYMER SCIENCE

Editors: H.-G. Kilian (Ulm) and G. Lagaly (Kiel)

Volume 89 (1992)

Trends in Colloid and Interface Science VI

Guest Editors:
C. Helm, M. Lösche, and H. Möhwald (Mainz)

 Springer-Verlag Berlin Heidelberg GmbH

ISBN 978-3-662-16077-0   ISBN 978-3-7985-1680-9 (eBook)
DOI 10.1007/978-3-7985-1680-9

ISSN 0340-255 X

© Springer-Verlag Berlin Heidelberg 1992
Chemistry editor: Dr. Maria Magdalene Nabbe; English editor: James Willis; Production: Holger Frey.
Originally published by Dr. Dietrich Steinkopff Verlag GmbH & Co. KG, Darmstadt in 1992

Softcover reprint of the hardcover 1st edition 1992

Type-Setting: Graphische Texterfassung, Hans Vilhard, D-6126 Brombachtal

Preface

This special issue contains the major portion of lectures and posters presented at a conference that combined the 35th biannual meeting of the Deutsche Kolloidgesellschaft and the 5th annual meeting of the European Colloid and Interface Society. It was held at the Johannes Gutenberg-Universität Mainz, FRG, September, 25—28, 1991, and brought together about 300 participants from 17 countries.

Although there was an emphasis on partially ordered systems, the following wide range of topics was covered in the program:

Particle and Lamella Interaction in Fluid Environments
Colloidal Particles: Size and Mobility
Rheology and Stability
Colloidal Suspensions under Stress
Surface Properties and Adsorption
Monolayers at the Air/Water Interface
Molecular and Collective Dynamic Properties
Phase Transitions and Phase Diagrams

The highlight of the conference was the lecture of the 1991 Nobel laureate P. G. de Gennes, who was awarded the Wolfgang-Ostwald Prize of the Deutsche Kolloidgesellschaft (designated to honor a lifetime achievement).

We are grateful to the ECIS International Committee for assisting us in selecting a quarter of the 170 original submissions for oral presentations. This required tough decisions which were based predominantly, but not totally on scientific grounds. Consequently, poster presentations received much time and space during the conference and were allocated the same amount of space as oral contributions in this volume.

Combining a meeting of national and international societies requires a balance of potentially conflicting interests. The selection process inevitably disappointed some, yet we hope that most participants felt that the program was evenly balanced between national interests and scientific fields. (We also ensured that our group is only sparingly represented in contributions to this volume).

We at the Institute of Physical Chemistry of the University of Mainz are grateful to the members of our group, especially those who, instead of preparing presentations of their own work, helped in organizing and holding the conference. We were delighted by compliments concerning the conference organization as well as about its scientific level. We hope that this has contributed to progress in science, as well as to international collaboration in this field. Especially, it was interesting to meet and talk with people from former Communist countries.

We also acknowledge the financial support of our corporate sponsors (Bayer AG, BASF AG, Coulter Electronics GmbH, Henkel KGaA, Hoechst AG, Hoffmann-La Roche AG, Hüls AG, LAUDA Dr. R. Wobser GmbH & Co. KG, Malvern Instruments GbmH, Partikel-Analytik HMS Elektronik, Riegler & Kirstein Ultrathin Organic Film Technology), institutions (Johannes Gutenberg-Universität Mainz, and the Royal Society of Chemistry), the two sponsoring Societies, and the Deutsche Forschungsgemeinschaft.

On behalf of the local organizers,

C. A. Helm, M. Lösche, H. Möhwald

Contents

Progress in Colloid & Polymer Science   Progr Colloid Polym Sci 89:1—8 (1992)

Modeling of solvation interactions
in non-polar dispersions of colloidal particles
using the liquid state theory of adhesive hard sphere mixtures

M. H. G. M. Penders and A. Vrij

Van't Hoff Laboratorium voor Fysische en Colloïdchemie, University of Utrecht, The Netherlands

Abstract: Colloidal particles dispersed in a non-polar solvent are modeled by a binary mixture of large spheres in a "solvent" of small spheres using the liquid state model of adhesive hard sphere mixtures. The discrete nature of the solvent molecules is explicitly taken into account. Solvation forces can be described fairly well using both solvent-solvent and solvent-solute interactions. By increasing the solvent-solvent interaction, keeping the solvent-solute interaction constant, the effective attraction between the large colloidal particles increases. The isothermal osmotic compressibility goes to infinity when the adhesive strength between the solvent molecules becomes very high and phase separation may occur ("poor" solvation). By increasing the solute-solvent interaction, keeping the solvent-solvent interaction constant, the effective repulsion between the large particles increases ("good" solvation). — When the solvent density is small (near the "critical" value), however, solvent-solute interactions may ultimagely lead again to effective attractions between the large spheres, if the adhesive strength between the solvent and solute particles is large enough. This phenomenon may be interpreted in terms of "bridge formation".

Key words: Solvation interactions; adhesive hard spheres; structure factors; colloidal dispersions; binary mixtures

1. Introduction

The action of solvent molecules plays an important role in the building up of interaction forces in colloidal systems containing particles stabilized by a protective surface layer of chain molecules. If the particle core is composed of material with a refractive index comparable to that of the solvent, the van der Waals-London attraction forces between the cores are small [1]. The interaction forces between the colloidal particles are then dominated by the chain-chain and chain-solvent interactions of the opposing protective layers of the touching particle surfaces.

If there is a preferential solvation of chain (ends) by solvent molecules, repulsive forces will already be felt before the bare chain segments are actually in contact, because the removal of solvent molecules requires work ("good" solvation).

If, however, there is a preferential solvation, not of solvent molecules, but of chain ends of the surface layers of opposing particles, effective attraction forces between the particles result. In other words: work is gained when chain/solvent contacts are replaced by chain/chain plus solvent/solvent contacts. Such forces are, in fact, directly measurable using macroscopic mica surfaces covered with chain molecules [2, 3].

In this paper, we elucidate such complex interactions with the help of a fluid state model in which the suspension is modeled as a two-component mixture of large and small spheres having certain adhesive interactions.

An important characteristic in the statistical description is that the solvent is not merely described as a continuous background, but possesses a discrete nature. The discrete nature of the solvent molecules was already considered by Chan et al.

et al. [6, 7], Jamnik et al. [8] and by our group [9, 10].

Henderson [5] gave an explicit expression for the solvent contribution to the force between colloidal particles using a hard sphere model. Hansen et al. [6, 7] used a binary hard sphere mixture and concluded that due to the presence of small solvent molecules the effective repulsion between the large particles decreases. According to them, phase separation may occur when the size ratio of the two species is less than 0.2, and the partial packing fractions of the two species are comparable.

Chan et al. [4], and Jamnik et al. [8] investigated the effects of solute/solvent size ratio on the solvent mediated potential of mean force between solutes at infinite dilution using a binary mixture of hard solutes dispersed in a "solvent" of hard spheres with surface adhesion (sticky spheres). Jamnik et al. found [8] that, at critical conditions of the model fluid, the solvation force between the macroparticles tends to vanish in parallel with the increasing compressibility of the fluid.

In this paper, we present results of model calculations of the osmotic isothermal compressibility and structure factors concerning binary mixtures of large particles dispersed in a "solvent" of small particles, taking both solvent-solute and solvent-solvent interactions into account. For the model calculations the Baxter theory of adhesive hard spheres [11, 12] is used; the calculations are based on the PY-approximation in the Ornstein-Zernike equation as worked out by Barboy [13, 14] and Perram and Smith [15, 16]. The direct interaction between the large spheres is neglected.

In section II the relevant equations for a binary mixture are given. In section III the influence of stickiness (solvent-solvent and solvent-solute interactions) on the osmotic compressibility and structure factor for a binary mixture of large spheres in a "solvent" of small spheres will be discussed, making use of the theory presented in section II. The results of model calculations concerning the solvation interactions in colloidal dispersions will be shown.

2. Theoretical background

We consider a binary mixture of large spheres with pair diameter d_{22} and volume fraction ϕ_2 in a "solvent" of small spheres with pair diameter d_{11} and volume fraction ϕ_1. The two-sphere mixture is considered to be in osmotic equilibrium with a one-

sphere system of "pure solvent" using a semipermeable membrane that is permeable to the small spheres, but impermeable to the large spheres. The volume fraction of the "pure solvent" is denoted with ϕ_1^0. In the case of $d_{11}/d_{22} \ll 1$ the following approximate relation between ϕ_1 and ϕ_1^0 can be given:

$$\phi_1 \approx \phi_1^0 (1 - \phi_2) . \tag{1}$$

In Eq. (1) ϕ_1^0 can be regarded as the volume fraction of the small spheres in the volume available once the volume occupied by the large spheres has been subtracted. The relation in Eq. (1) is exact for a binary hard sphere mixture with $d_{11}/d_{22} \to 0$ [6, 9].

In this paper the partial structure factor $S_{22}(K)$ (denoted as $S(K)$ in the rest of the paper) will be used to describe the interactions between the large particles (denoted as 2) dispersed in a "solvent" of small particles (denoted as 1). Here, K is the magnitude of the wave vector.

The structure factor at zero wave vector $S(K = 0)$ can be related to the osmotic isothermal compressibility $(\partial \rho_2 / \partial \Pi)_{\mu_1}$ (see, e.g., [9, 17]):

$$S(K = 0) = kT(\partial \rho_2 / \partial \Pi)_{\mu_1} , \tag{2}$$

with Π being the osmotic pressure of the mixture in osmotic equilibrium with the small spheres, ρ_2 the number density of particles 2, μ_1 the chemical potential of the "solvent", k the Boltzmann constant, and T the absolute temperature. For dilute dispersions $S(K = 0)$ can be written as:

$$S(K = 0) = 1/(1 + 2B_2\rho_2) \approx 1 - 2B_2\rho_2 , \tag{3}$$

where B_2 is the osmotic second virial coefficient.

To describe short-range interactions between particles the polydisperse adhesive hard sphere model can be used [13—16]. The adhesive interaction potential between particles i and j, $V_{ij}(r)$, which is a limiting form of the square well potential, is defined by

$$\frac{V_{ij}(r)}{kT} = \begin{cases} \infty & r \leqslant \sigma_{ij} \\ \ln\left[12\tau_{ij}(d_{ij} - \sigma_{ij})/d_{ij}\right] & \sigma_{ij} < r < d_{ij} \\ 0 & r \geqslant d_{ij} \end{cases} \tag{4}$$

with $d_{ij} - \sigma_{ij}$ allowed to be infinitesimally small, and τ_{ij} being the stickiness parameter for particles i and j, which goes to infinity in the case of hard sphere interactions.

The structure factor $S(K)$ is a function of the set of parameters $\{\lambda_{ij}\}$. In the case of hard sphere interactions λ_{ij} goes to zero. The parameter set $\}\lambda_{ij}\}$ and τ_{ij} are connected by the following relation:

$$\frac{\pi}{6} \sum_\gamma \rho_{\gamma\gamma} d_{\gamma\gamma}^2 (\lambda_{i\gamma}\lambda_{j\gamma} - 6[\lambda_{i\gamma} + \lambda_{j\gamma}] + 18)$$

$$- 12(1 - \xi_3) \left[\frac{\tau_{ij}\lambda_{ij}}{d_{ij}} - \frac{d_{ij}}{d_{ii}d_{jj}} \right] = 0 . \qquad (5)$$

In most cases the coefficients $\{\lambda_{ij}\}$ for given τ_{ij} can only be found numerically.

In the next section, we discuss the case of a binary mixture where both the solvent-solvent and solvent-solute interactions are taken into account ($\tau_{11} < \infty$, $\tau_{12} < \infty$). The direct attractions between the large particles are taken to be zero for simplicity ($\tau_{22} = \infty$). This implies that the emphasis on the interaction between the large (colloidal) particles is on the *indirect* interactions through the solvent. A more detailed description about the method of calculation of the structure factor can be found elsewhere [10].

3. Solvation interactions in binary mixtures

For a binary system of large colloidal spheres (denoted as 2) in a "solvent" of small spheres (denoted as 1) with solvent-solvent and solute-solvent interactions ($\tau_{11} < \infty$, $\tau_{12} < \infty$ and $\tau_{22} = \infty$) the following expressions for $S(K = 0)$, which is proportional to the osmotic isothermal compressibility, and the second virial coefficient B_2 can be given in the case of $d_{11}/d_{22} \to 0$:

$$\xi_3 = \phi_1 + \phi_2 , \qquad (7b)$$

and for B_2

$$B_2/V_{HS} = 1 + \frac{3}{1 - \phi_1^0}$$

$$\times \left\{ 1 - \phi_1^0 \left(\frac{3 + \lambda_{12}(\lambda_{12} - 6)}{1 + 2\phi_1^0 - \lambda_{11}\phi_1^0} \right) \right\} , \qquad (8)$$

with

$$V_{HS} = (\pi/6)d_{22}^3 . \qquad (9)$$

In the limit of τ_{11} and $\tau_{12} \to \infty$ (or $\lambda_{11} = \lambda_{12} = 0$) Eqs. (6) and (8) give the expression for a binary hard sphere mixture. In the case of $\lambda_{12} = 0$ the solvent-solute interactions disappear. By decreasing τ_{11} (or increasing the stickiness between the solvent molecules) "poor" solvation is simulated. In the limit of $\lambda_{11} = 0$ the solvent-solvent interactions disappear, and by decreasing τ_{12} "good" solvation is simulated. All three cases are described in more detail elsewhere [10].

To obtain both coefficients λ_{11} and λ_{12} Eq. (5) has to be solved. In this case a quartic equation in λ_{12} is found. Some results of model calculations for a binary mixture of large colloidal spheres in a "solvent" of small spheres with a pair diameter ratio $d_{22}/d_{11} = 160$ are given in Figs. 1—8.

In Figs. 1 and 2 $\ln[1/S(K = 0)]$ is plotted against the volume fraction of the large spheres ϕ (which equals ϕ_2) for a binary mixture of large hard spheres in a "solvent" of small adhesive hard spheres with $\phi_1^0 = 0.40$ at different τ_{11}-values ($\tau_{12} = \tau_{22} = \infty$). The volume fraction of the solvent ϕ_1 is taken to be equal to $\phi_1^0[1 - \phi_2]$ (see also Eq. (1)). The curve denoted with I, presented in Figs. 1 and 2, represents the $\ln[1/S(K = 0)]$ vs. ϕ plot for monodisperse hard spheres in a solvent, which is regarded as a continuous background. This curve

$$S(K = 0) = \frac{(1 - \phi_2)^2(1 + 2\xi_3 - 3\phi_2 - \lambda_{11}\phi_1)^2}{\left(1 + 2\xi_3 - \dfrac{\lambda_{11}\phi_1(1 + 2\phi_2 - \phi_1)}{1 - \xi_3} - \dfrac{\lambda_{12}(\lambda_{12} - 6)\phi_1\phi_2}{1 - \xi_3} \right)^2} , \qquad (6)$$

with

$$\phi_i = (\pi/6)\rho_i d_{ii}^3 \qquad (7a)$$

appears to be nearly linear over a large volume fraction range.

At $\tau_{11} = \infty$, i.e., in the case of a binary hard sphere mixture, the initial slope of the $\ln[1/S(K =$

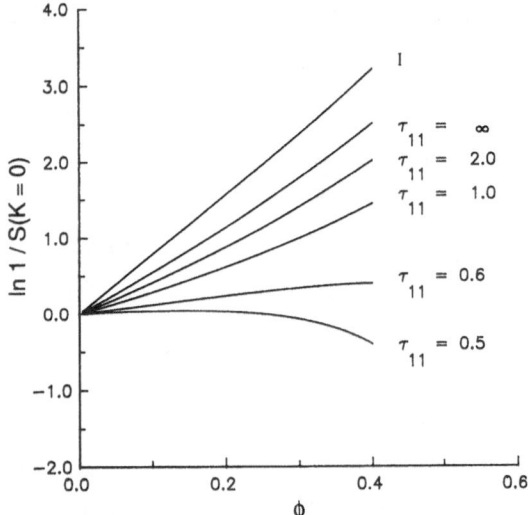

Fig. 1. $\ln 1/S(K = 0)$ vs. ϕ, the volume fraction of the large spheres for a binary mixture of large (colloidal) spheres and small (solvent) molecules (diameter ratio 160/1), with the volume fraction of "pure" solvent $\phi_1^0 = 0.40$ as a function of τ_{11}, the stickiness between the small particles. The curve denoted with I represents the $\ln 1/S(K = 0)$ vs. ϕ plot for monodisperse hard spheres dispersed in a solvent which is regarded as a continuous background

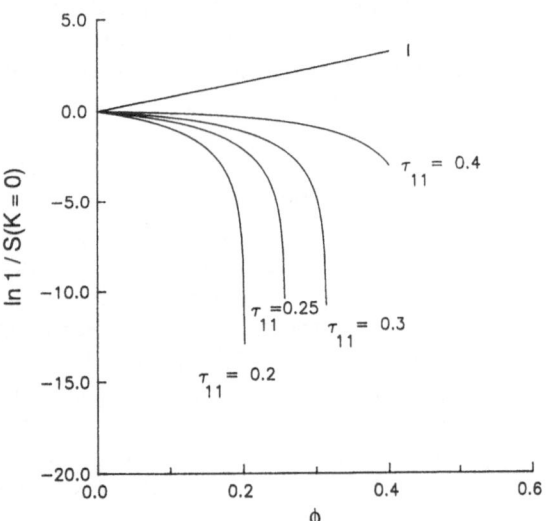

Fig. 2. $\ln 1/S(K = 0)$ vs. ϕ, the volume fraction of the large spheres, at low τ_{11}-values, the stickiness between the small particles, for the same binary mixture as used for Fig. 1. (I-curve defined in the legend of Fig. 1.)

0)] (which is equal to $\ln [(\partial [\Pi/kT]/\partial \rho_2)_{\mu_1}]$) vs. ϕ plot is smaller than the initial slope for the curve denoted with I (see Fig. 1). The initial slope is proportional to the second virial coefficient B_2. From

this slope it follows that the effective repulsion between the large particles decreases in the case of a binary hard sphere mixture due to the presence of small hard spheres, compared to the case of monodisperse hard spheres were solvent is considered as a continuous background. This is in agreement with what has been found before by Hansen [6, 7] and by our group [9, 10].

From Figs. 1 and 2 it follows that the initial slope of the $\ln [1/S(K = 0)]$ vs. ϕ plot decreases as τ_{11} decreases (or the stickiness between solvent molecules increases). For $\tau_{11} < 0.445$ there is an overall effective attraction ($B_2 < 0$) between the particles [10]. This is in good agreement with the plots in Figs. 1 and 2, from which it can be seen that the slope of $\ln [1/S(K = 0)]$ vs. ϕ plot becomes negative in the τ_{11}-range $0.4 < \tau_{11} < 0.5$. Below $\tau_{11} = 0.445$ there is an overall effective attraction (i.e., $B_2 < 0$) between the particles. From Fig. 2 it can be seen that at very low τ_{11}-values ($\tau_{11} \leq 0.30$) there exists a volume fraction of the large particles ϕ at which the osmotic isothermal compressibility [or $S(K = 0)$] goes to infinity and phase separation may occur.

In Fig. 3 the structure factor $S(K)$ is plotted as a function of τ_{11} at $\phi = 0.20$ for the same binary mixture as used for the plots in Figs. 1 and 2. At low K-values the increase of $S(K)$ is the largest at decreasing τ_{11} (or increasing stickiness between the solvent molecules). The first maximum of the $S(K)$ vs. K plot shifts to higher values at decreasing τ_{11}, indicating that the larger particles, on average, are closer together at higher stickiness between the solvent molecules. Similar $S(K)$ vs. K plots, obtained from SANS experiments, have been found by Duits et al. [18] for stearyl silica dispersions in benzene as a function of temperature.

From this model ($\tau_{11} < \infty$; $\tau_{12} = \tau_{22} = \infty$) it follows that, by increasing the stickiness between the solvent molecules, the larger particles are pushed closer toward each other, which eventually may lead to phase separation. One can say that the solvent molecules try to avoid contact with the surface of the colloid particle and prefer contacts among themselves (dewetting). Thus, this model can be used to simulate a binary mixture of large colloidal particles dispersed in a "poor" solvent.

In Fig. 4 $\ln [1/S(K = 0)]$ is plotted against ϕ at different τ_{12}-values ($\tau_{11} = \tau_{22} = \infty$) for a binary mixture of large hard spheres in a "solvent" of small hard spheres with $\phi_1^0 = 0.4$. From this figure it can be seen that the initial slope of the $\ln [1/S(K = 0)]$

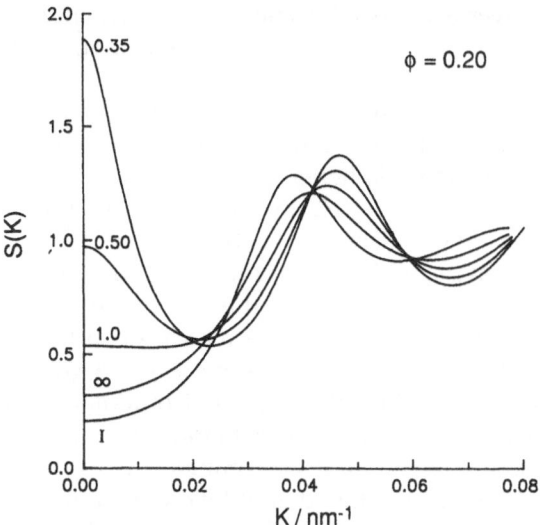

Fig. 3. Structure factor $S(K)$ vs. the wave vector K as a function τ_{11}, the stickiness between the small particles, at $\phi = 0.20$ for the same binary mixture as in Fig. 1. The values presented at the $S(K)$-curves correspond to the stickiness parameter τ_{11}. (I-curve defined in the legend of Fig. 1.)

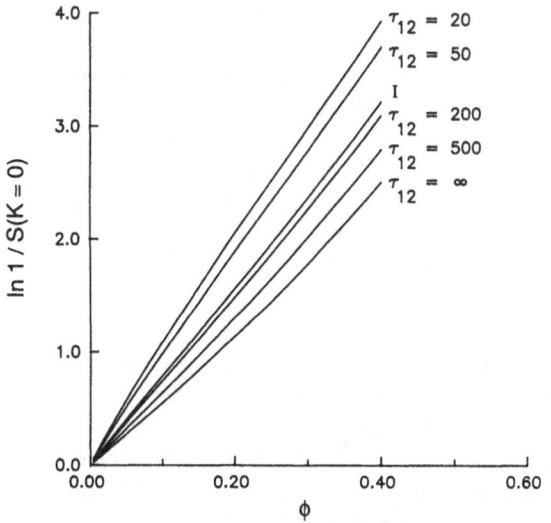

Fig. 4. $\ln 1/S(K = 0)$ vs. ϕ, the volume fraction of the large spheres, as a function of τ_{12}, the stickiness between the small (solvent) and large (solute) particles, at the same diameter ratio as used for Fig. 1 with $\phi_1^0 = 0.40$. (I-curve defined in the legend of Fig. 1.)

vs. ϕ plot increases when the stickiness between the solute and solvent particles also increases (or τ_{12} decreases). So the effective repulsion between the large particles increases.

At $\tau_{12} = 152$, B_2/V_{HS} gives the value of 4 (see [10]), the value found for monodisperse hard spheres where solvent is regarded as a continuous background. This is in good agreement with the plots in Fig. 4 where it can be seen that the curve of monodisperse hard spheres (denoted as I) lies between the curves with τ_{12} values 50 and 200.

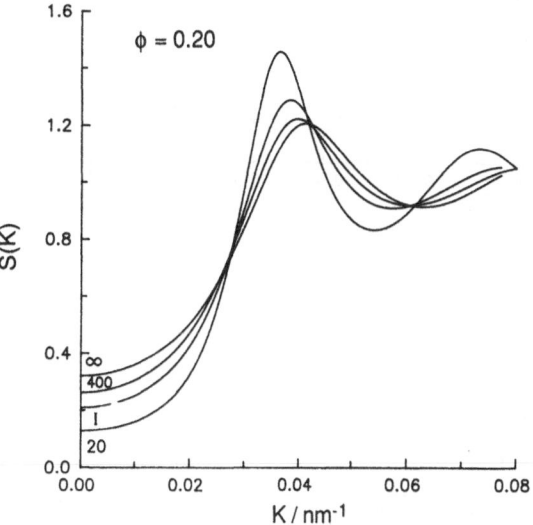

Fig. 5. Structure factor $S(K)$ vs. the wave vector K as a function of τ_{12}, the stickiness between small solvent molecules and large solute colloidal spheres, at $\phi = 0.2$ for the same binary mixture as in Fig. 4. The values presented at the $S(K)$ curves correspond to the stickiness parameter τ_{12}. (I-curve defined in the legend of Fig. 1.)

In Fig. 5 the $S(K)$ is plotted as a function of τ_{12} at $\phi = 0.2$ for the same binary mixture as used for the plots in Fig. 4. At small K-values the decrease of $S(K)$ vs. K plot shifts to lower values at decreasing τ_{12} indicating that larger particles on average are further away from each other at higher stickiness between solute and solvent particles.

From this model ($\tau_{12} < \infty$ and $\tau_{11} = \tau_{22} = \infty$) the conlusion can be drawn that by increasing the stickiness between solute and solvent particles the effective repulsion of the large particles increases. Therefore, this model can be used to simulate a binary mixture of large colloidal particles dispersed in a "good" solvent, where the large particles prefer to be surrounded by small solvent molecules instead of other large particles, resulting in an increase of the effective repulsion between the large particles ("good" solvation or wetting).

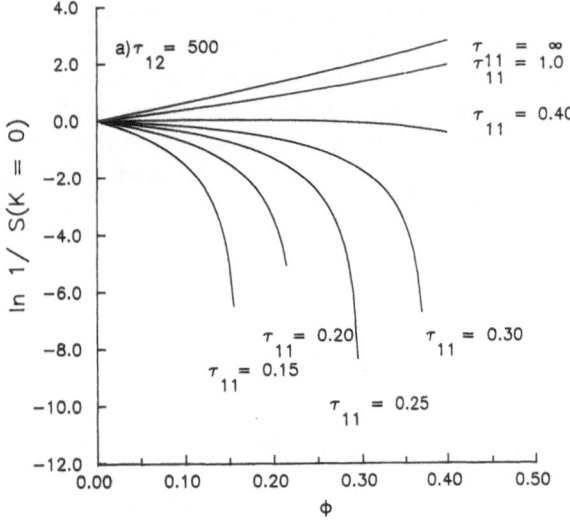

a

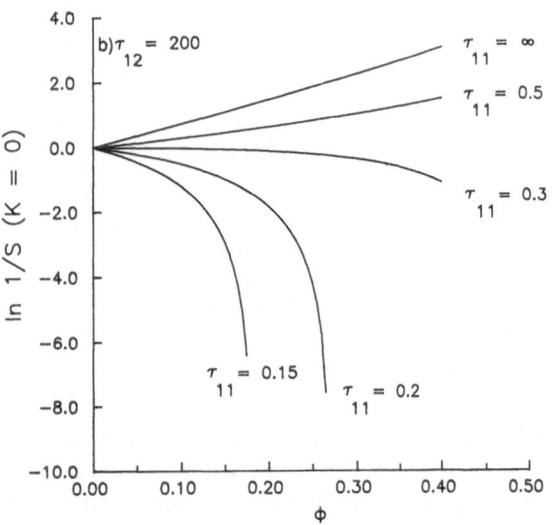

b

Fig. 6. $\ln 1/S(K = 0)$ vs. ϕ, the volume fraction of the large spheres, as a function of τ_{11}, the stickiness between the solvent molecules, at constant τ_{12} at the same diameter ratio as in Fig. 1, with $\phi_1^0 = 0.40$. a) $\tau_{12} = 500$; b) $\tau_{12} = 200$

For the calculated isothermal osmotic compressibility and structure factors presented in Figs. 1—5, only one type of interaction (solvent-solvent or solvent-solute) has been used; other types have been neglected. For the results of the model calculations given in Figs. 6—8, both the solvent-solvent

and solvent-solute interactions are taken into account ($\tau_{11} < \infty$ and $\tau_{12} < \infty$), which is more realistic.

In Figs. 6a—b $\ln[1/S(K = 0)]$ is plotted vs. ϕ at constant τ_{12}, but at different τ_{11}-values for a binary mixture of large spheres in a "solvent" of small spheres with $\phi_1^0 = 0.4$. By decreasing τ_{11} (or increasing the stickiness between the solvent molecules) the initial slope of the $\ln[1/S(K = 0)]$ vs. ϕ plot decreases and, thus, the effective repulsion between the large particles decreases. At sufficiently low τ_{11}-values ($\tau_{11} < 0.4$) there exists a volume fraction ϕ at which the osmotic isothermal compressibility (or $S(K = 0)$) goes to infinity and phase separation may occur. If we compare Figs. 1, 2 and 6a—6b with each other, then it follows that increasing the stickiness between the particles and solvent (decreasing τ_{12}) makes the solvent less poor, as expected on intuitive grounds.

In Fig. 7a—b and 8a, $\ln[1/S(K = 0)]$ vs. ϕ is plotted at constant τ_{11}, but different τ_{12}-values for a binary mixture of large spheres in a "solvent" of small spheres with $\phi_1^0 = 0.4$. By decreasing τ_{12} (or increasing the stickiness between solute and solvent molecules) the initial slope of the $\ln[1/S(K = 0)]$ vs. ϕ plot increases. Therefore, the effective repulsion between the large particles increases. At low τ_{12}-values, the particles prefer to be surrounded by solvent molecules, although the stickiness between the solvent molecules opposes this tendency when τ_{11} becomes smaller. If we compare Figs. 4, 7a—b, and 8a with each other, then it follows that increasing the stickiness between the solvent molecules (decreasing τ_{11}) makes the solvent less good, as expected.

To choose a reasonable value for τ_{11} the pressure concerning the adhesive solvent molecules has to be low. In that case the solvent mimics a condensed liquid with a low (atmospheric) pressure. It turns out that for $\tau_{11} = 0.12$, which is quite close to the critical value of $[2 - \sqrt{2}]/6$ (which ≈ 0.0976), this condition is reasonably fulfilled.

From Fig. 8a it follows that τ_{12} must be smaller than 30 to prevent phase separation. At $\tau_{12} \approx 20$, the initial slope of the $\ln[1/S(K = 0)]$ vs. ϕ plot is zero, meaning that the second virial coefficient $B_2 = 0$. Then the repulsive interactions are compensated by the attractive forces. At $\tau_{12} \approx 10$, the initial slope of the $\ln]1/S(K 3 0)]$ vs. ϕ plot approaches the initial slope of the HBMIX-curve, the case of a bidisperse hard sphere mixture, where the discrete nature of the solvent has been taken into

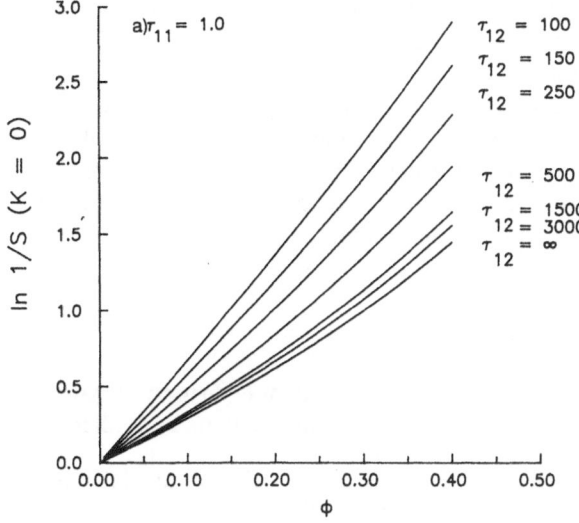

a

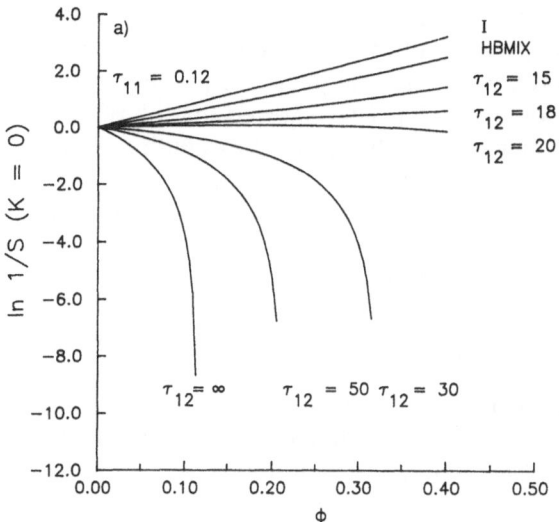

a

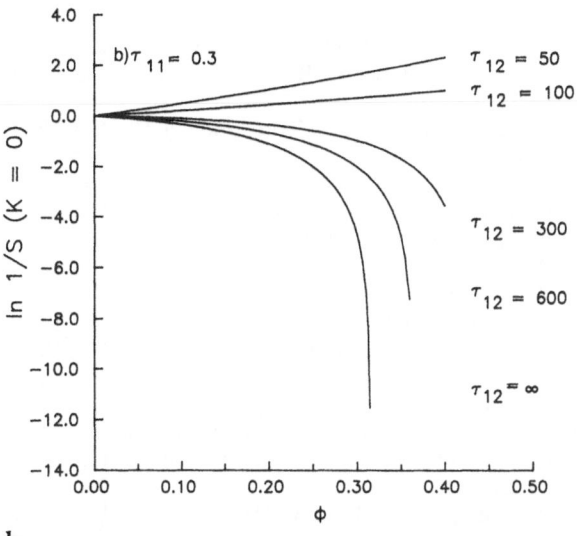

b

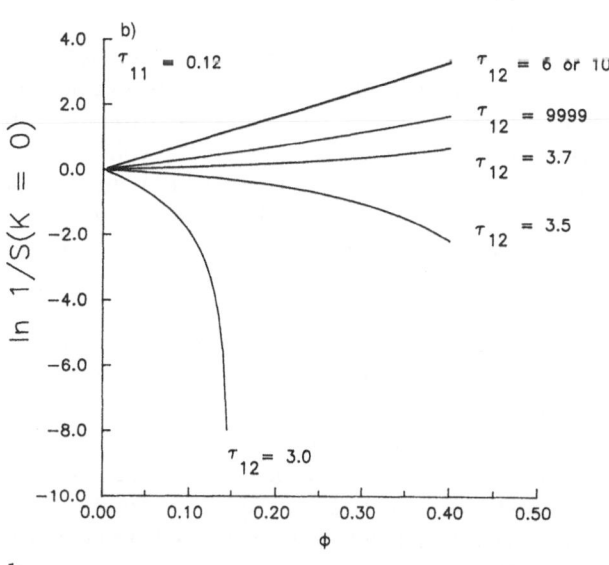

b

Fig. 7. $\ln 1/S(K = 0)$ vs. ϕ, the volume fraction of the large spheres, as a function of τ_{12}, the stickiness between solvent and solute particles, at constant τ_{11} at the same diameter ratio as in Fig. 1, with $\phi_1^0 = 0.40$. a) $\tau_{11} = 1$; b) $\tau_{11} = 0.3$

Fig. 8. $\ln 1/S(K = 0)$ vs. ϕ, the volume fractionof the large spheres, as a function of τ_{12}, the stickiness between solvent and solute particles, at $\tau_{11} = 0.12$, which is near the critical value, at the same diameter ratio as in Fig. 1. a) $\phi_1^0 = 0.4$; b) $\phi_1^0 = 0.12$ (critical volume fraction)

account. At $\tau_{12} \approx 6$, B_2/V_{HS} gives the value of 4. Thus, at the values of $\tau_{11} = 0.12$ and $\tau_{12} = 6$ the system behaves as if the colloidal particles are simple hard spheres in a continuous background.

In Fig. 8b $\ln[1/S(K = 0)]$ vs. ϕ is plotted at different τ_{12}-value for a binary mixture of large spheres in a "critical" solvent of small spheres at

$\tau_{11} = 0.12$ and $\phi_1^0 = 0.12$, instead of $\phi_1^0 = 0.40$ used before. From this figure it follows that at decreasing τ_{12} (or increasing stickiness between solvent and solute particles) the effective repulsion between the large particles increases (for $\tau_{12} > 8$). If $\tau_{12} < 8$, however, the effective repulsion decreases again and for $\tau_{12} < 3.7$ the effective attractive interactions between the large particles prevail, which eventually may lead to phase separation. This behavior is in accordance with the phenomenon described by Barboy et al. [14]. They showed that phase separation may occur in binary mixtures with a strong attraction between unlike molecules if $d_{22}/d_{11} > 1.32$. If $\lambda_{12} > 6$ (which is the case when $\tau_{12} < 3.7$) the structure factor at zero wave vector, $S(K = 0)$, goes to high values since $\lambda_{12}(\lambda_{12} - 6)\phi_1\phi_2/(1 - \xi_3)$ becomes positive.

From Fig. 8b it follows that at $\tau_{12} = 9999$ (which is near the case of zero stickiness between the large and small particles) there is a small effective repulsion between the large particles through solvation forces. Jamnik et al. [18] found in the case of $\tau_{12} = \infty$ that the solvation force between the large particles tends to vanish in the vicinity of the critical point, where virtually infinite compressibility of the solvent is expected.

The (indirect) strong attraction between the large colloidal particles dispersed in a "critical" solvent ($\tau_{11} = 0.12$ and $\phi_1^0 = 0.12$) caused by large (direct) attractive interactions between the solvent and colloidal particles ($\tau_{12} < 3.7$) may be explained in terms of "bridge formation". If ϕ_1, the volume fraction of the small particles is small, the surfaces of the large particles will not be fully covered with small particles. In that case, two not fully covered large particles may decrease their potential energy by sharing contacts of small particles, i.e., by forming bridges. The phenomenon of bridge formation has also been observed by Dickinson [19], who used Monte Carlo computer simulations for modeling a colloidal system of dispersed particles to which polymer molecules are added. If the decrease in potential energy is large enough the osmotic compressibility may increase, which may lead to phase separation.

4. Conclusions

With the adhesive hard sphere model for a binary mixture of large colloidal particles dispersed in a "solvent" of small spheres, solvation forces can be described by taking both solvent-solvent and solvent-solute interactions into account. By increasing the solvent-solvent interaction and keeping the solvent-solute interaction constant, the effective attraction between the large colloidal particles increases, which eventually may lead to phase separation ("poor" solvation). By increasing the solute-solvent interaction and keeping the solvent-solvent interaction constant, the effective repulsion between the large particles increases ("good" solvation). In the case of a critical solvent, attractions between the large particles dominate if the solvent-solute interaction becomes very high. This can lead to a phase separation, which may be interpreted in terms of "bridge formation."

References

1. Caljé AA, Agterof WGM, Vrij A (1977) In: Mittal KL (ed) Micellization, Solubilization, and Microemulsions. Plenum Press, Vol. 2
2. Herder CE, Ninham BW, Christenson HK (1989) J Chem Phys 90:5801—5805
3. Gee ML, Israelachvili JN (1990) J Chem Soc Faraday Trans 86:4049—4058
4. Chan DYC, Mitchell DJ, Ninham BW, Pailthorpe BA (1978) Chem Phys Lett 56:533—536
5. Henderson D (1988) J Colloid Interface Sci 121:486—490
6. Biben T, Hansen JP (1990) Europhys Lett 12:347—352
7. Biben T, Hansen JP (1991) Phys Rev Lett 66:2215—2218
8. Jamnik A, Bratko D, Henderson DJ (1991) J Chem Phys 94:8210—8215
9. Vrij A, Jansen JW, Dhont JKG, Pathmamanoharan C, Kops-Werkhoven MM, Fijnaut HM (1983) Faraday Discuss Chem Soc 76:19—35
10. Penders MHGM, Vrij A (1991) Physica A 173:532—547
11. Baxter RJ (1968) J Chem Phys 49:2770—2774
12. Watts RO, Henderson D, Baxter RJ (1971) Adv Chem Phys 21:421—430
13. Barboy B (1975) Chem Phys 11:357—371
14. Barboy B, Tenne R (1979) Chem Phys 38:369—387
15. Perram JW, Smith ER (1975) Chem Phys Lett 35:138—140
16. Perram JW, Smith ER (1976) Chem Phys Lett 39:328—332
17. Vrij A (1982) J Colloid and Interface Sci 90:110—116
18. Duits MHG, May RP, Vrij A, de Kruif CG (1991) Langmuir 7:62
19. Dickinson EJ (1989) J Colloid and Interface Sci 132:274—278

Authors' address:

Prof. A. Vrij
Van't Hoff Laboratory
University of Utrecht
Padualaan 8
3584 CH Utrecht, The Netherlands

Progress in Colloid & Polymer Science   Progr Colloid Polym Sci 89:9—19 (1992)

Monte Carlo investigations of the order-disorder transition in colloidal-sphere suspensions

F. S. Jardali and L. V. Woodcock

Department of Chemical Engineering, University of Bradford, U.K.

Abstract: The effects of varying the interparticle colloidal repulsion on the osmotic equilibrium freezing parameters of monodisperse sterically-stabilised colloidal-sphere dispersions have been investigated by Monte Carlo (NVT) simulations. — The model effective pair potential takes the form of a generalised soft sphere with a hard core, i.e.

$$\varphi_{ij} = \varepsilon \left[\frac{r_{ij}}{\sigma} - 1 \right]^{-n}.$$

Order-disorder thermodynamic transition parameters, for $n = 9$, 12 and 15 and $\varepsilon/kT = 10^{-9}$, 10^{-12}, and 10^{-15} respectively, have been determined by a development of the free volume method based upon a rigorous treatment of the cell theory of liquids. — Results are presented for the volume change (dilatancy) and excess entropy of melting compared to the limiting cases of the (infinitely) hard-sphere model and (zero-core) soft-sphere models. Comparisons are made with the experimental investigations of monodisperse latexes by Pusey and van Megen.

Key words: freezing parameters; colloidal-sphere dispersions; Monte Carlo; simulations; effective pair potential

1. Introduction

Colloidal dispersions of nearly monodisperse sub-micron particles are being increasingly used to investigate fundamental mechanisms of freezing and melting. Compared to atomic fluids, the time and distance scales are more amenable to laboratory investigations which allow growth rate, nucleation and two-phase equilibrium phenomena to be seen in great detail [1]. The early research in this area has been carried out on sterically stabilized colloidal suspensions which are "nearly hard spheres" [2]. These idealized suspensions are prepared with minimum surface charge and with a refractive index to match the suspending medium, thereby minimising attractive dispersion forces. In these circumstances the suspensions are the closest that presently be achieved to the widely studied and fundamentally important theoretical hard-sphere model, of which the freezing properties are now well-known from Monte Carlo [3] and molecular dynamics [4] simulations.

In order to develop effective pair-potentials to accurately model the slightly-soft steric interparticle repulsions, it is desirable to have knowledge of the effect of the softness and range of this surface repulsion on physical properties. In particular, on the osmotic pressure of the equilibrium freezing transition and the changes in volume and entropy on fusion.

Although the freezing properties of the inverse power soft-sphere model pair potential, defined by

$$\varphi_{ss} = \varepsilon \left[\frac{r_{ij}}{\sigma} \right]^{-n} \tag{1}$$

are known over a wide range of the softness exponent n [5,6], this model is inappropriate for sterically stabilized suspensions, but it may represent electrostatic stabilization repulsion in the limit of small relative core size.

A generalisation of the hard- and soft-sphere models has been suggested [7], which we here refer to as colloidal soft spheres corresponding to the pair potential

$$\varphi_{cs} = \varepsilon \left[\frac{r_{ij}}{\sigma} - 1 \right]^{-n}. \qquad (2)$$

In both (1) and (2), ε is a characteristic energy (ε/kT determines an "effective diameter"), rij is the distance between particle centres, σ is the hard-core diameter and n is the repulsive exponent.

Equation of state and rheological properties have been reported previously for this model for a range of n and ε values from MD and NEMD computations [8]. The objectives of the present computations are to determine the equilibrium freezing properties of the model colloidal-sphere suspension, defined by Eq. (2), compared to the hard- and soft sphere limiting models, and to compare the results with the experimental freezing properties of the monodisperse PMMA in CS_2/decalin as reported by Pusey and van Megen [2]. The comparison with experiment is, in effect, a test of the adequacy of the colloidal sphere model with a given softness parameter as an effective pair potential for these systems.

The method we use to obtain the relative free energies of the two phases utilises the free volumes determined from acceptance ratios in the Monte Carlo method as originally suggested 20 years ago by Gosling and Singer [9]. We show here also that although the method, in general, may be inexact, it becomes rigorous for the hard-sphere model, and is very accurate for soft-spheres which can be treated as having an effective hard-sphere diameter slightly larger than the core diameter. The procedure is expecially useful, therefore, for determining the freezing transition parameters of models for sterically stabilized colloidal suspensions.

2. The colloidal soft-sphere pair potential:

The realisation that the bulk thermodynamic properties are all primarily determined by the repulsive potential, which occurs on the close approach of particles, that the repulsion emmanates from the surface of the colloidal particle rather than its centre, and that the steepness of the repulsion by a number of factors (steric or electrostatic effects) has led to the introduction of a repulsive-sphere hybrid model, described as colloidal-soft sphere model and defined in Eq. (2).

Typically, the repulsion is steep at close particle approach, and the repulsion fades completely at about 1.35 σ. The short-range solvation forces at

high packing fractions and the screened Coulombic forces may all be accounted for in this potential. The limited number of variables associated with this potential, and the possibility of representing polydisperse systems, may be considered as additional advantages. The hard-sphere limit is approached in the limit of $\sigma \to \infty$/or $\varepsilon \to 0$ and the soft-sphere scaling is recovered when $\sigma \to$ or $r_{ij} \gg \sigma$. Unlike conventional soft spheres Eq. (1), two state variables, e.g. V and T, are generally required to specify the thermodynamic state of the colloidal soft-sphere system.

The value of m is given by $10^m = kT/\varepsilon$, and it determines the correspondence between system potential and kinetic energy. Values of m and n equal to 12, as used in this simulation, are such that the effective pair-potential broadly resembles that of a sterically stabilised, fairly short-ranged repulsion "skin" originating at the particle surface. Variation in the value of m and n is possible so as to produce a longer-ranged repulsion more appropriate to electrostatically stabilised systems (Fig. 1).

The effects of varying the exponent n at constant kT/ε upon the shape of the colloidal potential can be observed clearly from Fig. 2. An increase in the value of n would inevitably increase the repulsion

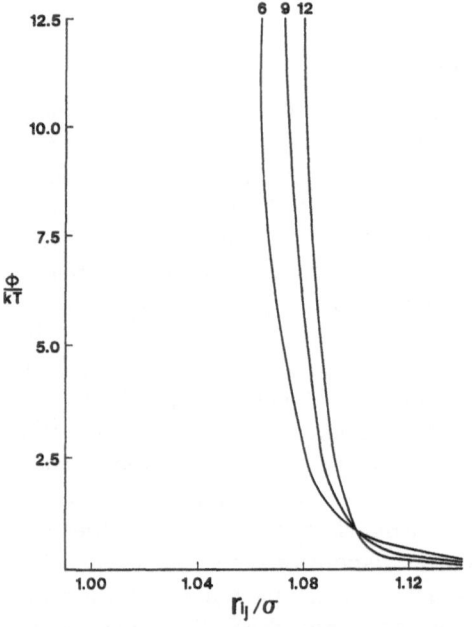

Fig. 1. The colloidal-sphere pair potential versus particle separation at different values of n and kT/ε: 6 and 10^6, 9 and 10^9, 12 and 10^{12} respectively

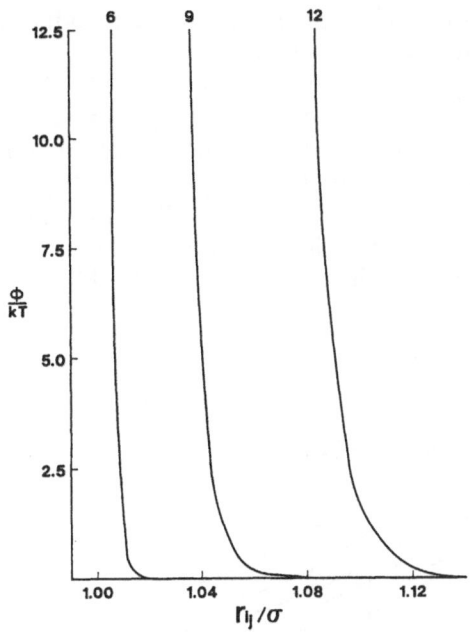

Fig. 2. The colloidal-sphere pair potential versus particle separation at constant $kT/\varepsilon = 10^{12}$, and at different values of n (6, 9, and 12)

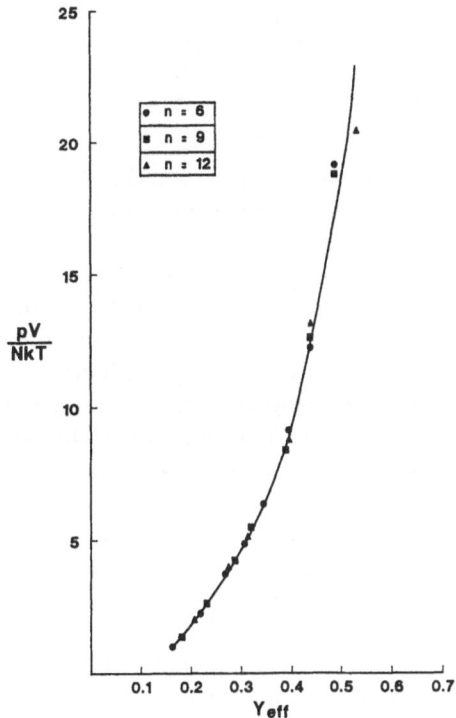

Fig. 3. The Carnahan-Starling (C-S) equation where the PV/NkT data of the amorphous states at $n = 6$, 9, and 12 are superimposed on the C-S curve at equivalent effective packing fractions

when compared at equivalent particle separation. Note that all the potentials in Fig. 1 cross the value of $\varphi = kT$ at $r/\sigma = 1.1$. This fact has been utilised in corresponding states scaling of this potential [8]. Remeniscent of the soft-sphere model [5, 6], a scaling feature of the colloidal-soft sphere potential allows the equations of state of constant volume systems to be predicted at different values of the repulsive exponent n, along any isotherm, from a knowledge of the equation of state along a single isotherm (kT/ε), at a particular value of n [8].

The colloidal-sphere potential may alternatively be written in the form of:

$$\varphi_{cs}(r_{ij}) = 10^{-m}kT\left(\frac{r_{ij}}{\sigma} - 1\right)^{-n};\qquad (3)$$

then at $r/\sigma = 1.1$, and at $m = n$ values, the value of the interparticle potentials is $\varphi = kT$. So r_{ij}, the separation of particle centres, can be replaced by σ_{eff}, the effective hard-sphere diameter, therefore, $\sigma_{eff} = 1.1\,\sigma$ for $m = n$ cases. For a general case, at $\varphi = kT$, the effective hard-sphere diameter is $\sigma_{eff} = (1 + 10^{-m/n})\,\sigma$, where m may or may not be equal to n. The osmotic pressure data, for example,

exhibited by the colloidal sphere system at ($n = 6$, 9, and 12) superimposes when compared at equivalent effective-hard sphere packing fraction [8]. These data coincide accurately with the data obtained for the hard-sphere equation of state for the amorphous system, [4] e.g. of Carnahan and Starling at equivalent packing fractions (Fig. 3). Thus, from a knowledge of n and m values, and the effective hard-sphere diameter, the osmotic pressure values can be predicted from hard-sphere equations of state.

3. Monte Carlo method

The ensemble averages investigated in the present work are the total pair potential energy $\langle\Phi(r_1 \ldots r_N)\rangle$, and the pressure values obtained in the form of compressibility factors PV/NkT, at various values of reduced densities, $\rho^* = N\sigma^3/V$, and calculated from:

$$\left\langle \frac{PV}{NkT} \right\rangle = 1 - \frac{\left\langle \sum\limits_{}^{N-1} \sum\limits_{}^{N} r\,d\varphi_{ij}/dr \right\rangle}{3NkT}. \qquad (4)$$

The derivation of the viral expression (Eq. (4)) can be seen in any textbook of statistical mechanics, e.g. by McQuarrie [10]; for a colloidal suspension at equilibrium p is the osmotic pressure.

Hopkins and Woodcock [8] have published some preliminary thermodynamic equation of state data for colloidal spheres at $kT/\varepsilon = 10^{12}$ by applying particulate dynamics simulations. They investigated the equilibrium properties of amorphous states for three values of n, 6, 9 and 12, and they, subsequently, parametrised the equation of state. The present work extends the investigation into the equation of state for the (NVT) amorphous and crystalline systems, and covers also the order/disorder transition regions at $kT/\varepsilon = 10^{12}$ and $kT/\varepsilon = 10^{18}$.

The present Monte Carlo computations assume isothermal conditions (kT/ε is constant) and are carried out on the colloidal soft-sphere pair potential (Eq. (2)), $\varepsilon_{cs} = 10^{-m} E_0$ is the repulsive exponent, where E_0 is the system unit kinetic energy which has been considered in the present case to be kT; $T^* = (kT/\varepsilon)$ is the reduced dimensionless temperature.

Simulations at values of the exponent n, ranging from 6 to 15, and at reduced temperatures, 10^{12} and 10^{18}, were carried out and the average pressures (PV/NkT) obtained at different dimensionless reduced number densities ($N\sigma^3/V$), where N is the number of particles in the simulation cell and V is its volume. These dimensionless properties, PV/NkT and $N\sigma^3/V$, are important since they make scaling properties easy to visualise. It is obvious from the data that the osmotic pressures (PV/NkT) increase with increasing packing fraction, and the increase in the values are much more pronounced in the amorphous states, especially at high densities.

This means that the system at such high pressures are not stable and that the system must have a transition to some ordered and more stable state, as does the hard-sphere model.

To construct the equation of state for the amorphous branch, eight state points were chosen. The values of 0.4, 0.5, 0.55, 0.60, 0.65, 0.70, 0.75, and 0.80 densities were considered for studying the variation of the osmotic pressure. Detailed tables of data giving the compressibility equilibrium Monte Carlo

data of the colloidal spheres at different reduced densities are given elsewhere [12].

The choice of the arbitrary displacement to give a minimal acceptance ratio (N_{accept}/N_{total}) was experimentally determined. Plots of the acceptance ratios versus the value of the displacement (Fig. 4) show that the value of DMAX = 0.07 was good enough. The conventional Monte Carlo (MC) method of Metropolis et al. was used [9, 10].

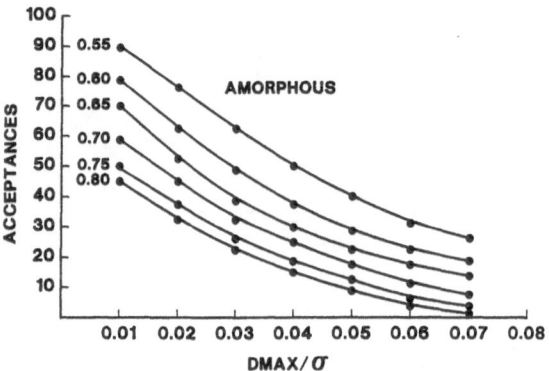

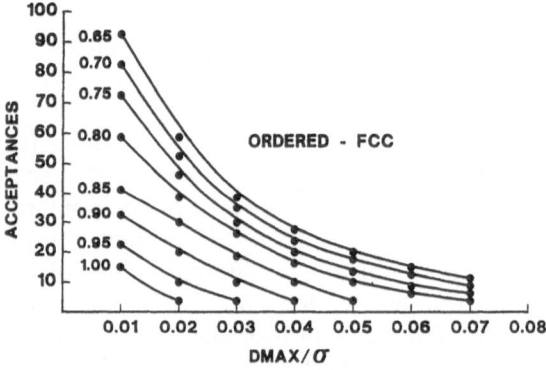

Fig. 4. Variation of the number of accepted moves at different values of DMAX; the top graph represents the amorphous systems, and the bottom one represents the ordered systems; the numbers correspond to the reduced densities

The number of moves were of the order of a few thousands for smaller density systems, going up to a few hundred thousand for the ordered systems at higher densities. Fewer moves were required by the systems at $n = 9$ and $n = 6$, owing to the fact that softer repulsion gives faster equilibration. The same conditions and reasons hold for the systems at $kT/\varepsilon = 10^{18}$ at n 3 15, 12 and 9; the equilibration time was the minimum for the $n = 9$. The initial lattices were simple cubic for the amorphous states and face-centered cubic (FCC) for the crystalline states for the initial runs on the systems $n = 6$ and $kT/\varepsilon = 10^{12}$. After a few thousand moves the system equilibrates and is ready for further sampling moves. Subaverages were sampled every thousand moves until the statistical uncertainties of the subaverages become very small. The accumulated acceptance ratios are related to the free energies and hence the freezing transition parameters as follows.

From statistical mechanics, the Helmholtz free energy of any fluid or crystal state, which is required to locate the crystal-fluid equilibrium melting transition, can be expressed in terms of the partition function:

$$A^\dagger = -kT \ln Q_N^\dagger , \qquad (5)$$

where $\dagger$ means that the free energy is defined in terms of the configurational partition function:

$$Q_N^\dagger = \frac{1}{N!} \int_v \dots \int_v \exp(-\Phi(r_1 \dots r_N)/kT] dr_1 \dots dr_N. \qquad (6)$$

Where $\int_v$ mean integration over all the volume of the system. Now based on the fact that the potential energy decays rapidly at very short distances (Figs. 1, 2) the configurational partition function can be expected to behave like that of hard-sphere models at high densities, where the configurational partition function can be expressed as a product of average individual particle configurational integrals, $q_1 \dots$. Therefore,

$$Q_N^\dagger = \langle q_1 \rangle^N , \qquad (7)$$

then,

$$A^\dagger = -kT \ln \langle q_1 \rangle^N . \qquad (8)$$

Since the configurational entropy:

$$S^\dagger = (\langle \Phi \rangle - A^+)/T ; \qquad (9)$$

therefore,

$$S^\dagger = Nk \ln \bar{v}_f , \qquad (10)$$

where v_f is the free volume, and in it can be expressed in terms of the acceptance ratio: $v_f =$ number of moves accepted/number of total moves (N_{accept}/N_{total});

$$v_f = v_D \frac{N \text{ accept}}{N \text{ total}} = \int_0^\infty f(v) dr , \qquad (11)$$

and $f(v)$ = acceptance ratio function, $dr = d$ (DMAX/σ), and $v_D = (2 \times \text{DMAX})^3$.

DMAX is chosen to be so large that there would be no chance of acceptances outside the displacement sphere/cube. It appears from Fig. 4 that the value of DMAX = 0.07, which is used in the present work, is equal to approximately one-fifth of the particle diameter. It is in contrast to the smaller value of 1/20 of the particle diameter which was suggested by Powles [11]. The choice of this value made the system spend more time to reach steady-state conditions, but it also made the systems give averages of lower standard deviations with low acceptances ratios.

The determination proceeds as follows: for a two-phase system to be in equilibrium, each phase must be at the same pressure for mechanical equilibrium, the same temperature for thermal equilibrium, and the same chemical potential, $(\delta G/\delta n_i) = 0$, for chemical equilibrium. This means that the molar Gibbs free energies of the two phases are equal at equilibrium,

$$G_f(P, T) = G_c(P, T) , \qquad (12)$$

and the Helmholtz free energies of the two phases are equal at equilibrium, for constant volume systems,

$$A_f(V, T) = A_c(V, T) . \qquad (13)$$

In other wors, $\Delta A = 0$ and $\Delta G = 0$ at equilibrium, for isobaric and isochoric conditions respectively.

It can be seen from Fig. 5 that $dG \ (= \int V dP)$ varies horizontally in a P versus V plot and $dA \ (\int P dV)$ varies vertically in the same plot. Now at P_1 and P_2, which are the two pressure values (one on the crystal branch and the other on the fluid branch) that correspond to the points of intersection with the branches at $\Delta A = 0$, the following relation holds,

$$\frac{1}{2} [V_0 - V_1][P_0 - P_1] = \frac{1}{2} [V_2 - V_0][P_2 - P_0] , \qquad (14)$$

where P_0 is the value of the pressure at $\Delta G = 0$, V_1 and V_2 are the points of intersection with the crystalline and fluid branches which represent the values of the volumes at melting and freezing

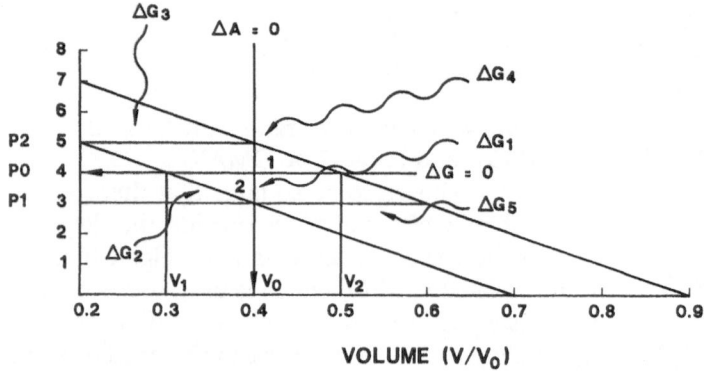

Fig. 5. A representative figure showing the locations of P_1, and P_2, the pressure values at the intersection of the $\Delta A = 0$, vertical line, with the crystal and the fluid branches respectively, and the volumes V_1 and V_2 at the intersection of the $\Delta G = 0$, horizontal line, with the crystal and the fluid branches respectively. P_0 is the pressure at $\Delta G = 0$, and V_0 is the volume at $\Delta A = 0$

respectively. So knowing P_1 and P_2, V_1 and V_2, and V_0, the value of P_0 can be determined; this is the value of the pressure where the two phases coexist at the same temperature and chemical potential.

Figure 5 may be considered a simple model representation of linear and parallel equations of state. In this case, however, the location of P_0, the constant pressure value at the transition, can be obtained simply by:

$$P_0 = \frac{1}{2}(P_1 + P_2),\qquad(15)$$

where P_1 and P_2 are obtained from the intersection of the $\Delta A = 0$, vertical line with the crystal and fluid linear branches respectively.

In case of having linear, but not parallel crystalline and fluid branches, an expression for P_0 can be obtained by rearranging the above equation (Eq. (15)), and so

$$P_0 = (P_1(V_0 - V_1) + P_2(V_2 - V_0))/(V_2 - V_1),(16)$$

which reduces to Eq. (15) in the case of parallel and linear crystalline and fluid branches. In fact, this equation contains three unknowns, but one should bear in mind that for any horizontal (ΔG) line (Fig. 5), the three unknowns, i.e. P_0, V_1 and V_2, are read simultaneously. Therefore, the best value of P_0 can be obtained by trial and error.

In real cases, however, to find a simple expression similar to Eq. (14) is difficult because the crystalline and fluid branches are always curves, and obviously not parallel. One way to determine P_0 numerically, therefore, is by finding expressionss of the equation of state of the fluid and crystal branches, then find the differences of the Gibbs free energies, in terms of the pressures, for the thermo-

dynamic cycles starting at (V_1, P_2) and (V_2, P_1) (Fig. 5), and hence, locate ΔG_0, which corresponds to P_0, by simple calculation. Alternatively, the "equal area rule", which is described below, can be used to locate P_0 with limited uncertainty where graphical integration is used rather than numerical calculation.

From the thermodynamic relationship

$$\frac{\Delta A}{NkT} = \frac{\Delta U}{NkT} - \frac{T\Delta S}{NkT},\qquad(17)$$

where $\Delta U (= \Delta\Phi)$ is the total potential energy, since the translational (kinetic) energies, $KE = 3/2\ NkT$, are equal in both phases.

i) From statistical thermodynamics using rigorous treatment of the cell theory of liquids due to Kirkwood (9, 10), the difference in entropy is related to the fractional number of acceptances:

$$\frac{\Delta S}{Nk} = Ln\ \frac{V_{\text{free}}^{\text{(Fluid)}}kT}{V_{\text{free}}^{\text{(crystal)}}kT}.\qquad(18)$$

The free volume at any state is related to the number of acceptances ratio and to the volume which is defined by the range of the Monte Carlo displacement, provided that DMAX is so large that there would be no chance of acceptances outside the displacement cube (or sphere)

$$V_{\text{free}} = (2\ \text{DMAX})^3\ \frac{N_{\text{Accept}}}{N_{\text{total}}},\qquad(19)$$

where D_{max}, is the maximum displacement, $\dfrac{N_{\text{accept}}}{N_{\text{total}}}$ is the acceptance ratio. Hence the difference in entropy can be seen as,

$$\frac{\Delta S}{Nk} = Ln\ \frac{N_{\text{Accept}}^{\text{(Fluid)}}kT}{N_{\text{Accept}}^{\text{(crystal)}}kT}.\qquad(20)$$

ii) The term $\Delta U = (U_{\text{fluid}} - U_{\text{crystal}})$ is the total internal energy difference. This thermodynamic average is equal to the difference between the average total potential energies of fluid and crystal respectively at equilibrium, since the kinetic energies are equal in both phases, these values are also averaged and recorded throughout the simulation.

iii) Knowing the values of the energy difference and the difference in entropy, the value of the difference between the two phases of the Helmholtz free energy is known at any one volume that correspond to two pressure values, which are the intersection of the vertical line (Fig. 5) with the fluid and crystalline branches. The next step is to plot the difference in Helmholtz free energy ($\Delta A/NkT$) versus the volume (the reduced volume, $V/n\sigma^3$, in the present case), and hence, the value of the reduced volume (V_0) at the minimum free energy would be read off (Fig. 6), which corresponds to equilibrium, i.e. at $\Delta A = 0$.

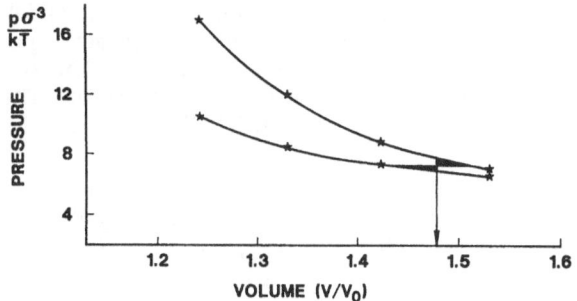

Fig. 7. Equation of state data illustrating matching of the areas to find the locations of the transition densities for the colloidal-sphere system $n = 12$, $T^* = 10^{12}$

fluid and the crystal branches respectively. Of these three sides, only the horizontal line, $\Delta G = 0$, has to be located exactly. The method of location is made by matching the areas of triangles 1 and 2, by shifting the horizontal line, upwards or downwards, to an optimum position. Matching the areas, however, involves some uncertainties, e.g. <2—3%, in measuring the areas, but the degree of uncertainty is much better than assuming linearity at the transition.

4. Results and Discussions

A summary of the coexisting volumes and densities as calculated by the present methods for the colloidal soft-sphere systems is given in Table 1. The following is an example of locating P_0 by the methods outlined above; from Table 1 and for $n = 12$ and $kT/\varepsilon = 10^{12}$ potential,

$$V_{\text{melting}}^{(\text{or } V_1)} = 1.415, \quad \text{and}$$

$$V_{\text{freezing}}^{(\text{or } V_2)} = 1.513,$$

and from Fig. 7 which corresponds to the reduced pressure values for the same potential,

$$P\sigma^3/kT_{\text{fluid}}^{(\text{or } P_2)} = 7.539, \quad \text{and}$$

$$P\sigma^3/kT_{\text{crystal}}^{(\text{or } P_1)} = 6.307.$$

Therefore, P_0 can be estimated by the following approximate methods

1) By assuming linear and parallel fluid and crystalline branches, i.e. from Eq. (3.23)

$$P_0 = \frac{1}{2}(P_1 + P_2) = 6.9235.$$

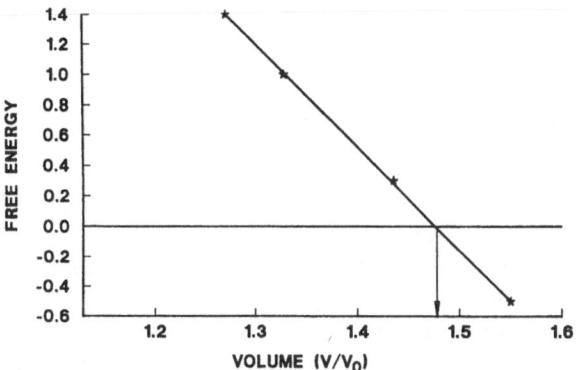

Fig. 6. Determination of V_0 that corresponds to the reduced volume at $\Delta A = 0$

iv) Now the location of the vertical (Helmholtz) free energy line is given by the positions of the values of P_2 and P_1 (Fig. 5), but still the values of V_1 and V_2 must also be known to locate the transition presusre P_0. The solution is to solve for that graphically by drawing a horizontal line, which represents the tie line at $\Delta G = 0$, and which crosses the vertical line at $\Delta A = 0$ (see Fig. 7).

From Fig. 5, one can see that the horizontal line, labelled as $\Delta G = 0$, crosses the vertical line, $\Delta A = 0$, to make two triangles, labelled as 1 and 2 on the diagram. The third sides of these triangles are, the

Table 1. Disorder/order transition data of colloidal-soft-sphere systems; $T^* = (kT/\varepsilon) = n$ is the repulsive exponent in the definition of the colloidal-soft potential; and m and f refer to melting and freezing respectively. The value of the bracketed $P\sigma^3/kT$ refers to the pressure at the transition

T^*	n	V^* melting $(1/\rho m^*)$	V^* freezing $(1/\rho f^*)$	$\Delta V/V_{melt}$ % $(P\sigma^3/kT)$
10^{12}	6	1.205 ($\rho = 0.8298$)	1.2750 ($\rho = 0.7843$)	5.8091 (6.33)
10^{12}	9	1.300 ($\rho = 0.7692$)	1.3816 ($\rho = 0.7238$)	6.2769 (6.50)
10^{12}	12	1.415 ($\rho = 0.7067$)	1.5130 ($\rho = 0.6609$)	6.9258 (7.05)
10^{18}	9	1.1975 ($\rho = 0.8351$)	1.2750 ($\rho = 0.7843$)	6.4718 (5.85)
10^{18}	12	1.925 ($\rho = 0.8386$)	1.2780 ($\rho = 0.7825$)	7.1698 (7.27)
10^{18}	15	1.2525 ($\rho = 0.7984$)	1.3500 ($\rho = 0.7407$)	7.7844 (8.10)

2) By assuming linear but not necessarily parallel, i.e. from Eq. (3.24); at $V_1 = 1.43$ and $V_2 = 1.547$

$$P_0 = (P_1(V_0 - V_1) + P_2(V_2 - V_0))/(V_2 - V_1)$$

$$= 7.12 .$$

3) By assuming the crystalline and the fluid branches to be linear at the transition; the expressions of their equations of state are:

$$P = -12.953 \, V + 24.869 \text{ for the crystal branch}$$

and

$$P = -12.538 \, V + 26.020 \text{ for the fluid branch,}$$

where P is the reduced pressure, i.e. P represents $P\sigma^3/kT$. Therefore, from Fig. 5

$$\Delta G_3 = \Delta G_1 + \Delta G_2$$

and

$$\Delta G_5 = \Delta G_1 + \Delta G_4 ,$$

where

$$\Delta G_1 = \Delta A(= 0) + \Delta PV_0$$

$$\Delta G_2 = \int_{P_2}^{P_1} V(P) dP \text{ (crystal)}$$

and

$$\Delta G_4 = \int_{P_2}^{P_1} V(p) dP \text{ (fluid) ,}$$

where $V^{(crystal)}_{(P)}$ is the equation of state of the crystal branch, and $V^{(fluid)}_{(P)}$ is the equation of state of the fluid branch. Solving for ΔG_2 and ΔG_4 by substituting the equation of state into these equations gives (in units of NkT)

$$\Delta G_2 = 2.3737 ,$$

and

$$\Delta G_4 = -2.4962 .$$

and

$$\Delta G_3 = (P_2 - P_1)V_0 + \Delta G_2 = 4.1887 ,$$

and

$$\Delta G_5 = (P_1 - P_2)V_0 + \Delta G_4 = -4.3122 .$$

Therefore, by joining (V_1, P_2) and (V_2, P_1) the slope of this line can be expressed as:

$$\frac{\Delta G_3 - \Delta G_5}{P_2 - P_1} = \frac{\Delta G_3 - \Delta G_0(= 0)}{P_2 - P_0} ,$$

4) Finally, the determination of the reduced pressure by graphical integration gives the value of: $P_0 = 7.05$. This final answer is probably the most accurate one.

The transition was expected to be near the $P\sigma^3/kT$, reduced pressure, region which was located by Hoover and Ree [3] for the hard spheres. Reduced densities equal to 0.65, 0.75, and 0.80 were considered for both fluid and crystalline systems. The data which are required for such calculations are given in [12] and Fig. 8 shows the locations of the transition regions. Figure 9 shows the PV/NkT data of $kT/\varepsilon = 10^{18}$ system versus the reduced densities. The $PVNkT$ data have been converted to (independent of volume) $P\sigma^3/kT$ reduced data in order to see the variation in the volume along a horizontal tie-line. Figure 10 shows a summary of the ratios of the change in the volume ($\Delta V/V_m\%$) across the transition at the two extremes of the coexistence regions. Plotting these ratios versus the exponent n, Fig. 10, and comparing it with the values stated by Hoover and co-workers [5, 6] for the soft-sphere potential in three dimensions, and

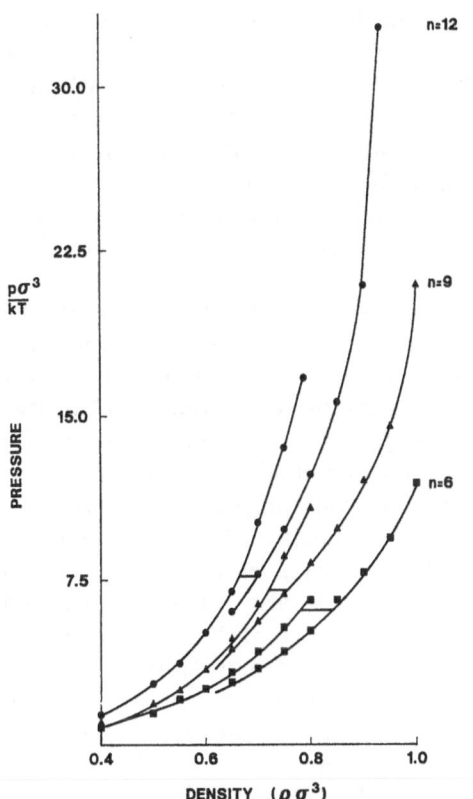

Fig. 8. Equation-of-state reduced pressure data ($P\sigma^3/kT$) for 3-dimensional monodisperse colloidal-sphere systems at $n = 6$, 9, and 12 at $T^* = 10^{12}$. The disorder/order transition regions are represented by straight lines joining the two states at equal Gibbs and Helmholtz free energy

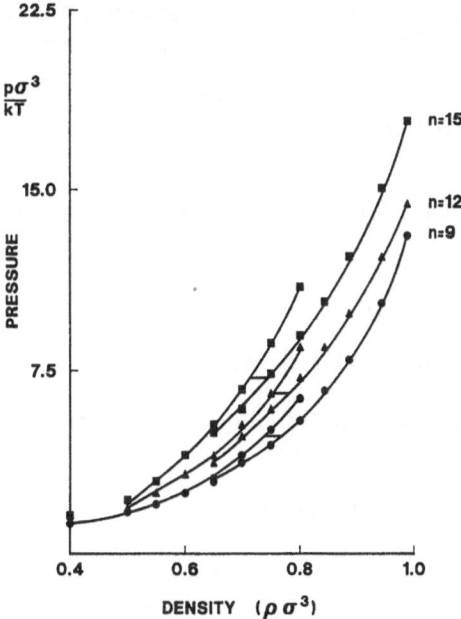

Fig. 9. Equation-of-state reduced pressure data ($P\omega^3/kT$) for three-dimensional colloidal-sphere systems at $n = 9$, 12 and 15 at $T^* = 10^{18}$. The disorder/order transition regions are represented by straight lines joining the two states at equal Gibbs and Helmholtz free energy

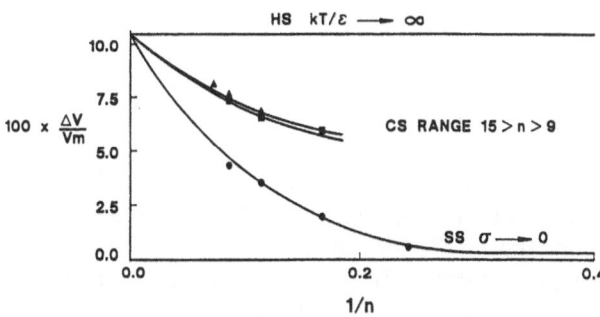

Fig. 10. Summary of the $\Delta V/V_{\text{melting}}$(%) values for colloidal soft-spheres at different values of the exponent n; where the top and the bottom lines represents the comparative behaviour of the hard- and soft-sphere models respectively

the value of 10.32%, which was predicted by Hoover and Ree [3] for the hard-sphere model $n = \infty$, it appears that the behaviour of the colloidal-soft spheres lies between these two extremes. The behaviour of the $kT/\varepsilon = 10^{18}$ model, as expected, is closer to the hard-sphere behaviour than the $kT/\varepsilon = 10^{12}$ model.

Pusey and van Megen [2] investigated the transition regions of ~305 nm PMMA dispersions by suspending the particles in equivalent refractive index mixture. After some time, the ordering of dense suspension altered the Bragg diffraction patterns which differ from disordered states, thus indicating a transition between the ordered and disordered states. By this method they were able to predict the $\Delta V/V_m$ ratio to be equal to 8.28% which can now be compared against the present colloidal-sphere results, i.e.,

$$\Delta V/V_m = \frac{1/0.407 - 1/0.4407}{1/0.4407} = 8.28\% \,.$$

At $\Delta V/V_m = 8.28\%$, the corresponding value of the exponent n, obtained from Fig. 10 is equal to $\cong 24.61$ at $kT/\varepsilon = 10^{12}$. Using the hardest CS potential, however, we obtain the best fit

$n \cong 20.0$ at $kT/\varepsilon = 10^{18}$ when expressed in terms of the effective colloidal soft-sphere potential, e.g.,

$$\varphi_{cs}(r_{ij}) = 10^{-18} kT \left(\frac{r_{ij}}{\sigma} - 1 \right)^{-20.0} .$$

The parameters at $kT/\varepsilon = 10^{18}$, give the effective diameter to be

$$\sigma_{eff} = (1 = 10^{m/n}) \sigma$$

$$\sigma_{eff} = (1 + 10^{-18/20}) 305 \cong 343 \text{ nm}$$

and the parameters are $kT/\varepsilon = 10^{12}$, however, give the effective diameter to be:

$$\sigma_{eff} = (1 + 10^{-12/24.61}) 305 \cong 404 \text{ nm}.$$

The value of 343 is closer to the value of 325.3 nm which they obtained by comparing the onset of their freezing with the hard-sphere value [3], i.e.

$$\sigma_{eff} = \sigma (y_f hs/y_f)^{1/3} ,$$

where y_f is the packing fraction at the onset of freezing of Pusey's system, i.e. y_f, and y_{fhs} is the Hoover and Ree [3] value of the packing fraction at the onset of freezing, i.e. $y_{fhs} = 0.494$. σ is the core diameter which is ~ 305 nm.

5. Conclusions

It may be concluded that the colloidal sphere potential can represent slightly soft interactions, and that a better description of the Pusey-van Megen system [2] could be obtained at higher reduced teperatures, i.e. slightly harder potentials, which are recommended to be used in the future.

The Metropolis-Rosenbluth-Teller Monte Carlo method has been employed to generate a Boltzmann-weighted distribution of condensed phase configurations to estimate the excess contributions to the excess osmotic thermodynamic properties. The equations of state have been determined at two reduced temperatures, 10^{12} and 10^{18}, for different values of n, e.g. $n = 6, 9, 12$ and 15. Data showing a comparison between results obtained for the compressibilities of the amorphous branches at $n = 12, 9$, and 6 at $kT/\varepsilon = 10^{12}$ for the colloidal-soft sphere model obtained via the Monte Carlo and the molecular dynamics methods and other details are given elsewhere [12].

The values seem to be in good agreement with each other, but the statistical uncertainties are higher for the present method than for the MD one.

Effective pair potentials may be determined from experimental osmotic pressure data on monodisperse systems by comparing with the colloidal spheres data by finding the best "n" value at any kT/ε using the parametrised equations of state given in [12].

Order/disorder transitions have been located by graphical integrations with limited uncertainty. From Fig. 10 which shows the location of the transition regions as functions of the values "n" at two different reduced temperatures, one can see clearly that the properties of colloidal soft-sphere models differ substantially from properties of both soft- and hard-sphere models. At higher reduced temperatures, the behaviour of the colloidal system becomes harder. It is also obvious that the behaviour of colloidal spheres is bounded by the extreme limits of the soft and hard spheres.

One other conclusion, which can be seen in Fig. 10, is that no matter what the value of reduced temperature, $\Delta V/V_m$ can never exceed the hard-sphere limit, i.e. 10.32%, the value which was predicted by Hoover and Ree [3] for the hard-sphere model.

The summarising diagram (Fig. 10) represents the alternative way of describing quantitatively the freezing properties, and hence determining the form of the effective pair potential that fits the experimental data, i.e. coexistence density data. Only by just knowing the values of the densities where the freezing and melting occur, the value of n can be obtained from the point of intersection with the isotherm.

Acknowledgements

We wish to thank the S.E.R.C. (UK) for the award of an Advanced Research Fellowship (to L. V. W.) and the Hairiri Foundation for the award of a Research Scholarship (to F. S. J.).

References

1. For a recent overview, see e.g. Oxtoby DW (1990) Nature 347:725–730
2. Pusey PN, van Megen W (1986) Nature 320:340–342
3. Hoover W, Ree F (1968) J Chem Phys 49:3609–3617
4. Alder BJ, Wainwright TE (1962) Phys Rev 127:359–361 see also Woodcock LV (1981) Ann NY Acad Sci 371:274–298

5. Hoover WG, Ross M, Johnson KW, Henderson D, Barker JA, Brown EC (1970) J Chem Phys 52:4931
6. Hoover WG, Ross M (1971) Contempary Physics 12:339
7. Woodcock LV (1989) Molecular Simulation 2:253—273
8. Hopkins AJ, Woodcock LV (1990) J Chem Soc Faraday Transactions 86:2593—2606
9. Gosling EM, Singer K (1973) J Chem Soc Faraday Trans II 69:1004
10. McQuarrie D (1976) Statistical Mechanics (Publishers Harper and Row)
11. Powles JG (1985) SERC CCP5 Quarterly Newsletter Daresbury Laboratory 18:24
12. Jardali FS (1991) Ph D thesis, University of Bradford

Authors' address for correspondence:

L. V. Woodcock
Dept. of Chemical Engineering
University of Bradford
Bradford, BD7 1DP, U.K.

Progress in Colloid & Polymer Science   Progr Colloid Polym Sci 89:20—24 (1992)

The Laplace equation and Winsor microemulsions

J. C. Eriksson, S. Ljunggren

Department of Physical Chemistry, Royal Institute of Technology, Stockholm, Sweden

Abstract: The genesis of microemulsion aggregates in the two- and three-phase regimes depends crucially not only on the magnitude of the oil/water interfacial tension γ but, in addition, on its curvature dependence at a constraint of constant chemical potentials. Hitherto, it has often been assumed that this rather obvious boundary condition would automatically imply that the hydrocarbon chain-packing density in the interfacial film, as well as the composition of the film would remain constant, irrespective of curvature. A more general approach is obtained, however, by employing a generalized Laplace equation for a surfactant-loaded interface of mean curvature H and Gaussian curvature K, and invoking the Helfrich bending free energy expression. On this basis, we derive what interfacial shapes are compatible with mechanical and physico-chemical equilibrium in the Winsor kind of microemulsions. Accordingly, when the spontaneous curvature H_0 is different from zero, spherical and cylindrical equilibrium shapes are possible. On the other hand, for $H_0 = 0$, we find that extended minimal surfaces with $H = 0$ can form if the Gaussian curvature bending constant $\bar{k}_c$ is positive.

Key words: Generalized Laplace equation; Winsor microemulsions; Helfrich curvature free energy expression; curvature-dependent interfacial tension; minimal surfaces

Introduction

The importance of the Laplace pressure at an interface Δp when applied to microemulsions stems from the fact that thermodynamic equilibrium requires Δp to be equal to zero when there is an excess oil or water phase present (Winsor cases) or when the sub-phases on both sides of the interface are identical (L_3 case). The basic reason for this is that two solutions with the same composition can have the same chemical potentials only when they are subject to the same pressure. Hence, for a large enough microemulsion droplet of equilibrium size and shape, formed in either the Winsor I (o/w + excess oil phase) or the Winsor II (w/o + excess water phase) regime, the pressure drop across the interface is zero. This condition may well be fulfilled, even when the interfacial tension γ is slightly larger than zero, if γ depends on curvature to some appreciable extent. In our previous treatments of (non-interacting) spherical and cylindrical micro-emulsion aggregates [1, 2] these fundamental notions were of great significance. In the present paper, we employ the most general form of the Laplace equation derived so far, following a surface-thermodynamic route, in order to develop a unifying framework for describing all kinds of Winsor microemulsion phases.

Expressions for the Laplace pressure

According to Melrose [3], Boruvka and Neumann [4] and, more recently, Markin et al. [5], surface thermodynamics yields the following equation for the Laplace pressure due to an interface of mean curvature H and Gaussian curvature K:

$$\Delta p = 2H\gamma - C_1(2H^2 - K) - 2C_2HK$$

$$- \frac{1}{2}\,\nabla_s^2 C_1 - K\nabla_s^* \cdot (\nabla_s C_2)\,, \qquad (1)$$

where ∇_s^2 is the Laplace-Beltrami operator in the surface and ∇_s^* is a special operator introduced by

Weatherburn (cf. [4]). The coefficients C_1 and C_2 are defined as follows (at constant temperature, T, and chemical potentials, μ_i):

$$C_1 = (\partial\gamma/\partial H)_K \tag{2}$$

and

$$C_2 = (\partial\gamma/\partial K)_H , \tag{3}$$

where γ is the interfacial tension, or to be more precise, the grand Ω-potential of the interface per unit area. This implies that, primarily, we do not attribute any mechanical significance to γ. Equation (1) holds for an arbitrary dividing surface, although γ and the coefficients C_1 and C_2 will have different values, depending on the choice of the dividing surface condition. In this context, we normally assume that the dividing surface is located at the hydrocarbon/water contact.

Now, in analogy with Helfrich's expression for the bending free energy [6], for a surfactant-loaded oil/water interface with low interfacial tension, we write

$$\gamma = \gamma_0 + 2k_c(H - H_0)^2 + \bar{k}_c K , \tag{4}$$

where $\gamma_0 = \gamma(H = H_0, K = 0)$ and H_0 is the spontaneous curvature, and where k_c and $\bar{k}_c$ are two (bending) constants relating to variations of the mean and Gaussian curvatures, respectively. We assume throughout that Eq. (4) refers to curvature variations occurring at constant T and μ_i (i.e., to an interface that is open in the thermodynamic sense), and that the same dividing surface condition is employed as in Eq. (1). Hence, we get

$$C_1 = 4k_c(H - H_0) \tag{5}$$

$$C_2 = \bar{k}_c , \tag{6}$$

and the following expression for the Laplace pressure consequently emerges:

$$\Delta p = 2H\gamma_0 - 4k_c(H - H_0)(H^2 + HH_0 - K)$$
$$- 2k_c\nabla_s^2 H . \tag{7}$$

This should constitute rather a satisfactory approximation, at least for curvatures of similar order as H_0. Accordingly, Δp is independent of the Gaussian curvature (or saddle splay) constant $\bar{k}_c$.

It is woth observing that the experimentally accessible case of a planar interface with $H = K = 0$ and an interfacial tension $\gamma = \gamma_\infty$ that is given by the expression

$$\gamma_\infty = \gamma_0 + 2k_c H_0^2 \tag{8}$$

serves as a reference here. In other words, we are actually considering changes of the interfacial tension relative to γ_∞, which are caused by curvature variations. Using Eq. (8), we obtain an alternative form of the expression for the Laplace pressure,

$$\Delta p = 2H\gamma_\infty - 4k_c[H(H^2 - K) + H_0 K]$$
$$- 2k_c\nabla_s^2 H . \tag{9}$$

Ou-Yang and Helfrich [7] recently proposed a detailed theory of the overall free energy of a vesicle membrane which resulted in the same expression for Δp as the above Eq. (7). In fact, their so-called shape equation also includes the additional term, $-2k_c\nabla_s^2 H$, which drops out for an interface of constant curvature.

For interfaces, which are actually minimal surfaces, the mean curvature H is zero by definition, and Eq. (7) reduces to

$$\Delta p = -4k_c H_0 K . \tag{10}$$

Thus, when physico-chemical equilibrium imposes a constraint of zero pressure difference between the two sub-phases and $k_c \neq 0$, it is a necessary requirement that the spontaneous curvature H_0 equals zero if minimal surface structures are to exist, since K can be zero only at the flat points, which is of no relevance in this connection.

For interfaces which deviate only to a minor extent from minimal surfaces, the mean curvature H is always a small quantity, irrespective of the length scale of a representative patch, whereas the Gaussian curvature K may vary considerably. For this class of interfaces, we can write, assuming H_0 to be equal to zero,

$$\gamma = \gamma_0 + 2k_c H^2 + \bar{k}_c K \tag{11}$$

and the corresponding expression for the Laplace pressure becomes

$$\Delta p = 2H[\gamma_0 - 2k_c(H^2 - K)] . \tag{12}$$

An expression for the Laplace pressure like Eqs. (7), (9) or (12), is nothing else than a minimum condition for the grand thermodynamic potential, $\Omega = \iint \gamma dS - p_i V_i - p_e V_e$, of the two-phase system in question when the interfacial tension obeys the usual Helfrich equation. The reason why this thermodynamic potential should be used rather than just the Helmholtz free energy is that we are dealing with an open system in the thermodynamic sense where physico-chemical equilibrium prevails.

Aggregation equilibrium

The second general condition of paramount significance which must be satisfied in order for an interface to exist within the realm of a macroscopic bulk phase at equilibrium is that the overall grand Ω-potential must necessarily be equal to its standard value $-pV$. Otherwise expressed, some additional mechanisms must operate which counterbalance $\Omega + p_i V_i + p_e V_e$ (that for a regular interfacial system equals γS, S denoting the interfacial area) for each one of the small two-phase systems that constitute the subunits of the microemulsion phase. Here, subscripts i and e refer to the two phases (oil, water) present in such a bulk phase. For Winsor I and II microemulsions the surface free energy $4\pi R^2 \gamma$ expended to form a droplet of radius R is counterbalanced by negative, entropy-dominated free energies, due to dispersing the droplet in the surrounding oil or water medium, and to shape fluctuations [1, 8, 9].

For infinite structures with non-interacting internal interfaces that may form when $\Delta p = 0$ and $H_0 = 0$ (Winsor III), we shall have to demand, however, that the overall excess surface free energy $\iint \gamma dS$ becomes equal to zero since there are no dispersion or fluctuation free energies to be separately invoked in this case. Interestingly, minimal surfaces offer novel possibilities to realize this condition for the important case when γ_0 is somewhat larger than zero. Note that, according to Eq. (8), γ_0 is equal to the interfacial tension γ_∞ of a planar, surfactant-loaded oil-water interface at the phase inversion where $H_0 = 0$.

The two conditions discussed above are actually the counterparts of i) the phase, and ii) chemical equilibrium conditions introduced by Hill in his treatment of aggregation of surfactants to form ordinary micelles [10].

Winsor microemulsions

Let us now consider Winsor I (o/w + excess oil phase), Winsor II (w/o + excess water phase), and Winsor III (middle phase + excess oil and water phases) microemulsions and apply the equilibrium condition $\Delta p = 0$. Implicitly, we then assume that we can neglect the interactions among the interfaces, i.e., that the bulk properties pertain in the central parts of the dispersed phase.

In the Winsor I and II cases, we have well-founded reasons to assume $H_0 \neq 0$ and spherical geometry ($K = H^2$) at equilibrium and, from Eq. (9), we then derive that

$$H_{eq}^s = \gamma_\infty/(2k_c H_0) \ . \tag{13}$$

Hence, in these regimes spherical microemulsion droplets of certain equilibrium sizes $R_{eq}^s = 1/H_{eq}^s$ are predicted, about which, however, fairly large size and shape fluctuations occur (cf. [1]). According to the stability analysis made by Ou-Yang and Helfrich [7] for the spherical case, we only need to spell-out the rather self-evident condition $\gamma \geqslant 0$ in order to secure that Eq. (13) represents a fully stable equilibrium corresponding to minimization of Ω.

Similarly, in the case of cylindrical geometry, from Eq. (9) we get

$$H_{eq}^c = \pm\sqrt{\gamma_\infty/2k_c} \ , \tag{14}$$

where further analysis shows that γ_0 must be equal to zero (and, hence, $\gamma_\infty = 2k_c H_0^2$, resulting in $H_{eq}^c = \pm H_0$) in order to allow the formation of infinitely long cylindrical aggregates. However, elongated aggregates of restricted length with a cylindrical or a string-of-bead-like contour may well form (even when γ_0 is somewhat larger than zero) by means of a size-fluctuation mechanism similar to that yielding rod-shaped surfactant micelles [2].

In line with a vast amount of experimental evidence, we can generally associate the three phase Winsor III regime with zero spontaneous curvature, i.e., $H_0 = 0$. In this case, it follows from Eq. (12) that the condition $\Delta p = 0$ can be written

$$H[\gamma_0 - 2k_c(H^2 - K)] = 0 \ , \tag{15}$$

showing that all minimal surfaces, which by definition are characterized by $H = 0$, $K \leqslant 0$, are compatible with physico-chemical equilibrium, irrespective of the values of γ_0, k_c, and $\bar{k}_c$. In other words, a periodic minimal surface structure is justified thermodynamically from the point of view of the phase equilibrium condition as a model of the average structure of a Winsor III microemulsion. Equally possible in this respect is, of course, a lamellar structure with $H = K = 0$.

It is to be expected that minimal surface structures based on single surfactant-loaded (monolayer) oil/water interfaces would generally accomodate about equal volumes of the oil and water phases in separate labyrinth network channels. By now, it is, in fact, rather firmly established that bicontinuous microemulsions (whether single-phase or middle-phase) usually fulfill such a volume condition [11, 12]. Moreover, it is known experimentally that γ_∞ of the flat oil/water interface mostly attains some ultra-low minimum value ($10^{-4} - 10^6$ Nm^{-1}) in the center of the middle phase region where $H_0 = 0$ and, hence, $\gamma_\infty = \gamma_0$, (cf. [13]). However, γ_∞ is always larger than zero. We note in passing that in this range with extremely low γ_∞-values, the flat, macroscopic interface becomes very rough as a result of large thermal fluctuations [14] and, consequently, we have to assume that thermal fluctuations modulate the structures formed by curved, internal interfaces in a corresponding fashion.

The above means that for each unit cell (subscript u) of the minimal surface structure there will always be a positive "stretching" free energy amounting to $\gamma_\infty S_u$, S_u denoting the corresponding interfacial area. Still, the excess of overall interfacial free energy associated with such a unit cell is characterized by $H = 0$, $H_0 = 0$, and $K \leqslant 0$, as obtained from Eq. (11), i.e.,

$$\Omega_u^s = \iint_u \gamma dS = \gamma_\infty S_u + \bar{k}_c \iint_u K dS , \qquad (16)$$

which may well become zero, as is definitely required for the formation of an extended internal interface within a macroscopic bulk phase. Noting that $\gamma_\infty > 0$, it is seen that $\bar{k}_c$ must also be positive in order to obtain $\Omega_u^s = 0$, since it follows from the Gauss-Bonnet theorem that $\iint K dS = 2\pi \chi_u^E$ where the Euler characteristic per unit cell χ_u^E is smaller than zero. Hence, we derive the following additional condition for the genesis of a minimal surface structure formed by non-interacting, surfactant-loaded oil/water interfaces:

$$\gamma_\infty S_u + 2\pi \bar{k}_c \chi_u^E = 0 . \qquad (17)$$

It is evident, however, that the condition expressed by Eq. (17) does not correspond to a stable equilibrium. This is so because, according to this equation, Ω_u^s does not have a minimum with respect to S_u-variations at the point where the two terms balance. To attain this necessary feature, we shall have to assume that some (repulsive) mechanisms enter into the picture in such a way that the average value of γ might be written

$$\langle \gamma \rangle = \Omega_u^s / S_u = \gamma_\infty + 2\pi \bar{k}_c \chi_u^E / S_u + C_u / S_u^2 , \qquad (18)$$

where C_u stands for a positive constant. As a matter of fact, such an expression results for a minimal surface upon extending the Helfrich expression, Eq. (11), to include a quadratic term in K and noting that $\langle K^2 \rangle / \langle K \rangle^2$ is a fixed number somewhat larger than 1. Apart from additional effects of pure bending at large $|K|$, repulsive monolayer interacting forces may also become important in this context. Hence, in order for $\langle \gamma \rangle S_u$ to become zero and, simultaneously, to be at a minimum, we find that the condition

$$S_u = -\pi \bar{k}_c \chi_u^E / \gamma_\infty \qquad (19)$$

has to be fulfilled, where γ_∞ becomes determined by $\gamma_\infty = \pi^2 \bar{k}_c^2 (\chi_u^E)^2 / C_u$. Equation (19) differs from Eq. (17) above by a factor 2. Inserting typical numerical values. $\bar{k}_c = 10^{-21}$ J, $\chi_u^E = -4$ (valid for a Schwarz P-type minimal surface structure) and $\gamma_\infty = 0.5$ s 10^{-5} Jm^{-2} yields $S_u = 2.51$ s 10^5 Å^2, which does seem to be of the right order of magnitude.

When $\bar{k}_c > 0$, this condition implies that the "stretching" free energy $\gamma_\infty S_u$ is counterbalanced by the "bending" free energy $C_u / S_u + 2\pi \bar{k}_c \chi_u^E$ gained by forming a doubly curved interface with $H = 0$ and $K \leqslant 0$.

When $\bar{k}_c < 0$, such a stretching/bending free energy compensation cannot occur and the main alternative left would be that $\gamma_0 = \gamma_\infty$ is reduced to zero by attractive surface/surface interactions resulting in the formation of a lamellar phase.

Finally, we note that our recent model calculations do, in fact, indicate that although $\bar{k}_c$ is mostly negative, it may attain positive values at certain solution states [15].

Conclusion

In the present paper, we have demonstrated how surface thermodynamics in a general version which accounts explicitly for the curvature dependence of the interfacial tension can be applied to Winsor-type microemulsions with negligible aggregate/aggregate interactions, employing Helfrich expressions to account for the free energy changes related to curvature variations. At full equilibrium, the Laplace pressure at a surfactant-loaded interface present in any one of the Winsor systems considered must be zero in order to comply with the requirements for mechanical and physico-chemical equilibrium. Consequently, only certain aggregate shapes and sizes are, in strictness, allowed. However, large fluctuations about the equilibrium positions are bound to occur because of the very low interfacial tensions encountered here.

As to the middle phase Winsor III regime (with excess phases of both oil and water), where the spontaneous curvature H_0 is equal to or very close to zero, we have furnished a rationale for modeling bicontinuous microemulsions of this kind by means of (disordered) extended minimal surface structures. A more complete account of our work in this direction will be published elsewhere [16].

References

1. Eriksson JC, Ljunggren S (1990) Progr Colloid Polym Sci 81:41
2. Eriksson JC, Ljunggren S (1991) J Colloid Interface Sci 145:224
3. Melrose JC (1968) Ind Eng Chem 60:53
4. Boruvka L, Neumann AW (1977) J Chem Phys 66:5464
5. Markin VS, Koslov MM, Leikin SL (1988) J Chem Soc Faraday Trans 2 84:1149
6. Helfrich W (1973) Z Naturforsch C28:693
7. Ou-Yang and Helfrich W (1989) Phys Rev A 39:5280
8. Overbeek JTG, Verhoeckx GJ, De Bruyn PL, Lekkerkerker HNW (1987) J Colloid Interface Sci 119:422
9. Borcovic M, Eicke H-F, Ricka J (1989) J Colloid Interface Sci 131:366
10. Hill TL (1963, 1964) Thermodynamics of Small Systems, Benjamin, New York
11. Billman JF, Kaler EW (1990) Langmuir 6:611
12. Kahlweit M, Strey R, Busse G (1990) J Phys Chem 94:3881
13. Aveyard R, Binks BP, Fletcher PDI (1989) Langmuir 5:1210
14. Langevin D (1988) Acc Chem Res 21:255
15. Eriksson JC, Ljunggren S (to be published)
16. Ljunggren S, Eriksson JC (1992) Langmuir, in press

Authors' address:

Prof. Jan Christer Eriksson
Dept of Physical Chemistry
Royal Institute of Technology
S-10044 Stockholm, Sweden

Progress in Colloid & Polymer Science   Progr Colloid Polym Sci 89:25—29 (1992)

Diffusion and interaction in gels and solutions

L. Johansson, C. Elvingson, U. Skantze[1]), and J.-E. Löfroth[1])

Department of Physical Chemistry, Chalmers University of Technology, Göteborg, Sweden
[1]) Department of Drug Delivery Research, Astra Hässle AB, Mölndal, Sweden

Abstract: The diffusion of monodisperse polyethylene glycols (PEGs) in gels and solutions of the polymers κ-carrageenan and polystyrene sulfonate (NaPSS) has been studied experimentally and theoretically. It was found that the diffusion quotient D/D_0 (ratio between diffusion coefficients of the PEGs in systems with and without polymer) was higher in gels than in solutions of κ-carrageenan at the same total polymer volume fraction. Also, D/D_0 of a PEG was lower in NaPSS solutions at low ionic strengths compared to the diffusion in solutions where NaPSS had a higher degree of flexibility (induced by added salt, which had no influence on D_0). Further, it was found that D/D_0 for each PEG was a convex declining function of the total polymer volume fraction, while at constant volume fraction the D/D_0 quotient was a concave function of the radius of the PEGs. — The results are discussed by means of a novel theory. The basic concepts and equations of this theory are outlined in the paper, and it is shown that the main hindrance to the diffusion of molecules like PEG is the result of sterical obstruction due to the presence of the polymer chains. The crucial parameters determining the diffusion are the size of the diffusing molecule, the polymer radius, the persistence length of the polymer, and the total polymer volume fraction.

Key words: Diffusion; interaction; gel; obstruction; theory

Introduction

In the pharmaceutical industry, polymers are often used as excipients in formulations, e.g., tablets, to control the release of the drug after administration. Thus, the transport of small molecules in polymer systems is a subject of great interest for pharmaceutical applications. However, it also deserves attention from a fundamental research point of view.

In our line of work, we have focused on the diffusion of solutes in polymer gels and solutions. Our strategy is based on the fact that polymer solutions constitute well-defined systems, and that fundamental knowledge about diffusion in these systems can be utilized to understand the diffusion in less well-defined gels.

In this contribution, we present results from studies we have carried out of small, monodisperse PEGs (polyethylene glycol) in polymer gels and solutions [1, 2]. Our results could, however, not be explained or predicted by existing theories. Therefore, a theoretical approach will be presented which was successfully applied to results in the literature, to Brownian dynamics simulations, and to our own data [3].

Experimental

Method

A homebuilt device was used; it measures the diffusion coefficient with a precision ~5%, and with good accuracy as compared to diffusion-NMR [1]. The experiments were carried out at 25°C. Results are presented as D/D_0, where D is the estimated diffusion coefficient in the polymer system, and D_0 denotes results from a system without polymer, but which can contain other added substances, e.g., salt or non-radioactive PEG.

Substances:

Radioactive, monodisperse PEGs, $300 < M_w < 4000$. Monodisperse fractions obtained with HPLC. Polydisperse samples from Berol Nobel and NEN.

κ-carrageenan in K^+-(gels above 1% w/w) or Na^+-form (solution) from Sigma.

NaPSS (polystyrene sulfonate, solution) from Polyscience.

The purification of the substances is described in [2].

Results and discussion

Figure 1 shows a typical result from studies of PEG diffusion in gels and in solutions of the same polymer. It was found that the diffusion was slower in the solution than in the gel, in which it has been suggested that double helices are formed at the gel-point [4]. Thus, in the gel-network the fibers are

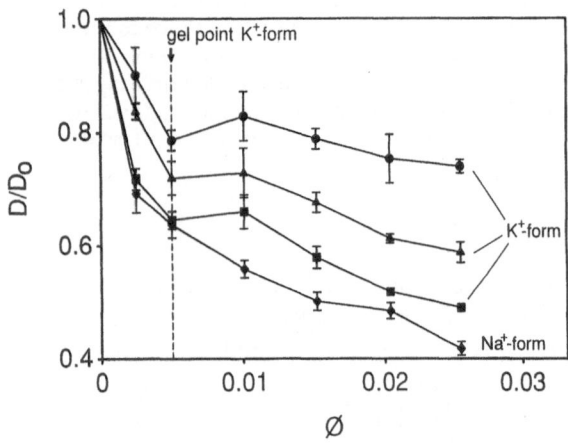

Fig. 1. D/D_0 vs. ϕ for PEGs of different molecular weights, 1118 (circles), 2834 (triangles), and 3978 (squares) in K^+-κ-carrageenan (gels above $\phi \sim 0.005$) and for the molecular weight 3978 (diamonds) in Na^+-κ-carrageenan (solution). For clarity, the results obtained with the PEGs of molecular weights 326, 678, and 1822 are not shown. The data for these molecules showed similar patterns. (Used by permission from [2])

thicker, which should lead to larger spaces between the fibers when compared to a solution at the same total polymer volume fraction. In other experiments it was found that there was no interaction between the PEG-molecules and the polymer. Further, when a slightly crosslinked gel of Na^+-κ-carrageenan

was used (on the average, three crosslinks per polymer chain, as estimated from rheological measurements) it was found that the diffusion was the same in the two Na^+-systems (crosslinked gel and solution), despite very different rheological behavior. Thus, the macroscopic viscosity was no measure of the diffusion in our systems. Instead, we attributed the faster diffusion of the PEGs in the K^+-κ-carrageenan system above the gel point to the larger spaces that become available for diffusion.

The conclusion above was supported by experiments in solutions of NaPSS, adjusted to different persistence lengths by added salt (up to 0.2 M) (see Fig. 2). Thus, in the more coiled system (at 0.2 M) the diffusion was faster than at lower ionic strength. A more coiled system indeed has larger spaces between the chains when compared to a less coiled system at the same total polymer volume fraction. In separate experiments, we did not detect any interaction between the PEGs and the NaPSS; neither was D_0 influenced by the amount of added salt. Although the viscosity of this system showed the qualitatively correct behavior (lower in the solution with the high salt concentration), we again attributed the faster diffusion in the coiled system to the larger spaces available for diffusion.

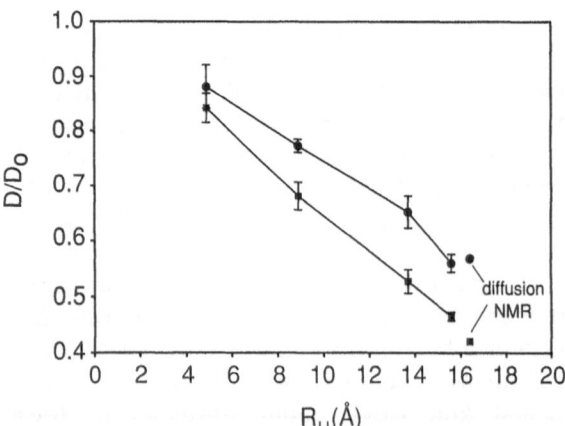

Fig. 2. D/D_0 vs. R_H for PEGs in solutions of NaPSS ($\phi = 0.012$) in water (squares), and in 0.2 M NaCl (circles). The result from measurements with diffusion-NMR, using a polydisperse sample of PEG 4000 in NaPSS solutions in D_2O, is included for comparison. (Used by permission from [2])

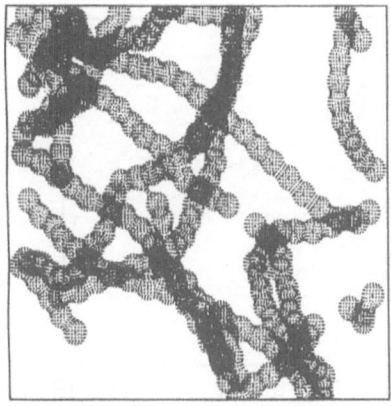 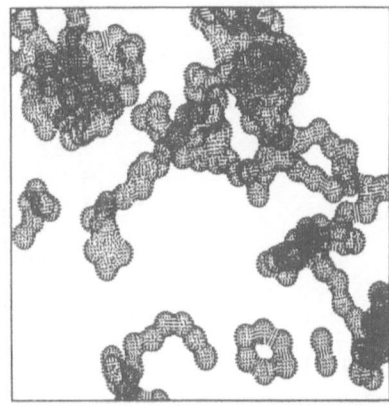

Fig. 3. Selected chain configurations from our network simulations projected in two dimensions with $\phi = 0.026$, $a = 3.3$ Å. The persistence lengths were 200 Å (left), and 10 Å (right). (Used by permission from [3])

Our results showed that the diffusion of the PEGs was not only determined by their hydrodynamic radii, R_H, and by ϕ, the polymer volume fraction, but also depended on the flexibility and the radius of the polymer, a, constituting the network. It should also be noticed that the curvature of D/D_0 vs. ϕ was convex, while D/D_0 vs. R_H was concave (see Figs. 1 and 2).

The only model presented so far in the literature that could correctly predict these behaviors without fitting parameters is the cylindrical cell model, modified to allow for a finite size of the diffusing molecule. However, this theory does not consider flexibility of the polymer chains. Thus, an appropriate theory was obviously lacking.

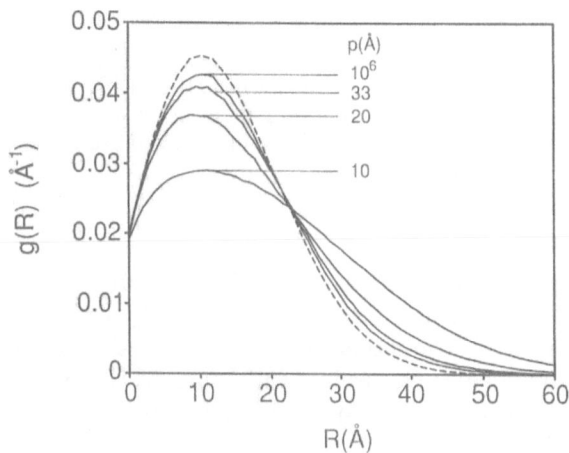

Fig. 4. $g(R)$ for networks composed of non-overlapping stiff chains ($p = 10^6$ Å), (dashed line) and of overlapping chains with different persistence lengths, p, (solid lines). In all cases, we used $\phi = 0.026$ and $a = 3.3$ Å. (Used by permission from [3])

Theory

In this contribution, we will only outline the main concept of our theory. A thorough description is given in [3].

Our results indicated that a crucial parameter describing the obstruction from the polymers was the space available for diffusion. Computer-generated networks were therefore constructed according to the discrete wormlike chain model with the chains composed of beads [5]. The results from two such simulations are shown in Fig. 3.

Intuitively, the diffusion should be faster in the more coiled system, due to the larger spaces available for diffusion. A quantitative measure of these spaces is given by their distribution, $g(R)$, which can be calculated numerically or analytically. In general, the numerical $g(R)$ is obtained by

calculating the distance of closest approach to a chain from points chosen at random in the network between the chains. Figure 4 shows the results when this approach was applied to networks of chains of different flexibility. A special case was described by Ogston, who gave the analytical form describing "spaces in a uniform random suspension of stiff fibers." We found that Ogstons expression [6], modified to allow for a finite polymer radius, could be used as a good approximation of $g(R)$, as long as $p/a > 10$, where p is the persistence length.

The approach to obtain the diffusion quotient D/D_0 used in this work is elucidated below, and summarized in Fig. 5:

a)
$$g(R) = \int_a^\infty f(b) \cdot g^{cc}(R)\, db \sim \sum_b f(b) \cdot g^{cc}(R)$$

b)
$$g^{cc}(R) = \frac{2(R+a)}{b^2} \cdot S\big(R\text{-}(b\text{-}a)\big)$$

c)
$$\frac{D(b)}{D_0} = \frac{1}{1+(R_s+a)^2/b^2} \cdot S\big(R_s\text{-}(b\text{-}a)\big)$$

d)
$$\frac{D}{D_0} = \int_a^\infty f(b) \cdot \frac{D(b)}{D_0}\, db$$

Fig. 5. Basic concepts and equations of the theory. $S(R - (b - a))$ is a step function $= 1$ when $R \leqslant b - a$, and $= 0$ when $R > b - a$. R_s is the solute radius. See text for further details

a) assume that the total space distribution $g(R)$ can be obtained as the sum of properly weighted space distributions in local models describing the chains. The weights are denoted $f(b)$, while $g^{cc}(R)$ is the local space distributions. See Eq. a) of Fig. 5.

b) use the cylindrical cell model to describe $g^{cc}(R)$, which then in each local description of a chain is determined by the polymer radius and the (local) concentration of the polymer by the cell radius b [7]. See Eq. b) of Fig. 5.

c) use the cylindrical cell model to describe local $D(b)/D_0$ quotients [7]. See Eq. c) of Fig. 5.

d) assume that the global D/D_0 can be obtained as the sum of properly weighted local diffusion ratios. The weights are the same as used to obtain the total space distribution. See Eq. d) of Fig. 5.

Applications of the theory

The theory, i.e., either the numerically calculated D/D_0 (obtained from our numerically calculated $g(R)$) or the analytical form (obtained from the modified Ogston theory) was applied to several experimental data. The input parameters were the

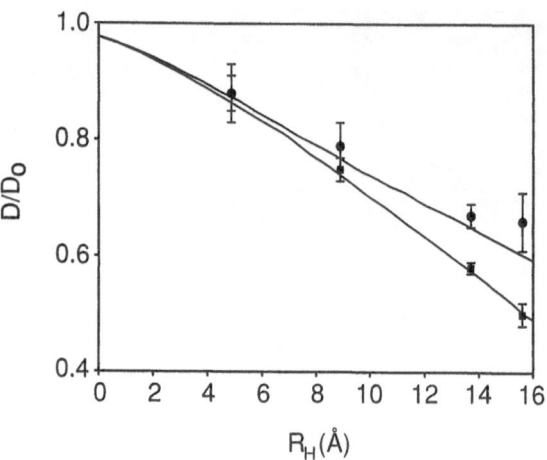

Fig. 6. D/D_0 vs. R_H from Brownian dynamics simulations in networks composed of stiff chains, $p = 10^6$ Å (squares), and flexible chains, $p = 10$ Å (circles). The solid lines are numerical predictions using our theory. (Used by permission from [3])

polymer radius, the radius of the diffusing molecule, the polymer volume fraction, and the persistence length of the polymers constituting the network. One application is presented in Fig. 6, which shows the theoretical predictions of D/D_0 from Brownian dynamics simulations. The simulations had been carried out in the computer-generated networks and are described in detail in [3]. The predicted D/D_0 were obtained with the numerical space distributions describing the networks.

Another application is presented in Fig. 7, which shows the theory applied to data by Ogston et al. [6], who studied the diffusion of albumin, $R_H = 35.5$ Å, in solutions of hyalouronic acid and dextran. It is seen that the theory (here the analytical part) successfully predicted the hyaluronic results. Since dextran is a flexible polymer, we used a numerically calculated space distribution. However, even in this case the prediction fairly well described the diffusion. The deviations from the dextran results can be attributed to the fact that dextran is a branched polymer. It should be emphasized that the same polymer radius was used for both hyaluronic acid and dextran. Thus, the results showed the effect of the different persistence lengths.

Finally, predictions of the diffusion of PEGs in the κ-carrageenan systems described above are shown in Fig. 8. The agreement was excellent for the

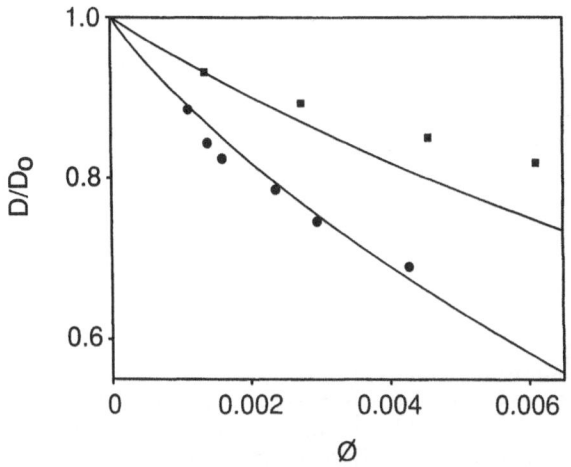

Fig. 7. D/D_0 vs. ϕ for experimental results [6] on the diffusion of albumin in solutions of hyaluronic acid (circles), and dextran (squares). The solid lines are our theoretical predictions using $a = 3.7$ Å for both polymers. (Used by permission from [3])

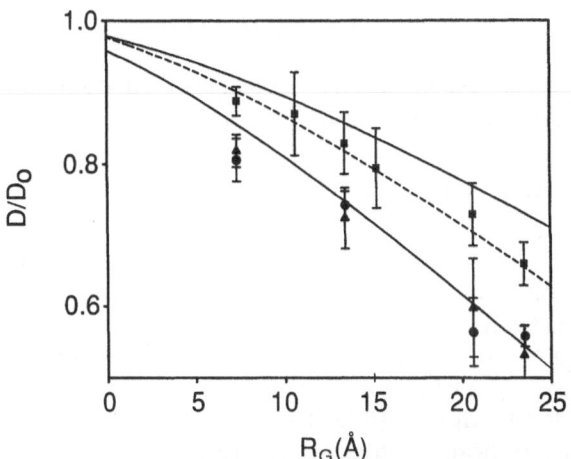

Fig. 8. D/D_0 vs. radius of gyration for experimental results [2] on the diffusion of PEG in κ-carrageenan: as a physically cross-linked gel (squares); as a solution (circles); and as a chemically cross-linked gel (triangles). The lines are predictions using our analytical theory [3]. The volume fraction ϕ was 0.01. (Used by permission from [3])

Na$^+$-systems (lower solid line), for which the polymer radius is known (coil conformation with a = 3.3 Å). The prediction for the K$^+$-system (upper solid line) was obtained using a double helix radius ($a = 5.1$ Å) for the polymer. Better agreement (dashed line) was obtained when we assumed that the

network was composed of 50% coils and 50% helices ($a = 4.2$ Å).

Conclusions

We have presented results from studies of the diffusion of PEG molecules in different polymer systems. The diffusion was shown to be governed mainly by the sterical obstruction effect from the polymers, with other interactions playing a minor role (e.g., hydrodynamics). We propose that our findings are general for solutes of the sizes used here that diffuse in polymer systems in the investigated concentration range. We also propose that the diffusion can be predicted with the present theory. We wish to emphasize the absence of scaling or fitting in the theory. However, if the theory is used with parameter fitting, the results indicated that the theory can be used to obtain structural information.

Acknowledgements

This work was financially supported by grants to LJ from Astra Hässle AB.

References

1. Johansson L, Löfroth J-E (1991) J Coll Interface Sci 142:116—120
2. Johansson L, Skantze U, Löfroth J-E (1991) Macromolecules 24:6019—6023
3. Johansson L, Elvingson C, Löfroth J-E (1991) Macromolecules 24:6024—6029
4. Nilsson S, Piculell L, Jönsson B (1989) Macromolecules 22:2367—2375
5. Hagerman PJ, Zimm BH (1980) Biopolymers 20:1481—1502
6. Ogston AG, Preston BN, Wells JD (1973) Proc R Soc London A 333:297—316
7. Nilsson LG, Nordenskiöld L, Stilbs P, Braunlin WH (1985) J Phys Chem 89:3385—3391

Authors' address:

Jan-Erik Löfroth, Associate professor
Drug Delivery Research
Astra Hässle AB
S-43183 Mölndal, Sweden

On mutual interactions in polycomponent surfactant systems

A. Šarić, R. Despotović, and S. Trikić

Department of Colloid Chemistry, Ruder Boškovic Institute, Zagreb, Croatia

Abstract: In order to confirm the models of mutual interactions in three- or four-component surfactant systems the experimental determinations of the electrophoretic mobility u_{+0-} (cm^2 s^{-1} V^{-1}) as a function of the molal ratios n/y, k/h, and x/y were carried out. The polycomponent models are $(n\text{Rhod}^+ + y\text{Cl2}-\text{SO}_4^-)^{0-} + (x\text{Blue}^+ + y\text{Cl2}-\text{SO}_4^-)^{+0-}$ and $(k\text{Cl2}-\text{NH}_3^+ + h\text{Fluo}^-)^0 + (x\text{Blue}^+ + y\text{Cl2}-\text{SO}_4^-)^0$ where n, y, x, k, and h correspond to molal fractions. The five chemically different surfactants were used to obtain polycomponent surfactant colloids. The behavior of all examined systems suggests nonuniform mutual interactions.

Key words: Colloids; electrophoretic mobility; surfactants

Introduction

The associates of diluted surfactant solutions possess typical colloid characteristics [1]. By mixing of cationic surfactant C^+ with the anionic one A^-, several typical phenomena occur: the turbidity, the surface tension, the conductivity, and the electrophoretic mobility change variously as functions of molal ratios of $[C^+]/[A^-]$ surfactants present in the solution [2]. On the basis of previously published data, it was reasonable to suppose the possibility of the formation of several colloid surfactant systems with various characteristics for the same chemical species used, if put into contact at different conditions [2, 3]. For example, the surfactants mixture of $C^+ + A^-$ possesses a positive, a zero, or a negative electrostatic sign, generally depending on the $[C^+]/[A^-]$ molal ratio [3]. By mixing of two different binary surfactant systems throughout the complex mutual interactions, several steps of transition equilibria appear. The analysis of the time-dependence of system properties leads us to the conclusion of substantial interchanges between colloid particles for both of the binary systems present. In order to confirm the model of complex structural changes in polycomponent surfactant systems, a series of experiments were carried out and the results discussed indicate the justifiableness of our hypothesis of possible mutual interactions in the observed systems. The behavior of all examined systems suggests nonuniform mutual interactions.

Experimental

Materials

Analar grade n-dodecylamine (Fluka) dissolved in 1:1 nitric acid (Merck) $\text{Cl2}-\text{NH}_3^+$, adsorption indicator Rhodamine 6G (B.D.H.) Rhod^+, methylene blue B (Merck) Blue^+, especially pure sodium n-dodecyl sulphate (B.D.H.) $\text{Cl2}-\text{SO}_4^-$, and adsorption indicator sodium fluoresceinate (Hopkin Williams) Fluo^- were used throughout the experiments. Twice distilled water was used.

Polycomponent systems

The following three- and four-component systems were prepared:

$$[(\text{Rhod}^+ + \text{Cl2}-\text{SO}_4^-)^0 + (\text{Blue}^+ + \text{Cl2}-\text{SO}_4^-)^+]^+ \quad \text{(I)}$$

$$[(\text{Rhod}^+ + \text{Cl2}-\text{SO}_4^-)^0 + (\text{Blue}^+ + \text{Cl2}-\text{SO}_4^-)^-]^- \quad \text{(II)}$$

$$[\text{Rhod}^+ + \text{Cl2}-\text{SO}_4^-)^- + (\text{Blue}^+ + \text{Cl2}-\text{SO}_4^-)^+]^- \quad \text{(III)}$$

$$[(\text{Rhod}^+ + \text{Cl2}-\text{SO}_4^-)^- + (\text{Blue}^+ + \text{Cl2}-\text{SO}_4^-)^0]^- \quad \text{(IV)}$$

$$[(\text{Rhod}^+ + \text{Cl2}-\text{SO}_4^-)^+ + (\text{Blue}^+ + \text{Cl2}-\text{SO}_4^-)^-]^- \quad \text{(V)}$$

$$[(\text{Cl2}-\text{NH}_3^+ + \text{Fluo}^-)^0 + (\text{Blue}^+ + \text{Cl2}-\text{SO}_4^-)^0]^- \quad \text{(VI)}$$

All the binary systems were prepared from total surfactant concentrations of 0.0010 mol dm^{-3} and were aged after mixing for 6000 s, before being used for mixing three- or four-component systems by the schemes (I) to (VI). The prepared polycomponent systems were kept at a constant temperature of 293 K using a Haake ultrathermostat. The systems' compositions are given in Table 1.

Table 1. The chemical composition of polycomponent systems given as the molal fractions of Cl2—NH$_3^+$ (k), Rhod$^+$ (n), Blue$^+$ (x), Cl2—SO$_4^-$ (y), and Fluo$^-$ (h):

System:	Molal fractions:			
	k	n	x	y
(I)		0.58		0.42
			0.80	0.20
(II)		0.58		0.42
			0.20	0.80
(III)		0.20		0.80
			0.80	0.20
(IV)		0.58		0.42
			0.50	0.50
(V)		0.80		0.20
			0.20	0.80
(VI)	0.35			0.65
			0.50	0.50

Table 2. The electrostatic sign z^{+-} and the electrophoretic mobility u_{+-} (cm^2 s^{-1} V^{-1}) for the polycomponent systems aged for t_A/s:

System:	z^{+-}	$u_{+-} \leftrightarrow 10^5$	t_A/s
(I)	+	3.8	6 000
(II)	−	5.0	600
(II)	−	8.2	6 000
(II)	−	12.7	60 000
(III)	−	2.5	6 000
(IV)	−	2.4	6 000
(V)	−	4.9	6 000
(VI)	−	1.9	6 000

Microelectrophoresis

The particle charge and the electrophoretic mobility u_{+0-} (cm^2 s^{-1} V^{-1}) of the surfactant colloid mixtures were determined using a double micro-electrophoretic cell after Smith-Lisse [4]. The results are presented as the average value of seven measurements of polycomponent systems aged for 60 000 s, and for 600 s and 600 000 s for system (II). The data are given in Table 2.

Results and Discussion

By mixing two binary systems, it is possible to prepare several different colloid systems [5]. Each of the binary colloid systems, including the four-component ones can be prepared with positive, negative, or zero electrostatic sign. The six examined three- and four-component systems generally showed three various results (Table 2). As a regular or model behavior for systems (I) and (II), the resulting electrostatic sign of the polycomponent mixture corresponds to the sign of the charged binary system in the mixture with the zero-charged binary colloid. It is of interest to note the time-dependence of the electrophoretic mobility (Table 1, system (II)). The observed phenomena are probably caused by the change of the inner structure between binary colloid particles, because for smaller particles the same electrostatic capacity causes an increase of electrophoretic mobility [6] that is expressed by strong time-dependence of u_{+0-} (cm^2 s^{-2} V^{-1}).

The second group of results was observed for the mixture of positively and negatively charged binary systems, as is presented for the three-component systems (III) and (V). The resulting electrostatic sign is negative for both systems with slowly mobile particles. For Cl2—SO$_4^-$ spherical associate structure the total negative charge z^- is proportional to the number x of surfactants ion, i.e., xz^{x-}, while for lamellar microcrystallite blocks of organic dye-stuffs particles Rhod$^+$ and Blue$^+$ correspond to yz^{y+} where y is smaller than the total number of surfactant cation ions, because of the shadowing effect [6]. The result is the negative electrostatic charge of polycomponent particles.

The third group of results shows inner structural changes in binary systems. The polycomponent model (IV) prepared by the mixing of two zero-charged binary systems has a negative electrostatic sign. The result observed indicates the formation of microcrystallites (Rhod^{z+}Blue^{z+})$^{2y+}$ with strong shadowing effect, causing the decrease of the outer electrostatic positive sign to Rhod^{z+} and Blue$^{z''+}$

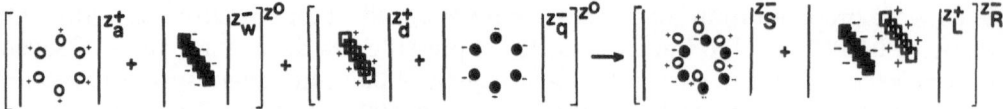

Scheme 1. The mutual interaction between two binary colloid surfactant systems (see Discussion)

where $z' < z$ and $z'' < z$ and $2y = x' + x''$. The result is the appearance of the negative electrostatic sign for polycomponent colloid particles according to the following scheme:

$$[(Rhod^{z+} + Cl2-SO_4^{z-})^0 + (Blue^{z+} + Cl2-SO_4^{z-})^0]^0$$

$$\rightarrow (Rhod'_+ Blue''_+ + 2\ Cl2-SO_4^{z-})^{r-},$$

where $r = 2z - z' - z''$. The prepared four-component system (VI) occurring when two binary systems are put into contact can be transformed according to the following scheme: if we symbolically indicate by open circles the cationic $(Cl2-NH_3)^z a^+$, by a filled square the anionic $(Fluo)^z w^-$, for the other pair, the cationic surfactant $(Blue)^z d^+$ by an open square, and the anionic surfactant $(Cl2-SO_4)^z q^-$ by filled circles, then the mutual interactions in four-component system (VI) can be presented by Scheme 1, or by the following equation

$$[(Cl2-NH_3)^{z_a^+} + (Fluo)^{z_w^-}]^0$$

$$+ [(Blue)^{z_d^+} + (Cl2-SO_4)^{z_q^-}]^0$$

$$\rightarrow [[(Cl2-NH_3)^{z_a^+} + (Cl2-SO_4)^{z_q^-}]^{(z_a^+ + z_q^-)z_S^-}$$

$$+ [(Blue)^{z_d^+} + (Fluo)^{z_w^-}]^{(z_d^+ + z_w^-)z_L^+}]^{z_R^-} \qquad (VIa)$$

where $z_S^- > z_L^+$ or $z_R^- > z_S^- + z_L^+$. The lamellar structures L or organic dye-stuffs and the spherical formations of aliphatic associates S are characteristic for the observed polycomponent surfactant agglomerates [7]. This means that, by spontaneous

processes of the inner structural changes in mixed surfactant solutions, the processes which essentially change the typical physico chemical characteristics take place. The results are well in accordance with the supposed model of mutual interaction [5] in polycomponent surfactant solutions, and offer answers about a series of the observed phenomena for different polycomponent surfactant mixtures in an aqueous medium.

References

1. Sepulveda I (1974) J Colloid Interface Sci 46:372—374; Despotović R, Despotović LjA, Filipović Vincekovi N, Horvat V, Mayer D (1975) Tenside Detergents 12:323—327
2. Malik WU, Verma SP (1966) J Phys Chem 70:26—29
3. Despotović R (1989) Proc 7th Yugoslav Symposium on Surface Active Agents, Vol I:107—117; Mayer D, Despotović R (1987) Tenside Surfactants Detergents 24:156—159
4. Smith ME, Lisse MV (1936) J Phys Chem 40:399—401
5. Despotović R, Šarić A, Trikić S (1991) Proc 12th Yugoslav Symposium on Electrochemistry, Igman pp 158—159
6. Despotović R (1977) Kemija u Ind 26:557—565
7. Tomašić V, Ph D Thesis, University Zagreb (1984) 1—146

Authors' address:

Prof. Dr. R. Despotović
Ruder Bošković Institute
P.O. Box 1016
41001 Zagreb, Croatia

Progress in Colloid & Polymer Science   Progr Colloid Polym Sci 89:33—38 (1992)

Osmotic pressure of highly charged swollen bilayers

Th. Zemb, L. Belloni, M. Dubois, and S. Marcelja*)

Service de Chimie Moléculaire Bât. 125, CE SACLAY 91191 Gif sur Yvette, France
*) Department of Applied Maths, Australian National University, Canberra, Australia

Abstract: This is a preliminary report about new osmotic pressure measurements in electrostatically stabilized highly swollen lamellar phases. We identify the stabilizing forces for different regimes: the equilibrium between dispersion and electrostatic forces explains the two coexisting lamellar phases while the lamellar to asymmetric disordered lamellar phase transition is entropically driven.

Key words: Osmotic pressure; lamellar phase; bilayers; electrostatics

Introduction

Our aim is to explain the presence of two coexisting lamellar phases in highly charged surfactant bilayers of low solubility such as those made of didodecyldimethylammonium bromide (DDAB), as well as the apparently paradoxical stability of the swollen lamellar phase in the presence of a salt content which is so high that the Debye length is one order of magnitude smaller than the period. What are the contributing factors to this lamellar liquid crystal stability? Are undulation forces relevant in the presence of salt? What sets the large but finite maximum swelling of these phases? Parsegian and coworkers have shown that this type of question can be answered when equilibrium structures occurring at fixed osmotic pressure equilibrium are identified using small-angle X-ray scattering [1]. In this paper, we extend osmotic pressure measurements to periods of the order of several hundred Å.

Description of the system:

The binary system under study is the (DDAB)/(water + KBr) solution. The evolution of DDAB microstructure in binary solution can be described as follows [2]. The area per surfactant is nearly independent of experimental conditions and is equal to 65+/—5 $Å^2$/molecule. The surface potential has been measured by Pashley et al. and is equal to 190 mV [3]. The molecular volume of DDAB is 784 $Å^3$/molecule; the bilayer thickness is 24 Å. The value of the elastic bending constant k_c of the DDAB bilayer, accessible from electrostatic calculations, has been estimated according to theoretical estimations to be ≈ 10 kT [4]. The value of the second bending constant k_c^* related to the Gaussian curvature is still unknown for this system, but is expected to be negative.

The succession of the structures found during a dilution experiment of DDAB is the following:

1) At temperatures higher than those for chain melting (290 K), a concentrated lamellar phase L_a' cannot be diluted to less than 75 wt% DDAB. This is the "condensed" lamellar phase L_a' with a periodicity of $D^* = 30.9$ Å in pure water, i.e., a water layer of only 6.9 Å between the charged DDAB bilayers [5].

2) Between 75 wt% and 28 wt% of DDAB, two lamellar phases (L_a' and L_a) coexist [5].

3) Between 28 wt% and 3 wt% DDAB, a pure lamellar L_a phase ("swollen lamellae") is formed. This phase is made of stiff bilayers under strong electrostatic repulsion and is identified by birefringence and at least three sharp Bragg peaks seen by small-angle neutron or X-ray scattering. The periodicity D^* in Å measured for these monophasic lamellar samples is given by $D^* = 24/\phi$, where ϕ is the volume fraction of surfactant.

4) At lower concentration (between 0.15% and 3 wt% DDAB), a metastable (lifetime a few months) spherulitic biphasic dispersion of $[L_a]$ crystallites in a disordered infinite bilayer (L_3) is obtained [2]. The metastable (months) L_3 flow birefringent solution shows a broad scattering correlation peak located around $q = \pi/\xi$, where ξ is the typical correlation length of the bilayers. The value of ξ is close to 300 Å, a typical size of the closed vesicles obtained in the solution [12]. This size ξ is concentration independent [2]: it is an *asymmetric* L_3 structure which does *not* swell as inverse volume fraction.

5) A diluted isotropic phase (Iso) is formed at DDAB concentrations less than 0.15% surfactant. The SANS scattering spectra of this diluted phase is shown in Fig. 1; the scattering of the Iso phase corresponds to a single bilayer with a very large correlation length ξ, of the order of 1600 Å, much larger than the Debye length set by pH and critical micellar concentration (CMC). This extremely intense scattering (orders of magnitude larger than those any small molecular sized agregate could produce!) can only be reconciled with giant close-packed vesicles or an asymmetric L_3 structure of very large correlation length ξ.

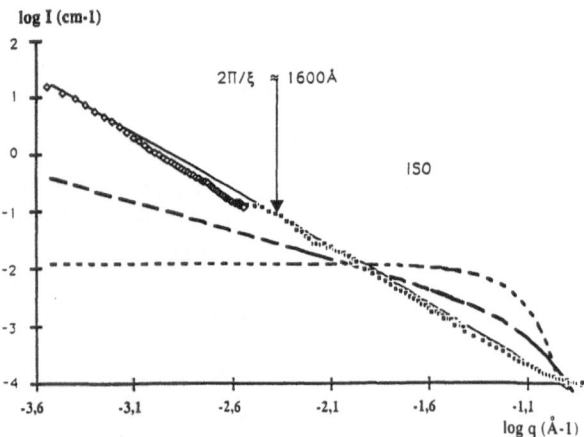

Fig. 1. SANS spectra of a 0.08 weight% (isotropic, not flow birefringent) aqueous solution of DDAB ($1.7 \, 10^{-3}$ M) obtained using the D11 setup (ILL), compared on absolute scale to the scattering which would be obtained if spherical (---) with the maximum admissible radius of $R = 34$ Å) or cylindrical (———) micelles ($R = 22$ Å, imposed by surface to volume ratio) were formed. Scattering calculated of the infinite *flat* bilayer is shown by the continuous line

At higher temperatures or in the presence of salt, one obtains the same succession of phases and microstructures, shifted however to higher surfactant concentrations. The maximum possible amount of salt content required to obtain the swollen lamellar phase is $4 \cdot 10^{-2}$ M [2].

Experimental method:

To set the osmotic pressure of the sample, we use the method pioneered by Parsegian and coworkers [1]; it consists in dissolving a neutral macromolecule (dextran) in the reservoir in order to reduce the chemical potential of the solvent.

The surfactant didodecyldimethylamonnium bromide (DDAB) was obtained from Kodak. Samples were dissolved in Millipore water in the case of H_2O, and D_2O from "Service des Molécules marquées, CEA Saclay" used as received.

The solutions are equilibrated at fixed osmotic pressure and ionic strength with a large reservoir containing Dextran 500 (Fine Pharmaceuticals) and KBr. No traces of Na^+ or K^+ were found in dextran using atomic adsorption. The reservoir volume is 10 to 100 times larger than the equilibrating sample. The calibration of osmotic pressure made by Véretout and Tardieu [6] was used. Since dextran is a non-charged, hydrophilic polymer, we have assumed that the osmotic pressure imposed by the dextran content is independent of the salt concentration (Parsegian, personal communication).

We have made systematic tests of permeability of the dialysis membranes brand (Spectrapor; Spectrum Lab), with molecular mass cutoffs ranging between 500 and 12000, corresponding to average radii of channels between 5 and 30 Å. A fixed concentration of DDAB was dialyzed against pure water and the concentration of DDAB outside the dialysis bag was checked using differential refractometry to quickly measure the evolution of the refractive index versus time; it was found that with a molecular mass cutoff higher than 5000, a low outflowing flux of DDAB gives artefacts in the measurements, because DDAB is not confined to the dialysis bag during the time needed for equilibration (weeks to months). Mass cutoffs lower than 5000 increase the equilibration times too much and it was decided to use 5000 as a cutoff value for the dialysis membrane. In any case, the slow outgoing flux of DDAB through the membrane is inhibited by the presence of added salt. This proves that the mechanism of DDAB exchange through the membrane occurs via monomer permeation: the salting-out reduces the concentration of DDAB monomer in equilibrium with the bilayers to a value much lower than the CMC, $5 10^{-5}$ M, thus reducing the permeation rate. The best results were obtained using a cut-off value of 8000, for which the radii of permeation holes are then equal to 20 Å. Measurable dextran passage through the membrane could only be detected after 6 months of dialysis. The total salt concentration reported in this paper includes the sum of the original K^+ and the small quantity of Na^+ released by the membrane.

Small-angle neutron scattering (SANS) was measured on the D11 camera (Institut Laue-Langevin, Grenoble) for periodicities D^* larger than 400 Å, requiring high resolution. Equilibration at the microscopic level was checked by measuring the sample at different times after preparation (up to 6 months). Other measurements were made at the PAXE setup (Orphée, Saclay) with a Guinier-Méring camera used with Cu Kα radiation allowing a range of Bragg spacings between 10 and 250 Å. At least two orders of Bragg peaks indexed 1 and 2 allow precise measurements of the periodicity D^* in the L_a phase. The Dextran concentrations required to perform the measurements reported in Fig. 2 vary from 1% to 20%. Without added salt, the pH of the DDAB solution is about 5 and the CMC is $5 \cdot 10^{-5}$ M. Therefore, we estimate an upper boundary for the Debye length (400 Å) to be due to these two types of ions contained in the thick water layers. The previous phase diagram study showed that the L_a phase exists up to $4 \cdot 10^{-2}$ M added salt [2].

Results

The osmotic pressure versus distance curves obtained at given salt concentrations are shown in Fig. 2. Without added salt, atomic absorption spectroscopy measurements reveal the presence of $5 \cdot 10^{-4}$ M Na^+ ions, 10 times more than the CMC. The pressures Π obtained are compared to the electrostatic repulsion calculated taking into account the salt ejection from the inside of the bilayers: for each salt concentration, the maximum swelling period corresponds to an osmotic pressure of the order of 300 Pa.

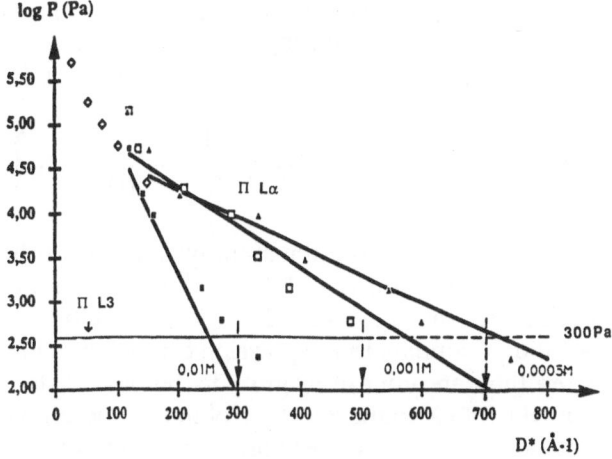

Fig. 2. Osmotic pressure versus periodicity D^* curves obtained with three different added monovalent salt (KBr) concentrations. The thick lines are the electrostatic contribution only. The arrows show the maximum swelling observed for these salt concentrations and the corresponding maximum swelling of the swollen L_a phase. Order-disorder transition pressure is of the order of 300 Pa for maximum swelling of the order of 200 to 600 Å. Thick lines are theoretical calculations of the electrostatic repulsion only. The arrows show the maximum swelling periods observed at $5 \cdot 10^{-4}$ M added salt (700 Å); 10^{-3} M added salt (500 Å) and 10^{-2} M added salt (300 Å)

We now have sufficient information to understand the origin of the minimum swelling δ^* and the maximum swelling Δ^* in the DDAB/water system, by numerically evaluating the forces between the bilayers.

Maximum swelling Δ^*:

The electrostatic repulsive pressure at fixed chemical potential (not concentration) of the added salt is calculated using the method described in [5], equivalent to the standard method described in [7]. At large spacings $D^* > 200$ Å, salt is ejected from the inside of bilayers, but the following asymptotic expression is valid [5]:

$$\Pi_{elec} = 64 \ kTC_s'\gamma^2 e^{-\kappa'(D^*-t)} \ ,$$

where D^* is the periodicity of the lamellar phase, $t = 24$ Å the apolar bilayer thickness and γ is close to 1. C_s' the concentration of salt and $1/\kappa'$ the screening length *in the reservoir.*

For the van der Waals pressure between bilayer stacks, we use the following expression, where D^* is also the period and t the bilayer thickness (24 Å):

$$\Pi_{vdW} = \frac{A}{6\pi} \cdot \left\{ \frac{1}{(D^* - t)^3} + \frac{1}{D^{*3}} - \frac{2}{(D^* + t)^3} \right\}.$$

For this first rough estimation, we have used the Hamacker constant $A = 1.3 \cdot 10^{-20}$ J which is imposed by the observed $L_a - L'_a$ equilibrium. Although popular, this expression can be wrong by more than an order of magnitude, because it does not take into account screening of the dipole-dipole interaction, ion-ion correlations or non-additive effects in a multilayer stack. In this first report, we have not introduced the rigorous calculation given in [13].

For the undulation pressure, we use the expression experimentally verified by Roux and Safinya and applied to fluctuation stabilized lamellar phases [8]:

$$\Pi_u = \frac{3\pi^2}{128} \cdot \frac{(kT)^2}{k_c \cdot D^{*3}}.$$

The numerical value of the elastic bending constant k_c is of the order of 10 kT [4]. As can be seen in Fig. 3, the strength of the undulation force is several orders of magnitude too low to participate in the lamellar stability in the case of DDAB. We known that $k_c \gg kT$ for DDAB since one observes several sharp Bragg peaks even at 300 Å spacing with electrostatically stabilized systems [2]. The three interaction pressures are compared in Fig. 3.

When water is added to the diluted lamellar phase L_a, the swelling (linear towards composition) corresponds to a decrease in the osmotic pressure. This decrease proceeds upon water addition until the osmotic pressure of the L_a phase is lowered enough to become equal to the osmotic pressure of the disordered L_3 phase. At that point, a microscopic biphasic suspension of $[L_a]$ crystallites, present as closed multilayers of lamellar phase (sometimes called "spherulites") emulsified in an L_3 phase is obtained [2]. Macroscopically, this sample is still a clear, unique, birefringent liquid which is not decomposed under centrifugation. Therefore, the total pressure at the maximum swelling when $D^* = \Delta^*$ provides a simple way to measure the pressure of the coexisting L_3 phase. At lower pressure and concentration, this is the system in

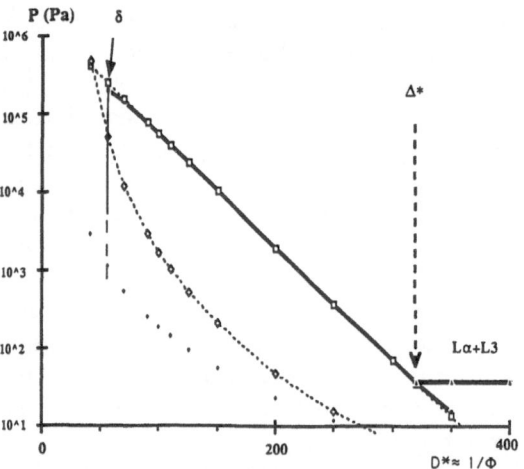

Fig. 3. Comparison of the electrostatic repulsive pressure (---), the Helfrich fluctuation force ($\cdots$), and the rough estimate of the attractive dispersion force ($\cdots \diamond \cdots \diamond \cdots$). The minimum swelling δ^* of the diluted lamellar phase L_a is set by equilibration of the attractive Van der Waals and the repulsive electrostatic forces while the maximum swelling Δ^* is obtained when the predominant electrostatic pressure is equal to the entropic term of the disordered bilayer structure (asymmetric L_3)

which giant vesicles in thermodynamic equilibrium have been seen by video-enhanced microscopy and scattering studies, leading to non-linear optical effects. When these closed multilamellar spherulites are diluted (as was shown a few years ago) part of the DDAB molecules are included in large, closed, singlewalled objects [12, 14, 15].

The osmotic pressure Π_t when the L_a and the L_3 phases coexist is set mainly by the electrostatic repulsion in the L_a phase, and equals the osmotic pressure at the transition point in the disordered open connected phase (L_3). We propose that the main term in the disordered lamellae phase is of entropic origin:

$$\Pi_\tau = [\Pi_{elec}]_{L_a} = [\Pi_{total}]_{L_3} = [\Pi_{config}]_{L_3} + [\Pi_{elec}]_L.$$

The coexisting L_3 phase exhibits a characteristic length $\xi = 300$ Å, nearly independent of concentration, as can be seen from its scattering behavior. Because the L_3 phase is asymmetric, composed of polydisperse closed bilayers in equilibrium with large open random bilayers evidenced by the scattering behavior at low q, we postulate that the configurational (entropic) contribution of the L_3 phase, approximated as [16]:

$$[\Pi_{\text{config}}]_{L_3} = \frac{\kappa T}{\zeta^3} ,$$

is predominant. The maximum swelling Δ^* of the L_a lamellar phase is then given at any salt content by the condition:

$$\Pi_{\text{elec}}(\Delta^*) \approx \Pi_{\text{total}}(\zeta) \approx kT/\zeta^3$$

$$[64 \cdot kTC'_s \cdot e^{-\kappa'\Delta^*}]_{L_a} = \left[\frac{kT}{\zeta^3}\right]_{L_3} \approx 300 \text{ Pa} .$$

The characteristic length ζ in the L_3 phase is known experimentally to be close to 300 Å, given by the position of the broad scattering peak in the flow birefringent L_3 phase. There is no attractive force keeping the lamellae together at the electrostatic pressure of 300 Pa, but there is competition between the high entropy L_3 phase and the pure L_a structure.

Minimum swelling δ^*:

We can use the same method to identify the origin of the L_a to L'_a transition: at periodicities of the order of 50 to 100 Å, the classical equilibrium between van der Waals and electrostatic repulsion appears: bilayer separations of less than 85 Å are unstable and collapse to a very concentrated bilayer, when only one or two water layers ($e = 6.9$ Å) remain between surfactants (L'_a phase). In Fig. 4, the electrostatic, Helfrich, van der Waals, and the observed steric "wall" at water thickness $e = 6.9$ Å are plotted together for the residual ion content 5 10^{-4} M, giving a Debye length of 135 Å. All the quantities needed to obtain these curves are known, except the Hamaker constant, which has been adjusted to $A = 0.7 \times 10^{-19}$ J. This adjustment is necessary in order to obtain the numerical value of the observed minimum water thickness in the L_a phase to $e \mapsto 80$ Å without added salt. $D^* = e + 24$ Å because bilayer thickness is constant and is equal to 24 Å in all our experiments even with added salt and has been measured by the Porod limit. The surface potential is 190 mV (surface charge 5 µC/cm² i.e. 300 Å²/e^-); the temperature 295 K, the dielectric constant 78, and the Debye length set by residual salt 135 Å. As can be seen in Fig. 2, the dispersion forces are of the same order of magnitude as the electrostatic repulsion for a value of the periodicity $D^* \approx 80$ Å, thus, a concentration

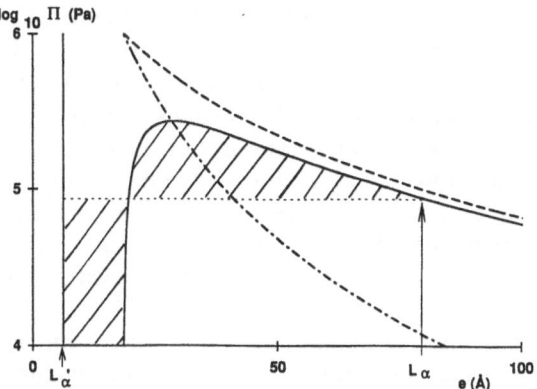

Fig. 4. Equilibrium between van der Waals and electrostatic pressures in the concentrated regime as function of the water layer thickness e evaluated for salt concentration KBr $\approx 5 \ 10^{-4}$ M. Parameters used in the calculation: Hamaker constant $A = 0.7 * 10^{-19}$ J, surface potential 190 mV, effective charge 5 µC/cm², area per charge 294 Å². The collapsed L'_a (minimum water layer thickness $e = 6.9$ Å) and the swollen L_a phase (water layer thickness ≈ 80 Å): sum of pressures due to electrostatic (---), van der Waals (—·—) and the added steric wall are plotted for water layer thicknesses e less than 100 Å. The total pressure is given by the thick line. The two arrows correspond to the two stable coexisting water layers corresponding to $e = 6.9$ Å (L'_a phase) and 80 Å (swollen L_a phase). The two distances correspond to the same pressure and free energy

of 28% DDAB. This is the order of magnitude of the minimum period δ^* observed for the diluted lamellar phase L_a.

A more precise evaluation of the forces involved in the range of the water layer thickness $e = 0$ to 100 Å is given in Fig. 3. It can be seen that two swelling periods, corresponding to the L_a and L'_a phases coexist with the same osmotic pressure, as has been observed in SAXS studies [5].

Conclusion

It is now possible to explain the maximum salt content that destabilizes the diluted lamellar phase. This is set by the maximum amount of salt C to realize the conditions $\delta^* = \Delta^*$, i.e., to satisfy, *simultaneously,*

$$[64kTC \cdot e^{-\kappa'\delta^*}]_{L_a} = \Pi_{L'_a} \approx 300 \text{ Pa} .$$

The maximum possible added salt for which the L_a phase is stable is set by the condition $\delta^* = \Delta^*$. this estimation is not very precise because of the low value of the water layer thickness (6.9 Å), assumed to be homogeneous in our estimation.

We have now explained the apparent paradox of the swollen lamellar phase with periodicity much larger than the Debye screening length. Using DDAB, several different regimes in the classification proposed by Joanny [9] and Roux [10] are now directly accessible for the first time using a binary solution. The maximum and minimum swelling behavior of the diluted L_a phase, according to this equilibrium mechanism, is similar to the one observed in dispersed clay systems by Langmuir [11]. This electrostatic repulsion between bilayers, compared to the negligible undulation and van der Waals forces, gives the evaluation of the pressure needed to allow an order-disorder transition between charged bilayers. This mechanism allows the coexistence of the two distinct lamellar phases in the DDAB phase diagram.

Acknowledgement

The initial ideas, the assistance of Barry Ninham in the preparation of this manuscript, as well as crucial suggestions about miscalculations widespread in the literature about dispersion and fluctuation forces are gratefully greatly acknowledged.

References

1. Parsegian VA, Rand RP, Fuller NL (1991) J Phys Chem 95:4777—4782
2. Dubois M, Zemb T (1991) Langmuir 7:1352—1360
3. Pashley RM, Mc Guiggan PM, Ninham BW (1986) J Phys Chem 90:5841—5845 and 90:1637—1642
4. Lekkerkerker HNW (1989) Physica A159:319—328; Fogden A, Mitchell DJ, Ninham BW (1990) Langmuir 6:159—162
5. Dubois M, Zemb T, Belloni L, Delville A, Levitz P, Setton R (1992) J Chem Phys 2278—2286
6. Véretout F, Tardieu A (1990) Eur Biophys J 17:61
7. Chan DY, Pashley RM, White LR (1980) JCIS 77:283
8. Roux D, Safinya CR (1988) J Phys France 49:307—318
9. Higgs P, Joanny JF (1990) J Phys Paris 51:2307
10. Roux D, this volume
11. Langmuir I (1938) J Chem Phys 6:873
12. Radlinska EZ, Ninham BW, Dalbiez JP, Zemb T (1990) Colloids and surfaces 46:213—230
13. Ninham BW, Parsegian VA (1970) J Chem Phys 10:664—674
14. Ninahm BW, Kachar B, Evans D (1984) Journal of Colloid and Interface science 100:287—301
15. Ninham BW, Talmon Y, Evans DF (1983) Science 221:1047—1048
16. Nelson D in Random fluctuation and pattern growth: Experiments and models, Stanley HE, Ostrowsky (eds) (1990) NATO ASI series: E157:193

Authors' address:

Th. Zemb
Service de Chimie Moléculaire
Bâtiment 125
F-91191 Gif sur Yvette Cedex, France

Progress in Colloid & Polymer Science

Progr Colloid Polym Sci 89:39—43 (1992)

Influence of proteins on the percolation phenomenon in AOT reverse micelles: Structural studies by SAXS

J. P. Huruguen[1]), Th. Zemb[2]), and M. P. Pileni[1,2])

[1]) Université P. et M. Curie, SRI, bâtiment de chimie-physique, Paris, France
[2]) C.E.N. Saclay, S.C.M., D.R.E.C.A.M., Gif sur Yvette, France

Abstract: The AOT water/oil microemulsions exhibit a percolation process which leads to the formation of aggregates when temperature or water droplets concentration increase. In the presence of cytochrome *c*, this aggregation process is more efficient. In order to know the influence of this protein on the structure of the aggregates so formed, we report structural studies by SAXS. The results show that the aggregates behave like polymer solutions in semidilute regime.

Key words: Reverse micelles; percolation; protein; biotechnology

Introduction

The system studied is composed of a dispersion of water droplets through an apolar media which is isooctane. The interface between oil and water is made up with a double-branched tail surfactant with an anionic polar head commonly named AOT. The average radius of these spheroidal water domains is controlled by the number of water molecules per AOT molecule [1], called w and given by:

$$w = [H_2O]/[AOT] .$$

Some groups have shown that these microaggregates interact via an attractive potential originating from the interpenetration of the surfactant tails [2]. This physical property has been taken into account in the interpretation of SAXS and SANS results [3, 4]. These attractive interactions are well described by a square well potential with a short-range tail equal to 3 Å equivalent to the maximum interpenetration of the surfactant tails, and a well depth equal to about $-kT$. This allows the formation of clusters of water droplets. The average size of the clusters and their number increase statistically with volume fraction of water and temperature, and diverge around the percolation threshold where an infinite aggregate exists (the so-called percolation transition) allowing the charge carriers to percolate through the system. The percolation threshold is estimated through electrical conductivity and permittivity measurements. On the other hand, the existence of attractive interactions interpret the low polar volume fraction and temperature values at which the percolation transition takes place, compared to the values that are expected from a system of "hard spheres" [5—8]. The percolation theory has been applied to these systems, and some critical exponents have been deduced [9—11]. Most of the results are in agreement with what is expected from theory. The discrepancy concerns the critical behavior of the electrical conductivity below the percolation threshold and is interpreted as arising from the dynamical character of the clusters which can break and rearrange in time under Brownian motion effects [6, 9, 12].

These water-in-oil microemulsions have the ability to serve as hosts of macromolecules in particular enzymes [13]. We chose to solubilize cytochrome *c* in the microemulsions. This protein is a water soluble hemoprotein with a small molecular weight (12400) having a gyration radius of 15 Å. This means that at any value of w larger than 10, the spontaneous size of the reverse micelle is larger than the protein. Cytochrome *c* is responsible for several electron transfer reactions across membranes.

Previously, we have developed a geometrical model tested by SAXS [1, 14] and a kinetic model [15] to determine the average location of low molecular weight proteins or enzymes at low volume fraction of water in reverse micellar systems. We demonstrated that cytochrome c is located at the interface and its presence increases the specific interface, $\sum (\text{Å}^2/\text{Å}^3)$, in the sample [14, 15].

It has been observed previously [16] that the solubilization of cytochrome c in reverse micelles induces a decrease in the percolation threshold as is shown in Fig. 1. Similar behavior has been observed by increasing temperature. In order to detect a structural evolution of the system in the presence of the protein, we decided to use SAXS, which was performed at Lure (Orsay).

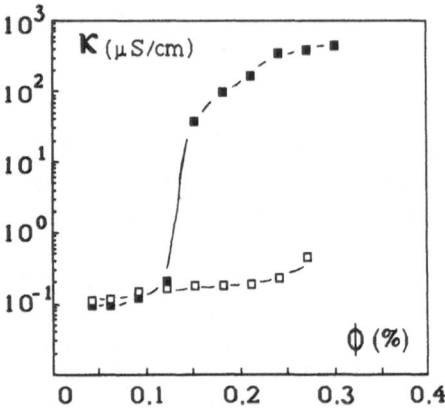

Fig. 1. Variation of the conductivity with the volume fraction of water in AOT-isooctane-water solution in the absence of protein (□) and in the presence of four cytochrome c molecules per water droplet (■). The w value is constant and equal to 40

Experimental section

Small angle x-ray scattering was performed at room temperature. We fixed the w value in order to fix the average size of the water droplets and the average number of cytochrome c molecules per water droplets, x ($x = 4$). This corresponds to 250 AOT per cytochrome c molecule. The investigations were done at different values of the polar volume fraction ϕ, corresponding to the relative volume of cytochrome c water solution in the microemulsion. The scattering intensity $I(q)$ is the product of two terms:

$$I(q) = S(q) \cdot P(q) , \qquad (1)$$

where $S(q)$ and $P(q)$ are the structure factor and the form factor of the scattering particles, respectively. The scattering intensity is given in arbitrary units. Therefore, in order to compare experimental data to theoretical simulations, the normalization is made using the invariant.

Results and discussion:

A) Results obtained in dilute solutions
   ($\phi = 7\%$ and 13%)

At low polar volume fraction ($\phi = 7\%$) the conductivity is very low (Fig. 1) and the percolation threshold is not reached. At higher polar volume fraction ($\phi = 13\%$) the system reaches the percolation threshold corresponding to the increase in the conductivity (Fig. 1). At such a polar volume fraction an infinite aggregate exists, allowing the charge carriers to percolate through the system.

Figures 2A and B show the scattered intensity $I(q)$ as a function of the wave vector q in a log-log plot for $\phi = 7\%$ and $\phi = 13\%$, respectively. A slope equal to -1 is observed over a large range of q values. This is characteristic of a cylindrical microscopic structure whose persistence length is larger than its cross-section radius R_c. The scattered intensity by a homogeneous cylindrical particule is given by [17]:

$$P_c(q) \sim q^{-1} \cdot [J_1(qR_c)/qR_c]^2 \qquad (2)$$

$J_1(x)$ is a first order Bessel function. Figures 2A and 2B show good agreement between the simulated curve of cylinder and the experimental data. From these data it can be concluded that the scattering of cylinder is observed at low volume fraction before and at the percolation threshold. This could indicate the formation of extended aggregates made up of water droplets connected by cytochrome c.

At low q values ($q < 10^{-2}$ Å^{-1}) there is no agreement between the scattered intensity and the simulated curve for a cylinder. This can be explained by the entanglement of the aggregates. To simulate this behavior at low q values, we use the "correlation hole" model [18] developed by de Gennes to describe the semidilute regime in polymer solutions where the polymers are entangled. The average mesh size ξ can be determined when the average distance between cylinders is smaller than their length. The intensity can be expressed as

$$I(q) = S(q) \cdot P_c(q) , \qquad (3)$$

where $P_c(q)$ is the form factor of the cylindrical particles, and $S(q)$ is the structure factor related to the lattice structure induced by the entanglement of the cylindrical particles, and given by [19]

$$S(q) \sim q^2 \xi^2 / (1 + q^2 \xi^2) . \qquad (4)$$

Figures 2C and D show the simulation obtained at $\phi = 7\%$ and $\phi = 13\%$, respectively. The values for the average mesh size ξ obtained from this method are equal to 310 Å and 260 Å for $\phi = 7\%$ and $\phi = 13\%$, respectively.

B) Results obtained at high polar volume fraction ($\phi = 24\%$ and 32%)

The correlations are important and lead to a distortion of the scattering curves. In order to ex-

tract the structure factor, we assumed the following hypothesis: the geometry of the particles is unchanged at various polar volume fractions. This corresponds to an unchanged form factor, $P_c(q)$. If we assume that for $q > 10^{-2}$ Å^{-1}, the scattering behavior is entirely controlled by the form factor $P_c(q)$, we have $S_{7\%}(q) = 1$. We divide the intensity at 24% and 32% by the intensity obtained at 7%:

$$I_{24\%}(q)/I_{7\%}(q) = [P_c(q) \cdot S_{24\%}(q)]/P_c(q) = S_{24\%}(q)$$

$$I_{32\%}(q)/I_{7\%}(q) = [P_c(q) \cdot S_{32\%}(q)]/P_c(q) = S_{32\%}(q) .$$

Figures 3A and 3B show the result obtained from this method at $\phi = 24\%$ and $\phi = 32\%$, respectively. The structure factor has the same behavior as one expected from a simple atomic liquid solution. So, we used the peak position q_{max} to determine the values of the average mesh size ξ:

$$\xi = 2\pi/q_{max} . \qquad (5)$$

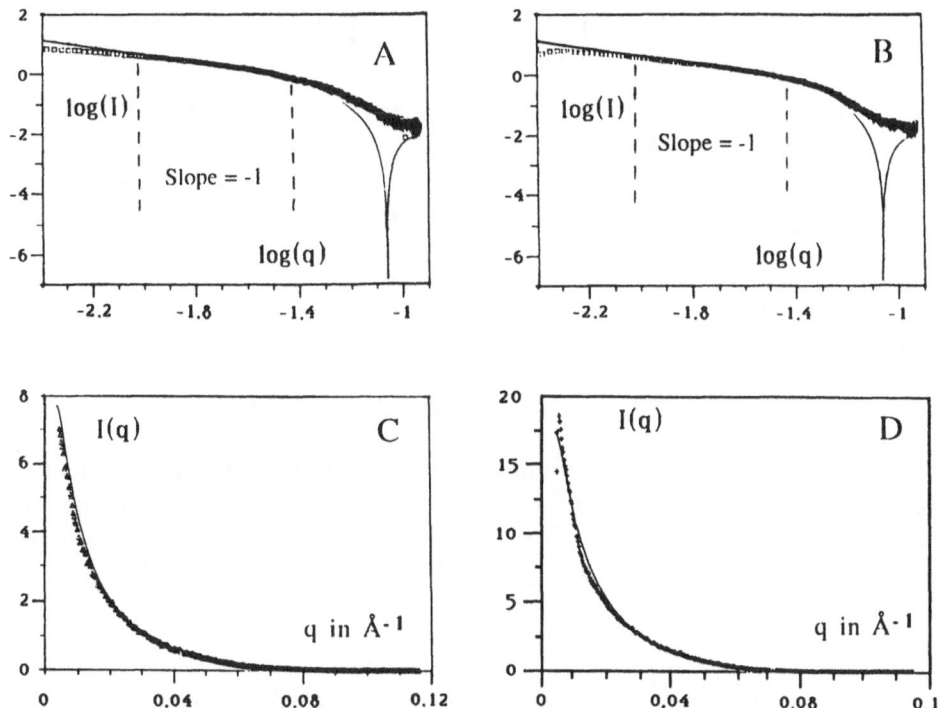

Fig. 2. Scattered intensity behavior versus wave vector in a log-log plot for a polar volume fraction equal to 7% (A) and 13% (B). The solid line represents the simulated curve of the scattering by a homogeneous cylinder ($R_c = 45$ Å). $x = 4$ and $w = 40$. The range of q values where a slope equal to -1 is observed is shown in this figure. Simulation of the scattering curve at $\phi = 7\%$ (C) and at $\phi = 13\%$ (B) using the correlation hole model described in the text. The average mesh size ξ is equal to 310 Å at $\phi = 7\%$ and to 260 Å at $\phi = 13\%$

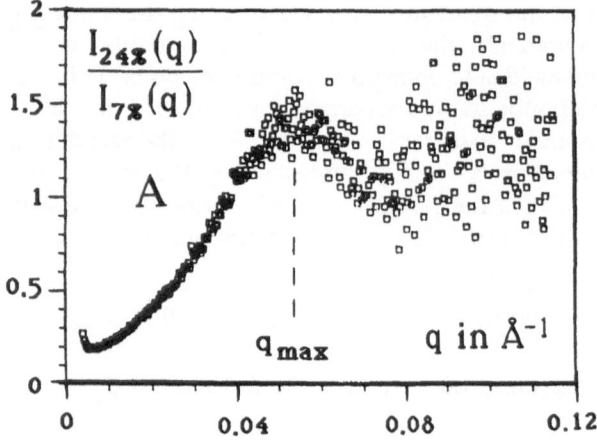

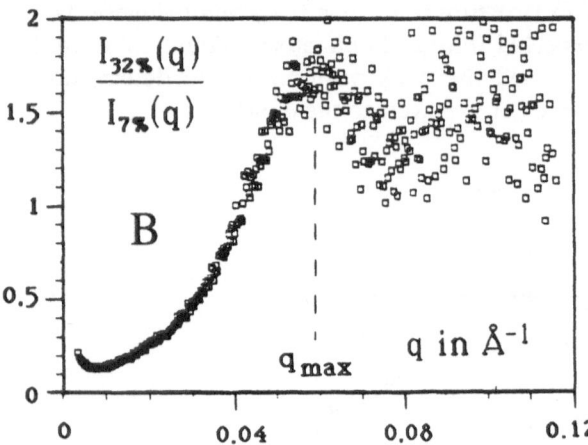

Fig. 3. Structure factor obtained by dividing the intensity at ϕ = 24% (A) and at ϕ = 32% (B) by the intensity at ϕ = 7%. The peak positions correspond, for q_{max}, to a value equal to 0.053 Å^{-1} at ϕ = 24% and to 0.059 Å^{-1} at ϕ = 32%. From Eq. (5), the average mesh size ξ is equal to about 118 Å and 106 Å, and at ϕ equal to 24% and 32%, respectively

For a polar volume fraction equal to 24% and 32%, we have an average mesh size ξ equal to 118 Å and 106 Å, respectively.

C) Geometric model

Assuming that the behavior of the AOT-isooctane-water-cytochrome c solutions is similar to that of polymer in semidilute regime, Table 1 shows a decrease of the distance, denoted ξ_{exp}, with increasing volume fraction of water.

Table 1. Comparison of the average mesh size obtained experimentally (ξ_{exp}) and from the geometric model (ξ_{geo})

ϕ	7%	13%	24%	32%
ξ_{exp}	310 Å	260 Å	118 Å	106 Å
ξ_{geo}	326 Å	236 Å	180 Å	156 Å

These distances can be determined using a geometrical model that assumes the aggregate to be a homogeneous cylinder. Its surface S_c and its volume V_c are given, respectively, by:

$$S_c = 2 \cdot \pi \cdot R_c l_v \quad \text{and} \quad V_c = \pi \cdot R_c^2 l_v . \quad (6)$$

R_c is the radius and l_v is the length of cylinder per unit volume. The total volume V_T, and the total surface S_T of cylinders are equal to nV_c and nS_c, respectively, where n is the total number of cylinders in the solution. Then, the ratio V_T/S_T is equal to $R_c/2$. Assuming that the total surface of the cylinder is due to the AOT molecules, the specific interface Σ, expressed per unit volume, is given by:

$$\Sigma(A^2/Å^3) = [AOT] \cdot N \cdot \sigma_{AOT} \cdot 10^{-27} . \quad (7)$$

σ_{AOT} is the surface area per polar head group and is found to be equal to 60 Å^2 [1]. For a given volume of solution V the total interface is: $S_T = V\Sigma$. Assuming that the total volume of the cylinders V_T is due to the volume of the water molecules V_{aq}, it is deduced by

$$V_T/S_T = R_c/2 = V_{aq}/V\Sigma = \phi/\Sigma . \quad (8)$$

Assuming that the entire interface between oil and water is only composed of the surfactant, the surface of the cylinder per unit volume S_c is equal to Σ, and it is deduced (Eqs. (6) and (8)) by

$$l_v = \Sigma^2/4 \cdot \pi \cdot \phi . \quad (9)$$

The average mesh size ξ, denoted ξ_{geo} for this geometric model, is the radius of the correlation tube containing a homogeneous cylinder from which the other cylinders are strongly repelled. From this model, a simple geometric calculation leads to

$$\pi \cdot (\xi_{geo}/2)^2 \cdot l_v = 1 ; \qquad (10)$$

and then

$$\xi_{geo} = 2(\phi)^{0.5}/\Sigma . \qquad (11)$$

Table 1 shows a decrease in the distance with increasing the volume fraction, which is in good agreement to the data obtained from simulation of the structure factor from the "correlation hole" model.

Conclusions

Because of its location at the interface, the addition of cytochrome *c* to water droplets induces an increase in the attractive interactions between the microemulsion aggregates which promote a percolation process. This usually takes place, using unfilled micelles, at larger volume fraction values and higher temperature. The percolation volume fraction in our experimental conditions ($T = 20\,°C$, $w = 40$) is close to 10%. The structural study by SAXS has been performed using the correlation hole model formulated in order to describe the polymer solutions in semidilute regime. This suggests that the aggregates so formed are highly extended and entangled. this probably means that cytochrome *c* is directly responsible for the aggregation process and must act as a connector between the water droplets.

References

1. Pileni MP, Zemb T, Petit C (1985) Chem Phys Lett 118:414
2. Lemaire B, Bothorel P, Roux D (1983) J Phys Chem 87:1023
3. Huang JS, Safran SA, Kim MW, Grest GS, Kotlarchyk M, Quirke N (1984) Phys Rev Lett, No 6, 53:592
4. Huang JS (1985) J Chem Phys, No 1, 82:480
5. Safran S, Webman I, Grest GS (1985) Phys Rev A, No 1, 32:506
6. Grest GS, Webman I, Safran S, Bug ARL (1986) Phys Rev A 33:2842
7. Bug ARL, Safran SA, Grest GS, Webman I (1985) Phys Rev Letters 55:1896
8. Kim MW, Huang JS (1986) Phys Rev A 34:719
9. Bhattacharya S, Stokes JP, Kim MW, Huang JS (1985) No 18, 55:1884
10. van Dijk MA (1985) Rev Lett, No 9, 55:1003
11. van Dijk MA, Joosten JGH, Levine YK, Bedeaux D (1989) J Phys Chem 93:2506
12. Laguës M (1979) J Phys (Paris) Lett 40:L331
13. Structure and reactivity in reverse micelles: Pileni MP (ed) Studies in physical and theoretical chemestry 65; Elsevier, Amsterdam, 1989
14. Brochette P, Petit C, Pileni MP (1988) 92:3505
15. Petit C, Brochette P, Pileni MP (1986) J Phys Chem 90:6517
16. Huruguen JP, Pileni MP (1991) Eur Biophys J 19:103
17. Guinier A, Fournet G (1955) in Small-Angle Scattering of X-rays; Wiley J (ed) Wiley and Sons, NY
18. de Gennes PG (1979) in Scaling Concept in Polymer Physics; Cornell University Press, London
19. Hayter J, Janninck G, Brochard-Wyart F, de Gennes PG (1980) J Phys Lett 41:L451

Authors' address:

M. P. Pileni
Université P. et M.Curie, SRST
Bâtiment F74, 6ème étage
4 place Jussieu
75230 Paris Cedex 05

Progress in Colloid & Polymer Science

Progr Colloid Polym Sci 89:44—48 (1992)

The use of methyl orange for the characterization of micelles in aqueous nonionic surfactant solutions

T. Sobisch

Department of Surface Active Substances, Central Institute of Organic Chemistry, Berlin*)

Abstract: As has been shown for the system poly(ethylene glycol) (PEG-600) — H_2O, a gradual change in the number of water molecules per oxyethylene (EO) unit leads to a corresponding change of the wavelength of the absorption maximum of methyl orange (MO). This effect has been used to investigate the alterations of the hydration state of oligoether chains in nonionic surfactant solutions. The measurements have been carried out with polydisperse preparations of *p*-(1,1,3,3-tetramethylbutyl)phenoxypoly(oxyethylene glycol) with an average EO-chain length varying from 3.1 (OPE-3.1) to 90 (OPE-90). Special attention has been focused on the influence of lower alkanols on solutions of OPE-9. The shift of the MO peak sometimes parallels the micellization of nonionic surfactants, but not necessarily. The main information obtained by the MO method is about the hydration state of the EO chains, i.e., their aggregation state. By varying the EO degree of the OPEs from low (3.1, 5.4) to medium (9, 10, 13, 18) to high values (70, 90), distinct alterations in the aggregation behavior have been observed. It has also been shown solubilization and the structure-breaking action of alkanols affect the hydration state of the oligoether chains.

Key words: Nonionic micelles; hydration; solubilization; additives; methyl orange; spectral shift

1. Introduction

Changes in the absorption spectra of water-soluble dye molecules have been widely used to provide information about micellar properties of surfactants. Most of the investigations done in this field have been limited to the determination of the critical micelle concentration (CMC), although for this purpose the applicability and precision of the method was often questioned.

In the case of the well-soluble dye methyl orange and nonionic surfactants, errors introduced by mixed micellization of dye and surfactant may be avoided. Moreover, as was shown by Klotz [1], methyl orange exhibits a solvatochromic behavior; hence, it can be used to trace the polarity of its environment.

The objective of the present study was to examine the applicability of methyl orange for investigating the micellization of nonionic surfactants. Therefore, a comparison with the results of the UV- and the surface-tension methods was made. Special attention was directed to the influence of surfactant concentration, EO chain length, and additives on the polarity of the oligoether shell of nonionic micelles.

2. Experimental

2.1. Materials

The nonionic surfactants used were ethoxylation products of recrystallized p-tert-octylphenol with an average EO chain length varying from 3.1 (OPE-3.1) to 90 (OPE-90). These materials were kindly synthesized by coworkers of our department according to [2].

Alkanols were purified as described in [3]. Methyl orange purchased from Feinchemie Sebnitz, as well

---

*) Present affiliation: Adlershofer Umweltschutztechnik- und Forschungsgesellschaft mbH, Berlin.

as PEG-600 supplied by Ferrak were used without further purification.

2.2. Absorption spectra

Absorption spectra of solutions of varying composition containing 10^{-5} M MO were measured relative to solutions without MO on a Specord M 40 spectrophotometer (Carl Zeiss Jena) at 25°C. The spectra were accumulated and smoothed to determine their maximum wavelength with high accuracy (± 0.5 nm), despite broad absorption bands.

2.3. Measurements with OPE-3.1 and OPE-5.4

The cloud points of OPE-3.1 and OPE-5.4 are below the measuring temperature, thus no real solutions can be obtained at 25°C in pure water. Aqueous dispersions of these OPEs were prepared by ultrasonication directly before measurement. To minimize scattering effects caused by the turbidity of the dispersions, the absorption spectra were determined by placing the absorption cells directly in front of the photomultiplier.

3. Results and discussion

3.1. PEG — water mixtures and aqueous OPE solutions

Figure 1 shows the solvatochromic behavior of MO in PEG-600 — water mixtures. The ratio of water molecules per EO group (H_2O/EO) was calculated on the basis of the composition (wt/wt) of these mixtures. A considerable shift of the maximum wavelength can be observed at values between 0 and 2 water molecules per EO group. At higher H_2O/EO values the absorption peak shifts only slightly towards the absorption peak characteristic of MO in pure water (463.5 nm).

Figure 2 comprises the shifts of the absorption peak for the OPEs investigated. Subsequent addition of OPEs causes the MO peak to shift from the absorption peak characteristic of pure water to a value which is somewhat higher than the maximum wavelength characteristic of pure PEG.

Surprisingly, addition of OPE-5.4 only begins to shift the MO peak at concentrations exceeding those necessary for OPE-18, whereas no shift has been observed for OPE-3.1 in the concentration range investigated.

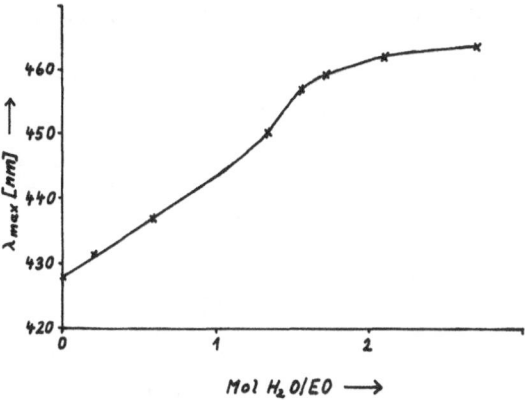

Fig. 1. Maximum wavelength of methyl orange in PEG-600 — water mixtures

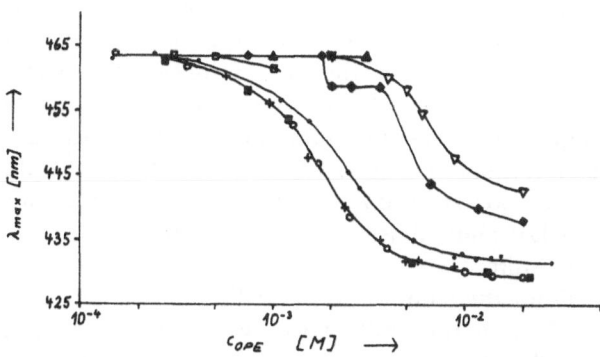

Fig. 2. Maximum wavelength of methyl orange in aqueous OPE solutions ▲ = 3.1; □ = 5.4; ○ = 9; ■ = 10; + = 13; ● = 18; ◆ = 70; ▽ = 90

Comparing the behavior of OPE-9, -10, and -13, no differences have been found.

In the case of OPE-70 a plateau has been observed at a wavelength of 458.6 nm corresponding to approximately 1.5 H_2O/EO (Fig. 1). Table 1 compares the results of the MO method with the results of surface tension and UV measurements [4]. The symbols c_s, c_{min}, c_1', c_2', c_1, and c_2 stand for the concentrations of the saturation at the air-water interface, the concentrations of the surface tension minima, and the approximate concentrations, where a pronounced shift of the UV or MO peak starts or ends. From this comparison the following can be seen.

For OPE-3.1 and OPE-5.4 the shift of the MO peak begins well above the concentrations of the satura-

Table 1. Comparison of results concerning micellization of OPEs obtained by the surface tension, UV-, and MO-methods

OPE	Surface tension		UV		MO	
	c_s	c_{min}	c'_1	c'_2	c_1	c_2
OPE-3.1	$1.1 \cdot 10^{-4}$	—	$5 \cdot 10^{-5}$	$5 \cdot 10^{-4}$	$>3 \cdot 10^{-3}$	
OPE-5.4	$1.2 \cdot 10^{-4}$	—	10^{-4}	$4 \cdot 10^{-4}$	$8 \cdot 10^{-4}$	
OPE-9		$2.5 \cdot 10^{-4}$	$1.5 \cdot 10^{-4}$	$1.5 \cdot 10^{-3}$	$5 \cdot 10^{-4}$	$4 \cdot 10^{-3}$
OPE-10		$2.8 \cdot 10^{-4}$	$2 \cdot 10^{-4}$	$1.5 \cdot 10^{-3}$	$5 \cdot 10^{-4}$	$4 \cdot 10^{-3}$
OPE-13		$2.9 \cdot 10^{-4}$	$2.5 \cdot 10^{-4}$	$2 \cdot 10^{-3}$	$6 \cdot 10^{-4}$	$5 \cdot 10^{-3}$
OPE-18		$3 \cdot 10^{-4}$	$3 \cdot 10^{-4}$	$2.5 \cdot 10^{-3}$	$7 \cdot 10^{-4}$	$5 \cdot 10^{-3}$
OPE-70		$5 \cdot 10^{-4}$			$2 \cdot 10^{-3}$	$9 \cdot 10^{-3}$
OPE-90		$8 \cdot 10^{-4}$			$3.5 \cdot 10^{-3}$	

tion at the air-water interface and those of the beginning shift of the UV peak. In the case of OPE-70 and OPE-90 the shift of the MO peak starts only at concentrations which are several times higher than the concentrations of the surface tension minima, whereas a shift of the UV peak could not be observed. If a c'_2 value could be obtained for the UV shift, it was always lower than c_2 for the MO shift.

From this comparison and Fig. 1, one can conclude that the shift of the MO peak sometimes parallels the micellization of nonionic surfactants, but not necessarily. The main information that can be obtained, however, is information about the hydration state of the EO chains, i.e., about their aggregation state. Therefore, the MO method indicates micellization or even demicellization via the change in the hydration state of the EO groups. It should be mentioned that, in micellar systems, the shift of the MO peak is obviously not an absolute measure of the hydration of the EO chains in general. On the one hand, the hydration of the EO groups in the oligoether shell changes with the distance from the benzene groups of surfactants. On the other hand, the MO molecules may be solubilized in the vicinity of the border of the hydrophobic core and the oligoether shell or somewhere else in the oligoether shell.

For the OPEs investigated, the following additional conclusions can be drawn.

In the case of OPE-3.1 and OPE-5.4 the EO chains of the surfactant phase dispersed in water remain in a highly hydrated state over a wide range of concentrations.

It seems that between 9 and 13 EO per molecule the hydration state of the oligoether shell of OPEs does not depend on the EO degree.

For OPE-9, -10, -13, and -18 there is still a remarkable change of the hydration state of the oligoether shell, even if the environment of the benzene groups of surfactants is only changed slightly. Thus, the aggregation in the hydrophobic core advances the aggregation in the oligoether shell. This effect is the more pronounced the longer the oligoether chains.

In the case of OPE-70 and -90, a change in the hydration state of the polyether chains takes place without an observable change of the environment of the benzene groups, hence, without significant aggregation of these parts of the molecules. Nevertheless, the aggregation of the polyether chains is strongly promoted by the relatively small hydrophobic part of the molecules, causing the shift for OPE-70 at a considerably lower concentration of EO groups than for OPE-90. The value of 1.5 H_2O/EO, corresponding to the plateau observed for OPE-70 but not for OPE-90, is believed to be characteristic of lamellar structures [5]. From all this, it can be seen that, unexpectedly, in the case of OPEs with high EO degree the changes in the aggregation state with varying concentration are highly sensitive to changes of the polyether chain length.

3.2. Effect of lower alcohols

Figure 3 shows the effect of the addition of alkanols on the MO absorption peak in $5 \cdot 10^{-3}$ M

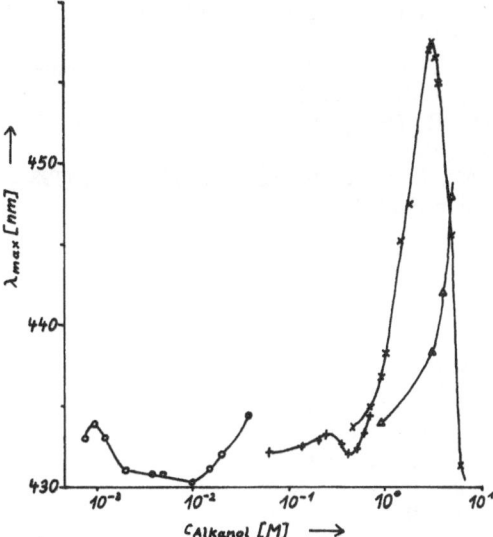

Fig. 3. Effect of alkanol addition on the maximum wavelength of methyl orange in aqueous solution of $5 \cdot 10^{-3}$ M OPE-9. $\triangle$ = methanol; $\times$ = 2-propanol; + = n-butanol; $\circ$ = n-hexanol

aqueous solutions of OPE-9. Addition of 2-propanol and methanol causes the MO peak to shift from the value characteristic of the aqueous $5 \cdot 10^{-3}$ M OPE-9 solution towards the value characteristic of water.

The addition of small amounts of n-butanol and n-hexanol also induces a shift towards higher values, but this amounts only to approximately 2 nm. The shift is then reversed and the maximum wavelength falls below the initial value. Whereas the reversal is very distinct in the case of hexanol, it is only slight for butanol. Approaching the alkanol concentration where the cloud point is depressed to 25 °C [3], the maximum wavelength is raised again.

The initial shift observed for n-butanol and n-hexanol addition may be explained by the assumption that the solubilization of small amounts of alkanols causes a loosening-up of the oligoether shell on sterical grounds. The uptake of greater amounts of alkanols, however, results in a dehydration of the palisade layer [6] and causes a reverse of the shift. The repeated increase of the maximum wavelength corresponds with the highly hydrated state of the EO chains of OPE-3.1 and -5.4 observed when they are dispersed in water at temperature above their cloud point (see 3.1.).

The shift caused by methanol and 2-propanol addition is due to the cosolvent action of these alkanols acting against the hydrophobic interactions in the oligoether shell and the hydrophobic core, thus giving rise to the hydration of the oligoether chains. By investigating the spectral shift in the UV region it could be confirmed that this is accompanied by a disaggregation of the benzene groups of surfactants. Hence, within certain limits, the MO method also allows to obtain information about the demicellization process. However, as demonstrated, many more conclusions can be drawn from the information provided about the changes in the hydration state of the oligoether shell.

In addition, at high concentrations of 2-propanol the MO traces the change in polarity of the mixed solvent due to its solvatochromic shift after passing the maximum.

4. Conclusions

The shift of the MO peak sometimes parallels the micellization of nonionic surfactants, but not necessarily. The main information which can be obtained by the MO method is that about the hydration state of the EO chains, i.e., about their aggregation state.

With the variation of the EO degree the following changes in the hydration state of the EO chains are observable.

In the case of OPEs with a cloud point below room temperature, the EO chains of the surfactant phase dispersed in water remain in a highly hydrated state over a wide range of concentrations. For OPEs with medium EO chain length (EO-9—OPE-18) there is still a remarkable change in the hydration state of the oligoether shell, even if the environment of the benzene groups of surfactants is only slightly changed. Thus, the aggregation in the hydrophobic core advances the aggregation in the oligoether shell. In the case of OPEs with long EO chains (OPE-70 and -90) the aggregation of the polyether chains is strongly promoted by the relatively small hydrophobic part of the molecules. Nevertheless, the change in the hydration state of the polyether chains takes place without an observable change in the environment of the benzene groups.

Addition of lower alkanols induces the following changes in the hydration state of OPE-9 micelles.

The solubilization of small amounts of alkanols causes a loosening-up of the oligoether shell. The uptake of greater amounts of alkanols, however, results in a dehydration of the palisade layer. The latter is reversed when the cloud point approaches room temperature. Whereas these solubilization effects dominate the hydration behavior when n-butanol or n-hexanol are added, the hydration is affected by 2-propanol or methanol due to their action as structure-breakers that cause gradual disaggregation of oligoether chains and demicellization.

References

1. Klotz IM, Burkhard RK, Urquhart JM (1952) J Phys Chem 56:77—85
2. Miller SA, Bann B, Thrower RD (1950) J Chem Soc 3623—3628
3. Sobisch T, Wüstneck R (1992) Colloids and Surfaces 62:187—198
4. Sobisch T (1989) Dissertation A, Academy of Sciences of the GDR, Berlin
5. Heusch R (1984) Ber Bunsenges Phys Chem 88:1093—1098
6. Ward AJI, Marie C, Sylvia L, Phillipi MA (1988) J Dispersion Sci 9:149—169

Author's address:

Dr. Titus Sobisch
Adlershofer Umweltschutztechnik-
und Forschungsgesellschaft mbH
Rudower Chaussee 5
O-1199 Berlin, FRG

Progress in Colloid & Polymer Science   Progr Colloid Polym Sci 89:49—52 (1992)

Structural order of Tobacco-Mosaic-Virus in aqueous solutions determined by static and dynamic light scattering

M. Hagenbüchle, C. Graf, and R. Weber

Universität Konstanz, Fakultät für Physik, FRG

Abstract: Aqueous solutions of rodlike tobacco mosaic virus (TMV) (L = 300 nm, d = 18 nm) below and above the overlap concentration c^* (c^* = 1 particle/length^3) are examined by means of dynamic light-scattering techniques (DLS) and compared to static light-scattering (SLS) results. There, samples of very low ionic strength show a liquid-like structure due to the electrostatic interaction of the charged rods, whereas at sufficiently high ionic strength the Coulomb interaction is screened and ordering disappears. The liquid-like structure at minimal ionic strength is examined in the concentration range 0.25 c^* — 1.4 c^*. A detailed comparison shows differences in the static structure factor $S(q)$ determined by SLS and DLS, respectively. Moreover, the transition from a liquid-like to a gas-like order is shown as a function of the salt concentration, i.e., the ionic strength. For the calculation of $S(q)$ from DLS-data, we have measured the apparent diffusion coefficient of TMV solutions where the Coulomb interaction is screened by added NaCl or Tris-HCl.

Key words: Tobacco mosaic virus; static and dynamic light scattering; ionic strength; particle concentration

1. Introduction

This article deals with light-scattering experiments on aqueous solutions of TMV. The virus particles are 300 nm long and their diameter is about 18 nm. Most of the work on TMV reported in the literature deals with dynamic properties in buffer solutions at low particle concentrations where the Coulomb interaction is totally screened. Since it is known from spherical particle systems that strong Coulomb interaction reveals interesting features, we have performed measurements on aqueous TMV solutions at very low ionic strength ($\leqslant 10^{-6}$ M) in the dilute and semidilute regime. These two concentration regimes are divided by the overlap concentration c^* defined as 1 particle/length^3 (c^* = 2.45 mg/ml).

2. Experimental

2.1. Tobacco-Mosaic-Virus

The original sample containing 46 mg/ml TMV in aqueous solution was a gift from Prof. C. Wetter (University of Saarbrücken). From this initial solution the various samples with different concentrations and ionic strengths were carefully prepared as described in detail in [1]. For concentration-dependent measurements at minimal ionic strength five saltfree samples were obtained by diluting the stock solution with deionized and highly purified water and adding a cleaned mixed-bed ion exchange resin. The concentration of these samples varied between 0.62 mg/ml (0.25 c^*) and 3.41 mg/ml (1.4 c^*). For the measurements with screened Coulomb interaction, samples were prepared by diluting with 100 mM Tris-HCl buffer, as well as with 10 mM NaCl.

All these samples were directly prepared in the scattering cells. To study the ionic strength dependence at a fixed particle concentration we had to use a different procedure, which was described in detail elsewhere [2, 3]. This technique allowed us a very precise adjustment of the ionic strength. The actual concentration of all samples was determined by their characteristic UV-absorption at λ = 260 nm.

2.2. Static Light Scattering

All static structure factors were measured with a commercial light-scattering apparatus (ALV, Langen, FRG) [2]. The temperature of the scattering cell was stabilized at $T = 21 \pm 1°C$. The measured intensity data were corrected by the angle dependence of the scattering volume and normalized by the concentration and the intensity of a reference sample (toluene). Then, the static structure factor $S(q)$ was calculated with the relation [1]

$$I(q) = P(q) \cdot S(q) , \tag{1}$$

where $P(q)$ denotes the formfactor for rodlike particles in the isotropic phase [4].

2.3. Dynamic Light Scattering

For dynamic light-scattering measurements we used the same experimental setup. But now a digital correlator (ALV 3000) formed the homodyne time correlation function $g_I(q, t)$ from the intensity data. With the measured background $\langle I(q) \rangle$ the normalized field correlation function $g_E(q, t)$ could be calculated by using the well-known Siegert relation [4]. Then, the first cumulant $K_1(q)$ and the apparent diffusion coefficient $D_{app}(q)$ were extracted from $g_E(q, t)$ according to

$$K_1(q) = D_{app}(q) \cdot q^2 = \lim_{t \to 0} -\frac{d}{dt} (\ln g_E(q, t)) . \tag{2}$$

The following equation had been originally derived and verified for systems of interacting spherical particles, but recently it was also claimed to be valid for weakly interacting rodlike particles [5, 6]:

$$D_{i,app}(q) = \frac{D_{0,app}(q)}{S(q)} , \tag{3}$$

where $D_{i,app}(q)$ is the short-time diffusion coefficient for interacting and $D_{0,app}(q)$ for noninteracting rods. For use in the next section it will be helpful to distinguish between the static structure factors defined by Eqs. (1) and (3), therefore, we call the static structure factor measured with SLS $S_S(q)$, whereas $S_D(q)$ stands for the structure factor determined by Eq. (3).

3. Results and discussion

According to Wilcoxon and Schurr [7] the apparent short-time diffusion coefficient of rodlike particles with screened Coulomb interaction is given by

$$D_{0,app}(q) = D_T + 2 \cdot (D_\parallel - D_\perp)$$

$$\cdot \left[\frac{1}{3} - F(q) \right] + 2 \cdot L^2 \cdot D_R \cdot G(q), \tag{4}$$

where D_T, $D_\parallel$, $D_\perp$ and D_R are the average translational, parallel translational, perpendicular translational, and rotational diffusion coefficients. Tirado and de la Torre [8, 9] calculated the friction coefficients for a rigid rod and derived the required diffusion coefficients from it. These theoretical considerations were summarized and presented by Stimpson and Bloomfield [10] in a form that is easy to apply. There, analytical expressions for the amplitudes $F(q)$ and $G(q)$ can also be found.

Wilcoxon and Schurr measured $D_{0,app}$ with a sample that contained 0.05 mg/ml TMV in 10 mM NaCl. Compared to the overlap concentration c^*, this is a very diluted sample. Since our measurements on TMV solutions at minimal ionic strength are in the concentration region around c^*, we first tested if there exists an influence on $D_{0,app}$ caused by the overlap of the particles in this concentration regime. Figure 1 shows the short-time diffusion coefficients $D_{0,app}(q)$ of TMV solutions with screened Coulomb interaction. The filled circles (●) apply to TMV at 3.85 mg/ml (1.57 c^*) in 10 mM NaCl. At this concentration the Coulomb interaction is not entirely screened by 10 mM NaCl as a slightly ascending $S(q)$ of this sample shows. Raising the ionic strength above 10 mM NaCl to improve the screening seems not feasible because TMV then forms large aggregates as also mentioned by Wilcoxon and Schurr. $D_{0,app}$ of this sample is therefore calculated by correcting the measured diffusion coefficients by the simultaneously measured $S(q)$. Sano [11] showed that the aggregation of the TMV particles depends on the species of the added salt, and found that Tris-HCl causes less aggregates than NaCl. Alternatively, we therefore measured the diffusion coefficients of TMV in 100 mM Tris-HCl at various concentrations below and above c^* (open symbols in Fig. 1). To our surprise, we find a concentration dependence of $D_{0,app}$. While $D_{0,app}(q)$ of the sample

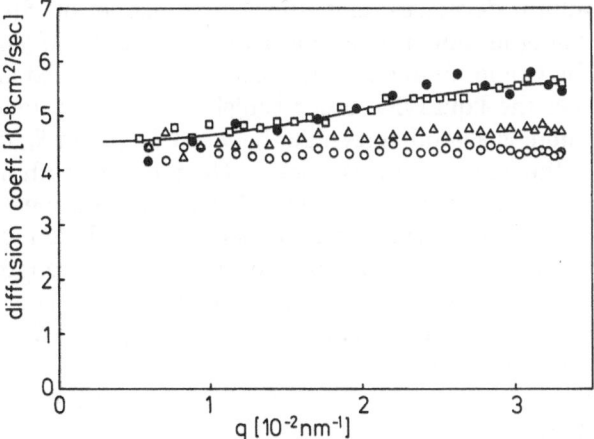

Fig. 1. Short-time diffusion coefficients $D_{0,app}(q)$ of TMV samples with screened Coulomb interaction versus the scattering vector q. The solid line gives the theoretical expression derived by Wilcoxon and Schurr; •) $c = 3.85$ mg/ml, 10 mM NaCl; □) $c = 0.6$ mg/ml, 100 mM Tris-HCl; △) $c = 2.27$ mg/ml, 100 mM Tris-HCl; ○) $c = 3.7$ mg/ml, 100 mM Tris-HCl

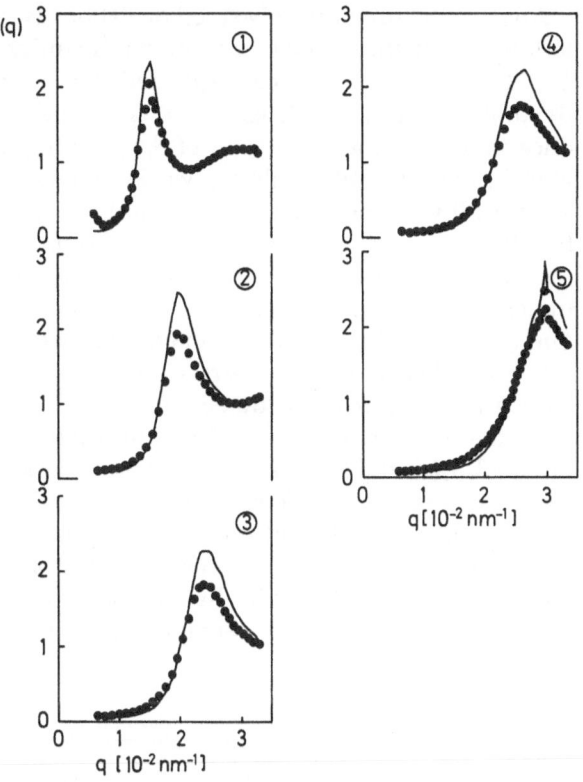

Fig. 2. Structure factor $S_D(q)$ (•) calculated from dynamic light scattering data compared with structure factor $S_S(q)$ (—) determined by static light scattering at minimal ionic strength. 1) 0.62 mg/ml; 2) 1.41 mg/ml; 3) 2.12 mg/ml; 4) 2.39 mg/ml; 5) 3.41 mg/ml

with the lowest concentration (□) closely follows the theoretical curve, the values in the high q-region decrease with increasing concentration, whereas $D_{0,app}(q)$ remains nearly constant in the low q-region. The static structure factor of these samples was constant over the whole q-region, demonstrating that the Coulomb interaction was sufficiently screened. Since, on the one hand, the data of TMV in NaCl solution show no concentration dependence and fit the theory, and on the other hand the well-known Doi-Edwards-theory for rods in semidilute solutions predicts a significant decrease of the rotational and parallel translational diffusion coefficients only for concentrations well above c^* [12], we assume that Eq. (4) is still valid in the examined concentration regime and ascribe the observed deviations in Tris-HCl solutions to the buffer. For the calculation of $S_D(q)$ from the dynamic light scattering data of the TMV solutions with strong Coulomb interaction according to Eq. (3), we therefore use the theoretical $D_{0,app}(q)$-values.

Dynamic light scattering on TMV solutions in the dilute and semidilute concentration regime at very low ionic strength is dominated by the Coulomb- and steric-particle interaction. Analogous to strongly interacting fd-virus solutions [13], the structure of the apparent short-time diffusion coefficient $D_{i,app}(q)$ is governed in a first approximation by the

reciprocal of the static structure factor $S(q)$. Using the theoretical values of $D_{0,app}(q)$ and the measured $D_{i,app}(q)$, we calculated $S_D(q)$ via Eq. (3) and compared $S_D(q)$ with $S_S(q)$, which was determined by static light scattering (Fig. 2). For the lowest concentration there is a rather good agreement between $S_D(q)$ and $S_S(q)$, whereas at higher concentrations $S_D(q)$ differs clearly from $S_S(q)$. The deviations occur in the region where the static structure factor exhibits its maximum. Graf et al. [14] found a similar behavior for fd solutions where, however, the deviations only occur at concentrations higher than $5\,c^*$. For these concentrations (also in the work of Schulz et al. [13]) it can be observed that $S_D(q)$ differs slightly from $S_S(q)$ in the region of the first maximum. The deviations in the low q-region observed by the same authors are in all probability due to aggregates or dust particles. In accordance with Graf et al., good agreement in the low q-region can be achieved by adequate centrifugation.

An explanation for the observed deviations is still lacking. Since these differences are on the order of 15—20%, they cannot be explained by hydrodynamic interactions alone, but it is to mention once more that the calculation of $S_D(q)$ via Eq. (3) is claimed to be valid only for weakly interacting systems. If we are looking at the highest concentration that was examined, we notice a sharp peak on top of the first maximum which indicates a phase transition from a liquid-like into a crystalline order due to the very high Coulomb interaction between the rods. Perhaps the observed deviations are related to the near phase transition.

For the concentration $c = 0.45$ mg/ml, we present in Fig. 3 $S_S(q)$ and $S_D(q)$ as a function of the ionic strength. Minimum ionic strength is reached at a conductivity of $\sigma = 0.65$ µS/cm. Here, $S(q)$ takes its maximum value. By adding small amounts of salt the conductivity and with it, the ionic strength, increases. Consequently, the height of the first maximum decreases and the whole slope of $S(q)$ smoothes down. At a conductivity of about 5 µS/cm the Coulomb interaction is extensively screened so that the liquid-like order vanishes almost entirely. There is a good agreement between $S_S(q)$ and $S_D(q)$ at the various ionic strengths. The ionic strength in the figure caption was calculated from the measured conductivities as described by Deggelmann et al. [3]. An examination of the ionic strength dependence at higher concentrations ($c > c^*$) where the significant deviations between $S_D(q)$ and $S_S(q)$ occur is planned.

Acknowledgement

This work has been supported by the Deutsche Forschungsgemeinschaft (SFB 306).

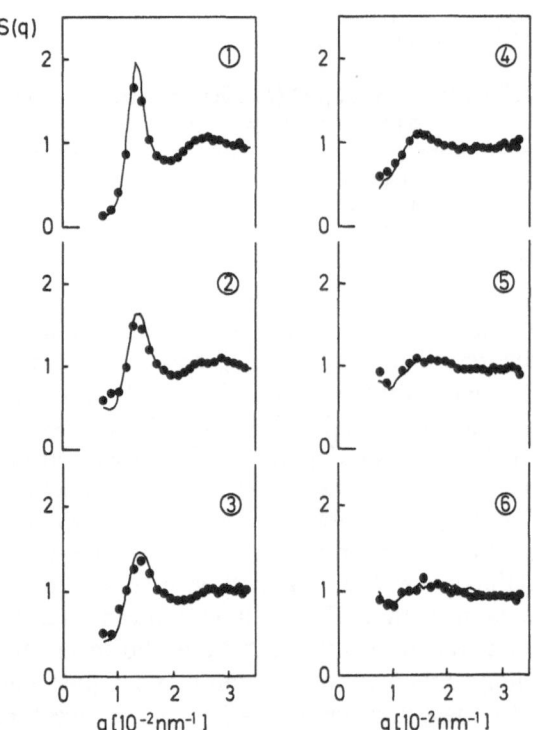

Fig. 3. Comparison of the static structure factor $S(q)$ at various ionic strengths I determined by SLS (—) and DLS (●), respectively. Particle concentration is 0.45 mg/ml. 1) $\sigma = 0.65$ µS/cm, I = 0.0023 mM; 2) $\sigma = 1.25$ µS/cm, I = 0.0116 mM; 3) $\sigma = 1.5$ µS/cm, I = 0.0155 mM; 4) $\sigma = 2.8$ µS/cm, I = 0.0358 mM; 5) $\sigma = 3.85$ µS/cm, I = 0.522 mM; 6) $\sigma = 5.4$ µS/cm, I = 0.0764 mM

References

1. Maier EE, Schulz SF, Weber R (1988) Macromolecules 21:1544—1546
2. Maier EE, Fraden S, Krause R, Deggelmann M, Hagenbüchle M, Weber R (1991) Macromolecules 25:1125—1133
3. Deggelmann M, Palberg T, Hagenbüchle M, Maier EE, Krause R, Graf C, Weber R (1991) J Coll and Int Sci 143:318—326
4. Berne JB, Pecora R (1976) In: Dynamic Light Scattering. John Wiley and Sons, Inc., New York
5. Weyerich B, D'Aguanno B, Canessa E, Klein R (1990) Faraday Discuss Chem Soc 90:245—259
6. Maeda T (1991) Macromolecules 24:2740—2747
7. Wilcoxon J, Schurr JM (1983) Biopolymers 22:849—867
8. Tirado M, de la Torre J (1979) J Chem Phys 71:2581—2587
9. Tirado M, de la Torre J (1980) J Chem Phys 73:1986—1993
10. Stimpson DI, Bloomfield VA (1985) Biopolymers 24:387—402
11. Sano Y (1987) J gen Virol 68:2439—2442
12. Doi M, Edwards SF (1986) In: The Theory of Polymer Dynamics. Oxford University Press, New York
13. Schulz SF, Maier EE, Weber R (1989) J Chem Phys 90(1):7—10
14. Graf C, Deggelmann M, Hagenbüchle M, Kramer H, Krause R, Martin C, Weber R (1991) J Chem Phys 95:6284—6289

Authors' address:

Martin Hagenbüchle
Universität Konstanz
Fakultät für Physik
Postfach 5560
7750 Konstanz, Germany

Progress in Colloid & Polymer Science   Progr Colloid Polym Sci 89:53—55 (1992)

A phase electric birefringence study of interdroplet attractions in water-in-oil microemulsions

M. Paillette

Groupe de Physique des Solides (URA 17/CNRS), Paris, France

Abstract: Phase electric birefringence measurements were performed in water/BHDC (hexadecylbenzyldimethyl ammonium chloride)/benzene microemulsion for molar water-to-surfactant ratio W_0 values between 16 and 20. The extent of negative and positive Kerr parts as W_0 varies is delimited. The two contributions appear simultaneously for $W_0 = 18$ and 19. The alternative Kerr modulus exhibits a sharp dip at $W_0 = 19$ as W_0 varies. From a recent model accounting for the negative electric birefringence, we yield a reasonable estimate for the penetration length of overlapping droplet pairs.

Key words: Kerr effect; BHDC; W/O microemulsions; interactions

Introduction

The three-component system H_2O, BHDC (hexa-decyl-benzyl-dimethyl ammonium chloride), benzene, shows from transient-(TEB) and phase (PEB) electric birefringence measurements that the Kerr constant B_k is negative for molar water-to-surfactant ratio W_0 values less than 15 [1] and positive for W_0 values larger than 22 [2]. This unexpected and interesting phenomenon seems common to all the attractive-type diluted W/O microemulsions, e.g., AOT/H_2O/Isooctane [3]. Two possible models to account for this behavior were proposed [4, 5].

In the investigated system, one would expect that a peculiar manifestation would be encountered in the W_0 values range [16—22]. The aim of the present study is to present the results obtained from PEB measurements in five series of samples with different W_0's [16—20].

Results and discussion

The preparation of the microemulsions and PEB measurement system were described in a previous paper [6].

The modulus $|B(2\omega)|$ and the phase angle $\varphi(2\omega)$ have been measured. The latter parameter plotted versus logarithm of the measurement frequency (1 kHz—10 MHz) is shown in Fig. 1.

The manifestation of a positive Kerr contribution corresponds to a phase variation range (0—90°), while the negative one corresponds to the range 180—270°. Figure 1 shows that the positive contribution extends up to $W_0 = 20$, and the negative one up to $W_0 = 17$. But the observed situation for $W_0 = 18$ and $W_0 = 19$ seems related to the simultaneous presence of these two opposite signs contributions.

Moreover, we verify that $|B(2\omega)|$ scales as φ^2, indicating only dimer contributions.

As the modulus remains constant near 1 kHz and down, Fig. 2 shows a log-log plot of the modulus Kerr constant at 1 kHz relative to the benzene Kerr constant (41 · 10^{-16} V^{-2} m) versus the parameter W_0. In the vicinity of $W_0 = 19$ there is a sharp dip of this ratio, in qualitative agreement with the hypothesis.

The presence of the dip reinforces the model proposed by Mayer [5], which is consistent with Lemaire's et al. approach [7].

The positive Kerr part comes from the electric field-induced alignment of the anisotropic neutral pairs of droplets leading to the contribution:

$$B_k^+ = +\Delta a_0 \cdot \Delta a_e / 45 \text{ kT},$$

where Δa_0 and Δa_e are the optical and electrical anisotropies of polarizability, respectively.

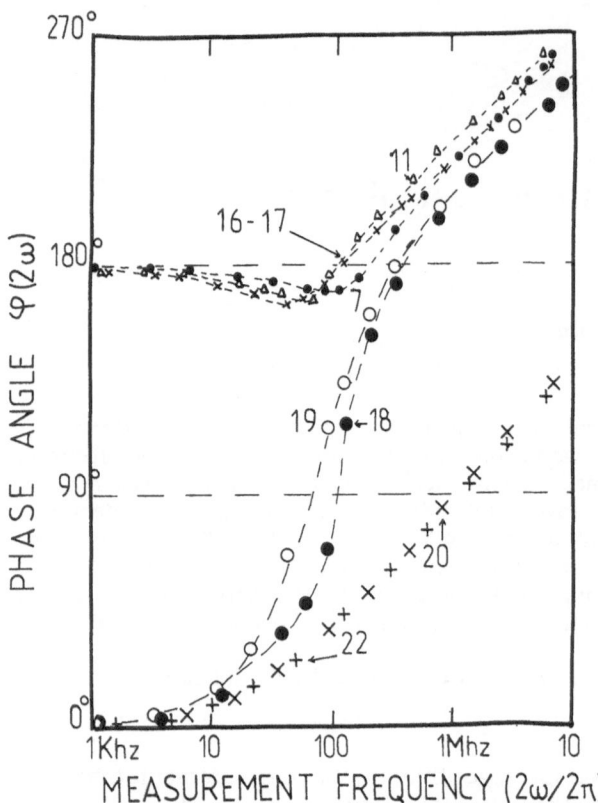

Fig. 1. Phase angle $\varphi(2\omega)$ as a function of the logarithm of the measurement frequency for different molar water-to-surfactant ratio values W_0 (from 7 up to 22). The lines are a guide for the eyes

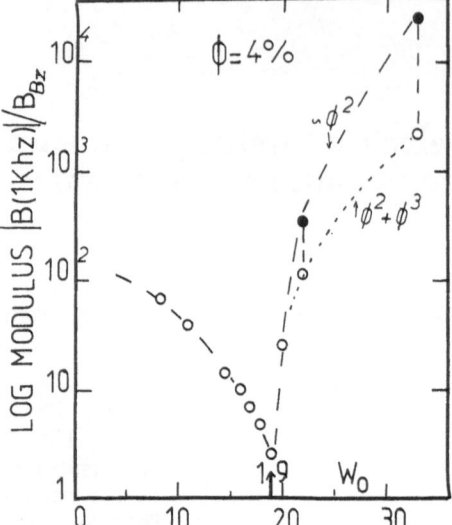

Fig. 2. Logarithm of the absolute value of the modulus $|B(1\text{ kHz})|$ relative to the benzene Kerr constant B_{BZ} as a function of the molar water-to-surfactant ratio values W_0 at volume fraction $\varphi = .04$. Note that the left side below $W_0 = 17$ corresponds to the negative part $(180° \leqslant \varphi(2\omega) \leqslant 270°)$

From T.E.B. measurements in the same system, the presence of the dip of the transient Kerr constant as W_0 varies is observed near $W_0 = 15$ [8].

The negative one is due to the electric field-induced orientation of the charge dumbbell-shape dimers where a charge q is assumed to be trapped along the circle where the droplets merge such that:

$$B_k^- = -\frac{\Delta a_0}{45\ kT} \cdot \frac{q^2}{4\pi\varepsilon_0} \cdot \frac{a}{kT}\left(R_h - \frac{a}{4}\right),$$

where a and R_h are the penetration length and the hydrodynamic radius, respectively.

When the positive and negative parts just balance, as observed for $W_0 = 19$, R_h known, we can estimate a.

For an optical polarizability anisotropy Δa_0:

$$\Delta a_0 = 2\ V_{\text{water core}} \cdot 3\left(\frac{R_h}{2R_h - a}\right)^3,$$

(we yield $a = 5\ \text{A}°$ and $1\ \text{A}°$ for $\Delta a_0 = 2V_{\text{water core}}$.)

Conclusions

The reasonable estimate for a from this rough approach is very promising for Mayer's model. From the fit of frequency-dependent data in progress, we hope to reinforce the model without an adjustable parameter.

Obviously, at the present stage, the basic idea of Mayer's model (i.e., assuming that one electron is "trapped" in the overlapping volume and, more precisely, is free to move along the circle where the droplets merge) remains an open question requiring verification.

References

1. Guering P, Cazabat AM (1983) J Phys Lett (Paris) 44:L601—607; Paillette M (1991) Progr Colloid Polym Sci 84:144—150
2. Guering P, Cazabat AM, Paillette M (1986) Europhysics Lett 2:953—960

3. Eicke HF, Markovic Z (1987) J Colloid Int Sci 79:151—158; (1982) Ibid 85:198—204; Hilficker R, Eicke HF, Hammerich H (1987) Helv Chim Acta 70:257—262
4. Van der Linden E, Geiger S, Bedeaux D (1989) Physica A 156:130—143
5. Mayer G (1990) Chem Phys Lett 168:575—578
6. Paillette M (1982) Opt comm 41:140—144; Paillette M, Guering P, Cazabat AM (1986) Opt Comm 60:244—250
7. Lemaire B, Bothorel P, Roux D (1983) J Phys Chem 87:1023—1028
8. Cazabat AM (1992) Adv Colloid Int Sci 38:33—42

Author's address:

Dr. M. Paillette
Groupe Physique des Solides (Tour 23)
2, Place Jussieu
75251 Paris Cedex 05, France

Progress in Colloid & Polymer Science   Progr Colloid Polym Sci 89:56—59 (1992)

Structure and properties of zwitterionic polysoaps: functionalization by redox-switchable moieties[1])

P. Anton, P. Köberle, and A. Laschewsky

Institut für Organische Chemie, Universität Mainz, FRG

Abstract: Redoxactive monomeric and polymeric surfactants containing viologen and N-alkylated nicotinic acid moieties were synthesized. These systems are potentially able to trigger reversible changes of self-organization by creation or removal of a charge via a redox reaction. Hence, they are investigated with respect to solubility, aggregation behavior, and their electrochemical properties in water. — All monomers, but only the viologen polymers are water-soluble and drastically decrease the surface tension of water. Critical micelle concentrations are observed for the monomers only. — The chemical reversibility of the redox reactions of the compounds in water was investigated using cyclic voltammetry. The redox reaction of N-alkylated nicotinic acid moieties proves to be only partially reversible, whereas the viologen moiety shows a reversible redox reaction for both monomers and polymers. Thus, viologen is a promising redox moiety to trigger reversible changes in self-organization of polysoaps.

Key words: Polymeric surfactant; polysoaps; redoxactive surfactants; self-organization; cyclic voltammetry

Polymeric surfactants may offer many interesting potential applications, e.g., in emulsion polymerization or in tertiary oil recovery. The characteristic properties of polymeric surfactants result from formation of micelle-like structures, because, due to hydrophobic interactions, the side chains undergo intra- and intermolecular aggregation.

The aggregation behavior of both monomeric and polymeric surfactants depends on concentration, temperature, and the applied pressure. In the case of surfactants bearing ionic groups, it can also be influenced by addition of electrolytes. These changes in self-organization are based on physicochemical variation of hydrophilicity influencing the interaction between the surfactants themselves and the solvent.

Alternatively, modification of the self-organization by chemical reactions may be considered, keeping the physical parameters of the system constant. Furthermore, much wider variations of properties should be accessible by chemical reactions. To achieve chemical switching, suitable functional groups have to be incorporated into the surfactants, which enable a marked change of the hydrophilic-hydrophobic balance. For such a purpose, addition or removal of charge in the surfactants would be most efficient.

In order to induce reversible changes of self-organization, functional groups like photo- and electroactive groups [1, 2] have been employed, with which charges can be created or removed without the addition of further reagents, if desired.

For polymeric surfactants, corresponding investigations on reversible changes of aggregation or solubility [3] are scarce, although the fixation of surfactants to a polymeric backbone should provide several advantages. For example, demixing of differently charged surfactant species during the transformation process should be avoided, and cooperative effects may be possible.

In order to explore the potentials of switchable polymeric surfactants, several monomeric and

---

[1]) Extended abstract of a poster presented by A. Laschewsky at the 5th European Colloid and Interface Society Conference, Sept. 25—28, 1991 in Mainz, FRG.

polymeric surfactants bearing different redox groups were synthesized and investigated [5]. Of particular interest was the influence of such functional groups on such parameters as solubility, surface tension, and self-organization behavior of the polymers in comparison to the corresponding monomers. For all systems studied the redox process should result in the formation or disappearance of charge, in order to strongly modify the hydrophilicity of the compound. In the optimal case, the redox processes are reversible in water, i.e., all redox states are stable. Further, the redox group should be small, so that it does not interfere with the self-organization of the surfactants. Hence,

we have chosen surfactants based on N-alkylated nicotinic acid and viologen for preliminary studies.

Scheme 1 shows the monomeric surfactants that were synthesized. The nicotinic acid derivatives 1—3 should allow the change from zwitterionic species to anionic ones. The viologen derivatives 4 and 5 should enable the transformation of cationic species via zwitterionic ones to anionic ones. The compounds bear acrylate and methacrylate moieties and can be homopolymerized by free radical polymerization. The non-functionalized surfactants 7, 8 and their corresponding polymers served as reference compounds to clarify the influence of the redox groups on surfactant behavior.

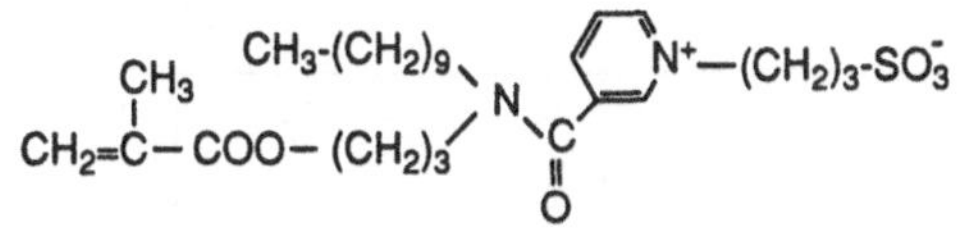

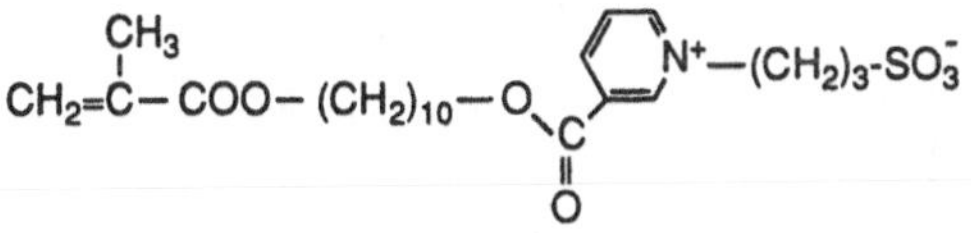

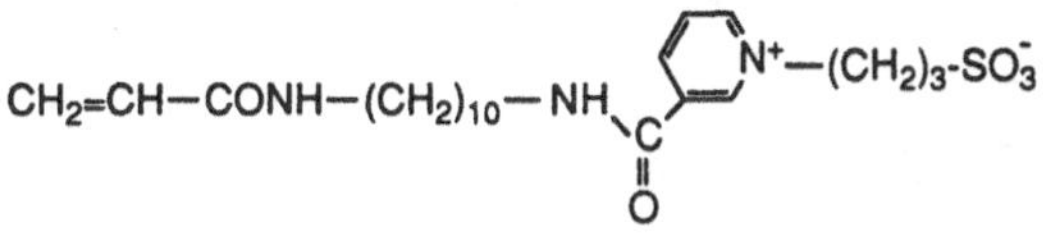

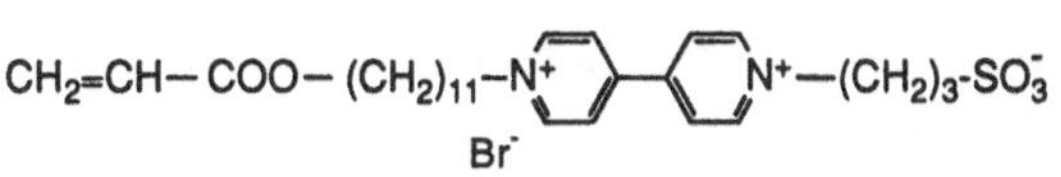

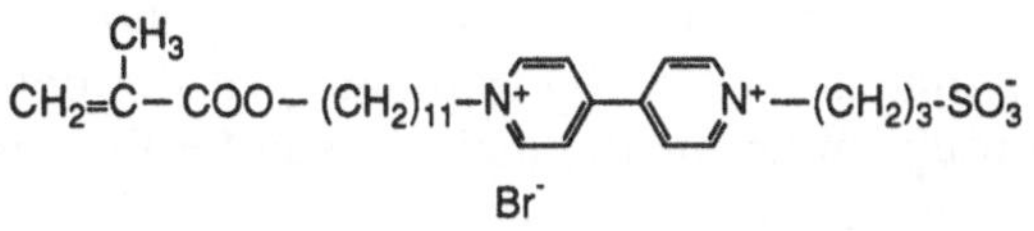

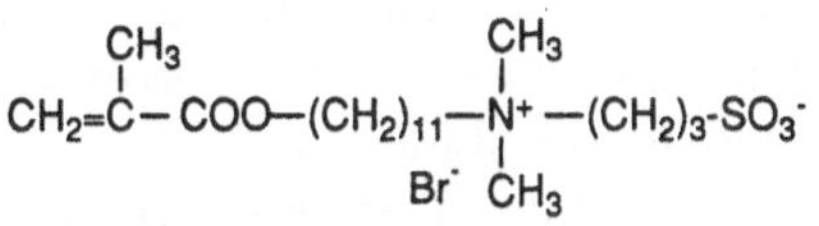

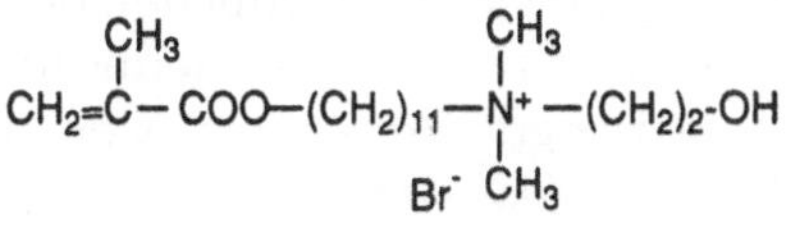

Scheme 1. Monomeric redoxactive surfactants

Polymerization of monomers 1—7 results in polymers of well-defined structure, as verified by IR, ^{1}H-NMR and ^{13}C-NMR [4, 5]. As standard methods fail, molecular weights of the polymers were estimated by end-group analysis. Nitrile-groups of the initiator azobis(isobutyronitrile) could neither be detected in the IR-spectra nor in the ^{13}C-NMR-spectra. Molecular weights of ca. 10^{15} D were found for chromophore-free poly6 and poly7, polymerized under standard conditions using a UV-active azoinitiator [7]. All results indicate the formation of polymers of considerable molecular weight.

Monomers 4—7 are soluble in cold water, monomers 1—3 are soluble in hot water only. With the exception of 3, all monomers dissolve in aqueous 0.1 M NaBr solution at room temperature. Polymers poly1-poly3 are not soluble in water or in brine, but in HCONH$_2$. Polymers poly4—5 are soluble in hot water only, but stay in solution when cooled to 20°C. Both polymers poly6 and poly7 are water-soluble.

Solution behavior of 1—3 and their polymers is probably governed by strong attractive forces of the aromatic moieties or steric effects [5]. Fixation of the surfactants to a polymeric backbone apparently seems to enhance these interactions, as poly1-poly 3 become insoluble in most solvents. Similarly, poly4 and poly5 are more difficult to dissolve than the monomers.

Surface tension measurements were used to investigate the aggregation behavior of the compounds. For the monomers, plots of surface tension vs. concentration are typical for low molecular weight surfactants. Comparing the values of the critical micelle concentration (CMC), values are lowest for the N-alkylated nicotinic acid derivatives 1 and 2 (CMC = 1.5×10^{-3} M and 1.1×10^{-3} M in 0.1 M aq. NaBr), whereas the viologen derivatives 4 and 5 have the highest values (CMC = 2.3×10^{-2} M and 0.9×10^{-2} M). Comparing the redox active compound 5 with its zwitterionic and cationic analoga 6 and 7 (CMC = 2.0×10^{-3} M and 5.9×10^{-3} M), the viologen moiety behaves more hydrophilic than both the quarternary ammonium group and the sulfobetain moiety (Fig. 1).

As reported for classical polysoaps [6], poly7 shows only a minor reduction of surface tension with increasing concentration (Fig. 2). In contrast, solutions of poly4, poly5, and poly6 notably decrease the surface tension, although to a lesser extent than their monomers 4—6. This effect is

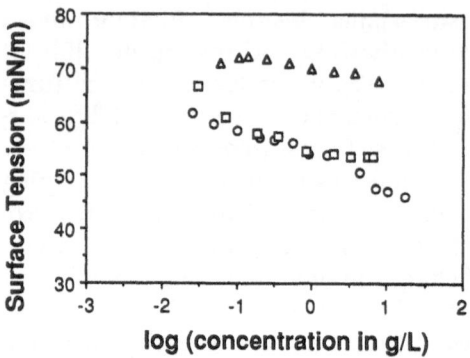

Fig. 1. Surface activity of monomers 5—7 at 25°C in water. □ = 5; ○ = 6; △ = 7

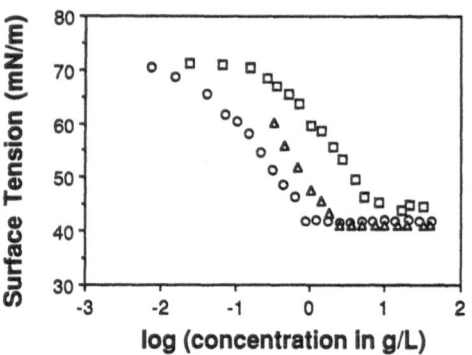

Fig. 2. Surface activity of polymers poly5-poly7 at 25°C in water. □ = poly5; ○ = poly6; △ = poly7

most pronounced for poly6. Its reasons are not yet clear, but seem to be characteristic for the zwitterionic moiety.

Interestingly, the surface activity of poly4 and poly5 is situated between those of the zwitterionic polymer poly6 and the cationic polymer poly7. This implies that the overall hydrophilicity of the polar head group cannot be the main reason for the surface activity of the polymers.

Cyclic voltammetry was employed to study the electrochemical properties of 1—5. Reduction potentials of 1—3 are close to the electrolysis of water at the mercury-covered electrode. Reoxidation potentials are shifted some hundred mV to more positive values. The reduction current i_{red} is larger than the oxidation current i_{ox} (Fig. 3).

These results indicate that reduction of 1—3 is not reversible in water and that the reduced products apparently undergo irreversible side reactions. Reversibility of redox-reactions of N-alkylated

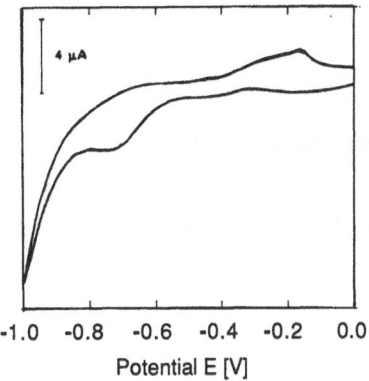

Fig. 3. Cyclic voltammogram of nicotinic ester 2 (c = 1.3 [g/l]) in 0.1 M KCl at 50°C, scan speed v = 100 [mV/s]. Working electrode: Hg-coated Pt, reference electrode: Ag/AgCl

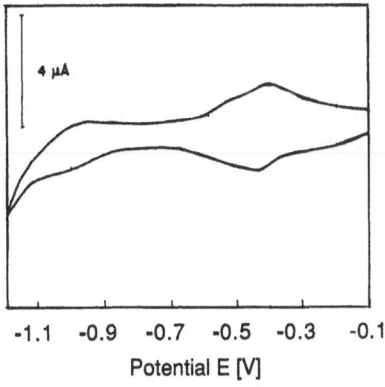

Fig. 4. Cyclic voltammogram of polymeric viologen poly5 (c = 1.4 [g/l]) in 0.1 M KCl at 25°C, scan speed v = 100 [mV/s]. Working electrode: Hg-coated Pt, reference electrode: SCE

show up. Additionally, chemical reduction with $Na_2S_2O_4$ and reoxidation by O_2 was monitored by UV spectroscopy. No hint of degradation of the reduced species in water was found.

Cyclic voltammograms of poly4 and poly5 differ noticeably from those of the monomers. The reduction current i_{red} is equal to the oxidation current i_{ox}, indicating that the redox reaction is reversible (Fig. 4). The redox potentials are shifted to more positive values compared to the monomers. Thus, reduction of the polymers is facilitated.

On reduction, monomers 4 and 5 adsorb at least partially onto the electrode, as the charge density is decreased and, therefore, the hydrophilicity is decreased. Fixation of the molecules to a polymeric backbone modifies this process. A detailed discussion of these differences will be given elsewhere [4].

References

1. Tazuke S, Kurihara S, Yamaguchi H, Ikeda T (1987) J Phys Chem 91:249
2. Saji T, Hoshino K, Aoyagui S (1985) J Am Chem Soc 107:6865
3. Okahata Y, Ariga K, Seki T (1986) J Chem Soc Chem Commun 73
4. Anton P, Heinze J, Laschewsky A (submitted)
5. a) Laschewsky A, Zerbe I (1991) Polymer 32:2070; b) ibid. 2081
6. Strauss UP (1961) J Phys Chem 65:1873
7. Anton P, Laschewsky A (1991) Makromol Chem, Rapid Commun 12:189

nicotinic acid derivatives in water is probably only guaranteed if catalyzed by enzymes.

Cyclic voltammograms of 4 and 5 show two separate reduction and oxidation waves. These are superimposed by adsorption of the reduced species onto the electrode, leading to sharp spikes. Multiple scans in fast succession using sweep rates v = 20—1000 mV/s do not change the form of the cyclic voltammogram, e.g., no degradation products

Authors' address:

Dr. A. Laschewsky
Institut für Organische Chemie
Universität Mainz
J. J. Becher-Weg 18—20
6500 Mainz, FRG

Progress in Colloid & Polymer Science
Progr Colloid Polym Sci 89:60—61 (1992)

Fractal aggregation of polystyrene spheres in the crossover region between DLCA and RLCA

D. Asnaghi, M. Carpineti, M. Giglio, and M. Sozzi

Physics Department, University of Milan, Italy

Abstract: We investigate both kinetics and fractal morphology of aggregating polystyrene latex, by means of a low-angle static light-scattering setup, covering two decades in scattering wavevectors. — The reactions are induced by adding various salt concentrations ranging from a value typical of diffusion limited colloid aggregation (DLCA) to one close to reaction limited colloid aggregation (RLCA). — The data show that the time evolution of the weight average molecular weight of the clusters is power law, with an exponent z varying in a continous fashion with the salt concentration c. — Measurements of the fractal dimension d_f are performed for each run. At high c (DLCA) the results are in good agreement with previous works: d_f is stable in time and equal to 1.65. For lower concentrations, however, d_f gradually decrases in time from 2, a value typical of RLCA, to lower values, close to the DLCA one. It should be pointed out that decreasing the salt concentration d_f becomes higher and more stable in time. — The data are interpreted consistently with both theoretical and simulation works presently available.

Key words: Colloids; aggregation; fractal; scattering; crossover

Many efforts have been devoted so far to the investigation of fractal aggregation of colloidal particles [1—3].

As a result, two limiting regimes with universality features have been identified: diffusion limited colloid aggregation (DLCA) and reaction limited colloid aggregation (RLCA). The former is characterized by a linear time evolution of the weight average cluster mass $\langle M \rangle_w$ and by aggregates with fractal dimension $d_f \sim 1.7$; the latter exhibits an exponential growth of $\langle M \rangle_w$ and leads to thicker aggregates: $d_f \sim 2.1$.

However, a few works [4—6] suggest that different regimes, which can be described as crossovers between RLCA and DLCA, should exist in addition to the other two wellknown classes, but experimental data are still lacking to characterize both kinetics and morphology during the crossover.

In order to investigate these new regimes, we studied salt-induced aggregation of polystyrene spheres 0.13 μm in diameter, suspended in a water — heavy water mixture, in order to match the solute density and to avoid differential sedimentation effects.

The salt $MgCl_2$ was added in concentrations starting from 30 mM, a value typical of DLCA, and then lowering in the attempt to move towards RLCA.

The data are collected by means of a low-angle static light-scattering setup, covering two decades in transferred momentum: $4 \times 10^2 \text{ cm}^{-1} \leqslant q \leqslant 4 \times 10^4 \text{ cm}^{-1}$.

From $I(q)$, we are able to extract the weight average molecular weight, the mean gyration radius, and the fractal dimension of clusters [7]. The experimental error in the determination of $\langle M \rangle_w$ is of about 5%. As to the fractal dimension, we can determine its value with a precision of about 2%.

As to the kinetics, we find that the weight average cluster mass grows according to a power law: $\langle M \rangle_w \propto t^z$. Furthermore, the exponent z shows a strong dependence on the initial conditions, varying in a continous fashion with the salt concentration c (Fig. 1). The data are in good agreement with the predictions of the analytical solution of the Smoluchowski equation [8], involving the homogeneity coefficient λ: $\langle M \rangle_w \propto t^z = t^{1/(1-\lambda)}$ with $0 \leqslant \lambda < 1$ (Fig. 2).

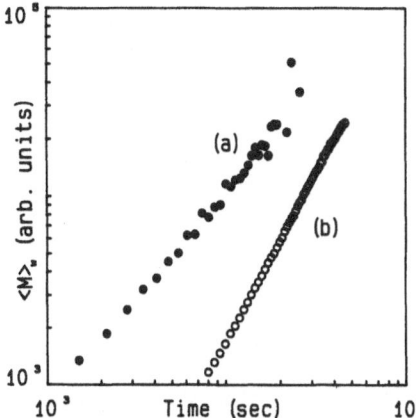

Fig. 1. Log-log plot of the average cluster mass as a function of time for two limiting salt concentrations: a) c = 30 mM (DLCA); b) c = 13.5 mM (intermediate regime)

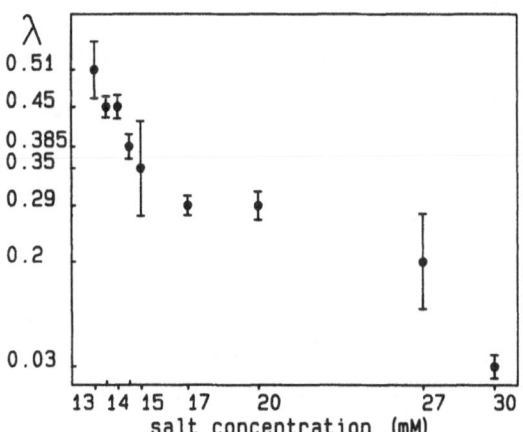

Fig. 2. Dependence of λ on the salt concentration

are more likely to stick on tips, leading to more tenuous structures. Actually, a crossover towards the DLCA morphology is taking place, which raises the question of whether such a transition should also be expected in the kinetics, at later times.

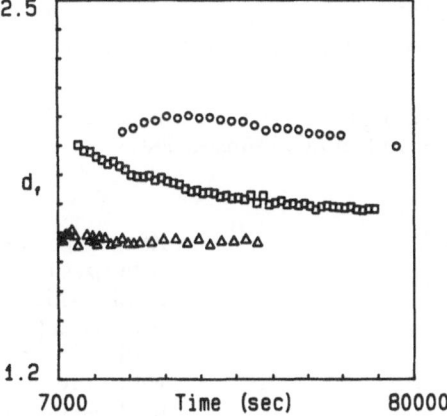

Fig. 3. The fractal dimension as a function at time for three salt concentration: △ c = 27 mM (DLCA); □ c = 17 mM (intermediate regime); ○ c = 14 mM (close to RLCA)

This last interpretation stimulates further theoretical and experimental work.

This suggests that these aggregation processes could be interpreted as self consistent, distinct new regimes, at variance with both DLCA and RLCA. Nevertheless, additional information about the clusters' fractal morphology can provide a different interpretation. We find that at high c (DLCA) the fractal dimension d_f is stable in time and equal to 1.65. On the contrary, for lower concentrations d_f surprisingly decreases during each run from 2, a value typical of RLCA, to lower values.

The variation of d_f with time (Fig. 3) can be explained as follows. As the aggregation proceeds, clusters grow larger so that the number of possible points of contact increases. Consequently, the effective sticking probability increases, too, and clusters

References

1. Lin MY, Lindsay HM, Weitz DA, Ball RC, Klein R, Meakin P (1990) Phys Condens Matter 2:3093—3113
2. Lin MY, Lindsay HM, Weitz DA, Ball RC, Klein R, Meakin P (1990) Phys Rev A 41:2005—2020
3. Lin MY, Lindsay HM, Weitz DA, Ball RC, Klein R, Meakin P (1989) R Soc Lond Sec A 423:71—87
4. Weitz DA, Huang JS, Lin MY, Sung J (1985) Phys Rev Lett 54:1416—1419
5. Kolb M, Jullien R (1984) J Phys (Paris) 45:L977—L981
6. Family F, Meakin P, Vicsek T (1985) J Chem Phys 83:4144—4150
7. Carpineti M, Ferri F, Giglio M, Paganini E, Perini U (1990) Phys Rev A 42:7347—7354
8. Van Dongen PGJ, Ernst HM (1985) Phys Rev Lett 54:1396—1399

Authors' address:

Prof. Marzio Giglio
Dipartimento di Fisica
Università degli Studi di Milano
Via Celoria, 16
20133 Milano, Italy

Progress in Colloid & Polymer Science   Progr Colloid Polym Sci 89:62—65 (1992)

Polymerization in microemulsion-size and surface control of ultrafine latex particles

M. Antonietti, S. Lohmann, and W. Bremser[1]

Institut für Physikalische Chemie der Universität Marburg, Marburg, FRG
[1] BASF Lacke & Farben AG, Münster, FRG

Abstract: Polymerization in microemulsion allows for the synthesis of latices with sizes of 10 nm $\leqslant R \leqslant$ 60 nm. In the system styrene/cetyltrimethylammoniumchloride/water, the particle size is controlled by the relative amount of surfactant to styrene and can be predicted by a simple three-phase picture of the microemulsion. Polymeric surfactants reveal a more complex behavior which prohibits an effective size control. — The particle surface is easily modified by addition of functional comonomers or additives which are incorporated in the interface. The additives stabilize or poison the original microemulsions. The results of the sucessful modification are highly reactive particles with a larger inner surface which can be used for different purposes. For example, the synthesis of a material with the ability of selective ion binding is described.

Key words: Polymerization in microemulsion; latex size; surface functionalization; surface "poisoning"; synergistic stabilization

1. Introduction

The polymerization in microemulsion is a quite new and versatile technique (see, e.g., [1—5]. It results in particies with a well-defined size and surface structure. Contrary to standard emulsion polymerization, the droplet size in microemulsions is thermodynamically controlled by the amount and character of the surfactant [4]. Such latices are narrowly distributed and have sizes of 10 nm $< R < $ 60 nm.

The dependence of the droplet size on the relative amount of surfactant is described for two surfactant systems which form microemulsions with styrene and water: cetyltrimethylammoniumchloride (CTMA) and a nonionic surfactant, Lutensol AT 50. The data are related to simple geometrical models which account for the surfactant geometry within the interface.

Due to the extremely large inner surface of such dispersions (up to 100 m^2/ml !), the examination of the droplet surfaces and their potential modification is of scientific and technical interest. This func-tionalization is conveniently performed by the addition of functional comonomers or even non-polymerizable dopants [5]. The resulting changes of the primary microemulsion and the efficiency of surface modification is the topic of the second part of this communication.

2.1. Size control of microemulsion droplets

In general, we are dealing only with microemusions in the composition range with oil contents less than 30%, where we expect spherical microdroplets dispersed in the water phase. A polymerization reaction in such microemusions is comparably simple to perform: Monomer, water, and surfactant are mixed until the equilibrium structure has settled; afterwards, the polymerization is started by a monomer soluble initiator. In the case of CTMA, the microemulsion structure is not affected by polymerization, and the micelles keep their size. For more details concerning the polymerization reaction and the characterization of the polymerized latex particles, the interested reader is referred to the literature [4, 5].

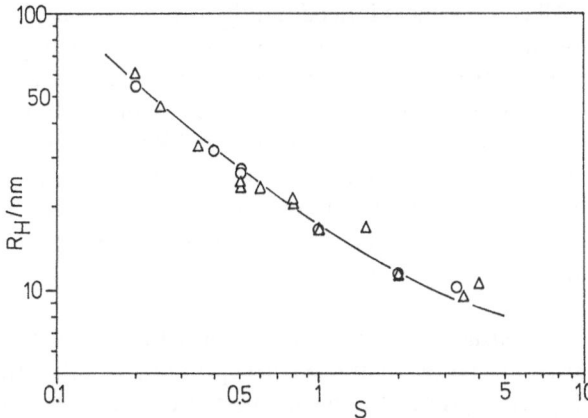

Fig. 1. Radius of microparticles R in dependence of the relative amount of surfactant S for CTMA (o) and the C_{12}-derivative, DTMA ($\triangle$): The line is a fit of the data with Eq. (1) and b = 3.5 nm

Figure 1 recapitulates the size of the resulting latex particles for the system styrene/CTMA/water in dependence of the mass ratio of surfactant to monomer S. We observe that, within the limits of experimental error, a simple relationship holds over more than one decade of S. Below $S \approx 0.3$, the microemulsions become unstable, and standard emulsions are obtained. In the region of stability, the droplet size distribution is quite narrow (standard deviations between $0.1 < \sigma < 0.3$, depending on the size), as seen by dynamic light scattering and electron microscopy [4]. It must also be emphasized that these latices are about one order of magnitude smaller than particles prepared by standard emulsion polymerization.

The relation of droplet radius R and composition can be described by very simple geometric arguments. In this model, the monomer droplets are assumed to be uniform and to be covered by a surfactant interlayer with a constant thickness b which separates the oil- and water-phases. This "three-phase"-approach surely oversimplifies the real situation. However, the examination of the phase compositions of similar microemulsions with a combination of different techniques reveals that mixing of the three components on a molecular level only occurs in minor amounts [6].

The macroscopic weight ratio of surfactant to monomer S is reflected in the volume ratio of inner oil core to the spherical shell of surfactant for each droplet. We derive:

$$S = \frac{m_{surf}}{m_{oil}} = \frac{\rho_{surf}}{\rho_{oil}} \cdot \frac{4/3\, \pi R^3 - 4/3\pi (R-b)^3}{4/3\pi (R-b)^3}$$

$$\approx \frac{R^3}{(R-b)^3} - 1$$

$$\Rightarrow R \approx b \cdot (1 - (1+S)^{-1/3})^{-1} , \; (1)$$

where ρ_{surf} and π_{oil} are the densities of the surfactant and oil phases, respectively, and R is the micelle radius.

Equation (1) is successfully used in Fi.g 1 to describe the experimental data; the only adjustable parameter b is obtained as $b \approx 3.5$ nm. One can relate this value to the geometric length of the surfactant molecule of $b' = 2.7$ nm. The difference is within a reasonable range, and one can picture the droplets of this system to be covered by a monomolecular layer of surfactant molecules.

We cannot expect, however, that all microemulsions behave similarly. On the one hand, we can imagine surfactant interfaces which stabilize the droplets with surfactant densities being less than densely packed. Here, b values less than the molecular length are obtained. On the other hand, the layer thickness may depend on composition, and a different size dependence is obtained.

Both effects are seen in the case of a polymeric surfactant, Lutensol AT 50 (a modified polyethyleneoxide), the data of which are presented in Table 1 [7].

Table 1. Characterization of latices made with Lutensol AT50 and styrene. r_H (meas.) is the experimentally determined hydrodynamic radius of the latices, and u_2 (90°) the second cumulant of dynamic light scattering at 90° scattering angle (a measure for the polydispersity)

S	r_H/nm	$u_2(90)$	b/nm
0.01	92	0.13	0.3
0.02	72	0.06	0.5
0.14	57	0.02	2.5
0.56	39	0.04	5.4

Here b is an increasing function of S, which is attributed to a "folding-up" of the polymer chains with increasing concentration. Such a behavior is known from polyethyleneoxide at planar interfaces [8]. In addition, the polymeric surfactant stabilizes the droplets already at lower S-values. The calculated layer thicknesses b in this region are as small as $b = 0.5$ nm, which can be interpreted as an incomplete coverage of the spheres or, alternatively, a nearly planar arrangement of the chains along the surface.

2.2. Surface functionalization of microemulsion droplets

With the technique of polymerization in microemulsion, the surface of the resulting latex particles can conveniently be modified by addition of functionalized comonomers or other dopants. These molecules are mainly incorporated in the interface, if their polarity is between the one of monomer and water or if they are even amphiphiles. On the other hand, certain mixing rules have to be considered since it is not evident that the interface accepts the dopant. A 1:1 mixture of CTMA and DTMA, for instance, does not produce a stable microemulsion, although each of the components does.

Therefore, we have tried to modify the system styrene/CTMA/water ($S = 1$) by incorporation of seven different comonomers (10% relative to styrene): methylmethacrylate (MMA), hydroxyethylmethacrylate (HEMA), dimethylaminoethylmethacrylate (DAMA), glycidylmethacrylate (GMA), potassium styrenesulfonate, diethyl(4-vinylbenzyl) amine (DVBA), and 2-vinylpyridine. The latter three substances destabilize the microemulsions: only very large and polydisperse latex particles are obtained. The methacrylate derivatives, however, result in stable microemulsions which are characterized in Table 2 (taken from [5]). For comparison, Table 2 also contains the data for pure styrene.

All comonomers indeed modify the microemulsion and the interface structure, as seen by the changed particle size and polydispersity. The addition of MMA and DAMA results in smaller which are also narrowly distributed. The increase of interface area is larger than the geometric requirements of the comonomer. Within the simple geometric picture, the droplets are covered by a stable, less dense surfactant layer consisting of surfactant and comonomer. We call this effect "synergistic stabilization".

In contrast, microparticies made with HEMA and GMA are larger than the corresponding ones made of pure styrene. In these cases, the comonomers seem to "consume" a part of the surfactant, and a thicker interface is required for an effective droplet stabilization. We call such behavior a beginning "poisoning" of the interface. The final state of this poisoning is the complete break of the microemulsion, as detected for the three other comonomers.

It is, a priori, not evident, at least not from the chemical structure, if the addition of a comonomer results in synergistic stabilization or surface poisoning. From these data, no general rule for a successful surface management can be given.

We have also examined the influence of polystyrene-poly(4)vinylpyridine blockcopolymers on the stability of these microemulsions. It turned out that for a successful introduction of the blocks, the pyridine units have to be protonated with *HBr*. Some data of the experiments with a very small block ($M_w \approx 3000$ g/mol, $x_{VP} = 54\%$), are also included in Table 2.

At low amounts of CTMA, the introduction of the block-copolymers actually stabilizes the microemulsions: the solubilization limit is increasing, and the droplets are smaller than corresponding ones made with pure CTMA. When the amounts of surfactant

Table 2. Characterization of latices made with functional monomers and styrene. The surfactant is in all cases CTMA. For definition of the abbreviations see the text

Monomer	S	Additive	r_H(meas.)/nm	$u_2(90°)$
Styrene	1	—	16.4	0.09
Styrene	1	10% MMA	14.2	0.04
Styrene	1	10% DAMA	12.5	0.04
Styrene	1	10% HEMA	24.0	0.03
Styrene	1	10% GMA	24.2	0.08
Styrene	0.2	10% Bloco1	27.1	0.04
Styrene	0.2	50% Bloco1	36.0	0.07

and blockcopolymer balance, the microemulsions become ill-defined; just in the case of an excess of block copolymer does the system become well defined again. In this region, comparably large droplets are produced. No microemulsions are obtained when block copolymers are used without CTMA.

The polymerized microemulsion particles posess polyvinylpyridinium chains protuding from their surface; these chains are "anchored" by means of the styrene blocks which sit in the polystyrene cores. Such a dispersion can be used for selective ion binding: transition metals are bound at the surface via complexation, whereas "normal" cations as sodium or calcium remain unattached. In our opinion, this example illustrates very vividly the possibilities of chemically modified microparticles.

3. Conclusion

It has been shown that the system styrene/ CTMA/water forms an "ideal"microemulsion in terms of a simple geometric model where the droplet size is a function of the amount of surfactant. Contrary to that, mixtures with the nonionic polymeric surfactant Lutensol AT 50 obey a more complex behavior which prohibits an effective size contrel.

These microemulsions are easily functionalized by addition of functional methacrylates or nonpolymerizable dopants such as block copolymers. The stability of the mixed microemulsions is very sensitive to the nature of the dopants in the interface layer: MMA and DAMA improve the droplet size and polydispersity, whereas HEMA and GMA

apparently reduce the surfactant effectiveness. A poisoning of the surface and a synergistic stabilization are very close to each other. Because of these sensitivities, the introduction of functionalized cosurfactant for the surface management of polymerized microemulsions must be carefully performed.

The results of these efforts are highly reactive particles which could be used for catalysis, for instance. It must be emphasized that most functionalization reaction strategies known for planar surfactant structures can also be applied to microemulsions, but with a "synthesized" surface area being 10 times larger.

References

1. Leong YS, Riess G, Candau F (1981) J Chem Phys 78:279
2. Candau F (1986) Encycl Polym Sci Eng 9:718
3. Atik SS, Thomas JK (1981) J Am Chem Soc 103:4279 and (1982) ibid. 104:5868
4. Antonietti M, Bremser W, Müschenborn D, Rosenauer C, Schupp B, Schmidt M, Macromolecules, accepted
5. Antonietti M, Lohmann S, van Niel C, Macromolecules, submitted
6. Guo JS, El-Asser MS, Sudol ED, Yue HJ, Vanderhoff JW (1990) J Coll & Interfac Sci 140:175
7. Antonietti M, Bremser W, Schmidt M (1990) Macromolecules 23:3796
8. Takahashi A, Kawagushi M (1982) Adv Polym Sci 46:1

Authors' address:

Prof. Dr. M. Antonietti
Inst. f. Physikalische Chemie
Phillips Universität Marburg
Hans Meerwein Str.
D-3550 Marburg

Progress in Colloid & Polymer Science   Progr Colloid Polym Sci 89:66—70 (1992)

Interconnection of microemulsion droplets with block-copolymers

Z. Zhou, R. Hilfiker, U. Hofmeier, and H.-F. Eicke

Institut für Physikalische Chemie der Universität Basel, Switzerland

Abstract: Static and dynamic light-scattering measurements were performed on water/AOT(sodium di-2-ethylhexylsulfosuccinate)/i-octane(2,2,4-trimethylpentane) w/o-microemulsions with various amounts of block-copoly(oxyethylene/isoprene/oxyethylene). We could show that the blockpolymer causes an aggregation of the microemulsion droplets. The mean size and polydispersity of the aggregates increase as the number of copolymer molecules per nanodroplet is increased. When the formal ratio of nanodroplets and copolymer molecules equals one, the mean aggregation number is three. The relative steepness of the increase of molar mass and hydrodynamic radius with increasing copolymer concentration suggests that the aggregates formed have an elongated shape.

Key words: Microemulsion; copolymer; light scattering; network; AOT

1. Introduction

In previous publications, we have studied network formation which is induced by copolymers of the ABA type (where *A* is a hydrophilic and *B* a hydrophobic block) in water-in-oil microemulsions (see Fig. 1), by conductivity, viscosity [1, 2], and transient electric birefringence (TEB) [3] measurements. We demonstrated that the initial step of gelformation is the formation of dimers of two microemulsion droplets connected by a copolymer molecule. The average distance of the droplets within a dimer could also be estimated. However, no further information on the subsequent growth of the dimeric subunits could be obtained. Scattering techniques such as small angle x-ray scattering (SAXS), time-averaged intensity (LS), and dynamic light-scattering (DLS) measurements have proved useful for the characterization of the aggregation behavior in colloidal systems. SAXS measurements

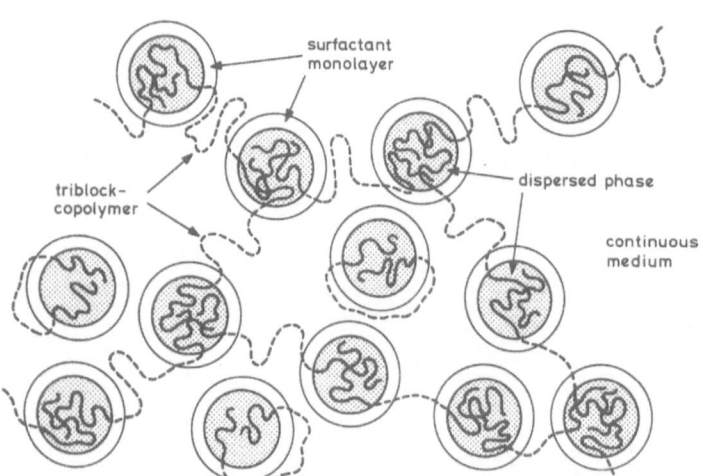

Fig. 1. Possible structure of a microemulsion-copolymer network

are most sensitive to small structures (< 30 nm), whereas LS and DLS measurements are also sensitive to larger aggregates. Consequently, SAXS measurements probe sensitively the shape and dimensions of dimeric aggregates of microemulsion droplets, whereas light scattering techniques probe larger aggregates as well.

2. Experimental

Materials

AOT from Fluka, Switzerland, was purified by dissolving it in a methanol-charcoal slurry. The suspension was filtered and the solution was dried in vacuo. Water was de-ionized and doubly distilled, and iso-octane was of the highest grade commercially available from Fluka. The triblock copolymer was synthesized and characterized at the Laboratoire de Chimie Macromoleculaire, ENSCM, Mulhouse, France [3, 4]. The copolymer which was used in our experiments had the following characteristics, i.e., M_n^{POE} (number average molar mass of the polyoxyethylene part) = 13.05 kg mol^{-1}, M_n^{PI} = 15.95 kg mol^{-1}, M_w^{cop} (weight average molar mass of the copolymer) = 33 kg mol^{-1}, and M_w^{cop}/M_n^{cop} = 1.14.

The microemulsion-copolymer systems were prepared by first mixing appropriate amounts of water, AOT, and iso-octane. The copolymer was then added to the microemulsion. The solution was stirred for 3 to 4 days with a magnetic stirrer in order to assure complete dissolution of the copolymer.

The ratio w_0 = [H_2O]/[AOT] controls the size of the nanometer-sized water droplets (ND). The weighed-in copolymer concentration (c_0 = mass of copolymer/{mass of water + AOT}) was converted to a more graphic variable, i.e., the number of copolymer molecules per nanodroplet (r_{cop}). To convert c_0 to r_{cop}, it has to be multiplied by the ratio of the molar masses of ND and copolymer.

Light scattering

A commercial light scattering goniometer (SP-86, ALV, Langen) with an ALV-3000 digital correlator was used with an argon ion laser operated at 496.5 nm. The temperature of the sample cell was set to 305.2 K ± 0.05 K. The incident light was vertically polarized and the intensity of the unpolarized scattered light was detected.

Refractive index increment

The refractive index increment $(\partial n/\partial c)_T$ was determined using a Brice-Phoenix differential refractometer. Measurements were made at 305.2 K and λ_0 = 436 and 546 nm. From the measured values, the refractive index indrement at 495.8 nm was obtained by interpolation.

3. Results and discussion

Static and dynamic light-scattering measurements were performed on microemulsion-copolymer solutions with w_0 = 19.6 and r_{cop} = 0.0, 0.25, 0.51, 0.76 and 1.08, respectively. The range of concentrations (c = grams of AOT + water + copolymer per cm^3 of solution) was varied from 3.9×10^{-2} to 1.63×10^{-1} gcm^{-3}.

After extrapolation to zero angle the weight average molar mass (M_w) of the NDs or aggregates can be determined by using Debye's formula, i.e.,

$$\frac{Kc}{R'_{\Theta \to 0}} = \frac{1}{M_w} + 2A_2c ,\tag{1}$$

where $K = 4\pi^2 n^2(\partial n/\partial c)^2/(N_A\lambda_0^4)$. R'_Θ is the excess Rayleigh ratio (i.e., the Rayleigh ratio of the solution minus the Rayleigh ratio of the solvent); Θ is the scattering angle, A_2 the second virial coefficient, n the refractive index of the solution, N_A Avogadro's constant, and λ_0 the wavelength of the incident radiation in vacuo.

If the system consists of a mixture of species with different refractive index increments, the apparent molar mass is obtained when Eq. (1) is applied [5]. In oder to calculate the weight average molar mass from the apparent molar mass, scattering experiments have to be performed in a series of solvents with different refractive indices, but otherwise "identical" properties. Since we are only concerned with the molar mass in a semi-quantitative way, and since the refractive index increments of single NDs and ND-copolymer aggregates are not too different, we neglect the difference between apparent and weight average molar mass.

Table 1. Results from static and dynamic light-scattering measurements

r_{cop}	$(\partial n/\partial c)_{305K}$[a]/ 10^{-2} cm^3 g^{-1} (496 nm)	M_w[a]/ 10^5 gmol^{-1}	M[b]/ 10^5 gmol^{-1}	R_{hs}[b]/ nm	D_0/ 10^{-7} cm^2 s^{-1}	R_h/ nm	σ[c]
0	1.94	2.26	2.14	3.80	12.0	4.25	.15
.25	2.24	3.10	3.29	4.99	8.80	5.80	.30
.51	2.56	4.46	4.41	6.70	6.86	7.44	.35
.76	2.88	6.21	6.33	8.12	5.42	9.41	.40
1.08	3.28	9.52	9.28	10.5	4.30	11.9	.42

[a]) From Eq. (1);
[b]) From Eq. (4);
[c]) $\sigma = \sqrt{\mu_2/\bar{\Gamma}^2}$.

The logarithm of $R'_{\Theta\to 0}/c$ was plotted against the concentration of solute and $(R'_{\Theta\to 0})_{c\to 0}$ was derived by extrapolating the initial linear portions of the semi-logarithmic plots to infinite dilution. An angular dissymetry of the R'_Θ-values was only noticeable for the solutions with $r_{cop} = 1.08$. The dissymmetry was too small, however, to obtain reliable values of the radius of gyration. From $(R'_{\Theta\to 0})_{c\to 0}$, the weight average molar mass of the interconnected nanodroplets was calculated for the various r_{cop}-values (see Table 1). The measured refractive index increments are also listed in Table 1.

The excess Rayleigh ratio is related to the osmotic compressibility by the following expression, i.e.,

$$R'_{\Theta\to 0} = 4\pi^2 n^2 (\partial n/\partial c)^2 N_A^{-1} \lambda_0^{-4} k_B T c (\partial c/\partial \Pi)$$

$$\sim c(\partial c/\partial \Pi) , \qquad (2)$$

where k_B is the Boltzmann constant and Π the osmotic pressure. A maximum in R'_Θ vs. c plots is observed when the inverse osmotic compressibility increases faster with increasing concentration than the concentration itself. One may argue that the hard sphere repulsion makes the main contribution to the osmotic compressibility. Considering the hard sphere model, $(\partial c/\partial \Pi)$ can be related to the volume fraction (Φ) of spheres using an approximate formula of Percus and Yevick [6], i.e.,

$$\left(\frac{\partial c}{\partial \Pi}\right) = \frac{M}{RT} \frac{(1 - \Phi_{hs})}{(1 + 2\Phi_{hs})^2} , \qquad (3)$$

where $\Phi_{hs} = 4\pi R_{hs}^3 N_A c/(3 M)$ with R_{hs} and M being the hard-sphere radius and the molar mass of the

microemulsion droplet or nanodroplet-copolymer complex, respectively. As will be discussed later, the complexes are characterized by their broad size distribution. For polydisperse spheres, Pusey et al. [7] have introduced a correction term which contains the χ-th moments ($\mu_\chi = \langle R^\chi\rangle/\langle R\rangle^\chi$) of the radius distribution function, i.e.,

$$R'_{\Theta\to 0} = KcM \frac{(1 - \Phi_{hs})^4}{(1 + 2\Phi_{hs})^2} \left\{ 1 + 6\Phi_{hs} \frac{1 + 2\Phi_{hs}}{(1 - \Phi_{hs})^2} \right.$$

$$\times \left[1 - \frac{\mu_4 \mu_5}{\mu_3 \mu_6} \right] - \frac{9\Phi_{hs}^2}{(1 - \Phi_{hs})^2}$$

$$\left. \times \left[1 - \frac{\mu_4^3}{\mu_3^2 \mu_6} \right] \right\} . \qquad (4)$$

By assuming a Gaussian size distribution for the nanodroplet-copolymer complexes and taking the standard deviation of the distribution as the square root of the variance ($\mu_2/\bar{\Gamma}^2$) obtained from dynamic light-scattering measurements, Eq. (4) can be used to fit the intensity data and to determine R_{hs} and M, both of which are dependent on r_{cop}. The solid curves shown in Fig. 2 are calculated by using Eq. (4) with the values of R_{hs} and M, as given in Table 1. It is interesting to note that, while the fitting of the data of the pure microemulsion ($r_{cop} = 0$) is rather satisfactory over the entire concentration range, the theoretical fits using Eq. (4) for the complexes ($r_{cop} > 0$) exhibit large deviations at high concentrations as compared to the experimental data. Such discrepancies become even more pronounced and occur at lower concentrations of the

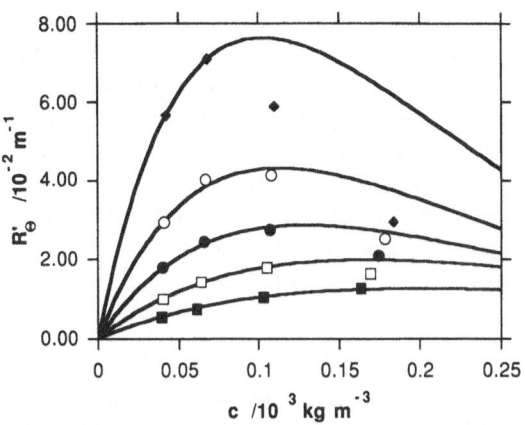

Fig. 2. Plots of the excess Rayleigh ratio vs. solute concentration at different number ratios of triblock copolymer to nanodroplet (r_{cop}). r_{cop} = 0 (solid squares), 0.25 (open squares), 0.51 (solid circles), 0.76 (open circles), and 1.08 (diamonds). System: water/AOT/i-octane/blockcopoly-(oxyethylene/isoprene/oxyethylene), T = 305.2 K. The solid lines are calculated from Eq. (4) with the values of M and R_{hs} as given in Table 1

dispersed phase when r_{cop} is increased. For some water-in-oil microemulsions both the hard sphere repulsion and van der Waals attractive terms have to be taken into account to interpret the scattering data over a large concentration range [8—11]. In our case, however, the inclusion of an attraction term makes the above deviations even larger. A reasonable explanation is the assumption that even for constant r_{cop}, the size of the nanodroplet-copolymer complexes may vary with microemulsion concentration.

Note that the molar masses of the complexes obtained by the two different approaches (Eq. (4) and Eq. (1)) are in good agreement. This might be somewhat fortuitous. We believe that, in principle, the first approach gives more reliable results since the hard sphere model is only a rough approximation for the nanodroplets interconnected by more or less flexible polyisoprene chains. Nevertheless, the results clearly demonstrate the interlinking effect of the triblock copolymers.

From SAXS experiments [12] we conclude that the radius of a nanodroplet which contains a polymer chain is about 11% larger than the radius of a polymer-free nanodroplet, i.e., an n-mer has a 1.37 times n larger molar mass than a ND in the copolymer free microemulsion. Therefore, for a r_{cop}-ratio of about 1, the mean aggregation number of nanodroplets is about 3.

$$G^{(2)}(q, \tau) = A(1 + \beta \mid g^{(1)}(q, \tau) \mid^2) , \qquad (5)$$

with $q = (4\pi n/\lambda_0)\sin(\Theta/2)$ the scattering vector, τ the delay time, A the background, and β the coherence factor. The measured electric field correlation function $|g^{(1)}(q, \tau)|$ was fitted with a cumulant expansion [13].

$$g^{(1)}(q, \tau) = \exp\{-\bar{\Gamma}(q)\tau + 0.5\mu_2(q)\tau^2 + ...\} ; \qquad (6)$$

$\bar{\Gamma}$ is the mean linewidth and $\mu_2/\bar{\Gamma}^2$ is the variance of the linewidth distribution function. The cumulants fit is applicable if the distribution is not too broad ($\mu_2/\bar{\Gamma}^2 < 0.4$). From $\bar{\Gamma}$ the apparent z-average translational diffusion coefficient ($\bar{D}$) is obtained, i.e.,

$$\bar{D} = \bar{\Gamma}/q^2 . \qquad (7)$$

By extrapolating the values of the apparent diffusion coefficient to zero concentration (D_0), the hydrodynamic radius (R_h) can be calculated by using the Stokes-Einstein relation

$$R_h = \frac{k_B T}{6\pi\eta D_0} , \qquad (8)$$

where η is the solvent viscosity.

We find that in the whole r_{cop}-range the measured diffusion coefficients were independent of the scattering vector. As far as the bridging effect between the nanodroplets by a triblock copolymer chain is concerned, we are mainly interested in the variation of the diffusion coefficient at infinite dilution D_0 vs. the amount of copolymer. In this way, we are able to eliminate the effect of interdroplet interactions and to deduce information about the growth of the complexes with increasing r_{cop}. The corresponding D_0 and R_h data are given in Table 1. The average hydrodynamic radii of the complexes become increasingly larger as r_{cop} is increased. Note that for all r_{cop}-ratios, the hard sphere radius is somewhat smaller than the hydrodynamic radius corresponding to a difference of 0.7 to 1.4 nm.

The standard deviation of the radius distribution function is estimated as the square root of the average variance for each r_{cop}-value (also listed in Table 1). It increases with increasing r_{cop}. This appears to be reasonable, since the aggregation of the nanodroplets has to be described by a multiple equilibrium.

The diffusion coefficient of polymer solutions at infinite dilution scales with the molar mass, i.e.,

$$D_0 \sim M^{-a_D} . \tag{9}$$

For theta conditions polymer coils have an a_D value of 0.5, and for good solvents a_D is intermediate between the ideal value of 0.5 and the Flory exponent 0.6 [14], while a_D of rodlike polymers is about 1 and for hard spheres 1/3. Therefore, from the molar mass dependence of the diffusion coefficient, we can extract information about the shape or conformation of the nano-droplet-copolymer complexes.

Figure 3 shows a double logarithmic plot of D_0 vs. M_w for the aggregates studied. From the slope, we obtain $\langle a_D \rangle = 0.71 \pm 0.04$. It should be noted that the scaling exponent $\langle a_D \rangle$ deduced from Fig. 3 may not be exactly the same as defined in Eq. (9), because the complexes studied are polydisperse, i.e., the D_0 and M_w-values listed in Table 1 are z-averages and weight-averages by nature, while Eq. (9) applies to monodisperse fractions. Nevertheless, the above $\langle a_D \rangle$ value suggests that the aggregates consisting of interconnected nanodroplets have a more linear than spherical shape [15].

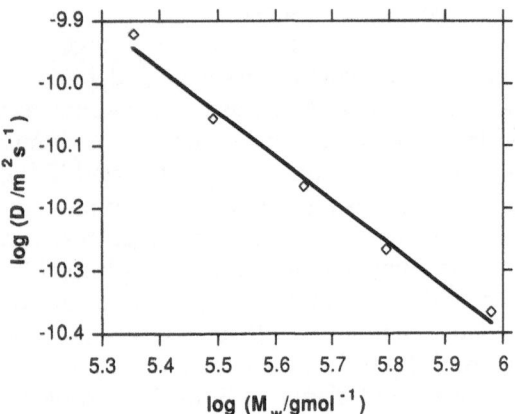

Fig. 3. Log-log plot of diffusion coefficients at infinite dilution vs. weight average molar mass for the water/AOT/i-octane/copolymer-system. $T = 305.2$ K

4. Conclusions

Static and dynamic light-scattering results demonstrate clearly the continuous growth of nanodroplet clusters with increasing amount of ABA-block copolymers. Both the hydrodynamic radius and the molar mass increase monotonically. The size polydispersity of the aggregates as determined by dynamic light-scattering measurements increased with increasing amount of copolymer. It seems plausible to assume that the aggregation of the nanodroplets is governed by multiple aggregation steps between aggregates of 2, 3, 4, etc., nanodroplets and monomers.

Acknowledgement

The authors are grateful to the Swiss national Science foundation for financial support of this work. One of us (R.H.) acknowledges generous support by the Treubel Foundation.

References

1. Eicke H-F, Quellet Ch, Xu G (1989) Colloids and Surfaces 36:97
2. Quellet Ch, Eicke H-F, Xu G, Hauger Y (1990) Macromolecules 23:3347
3. Hilfiker R, Eicke H-F, Steeb Ch, Hofmeier U (1991) J Phys Chem 95:1478
4. Xu G, PhD-Thesis, University of Basel, Switzerland, 1990
5. Benoit H, Froelich D, in "Light Scattering from Polymer Solutions" (Huglin MB, Ed), Academic Press, New York, 1972
6. Percus JK, Yevick GJ (1958) Phys Rev 110:1
7. Pusey PN, Fijnaut HM, Vrij A (1982) J Chem Phys 77:4270
8. Hou JM, Kim M, Shah DO (1988) J Colloid Interface Sci 123:398
9. Brunetti S, Roux D, Bellocq AM, Fourche G, Bothorel P (1983) J Phys Chem 87:1028
10. Calje AA, Agterof WGM, Vrij A, in "Micellization, Solubilization and Microemulsions" (Mitteal KL, Ed), Vol 2, p 779, Plenum, NY 1977
11. Bedwell B, Gulari E (1984) J Colloid Interface Sci 102:88
12. Hilfiker R (1991) Ber Bunsenges Phys Chem, 95:1227
13. Koppel DE (1972) J Chem Phys 57:4814
14. Schaefer DW, Hans CC, in "Dynamic Light Scattering: Applications of Photon Correlation Spectroscopy" (Pecora R, Ed) p 181, Plenum, NY 1985
15. Struis RPWJ, Eicke HF (1991) J Phys Chem 95:5989

Authors' address:

Prof. H.-F. Eicke
Institut für Physikalische Chemie
Universität Basel
Klingelbergstrasse 80
CH-4056 Basel, Switzerland

Progress in Colloid & Polymer Science   Progr Colloid Polym Sci 89:71—76 (1992)

Direct measurements by light scattering of the self diffusion in dense macromolecular solutions

D. Lombardo[1]), F. Mallamace[1]), N. Micali[2]), C. Vasi[2]), and F. Sciortino[3])

[1]) Dipartimento di Fisica dell'Universita' di Messina, Italy
[2]) Istituto di Tecniche Spettroscopiche del C.N.R., Messina, Italy
[3]) Center of Polymer Studies and Department of Physics, Boston University, Massachusetts, USA

Abstract: We report the study of the self-diffusion coefficient for strongly interacting polystyrene latex particles in a water solution, where aggregation processes are present. The reported data, obtained by measuring the contribution in the intensity-intensity correlation function of the occupation fluctuation number of particles in the scattering volume, agree with the current models of kinetic growth for random aggregation.

Key words: Diffusional modes; "light scattering"; colloidal aggregation

The study of diffusional motion in colloidal systems is essential in many fields of science in order to obtain a clear explanation about the dynamics of these systems [1] and, in particular, when interaction effects among the dispersed particles are the relevant phenomena or when aggregation processes take place. Such aggregation phenomena, depending on the type of the dispersed particles (ionic or non-ionic), are due to the particular interparticle potential that, as is well known, for a colloidal system is originated by two different contributions: a repulsive one (hard core or shielded Coulombic for charged particles) and an attractive one (Yukawa or London — van der Waals) [2].

The properties of such aggregation processes and, in particular, their growth kinetics, or rather the underlying mechanism of cluster formation from small isolated subunities, is a subject of considerable interest in ionic colloids as well as in non-ionic suspension like micelles and microemulsions. In such systems aggregation can be studied by changing different thermodynamic variables such as temperature, concentration or strength of interaction. Ionic colloids such as polystyrene latex particles suspension in water are well described in terms of the Derjaguin-Landau-Verwey-Overbeek potential, which, as is well known, shows two minima with a barrier whose form depends on the particle size and, particularly, on the ionic strength

[3]. When particles are able to overcome the barrier, the primary minimum is reached, giving rise to the so-called irreversible flocculation. For particles which are in the secondary minimum the reversible flocculation is observed (metastable or colloidal stable state). Those mechanisms of aggregation are different and give rise to different kinetics; in the particular case of the irreversible flocculation, we can distinguish between a reaction-limited aggregation and a diffusion-limited-cluster-cluster-aggregation as a function of ionic strength. Furthermore, micellar systems or microemulsions in the water-in-oil stable phase, usually described as Yukawa systems, originate percolation phenomena with a well-defined threshold that depends on temperature as well as on the concentration of the dispersed phase [4, 5]. Above the threshold a dynamic percolation regime exists where the droplets show an increasing connectivity with the formation of fractal clusters [5].

Light-scattering techniques, elastic and quasi-elastic, are powerful tools for the study of such structures. In general, a quasi-elastic light-scattering experiment using the photon correlation spectroscopy allows to measure both the self and mutual diffusion coefficients. The self-diffusion contribution describes the true motion of the particle within the fluid system, determining the rate at wich matter is transported through the system. The latter quantity, usually obtainable from NMR or tracer experiments, can be related and compared with observable thermodynamic quantities like viscosity. At present, while many studies have been performed in order to analize the mutual dynamics of colloidal particles within structural networks, the self dynamics is an open question with many unknown properties which may be the subject of considerable interest.

In this paper, we measure the self-diffusional dynamics by means of light-scattering photon counting spectroscopy, and report some experimental results for the self-diffusion coefficient of strong interacting colloids (water suspension of polystyrene latex spheres for different values of the solution ionic strenth) where aggregation processes can be observed. In particular, we discuss the obtained results on the basis of proper theoretical models for light scattering and in terms of the current models on the kinetic of growth due to random aggregation [2].

The scattering theory [1] gives the following expression, for the measured homodyne correlation function of the intensity fluctuation:

$$f_2(\vec{k}, \tau) = \langle N \rangle^2 (|g_1(\vec{k}, \tau)|^2 + 1) + \langle \delta N(0) \cdot \delta N(\tau) \rangle , \quad (1)$$

where $g_1(\vec{k}, \tau)$ represents the first-order correlation function of the dielectric tensor fluctuations; $\vec{k}$ is transferred wave vector, and τ the delay time; $\delta N(\tau)$ gives the deviation of the number of particles from the average number in the scattering volume. In a translational-diffusional model the first term of Eq. (1) represents the "coherent" contribution of the correlation function due to the time-dependence of density fluctuations, whose decay rate depends on k. The second one, is the correlation of the occupation fluctuation number of particles in the scattering volume, i.e., the time correlation of the deviation of the mean number of scattering particles in the scattering volume; its decay rate, proportional to the self-diffusion coefficient (k independent), depends on the intensity profile of the exciting source, as well as on the geometry of the scattering volume. The amplitude of the latter contribution (occupation fluctuation number) in the $F_2(\vec{k}, \tau)$ is proportional to $\langle N \rangle^{-1}$; thus, in a conventional experiment this contribution is negligible, since the number of scattering particles within the scattering volume is very large.

As pointed out by Pusey et al. [6] for an interacting colloidal system with narrow polydispersity, the normalized correlation function $g_1(\vec{k}, \tau)$ can also yield information about the self-diffusion motion. This correlation function, can be written in analogy with the intermediate scattering function for neutron scattering:

$$g_1(\vec{k}, \tau) = X \exp(-\Gamma_C \cdot \tau) + (1 - X) \exp(-\Gamma_S \cdot \tau), \quad (2)$$

with $\Gamma_C = k^2 D_C$ and $\Gamma_S = k^2 D_S$; D_C represents a cooperative diffusion coefficient, obtained in terms of the velocity correlation function [7],

$$D_C = \frac{1}{3 S(k)} \int_0^\infty dt N^{-1} \sum_{i,j=1}^N \langle v_i(t) v_j(0) \rangle , \quad (3)$$

while for the self-diffusion coefficient D_D, we have

$$D_S = \frac{1}{3} \int_0^\infty dt \langle v_i(t) v_i(0) \rangle , \quad (4)$$

v_i being the velocity of the i-th particle and $S(k)$ the structure factor.

The term X in Eq. (1) is connected with the scattering amplitude A (polydispersity in the particle size or in the particle charge [6]):

$$X = \frac{|\langle A(k) \rangle|^2}{\langle |A(k)|^2 \rangle} . \quad (5)$$

In general, X is a complicated function of the transferred wavevector k, interparticle interaction and concentration. For size polydispersity we can substitute $A(k)$ with $F(k)$ (particle form factor) and, in this case, $\langle ... \rangle$ represents an average weighted by the distribution of particle size [6]. Therefore, in this frame work, the experimental first-order correlation function also gives information about the self particle motion.

On the other hand, the normalized contribution of the occupation number fluctuation in the measured intensity correlation function can be written [1] in terms both of the spatial Fourier Transform of the intensity profile of the incident light, $I(k) = F.T. (I(r))$, and of the Fourier Transform of the correlation function of the concentration fluctuations $g_1(k, \tau)$ as:

$$g_N(\tau) = \frac{\langle \delta N(0) \cdot \delta N(\tau) \rangle}{\langle (\delta N(0))^2 \rangle} = \frac{\int dk^3 |I(k)|^2 g_1(k, \tau)}{\int dk^3 |I(k)|^2 g_1(k, 0)} . \quad (6)$$

In particular, in our translational-diffusional model [1], we have $g_1(k, \tau) \propto \exp[-k^2 D_S \tau]$. Considering a Gaussian illuminating field (Gaussian spatial profile), characterized by σ_2 parallel to z-axis (and the propagation direction $\vec{k}$, which can be selected by proper experimental conditions) by and σ_1 in the x, y coordinates, we have:

$$I(r) = I_0 \exp\left(-\frac{(x^2 + y^2)}{2\sigma_1^2}\right) \exp\left(-\frac{z^2}{2\sigma_2^2}\right), \quad (7)$$

whose spatial Fourier Transform is

$$I(k) = F.T. \ (I(r))$$

$$= I_0 (2\pi)^{3/2} (\sigma_1^2 \sigma_2) \left\{ \exp\left(-\frac{1}{2}(k_x^2 + k_y^2)\sigma_1^2\right) \right.$$

$$\left. \cdot \exp\left(-\frac{1}{2} k_z^2 \sigma_2^2\right) \right\}. \quad (8)$$

Considering two characteristic times, respectively, the time for a particle to diffuse out of a two-dimensional region of the size of the incoming beam diameter $\tau_1 = (\sigma_1^2/D_S)$, and the time that a particle takes to traverse a distance equal to the one-dimensional distance defined by the experimental geometry $\tau_2 = (\sigma_2^2/D_S)$, we can calculate the corresponding correlation function:

$$g_N(\tau) = (1 + \tau/\tau_1)^{-1} \cdot (1 + \tau/\tau_2)^{1/2}, \quad (9)$$

which, in the case of $\sigma_2 \gg \sigma_1$ (usual experimental conditions), becomes

$$(g_N(\tau))^{-1} = 1 + \frac{D_S}{\sigma_1^2} \tau. \quad (10)$$

This expression directly gives the contribution of the self-diffusion that, as previously said, can be observed for a suitable number of particles in the scattering volume, since their scattering amplitude is proportional to $\langle N \rangle^{-1}$. In particular, a measurement of this contribution [1] can be easily obtained for $\langle N \rangle \sim 10^5$; in this case the dominant contribution in $F_2(\vec{k}, \tau)$ comes from $(\langle N \rangle^2 (|g_1(\vec{k}, \tau)|^2 + 1))$.

On the basis of the concepts shown above, we present different data obtained through an experimental apparatus that allows for a separate determination of the self-diffusion coefficient from both terms in Eq. (1), using the same samples and the same experimental conditions. Specifically, we report the data obtained for a water suspension of latex polystyrene particles, with a radius of $R_0 = 900$ Å and a standard deviation of 55 Å, at very low fractional volume of the dispersed phase, $\phi = 5 \cdot 10^{-5}$, where aggregation phenomena take place. The samples were filtered and dialyzed in order to remove any contaminating substances. The obtained solution is electrically neutral and contains the charged polystyrene particles with their respective counter ions. The aggregation is obtained, by changing the ionic strength [2]. The appropriate amount of a univalent-univalent electrolyte (NaCl) as flocculating agent, at the different concentrations of 0.04 M and 0.4 M, was used. The measurements were made at constant temperature $T = 22\,°C$ in an optical transparent with a temperature control better than $10^{-2}\,°C$.

In Fig. 1 we show the experimental set-up. The exciting light from an Ar^+ laser ($\lambda = 5145$ Å), is collected into the sample S by a lens L_1 which is coupled with the telescope T in order to minimize the lens' focus spot size σ_1 (which in the present experiment is 0.032 mm, as measured through a calibration with a system of known D_S) and, consequently, gives the number of scattering particles into the scattering volume $\langle N \rangle$ [8]. The scattered light is collected by the lens L_2 with a numerical aperture n.a. variable by means of a diaphragm VP. In particular, the numerical aperture gives us the possibility to observe separately both the density-density correlation function (very small n.a.) and the contributions due to $g_N(\tau)$ (higher n.a.); moreover, an intermediate value of n.a. allows the observation of both the contributions. The pinhole P on the photomultiplier gives σ_2 (0.025 mm in our case). The photon correlation function is measured using a 4700C Malvern digital correlator. The optical apparatus, sample cell, thermostated bath, etc., were designed in order to have flare-free measurements.

The first set of experiments has been performed in order to test the general validity of the Pusey odel [6] by directly comparing the self-diffusion coefficient measured considering the two distinct contributions reported in Eqs. (2) and (10), i.e., through the long-time contribution to the density-density correlation function and through the occupation fluctuation number contribution, respectively. The Pusey model shows a general validity, even for systems with such complex dynamics as ours. In particular, the self-diffusion coefficient measured

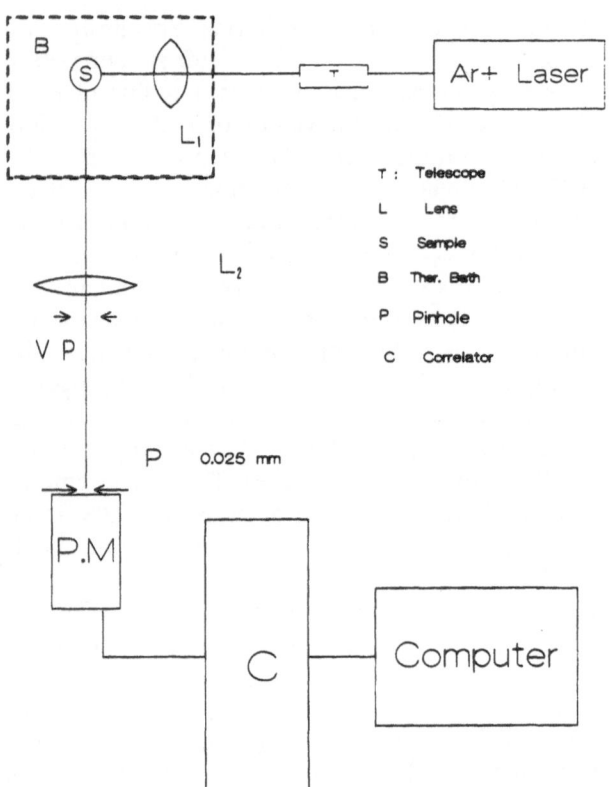

Fig. 1. Experimental set-up. Lens L is coupled with the telescope T in order to minimize the lens focus spot size σ_1 (see text). The scattered light is collected by the lens L_2 with a numerical aperture (n.a.) variable by means of a diaphram VP. The pinhole P *on the photomultiplier gives* σ_2 (see text)

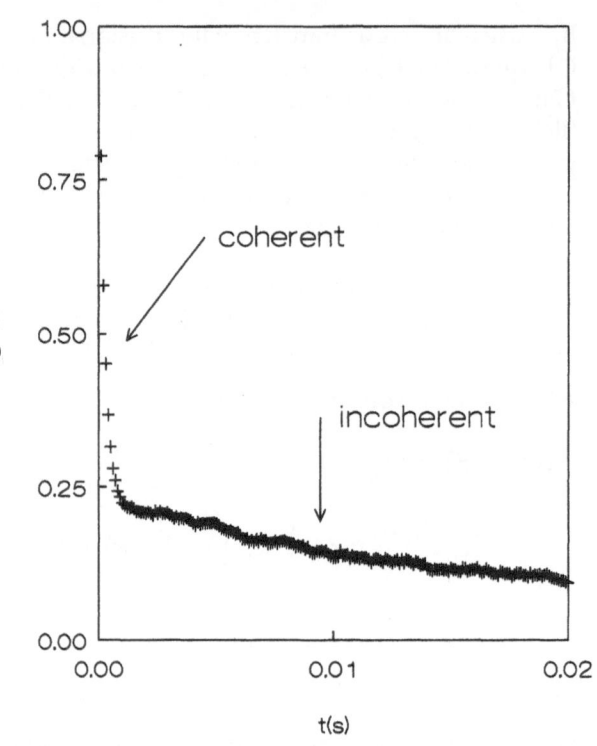

Fig. 2. Normalized correlation function $g_2(k, \rho)$ for the salt-free colloidal solution. It is possible to distinguish the two different contributions coming from $(|g_1(\vec{k}, \tau)|^2 + 1)$ and $g_N(\tau)$, named "coherent" and "incoherent", respectively

from the long-time contribution in the density-density correlation function ($g_1(k, \tau)$) and the data obtained from the occupation number fluctuation ($g_N(\tau)$) give analogous results. Those results are reported in a separate paper [9]. In this work, we will show how the obtained data for the self-diffusion coefficient in aggregating polystyrene water suspensions agree with the current results on the kinetic aggregation models [1]. Also, we show that the study of this quantity (D_S), in terms of the $g_N(\tau)$ contribution to the measured intensity correlation function, constitutes a powerful mean to investigate such complex systems.

In Fig. 2, we show for the latex suspension without salt the obtained normalized correlation function $g_2(k, \tau)$ for an intermediate n.a. value. It is obvious that the two different contributions coming from $(|g_1(\vec{k}, \tau)|^2 + 1)$ and $g_N(\tau)$ are easy to distinguish. These contributions are named "co-

herent" and "incohrent", respectively. In Fig. 3, we report three different normalzed correlation functions measured 3 min after salt addition; the salt concentrations used are $c = 0$ M, (curve A), $c = 0.04$ M (curve B), and $c = 0.4$ M (curve C), respectively. In Fig. 4 is shown the best fit with Eq. (10) (straight line) of the same experimental correlation functions reported in Fig. 3; as can be seen, we have a very good agreement with the obtained data. The measured D_S values are: $4.9 \cdot 10^{-8}$ (cm^2 s^{-1}) for $c = 0$ M (curve A of Fig. 3) reported here in the inset, $1.3 \cdot 10^{-8}$ (cm^2 s^{-1}) for $c = 0.04$ M (curve B), and $7.5 \cdot 10^{-9}$ (cm^2 s^{-1}) for $c = 0.4$ M (curve C).

Since the measurement was performed at a very low volume fraction of the dispersed phase, we can measure the correlation length ξ of the particles in the system via the Stokes-Einstein equation. The obtained values are larger than the hydrodynamic radius of the particle, $R_0 = 900$ Å, and reflect the strong interaction between the particles within the structural network. We observe that the self-

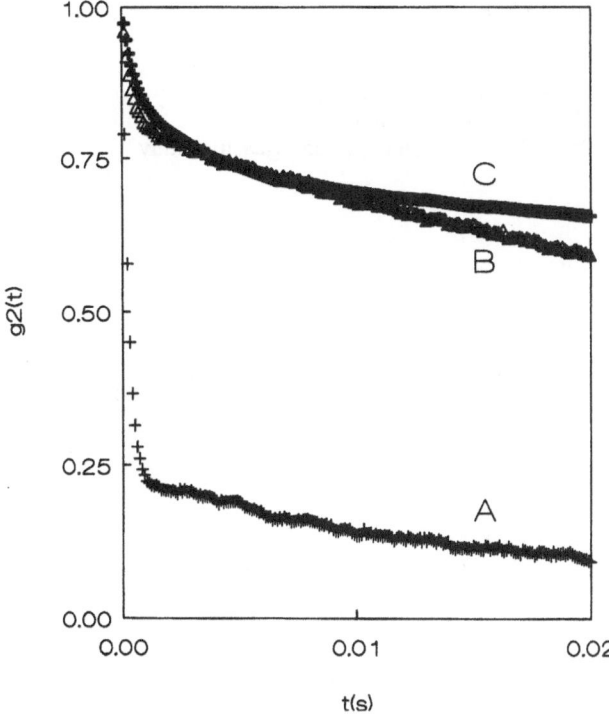

Fig. 3. Tree different normalized correlation functions measured 3 min after salt addition. A, B, and C indicate the following salt concentrations of the colloidal solution $c = 0$ M, $c = 0.04$ M, and $c = 0.4$ M, respectively

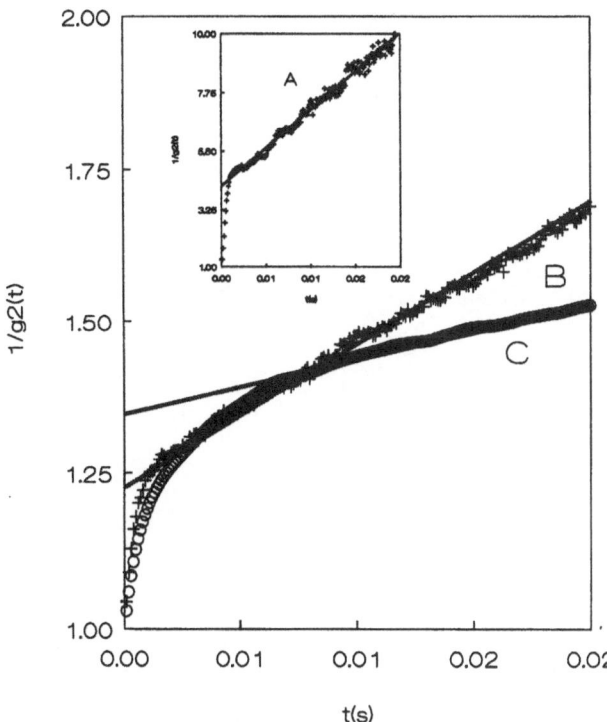

Fig. 4. The best fits with Eq. (10) (straight line) to the experiment correlation functions shown in Fig. 3

dynamics of the particles are hindered because of the interaction between the particles within the aggregating structure. It has been shown that such a structural arrangement has a fractal character with well-defined growth kinetics [10]. Morevoer, in the presented experiments agree very well with previous data [11] obtained from convention elastic and quasi-elastic light scattering (by means of the analysis of the mean decay rate in the density correlation function). In particular, the time-dependence of the correlation length ξ, described using scaling arguments in a numerical solution of the Smoluchowski equation, is $\xi \sim D_S^{-1} \sim t^{1/d_f}$ (d_f being the fractal dimension [12]), and our results numerically agree with such a picture.

In conclusion, we have performed the study of the self diffusion motion in strongly interacting colloidal particles, using an experimental set-up which allows for such measurements in a direct way. A comparison of the obtained data with the long-time contribution of the density-density correlation function reveals a good agreement. Furthermore, we are able to study the self-dynamics in a colloidal suspension where aggregation processes are present. In this case all the obtained results agree with the theoretical model proposed and explain the kinetics of such processes.

References

1. Berne BJ, Pecora R (1976) Dynamic light scattering, Wiley, New York
2. Stanley HE, Ostrowsky N (1986) On Growth and Form, Nijhoff, Dordrecht
3. Derjaguin BV, Landau L (1941) Acta Phys Chim Debricina 14:633; Verwey EJ, Overbeek GJT (1948) Theory of the Stability of Lyophobic Colloids, Elsevier, Amsterdam
4. Cametti C, Codastefano P, Tartaglia P, Rouch J, Chen SH (1990) Phys Rev Lett 64:1461
5. Magazu' S, Majolino D, Maisano G, Mallamace F, Micali N (1989) Phys Rev A 40:2643.4
6. Pusey PN, Fijnault HM, Vrij A (1982) J Chem Phys 77:4279; Pusey PN (1985) in: Physics of Amphiphiles: Micelles, Vesicles and Microemulsions, Degiorgio V, Corti M (eds) North-Holland, Amsterdam. p 152 (and references therein)
7. Hess W, Klein R (1983) Advances in Phys 32:174
8. Born M, Wolf E (1980) Principles of Optics, Pergamon, New York
9. Lombardo D, Mallamace F, Micali N, Vasi C, Sciortino F (to be published)

10. Bolle G, Cametti C, Codastefano P, Tartaglia P (1987) Phys Rev A 35:837
11. Majolino D, Mallamace F, Migliardo P, Micali N, Vasi C (1989) Phys Rev A 40:4665
12. Cametti C, Codastefano P, Tartaglia P (1987) Phys Rev A 36:4913

Authors' address:

Prof. F. Mallamace
Dipartimento di Fisica
Universita' di Messina
98166 Vill. S. Agata C.P. 55, Messina, Italy

Progress in Colloid & Polymer Science　　　　　Progr Colloid Polym Sci 89:77—81 (1992)

Micro-phase separation in cross-linked gels: Depolarized light-scattering results

F. Mallamace[2]), N. Micali[1]), C. Vasi[1]), R. Bansil[3]), S. Pajevic[3]), and F. Sciortino[3])

[1]) Istituto di Tecniche Spettroscopiche del C.N.R., Messina, Italy
[2]) Dipartimento di Fisica dell'Universitá di Messina, Messina, Italy
[3]) Center for Polymer Studies and Department of Physics Boston University, Massachusetts, USA

Abstract: We report depolarized light-scattering measurements on methyl-methacrylate (MMA) gels crosslinked with ethylene-dimethylacrylate (ED-MA). The study is performed varying the crosslinking amount of the gels from 0 to 6%, in the range $0-150$ cm^{-1} of the scattered light-spectrum. The obtained data confirm that the micro-phase separation phenomenon takes place in our system.

Key words: Gels; cross-linking; segregation; depolarized scattering

In recent years a great attention has been paid the properties of complex supramolecular aggregates such as dense polymeric fluids, strongly interacting colloids, polyelectrolytes, and gels. These systems represent one of the most interesting states of aggregation within the field of condensed matter and at present are considered, by researchers in physics, chemistry, and biology, to be of fundamental importance, both for pure science and for technological applications. Many theoretical and experimental works [1] have been performed in order to clarify the structural and dynamical properties of such systems, in particular, viscosity and light-scattering studies have been used to investigate the singularities near the gelation threshold. As is well known, a gel is a jellylike binary material which is composed of a crosslinked polymer network with a solvent occupying the pores in the network; therefore, the knowledge of the translational and rotational motion of solvent, polymer molecules, and segments of the latter has been important in elucidating the whole structure of the system [1, 2]. Light-scattering spectroscopy constitutes one of the most useful experimental methods for the investigation and the characterization of the properties of crosslinked gels and, in particular, for the sol-gel transition. Quasi-elastic light-scattering studies, measuring the time-dependence of the intensity correlation function, have established the existence of a collective diffu-sional mode with a diffusion coefficient inversely proportional to the hydrodynamic correlation length in the network [3], while polarized Rayleigh-Brillouin spectroscopy has established the existence of well-defined visoelastic properties above and below the gelation threshold [4]. Several investigators studied the dependence of the diffusional modes as a function of different parameters like concentration, temperature of the gel, and extent of crosslinking. A critical line, characterized by the divergence of the pair connectedness function, separates the "concentration of polymer" versus "concentration of crosslinker" phase diagram into a sol and a gel phase. Percolation models explain the properties of the sol-gel transition and give the theoretical background in order to explain the behavior of the relevant physical quantities at this "critical point". The sol-gel transition can be triggered at fixed total monomer concentration by increasing either temperature or concentration of the crosslinking molecules; at this point the viscosity of the solution exhibits a divergence, then, beyond the gel transition the system displays elastic behavior. The structure of the system at this threshold can be described by a power-law distribution of clusters with no characteristic length, and the corresponding scaling behaviors have been verified by several different studies.

In comparison, much less is known above the threshold about the scaling behavior or, in general, about many physical properties at high or very high crosslinking content, because of complicatioans due to the heterogeneities in the gel. In fact, even in a very good solvent the branched polymers generated in the crosslinking process will tend to segregate as the number of crosslinking molecules goes beyond the critical percolation point. In any case, macroscopic segregation is prevented by the existence of the spanning percolating network. As a result, only a microscopic segregation process takes place, with a corresponding building up of localized heterogeneities in the gel structure. This phenomenon, predicted by de Gennes [2], is known as micro-synerisis or microphase separation, and is also the major problem in making homogeneous interpenetrating networks. An example is the polyacrylamide gels that become turbid at high content of crosslinking molecules, in particular, their swelling capacity and pore size distributions are non-monotonic functions of crosslinking content. Computer simulations also showed the existence of spatial correlations in the distribution of crosslinking monomers [5], while Raman spectra [6] showed that the distribution of the crosslinking was non-uniform, and the clusters of the crosslinking appear to form at high crosslink concentration.

Similarly, linear polystyrene incorporated in a styrene-divinyl benzene gel exhibits phase separation as a function of crosslink content, even though the monomers involved are compatible and miscible. Recent theoretical works support the hypothesis that such segregation phenomena may occur in polymeric gels, in particular, for high crosslinking concentrations. Since the preexisting crosslinking network hinders macroscopic segregation, such a phase separation would take place a microscopic scale.

At present, not much is known about the structure of such microphase separated gels, although a recent Brillouin light-scattering experiment for the first, time gave evidence of such effects [7]. In particular, it has been observed that the wavevector dependence of the measured phase velocity $V(k)$ changes upon increasing the crosslinking content. More specifically, for high concentrations of crosslinking, maxima and minima in $V(k)$ have been observed. The k position of these minima and maxima have been correctly associated with the characteristic size of the more or less dense

crosslinked regions respectively. In fact, we can expect a higher (solid-like) sound velocity in the highly crosslinked regions, and a slower (liquid-like) sound velocity in the less cross-linked regions. Although such reported data do not extend over a very large k-range, the whole trend in the position of the minimum, as a function of k, suggests that the size of the less dense regions increases as the crosslink content increases. Correspondingly, the characteristic size of the high crosslink regions seems to decrease. This is consistent with the possibility that increasing the concentration of the crosslinking molecules favors a further clustering of the polymers [2]. Since a spanning network is already present in the gel, distribution of additional crosslinking clusters can only be accomplished by an exclusion of the solvent from the already existing clusters, with a corresponding reduction of the cluster sizes and an increased size of the low density "liquid like" islands in the gel. The total scattered intensity data at a fixed scattering angle (90°) also support the Brillouin data interpretation: in fact, on increasing the crosslinking content a marked increase of the intensity, has been observed, in agreement with the corresponding increase in the difference of population of high and low crosslinked regions. This interpretation is further supported by computer simulation [5]; in fact, using the kinetic gelation model, it has been found that at higher crosslinking content the distribution of crosslinking monomers became non-uniform, indicating the presence of some aggregations. Furthermore, in polyacrylamide gels, Raman spectra showed the presence of clusters of crosslinking monomers, and the decreasing in their swelling capacity (increasing the crosslinks) is well known and consistent with the suggestion from Brillouin data that solvent is excluded from regions of high crosslinking content.

Having such indications, we report here the results of depolarized Rayligh light-scattering measurements in the same gels and at the same concentrations as used in the Brillouin light-scattering experiments, where the solvent-exclusion phenomenon is observed. Modern developments in optical technology have shown that this experimental technique constitutes one of the most useful methods for the investigation of the rotational motions of polymers molecules and their segments, especially, in such complex fluids as ours [8]. Systems such as interacting supramolecular aggregates, dense and supercooled liquids, and polymeric solutions are characterized by strong viscoelastic behavior due to strong molecular configurational rearrangements and segmental motion of flexible polymer chains [9], in which the rotational dynamics plys the main role. If, as previously proposed, the structure and the dynamics of our gels are determined by the crosslinking content, a question of particular interest is to verify the behavior of freely rotating terminal groups of the polymer molecules of the gel and, if possible, the rotational dynamics of the solvent.

We study the depolarized light-scattered spectrum performing the measurements with a double pass double monocromator (DMDP 2000) SOPRA spectrometer that allows a very good maximization in the stray-light rejection. The exciting source is the 5145 Å line from a Ar$^+$ laser (Spectra Physics 2020) operating at a mean power of 1 watt. All the experiments are performed at constant temperature T = 17 ± 0.02 °C, using a fluid-filled thermostated scattering cell. The matching fluid has the same refractive index of the scattering cell in order to avoid unwanted stray-light effects. The scattering angle in the classical scattering geometry is 90°. In any case, the experimental conditions are the same as those used in the Brillouin measurements.

The experiments were performed on methylmethacrylate (MMA) gels crosslinked with ethylene-dimethylacrylate (EDMA) prepared with a well-defined procedure [3]. The gels were made by free radical copolymerization of MMA and EDMA in dioxane with ABIN as the initiator. Sealed vials containing 12% MMA + EDMA in dioxane were polymerized for 80 h in a 50 °C bath. The volume fraction of EDMA in the total monomer mixture of MMA + EDMA was varied from 0 to 6%. The sample with 0% EDMA corresponds to a linear polymer solution. All other samples were gels, in the phase beyond the sol-gel threshold. For comparison we also studied, the pure solvent (dioxane) and a solution of EDMA.

In order to have more information about the observed micro-phase separation phenomenon and a more direct comparison with our depolarized data, we report the group sound velocity in Fig. 1. It was obtained as the k derivative of the dispersion curves $V_g = (\partial \Delta \omega / \partial k)$, as a function of the crosslinking contents, for different scattering angles: θ = 90°, 115°, 135°, and 150°. Such a quantity, as is well known, represents the propagation of the sound wave packet in the system and can be directly connected with the structural properties of the medium [10]. In fact, the coupling between the physical pro-

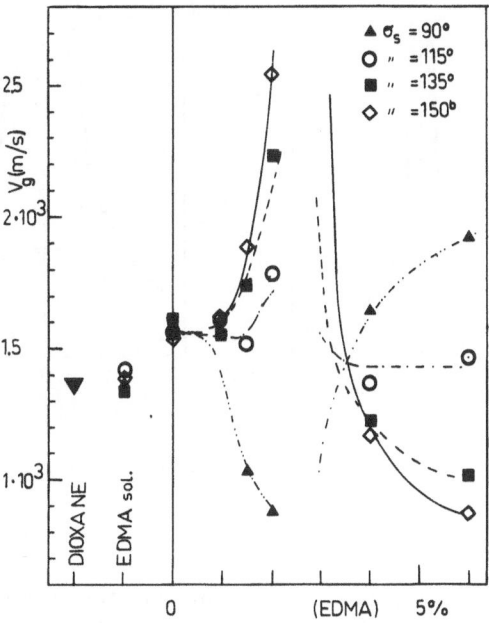

Fig. 1. The group velocity $V_g = (\partial \Delta \omega / \partial k)$ as a function of the crosslink contents and for the different scattering angles: θ = 90°, 115°, 135°, and 150°

perties of the system and the probe, selects the proper wavector (or frequency) and gives details about the interactions and structures. Consequently, when dispersion effects are present, a crossover wave vector (or frequency) exists such that the behavior of the system can be interpreted as mainly elastic for short times and large k, and as fluid for long times and small k. In fact, positive dispersions in the wave vector give information on a more solid-like behavior of the system, while negative dispersion is typical for a liquid-like system. Such a structural behavior is reflected by definite maxima or minima in V_g measured at different k. The reported promote such a point of view, in fact, for the lowest wave vector (θ = 90°) we have a minimum in V_g, while for the largest $k1$ (θ = 135 and 150°) we have the most pronounced maxima; an intermediate situation is observed for the k value corresponding to θ = 115°. Such a result confirms our structural model and can be definitely rationalized in comparison with the depolarized light-scattering data.

In Fig. 2, we report a characteristic spectrum of the depolarized scattering in the frequency range —30 — +30 cm^{-1} for the sample with 1% of crosslinking content. The continuous line represents

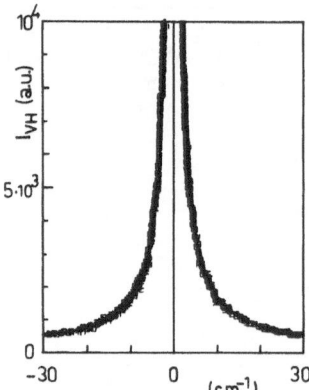 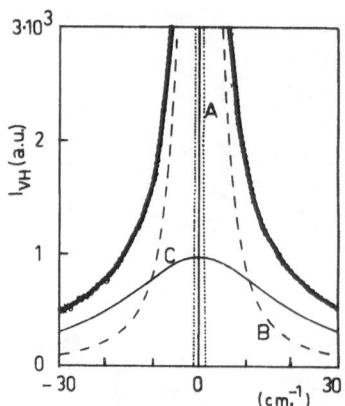

Fig. 2. Left: depolarized scattering in the frequency range $-30 - +30$ cm^{-1} for the sample with 1% of crosslinking content. The continuous line represents the best fit with the instrumental resolution (0.15 cm^{-1}) and two curves. Right: details of such fits are shown: curve A represents the instrumental resolution, while curves B (narrower) and C (broader) indicate the two significant physical contributions

the best fit, with the instrumental resolution (0.15 cm^{-1}) and two Lorentzian curves. In Fig. 2 (right) the detail of such a fitting procedure (curve A represents the instrumental resolution, while curves B (narrower) and C (broader) are the two spectral contributions of the system) is shown. By comparison with the spectra of pure solvent, we ascribe the narrow contribution to the solvent molecules; while the comparison with the spectra of the sample with 0% EDMA, which corresponds to a linear polymer (MMA) solution, allows us to associate the broad contribution to the freely rotating terminal group of the MMA polymer.

A general trend of the data obtained from such a fitting procedure is that the half-width at half maximum (HWHM) of both Lorentzians B and C, is constant for the different crosslinking contents, and is ~ 3 cm^{-1} and ~ 18 cm^{-1} for B and C, respectively. In contrast the behavior for the integrated intensity is very different (related to the population of the scatterers) [10]. In particular, we find that, for all the samples, I_B is constant while I_C changes significantly by changing the crosslinking contents. This result is shown in Fig. 3, additionally shown is the ratio $I_C/(I_B + I_C)$. As can be easily observed, I_C decreases slowly up to the gel with 3% of EDMA, and exhibits a marked jump immediately above this concentration value.

In conclusion, the whole behavior of the measured quantities confirms the model proposed for the microphase separation. For samples with cross-linking contents higher than 3%, Brillouin data give information about the existence of well-defined solid- and liquid-like islands (hetero

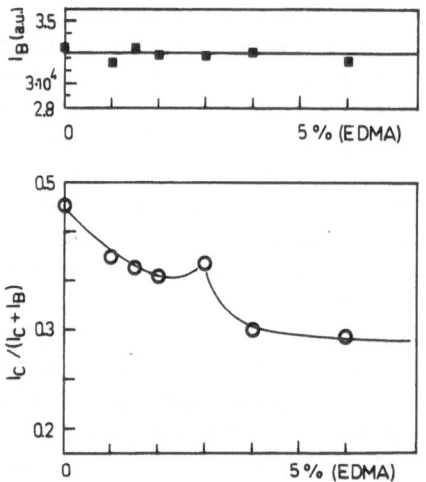

Fig. 3. The integrated intensity (population of the scatterers), I_B (solvent contribution), and ratio $I_C/(I_B + I_C)$ as a function of the crosslinking contents (EDMA); I_C is the contribution to the spectrum from freely rotating polymer (MMA) terminal groups

geneities in the gel structure) with an extent of about several hundred Angstroms (as confirmed by elastic scattering data). Depolarized light-scattering data show (in the same concentration region of EDMA) a change in the freely rotating terminal groups of MMA polymer and confirms "consequently, the solvent-exclusion phenomenon due to further clustering of the polymers, as proposed theoretically by de Gennes [2].

References

1. Ferry JD (1970) Viscoelastic Properties of Polymers, Wiley New York
2. de Gennes PG (1979) Scaling Concepts in Polymer Physics, Cornell University Press, Ithaca
3. Nishio I, Reina JC, Bansil R (1987) Phys Rev Lett 59:684; Bansil R, Pajevic S, Konak C (1990) Macromolecules 23:3380
4. marqusee JA, Deutch JM (1981) J Chem Phys 75:5239
5. Bansil R, Willings M, Herrmann HJ (1986) J Phys A 19:L1209
6. Bansil R, Gupta MK (1991) Polym Lett. In press
7. Mallamace F, Micali N, Vasi C, Bansil R, Pajevic S, Sciortino F (1991) J de Physique (in Press)
8. Huang YY, Wang CHJ (1975) Chem Phys 64:4748
9. Fytas G, Rizos A, Floudas G, Lodge TG (1990) J Chem Phys 93:5096
9. Brillouin L (1960) "Wave propagation and Group Velocity" (Academic Press New York)
10. Fabelinskii IL (1968) "Molecular Scattering of Light" (Plenum New York)

Authors' address:

Dr. Cirino Vasi
Istituto di Tecniche Spettroscopiche del C.N.R.
98166 Vill. S. Agata
C.P. 55
Messina
Italy

Progress in Colloid & Polymer Science   Progr Colloid Polym Sci 89:82—86 (1992)

Evidence by light scattering of long-range structures connected with the percolation transition in water-decane-AOT microemulsions

D. Lombardo[1]), F. Mallamace[1]), D. Majolino[1]), and N. Micali[2])

[1]) Dipartimento di Fisica dell'Universita' di Messina, Italy
[2]) Istituto di Tecniche Spettroscopiche del C.N.R., Messina, Italy

Abstract: We report extensive light-scattering measurements (elastic, quasielastic, and Brillouin) performed in dense micellar or microemulsion systems. The overall analysis of the data, in terms of a static effective-medium approach, indicates that the observed dynamic and structural effects are related to a dynamical percolation regime.

Key words: Percolation; fractal structures; microemulsions; dense systems; slow dynamics

Micellar systems and microemulsions in the water-in-oil phase are supramolecular liquid aggregates, which are thermodynamically stable in a wide range of temperatures and concentrations. They, constitute an interesting field of research due to its various chemical-physical properties. In past years, many researchers viewed such colloidal structures with great interest because they constitute a model system for the investigation of the dynamics of interacting particles in suspension and for the study of different solid-state structural phenomena [1]. Compared with a real system, they exhibit larger dimensions, and it is possible to change with relative case the thermodynamic conditions. Therefore, a large number experimental techniques can be applied in order to test the current theoretical models. In particular, the dilute and the concentrated phases can be treated in the same way as in molecular liquids, as far as the equilibrium properties are concerned, and the theories of the liquid state can be applied to analyze these suspensions.

The critical behavior [1], the percolation phenomenon [2], and the fluid-glass transition for high-volume fractions of the dispersed phase [3] are therefore the most studied processes in our supramolecular aggregates. The latter process has been intensively studied by means of different experimental techniques such as elastic and quasielastic light-scattering experiments, small-angle neutron scattering (SANS) [1, 4], viscosity [5], ultrasound [6], and hypersound [7, 8]. The data analysis shows a common behavior with that of the corresponding phases of real, systems especially, it has been possible to verify the findings of the recently developed mode-mode coupling theories on the dynamics of dense supercooled liquids or glass-forming systems, in particular, the slowing-down of the diffusional modes, or structural arrest.

The percolation transition observed in many different physical quantites, such as sound velocity [7, 8], ultrasound absorption [6] and electrical conductivity [2, 9, 10] represents another interesting property of micelles and microemulsions in which the interaction, originated by a well-defined interparticle potential form (repulsive hard-core plus an attractive Yukawa tail [4]) constitutes the driving mechanism [9, 10]. In particular, for water in oil microemulsions, i.e., the stable phase where the system is made of a dispersion of surfactant-coated water droplets in a continuous oil medium, electrical conductivity measurements show that at high temperatures and for different concentrations, the system changes from an insulator to a very good conductor. Crossing the percolation threshold [10], accurately determined by electrical conductivity data, the value of this quantity increases by several (six) orders of magnitude. The overall behavior, as a function of volume fraction ϕ and T, of the percolation has been clarified by means of electrical conductivity and dielectric constant measurements [9, 10]. Above the threshold a static percolation picture holds, below a dynamic one. For small volume fraction ϕ and within a large temperature interval a fluctuating charge model, holds while a power law behavior dominates above and below the percolation threshold. The overall percolation behavior is accounted for by a theoretical model that assumes an interdroplet potential consisting, as previously said, of a repulsive hard core plus a small but long-range attractive Yukawa tail with temperature-dependent parameters. In the dynamic percolation regime the droplets show an increasing connectivity accompanied by the building of transient fractal clusters due to the appreciable attractive interaction among the droplets [11]. Such a structural picture, in agreement with the conductivity date has been clearified by means of light-scattering experiments.

Starting from the ultrasound and hypersound data obtained in inverted micelles and water-in-oil microemulsions with the same surfactant and oil suspension, the object of the presented work is to give evidence of such long-range fractal structures in terms of elastic and quasi-elastic light-scattering measurements, and to clearify from a structural point of view the entire percolation process. In this respect, we report preliminary light-scattering data on these systems obtained by changing ϕ, T, the transferred wavevector k and the suspending oil. Quasielastic light-scattering data obtained by photon-correlation spectroscopy give us the information that our systems obey a well-defined dynamic with a scaling behavior typical for fractal percolating clusters. Furthermore, we fine that the measured density-density correlation function has a well-defined form that agrees with the result of a rigorous theory recently developed by Tartaglia et al. [12]. In particular, we obtain a measured correlation function that, while initially decaying exponentially, evolves continuously into a stretched exponential form for long times.

For the experiments performed with inverted micelles the micelles consists of spherical aggregates of AOT (AOT is di-2-ethylhexylsulfosuccinate) suspended in decane, with a mean radius of about 15 Å; this size remained constant as ϕ is varied (in the range 0—0.5). The microemulsions were studied in the one-phase region w/o of the AOT-water-decane microemulsion at a constant molar ratio of water to AOT, $X = 40.8$, in order to have droplets with a constant water core radius ($R_0 \sim 50$ Å) [1]. The samples were prepared according

to a well-established procedure [1] covering the ϕ range 0—0.75. A computer-controlled goniometer was used in a classical scattering geometry for the intensity meaurements; a full-correlator was employed for the quasi-elastic light scattering. The elastic properties were studied measuring the speed of the longitudinal sound wave of the fluid. This quantity in obtained from the Brillouin doublet, according to well-established experimental procedures [8]. The scattering experiment was made at a scattering angle of 90°, using as excitation light source an Ar^+ laser operating at 5145 Å, so that the scattering wavevector k is $\sim 2.2 \cdot 10^5$ cm^{-1}. The data are therefore taken in the frequency range 4.6 GHz to 5.2 GHz, and the current accuracy is of the order of 1%.

In Fig. 1, we report the measured hypersound velocity as a function of ϕ for different temperatures (for comparison, ultrasound values are shown for $T = 25$ and 35 °C [6]) of AOT microemulsions. The velocity data for AOT micelles are reported together with the microemulsion data in Fig. 1b ($T = 25$ °C) and Fig. 1c ($T = 30$ °C). For AOT micelles, crosses refer to ultrasonic data while rhombs refer to hypersound. The marked viscoelasticity, strongly dependent on volume fraction, temperature and suspending oil of different carbon chain lengths (same samples prepared in hexane) do not show any dispersion, even for very high ϕ, while micelles and microemulsions prepared in hexadecane exhibit a marked increase in the sound velocity V and, therefore, viscoelasticity, even at the lowest accessible frequencies (ultrasound) for $\phi \simeq 0.4$. Therefore, such a velocity dispersion can be related to the connected network formed by the droplets aggregation, and is due to the attractive part of the interparticle potential. The dynamics of such structures reflect a well-defined time behavior with a characteristic frequency dependence. For long time scales (low frequencies) the system is a collection of non-interacting spheres in which the shear stresses are relaxed; particles behave like isolated spheres undergoing Brownian motion and the observed sound velocity mainly reflects the physical behavior of the suspending fluid. For short times the solid-like network is able to support shear stresses and exhibits a finite elastic modulus. By increasing concentration, the system rigidity (and therefore the sound velocity) increases. In particular, it has been observed that the rigidity of this network exhibits power-law scaling with the volume fraction, consistent with rigidity percolation [7].

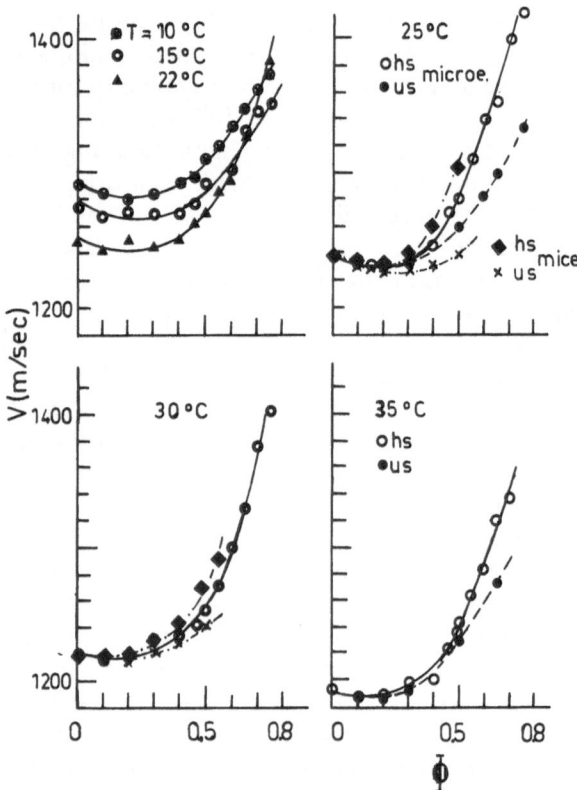

Fig. 1. The hypersound velocity V as a function of ϕ for different temperatures in AOT microemulsions and micelles (for microemulsions ultrasound values [6] for T = 25 and 35°C are also reported). In Fig. 1b (T = 25°C) and 1c (T = 30°C) data are reported together that refer to AOT microemulsions (circles: hypersond data, dots: ultrasound data) and AOT micelles (crosses: ultrasonic data, rhombs: hypersound data)

On this basis, we performed elastic and quasi-elastic light-scattering measurements considering that, for systems in which aggregation processes are present, scaling arguments and kinetic models for random aggregation give as a function of the wave vector the measured intensity $I(k)$ and the mean linewidth $\langle \Gamma(k) \rangle$. The scattered intensity can be written as $I(k) \sim P(k) S(k)$, where $P(k)$ is the form factor (water droplets in our case), and $S(k)$ represents the interparticle structure factor. Since the size of the particles R_0 is very small in comparison with the scattering wavelength, it turns out that $I(k) \propto S(k)$. In systems that originate self-similar aggregates the mass M scales [13] with the radius R according to a power law: $M \propto R^D$, where D is the fractal dimension directly connected to the peculiar aggregation kinetics. Consequently, the scattered intensity is

related to the transferred wavevector k through the simple relation:

$$I(k) \propto k^{-D} . \tag{1}$$

Furthermore, taking into account that the clusters of droplets with radius R_0 have a finite range of correlation, ξ, and, therefore, introduce into the calculation an exponential cut-off factor $\exp(-r/\xi)$ (analogous to the one used in critical phenomena) the structure factor $S(k)$ becomes, as shown by Chen and Teixeira [14],

$$S(k) = 1 + \frac{1}{(kR_0)^D} \frac{D\Gamma(D-1)}{\left(1 + \frac{1}{k^2\xi^2}\right)^{(D-1)/2}}$$

$$\times \sin[(D-1)\mathrm{tg}^{-1}(k\xi)] . \tag{2}$$

This allows for the determination of D and ξ by measuring $I(k)$ ($\Gamma(x)$ is the gamma function). It has been shown that in the range $1/\xi \ll k \ll 1/R_0$, we again have $S(k) \sim (kR_0)^{-D}$ and $I(k) \sim k^{-D}$. Outside these limits, i.e., $k\xi \gg 1$ and $kR_0 \ll 1$, D cannot be determined. In any case, the k region, in which Eq. (1) holds, is strongly dependent on the ratio ξ/R_0, and a good determination of D is obtainable for $\xi/R_0 \sim 50$. In that case, the obtained values of D can be directly connected to the peculiar aggregation kinetics; for example, values of $D = 1.75$ and $D = 2.5$, respectively, indicate a cluster-cluster diffusion limited aggregation or the percolation aggregation. Experimentally, our angular range is $30 \leqslant \theta \leqslant 120°$, corresponding to a k range of $7.19\ \mu\mathrm{m}^{-1} \leqslant k \leqslant 24.1\ \mu\mathrm{m}^{-1}$. Such a k interval, available with the use of the light scattering, is too narrow to accurately define a fractal behavior; usually, the region of self-similarity for fractal system is observed on a larger scale of lengths. In fact, such a k interval, is even narrow for the application of Eq. (2), which requires k ranges of several orders of magnitude. Nevertheless, it is sufficient, utilizing Eq. (1) to point out the existence of a scaling behavior. In Fig. 2a (T = 22°C) are reported the intensity profiles ($\ln I(k)$ vs. $\ln k$) for some ϕ of our system. As can be seen, the measured intensity for low ϕ is independent of k, and shows that if clusters are present they have dimensions of some hundreds of Å. For $\phi = 0.56$, we observe a well-defined k-dependence and the best fit of the experimental data with Eq. (1) (solid line) gives a fractal dimension $D \simeq 2.5 \pm 0.05$. For $\phi > 0.56$, we observe a sharp fall in D. In

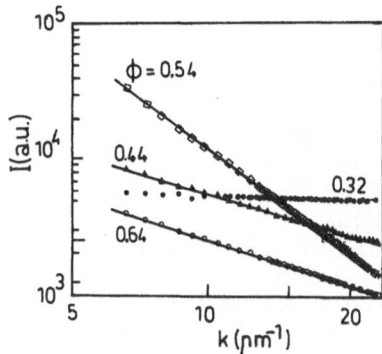

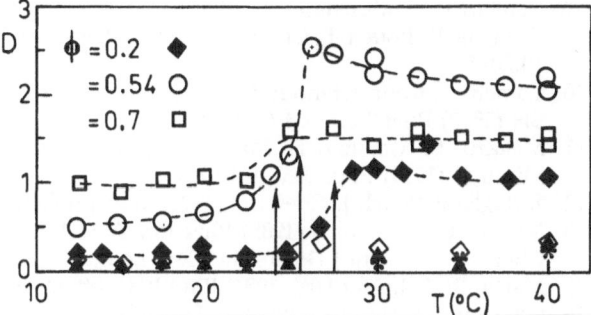

Fig. 2. a) Intensity profiles from elastic light-scattering measurements in AOT microemulsions, $\ln I(k)$ vs. $\ln k$, for various ϕ at $T = 22°C$. b) The obtained fractal dimension D as a function of T; arrows indicate the temperature values of the percolation threshold for the studied ϕ

Fig. 2b, we show for the same concentrations the extracted fractal dimension d as a function of T; arrows indicate the temperature values of the percolation threshold for the studied ϕ [10]. As a result, we observe that the measured D gives us a detailed mapping of the percolation effects in our system. In addition, application of Eq. (2) to our data indicates that below the threshold we have clusters of finite extension, while above it structures are present that span the entire scattering volume. The entire picture shown from these data confirm that the observed structural effects are directly connected with the percolation, i.e., in agreement with the current theoretical models.

The mean linewdith $\langle \Gamma(\vec{k}) \rangle$, obtained in a QELS experiment as the first cumulant of the measured density-density correlation function, depends on the center-mass motion or on cluster internal modes. It has been shown that [15]: $\langle \Gamma \rangle =$ $k^2 D_M F(kR_M)$, where D_M is the average diffusion coefficient and R_M is the average radius of gyration of the clusters. For $kR_M \ll 1$ (i.e., in the so-called Guinier regime), one obtains $F(kR_M) = 1$, the linewidth data must show the known k^2 dependence $\langle \Gamma \rangle = D_M k^2$ and D_M represents the aggregate diffusion constant. For $kR_M \gg 1$ (Porod regime) the same scaling arguments for self-similar objects as in the calculation of the intensity [16] give $F(kR_M) \sim kR_M$ and, in this limit of large kR_M, the mean linewidth becomes independent of the correlation range: $\langle \Gamma \rangle \sim k^3$. Although such a behavior has been recently confirmed [11] out information about a limited portion of the measured correlation function is given (in fact, the first cumulant represents only its initial decay rate), but it is important to obtain information on the overall function. In fact, the long time behavior in the measured correlation function can give details on the relaxational behavior of the system similar to that observed in other disordered complex systems as dense fluids, polymers, and gels, and verifying thoroughly the measured quantities with the percolation theory. In a recent work of Tartaglia et al. [12] for dense systems forming fractal percolation clusters, a rigorous theory of the scattering was developed that well accounts for the properties of the scattered intensity and for the density-density time correlation function. In particular, they obtained the following complete analytical form for the correlation function which, while initially decaying exponentially, evolves continuously for long times into a stretched exponential and explains the observed relaxation phenomena:

$$F(x, v) = \frac{(1 + x^2)^{D(3-\tau)/2}}{I(x)}$$

$$\times \int_0^{h^2((1+x^2)/x^2)D/2} (z^{2-\tau} e^{-z - v(1+x^2)^{1/2} z^{-1/D}}) dz$$

$$+ \frac{\sin[(D-1)\pi/2]}{(D-1)} \frac{(x/h)^{-D}}{I(x)}$$

$$\times \int_{(x/h)^{-D}}^{\infty} (a^{1-\tau} e^{-z - vz^{-1/D}}) dz , \qquad (3)$$

where we have $v = D_1 k^2 t/s^{1/D} = D_1 R_0 h k^2 t/\xi$, $x = k\xi$, $h = [D(D + 1)/6]^{1/2}$ and $\Gamma_c^* = \Gamma_c/D_1 R_0 k^3$; D_1 is the particle diffusion coefficient, τ a polydispersity exponent for percolation ($\tau = 2.2$ [17]), and $I(x)$ is the scattered intensity. Using the dimensionless quantity $\Gamma_c t = \Gamma_c^*(x/h)v$ the function can be calculated and shows exponential decay for short times with the decay time $1/\Gamma_c$, it evolves into a stretched exponential $\exp[-(\Gamma_s t)^\beta]$ with $\Gamma_s \sim k^2$ and $\beta = D/(D + 1)$. Figure 3 shows a fit a measured correlation function and good agreement with the present model (within the experimental errors is obtained. The best fit exponent $\beta \sim 0.7$ agrees well with the percolation predictions.

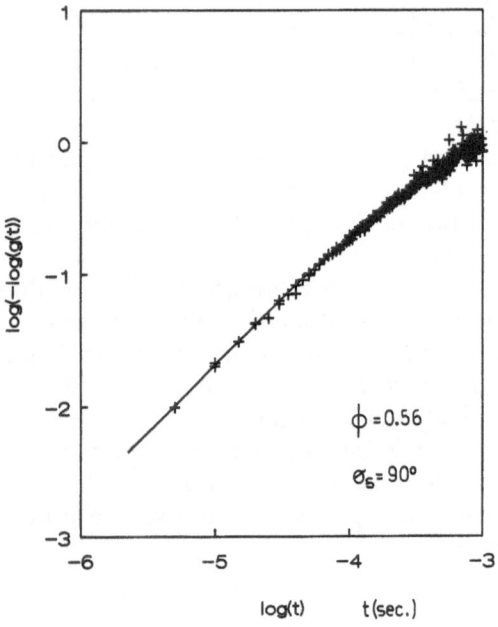

Fig. 3. The complete measured correlation function; the continuous line is a fit to Eq. (3)

In summary, we have shown that percolation concepts can be successfully used in order to explain light-scattering results from dense interacting colloidal fluids such as micelles and microemulsions; in particular, dynamical and structural properties of such complex systems can be investigated.

References

1. Kotlarchyk M, Chen SH, Huang JS, Kim MW (1984) Phys Rev Lett 53:941; (1984) Phys Rev A 29:2054
2. van Dijk MA (1985) Phys Rev Lett 55:1003
3. Chen SH, Huang JS (1985) Phys Rev Lett 55:1888
4. Sheu E, Chen SH, Huang JS, Sung JC (1989) Phys Rev A 39:5867
5. Majolino D, Mallamace F, Micali N, Venuto S (1990) Phys Rev A 42:7330
6. Cametti C, Codastefano P, D'Arrigo G, Tartaglia P, Rouch J, Chen SH (1990) Phys Rev A 42:3421
7. Ye L, Weitz DA, Ping Sheng, Bhattacharya S, Huang JS, Higgins MJ (1989) Phys Rev Lett 63:262
8. Mallamace F, Micali N, Vasi C, D'Arrigo G (1991) Phys Rev A 43:5710
9. Cametti C, Codastefano, Di Biasio AP, D'Arrigo G, Tartaglia P, Rouch J, Chen SH (1989) Phys Rev A 40:2013
10. Cametti C, Codastefano P, Tartaglia P, Rouch J, Chen SH (1990) Phys Rev Lett 64:1461
11. Magazu' S, Majolino D, Maisano G, Mallamace F, Micali N (1989) Phys Rev A 40:2643
12. Tartaglia P, Rouch J, Chen SH (1991) to be published
13. Martin JE, Ackerson J (1985) Phys Rev A 31:1180
14. Chen SH, Teixeira J (1986) Phys Rev Lett 57:2583
15. Martin JE, Schaefer DW (1984) Phys Rev Lett 53:2457
16. Schaefer DW, Han CC (1985) in: Dynamic Light Scattering, Percora R (ed) Plenum New York
17. Stauffer D (1979) Phys Rep. 54:1

Authors' address:

Prof. F. Mallamace
Dipartimento di Fisica
Universita' di Messina
98166 Vill. S. Agata
C.P. 55
Messina
Italy

Progress in Colloid & Polymer Science   Progr Colloid Polym Sci 89:87—88 (1992)

Solubilization of a water-insoluble dye:
A light-scattering study

B. Herzog and K. Huber

Ciba-Geigy AG, Dyestuffs and Chemicals Division, Basel, Switzerland

Abstract: In aqueous solutions of ionic surfactants above critical micelle concentration (CMC) spherical micelles form. Adding salt causes screening of their charges. This may result in one-dimensional growth, leading to micelles with wormlike shape. Dye-containing micelles in the presence of salt can adopt a similar structure. — The static light-scattering technique was used to get information about the structure of dye-containing micelles of the cationic surfactant tetradecyl-trimethylammoniumbromide (TTAB). The following results were obtained: At the same salt concentration dye-containing micelles are much larger than those without solubilized dye. Obviously, dye and salt act synergistically on the growth of wormlike micelles. — The flexibility of the dye-containing wormlike micelles is higher when the salt concentration is increased. This, in turn, can be explained by the screening of the charges, which decreases chain-stiffness of the micellear aggregates.

Key words: Solubilization; surfactant; micelles; static light scattering; persistence length

Introduction

In aqueous solutions of ionic surfactants above critical micelle concentrating (CMC), spherical micelles form. Adding salt causes srceening of their charges. This may result in one-dimensional growth, leading to micelles with wormlike shape [1]. Dye-containing micelles in the presence of salt can adopt a similar structure.

In this work, aqueous solutions of the cationic surfactant (TTAB) were used for solubilization of a water-insoluble dye (Fig. 1) at different KBr-concentrations. Clear solutions were prepared for static light-scattering measurements. The scattered intensity was measured at 19 angles simultaneously. Light of a kyrpton ion laser was used (λ = 647.1 nm), which does not interfere with the absorption spectrum of the dye.

TTAB-Micelles without dye

From static light-scattering measurements apparent inverse molecular weights of particles can be obtained. In the case of pure TTAB-micelles without

salt, those data could be extrapolated to zero-concentration (Table 1). In the presence of KBr, however, the apparent inverse molecular weights decrease to a minimum as surfactant concentration increases. In the low-concentration range molecular weights were evaluated under the assumption of negligible second virial coefficients. The highest molecular weights at 0.25 and 0.5 M KBr are listed in Table 1.

Dye-containing TTAB-micelles

In the absence of salt the molecular weight of dye-containing micelles could be obtained by extrapolation to zero-concentration (Table 2). The values of apparent inverse molecular weight measured in KBr-solutions behave similar to corresponding values where no dye is present. An increase of molecular weight with concentration is observed. Maximum values are shown in Table 2. From comparison of Tables 1 and 2, it can be concluded that solubilization of the dye leads to further growth of the micelles.

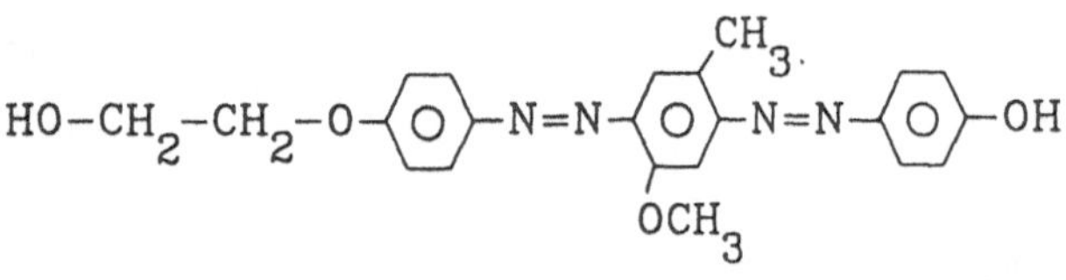

Fig. 1. Structure of the dye

Table 1. Molecular weights of pure TTAB-micelles at different KBr-concentrations

$c(KBr)/M$	M_w/g/mol
0	$2.0 \cdot 10^4$
0.25	$7.3 \cdot 10^4$
0.5	$2.2 \cdot 10^5$

Table 2. Molecular weights of dye-containing TTAB-micelles at different KBr-concentrations

$c(KBr)/M$	M_w/g/mol
0	$4.5 \cdot 10^5$
0.25	$3.6 \cdot 10^6$
0.5	$6.8 \cdot 10^6$

Properties of dye-containing micelles in the presence of salt

A) Overall dimensions

In a certain range of concentration the radius of gyration increases, as well as the molecular weight. An evaluation according to Benoit and Doty [2] was applicable which demonstrates consistence with wormlike structure of the dye-containing micelles. Experimental data at 0.25 M KBr were best fitted with a persistence length of 391 nm assuming a monodisperse distribution of micellar aggregates. In the case of 0.5 M KBr, however, fit-quality was increased if polydispersity was taken into account (Schulz-Flory-distribution, $M_w/M_N = 2$). A persistence length of 71 nm was obtained.

B) Q-dependent particle scattering functions

Particle scattering functions for wormlike chains were calculated according to Koyama [3]. For both salt concentrations experimental scattering functions were evaluated at three different molecular weights. At 0.25 M KBr good agreement of theory and experiments is obtained, assuming monodispersity and a persistence length of 391 nm. In the case of 0.5 M KBr, theory and experiments agree best with a polydispersity of $M_w/M_N = 2$ and a persistence length of 71 nm. Hence, particle-scattering functions led to the same results as obtained from overall dimensions.

Conclusion

With an increase of KBr-concentration, TTAB-micelles grow. This effect is greatly enhanced when the dye is solubilized.

In the presence of salt, dye-containing micelles behave like wormlike chains. As indicated by the overall dimensions and by the particle-scattering behavior, an increase of the salt concentration leads to a higher flexibility and polydispersity of the dye-containing wormlike micelles.

References

1. Imae T, Ikeda S (1986) J Phys Chem 90:5216
2. Benoit BH, Doty P (1953) J Phys Chem 57:958
3. Koyama R (1973) Phys Soc Jpn 34:1029

Authors' address:

Bernd Herzog
Ciba-Geigy AG
K-420.504
CH-4002 Basel, Switzerland

Progress in Colloid & Polymer Science   Progr Colloid Polym Sci 89:89—94 (1992)

Interdiffusion in polymer mixtures

A. Z. Akcasu, G. Nägele, and R. Klein

Fakultät für Physik, Universität Konstanz, FRG

Abstract: It is shown that the two exponential modes in the dynamic scattering intensity obtained by scattering experiments from ternary polymer mixtures cannot, in general, be identified with the interdiffusion and cooperative diffusion processes, contrary to recent suggestions in the literature. Conditions under which such an identification is possible are obtained. General expressions for the interdiffusion and cooperative diffusion coefficients associated with a given pair of components a and b in a ternary, or, in general, multicomponent polymer mixture are obtained in terms of the partial mobilities $\mu_{\alpha\beta}$ and partial scattering functions $S_{\alpha\beta}(q)$ of the a and b polymers. In the framework of the random phase approximation, a new expression of the interdiffusion coefficient in an incompressible ternary mixture of a and b homopolymers in a matrix of c homopolymers is found in terms of the tracer diffusion coefficients of the three species. It is shown that the results of the controversial "fast mode" and "slow mode" theories are obtained from this new expression as two limiting cases.

Key words: Interdiffusion; identification of modes; light scattering; polymer mixtures; random phase approximation

1. Introduction

The dynamic scattering function $I(q,t)$ in a ternary incompressible mixture of homopolymers consisting of species a and b in a matrix of homopolymers c, or, in a binary solution of a and b polymers it can be represented in the diffusion limit by a superposition of two exponentially decaying modes (bimodal relaxation) as

$$I(q,t) = I_+(q)e^{-q^2 d_+ t} + I_-(q)e^{-q^2 d_- t} . \qquad (1)$$

In the case of imcompressible ternary mixtures, the bimodal relaxation is a consequence of the incompressibility constraint, which enables one to express the dynamics of one of the components, identified as the "matrix", in terms of the dynamics of the other two components. In the case of binary polymer solutions, the bimodal relaxation arises from the large difference between the relaxation times characterizing the dynamics of the solvent and the polymers. Since the solvent relaxes much faster than the polymers, one is allowed to invoke the diffusion limit, and to eliminate the dynamics of the solvent completely. In a ternary mixture, the relaxation time of the matrix, which plays the role of the solvent in a binary solution, is comparable to those of the other two components. In this sense, the ternary mixtures and binary solutions of polymers correspond to two opposite limits, although they both are described by bimodal relaxation.

This paper is concerned with the identification of the two modes in a bimodal system, which may be a binary polymer solution, or a ternary polymer mixture. It has been suggested in recent literature [1, 2] that the two observed modes can be associated with the interdiffusion and cooperative diffusion processes in the mixture, and that the decay constants d_+ and d_- can be exactly identified as the cooperative and interdiffusion coefficients. It will be shown that this identification is in general not valid. For this purpose, a general definition of the interdiffusion and collective diffusion coefficients in mixtures is given and its relation to the observed decay constants is discussed. We will also address the question of whether it is possible to infer the interdiffusion and cooperative diffusion coefficient from a single dynamic scattering experiment on a labeled component.

2. Bimodal relaxation and normal modes

The number densities of the components are expressed as a column vector $\vec{\rho}(\vec{q}) =$ column $[\rho_a(\vec{q}),$ $\rho_b(\vec{q})]$, where $\rho_a(\vec{q}) = \sum_{j=1}^{N_a} \exp(i\vec{q} \cdot \vec{R}_j^a) - N_a \delta_{\vec{q}0}$ are the incremental number densities in Fourier space. Here, $\vec{R}_j^a$ is the position vector of the j-th monomer of species $a = (a, b)$, $\vec{q}$ is the wave vector, and N_a is the total number of monomers of species a. The dynamic scattering matrix $S(q, t)$ is defined by $S(q, t)$ $= \langle \vec{\rho}(\vec{q}, t) \vec{\rho}^\dagger(\vec{p}, 0) \rangle$, where $\dagger$ denotes hermitian conjugation, and the bracket implies an ensemble average.

The dynamic scattering intensity can be expressed in terms of the dynamic scattering matrix as

$$I(q, t) = \vec{a}^\dagger S(q, t) \vec{a} , \qquad (2)$$

where $\vec{a} = [a_a, a_b]^T$, and $a_a = \bar{a}_a - \bar{a}_c$ is the excess scattering length of species a. The scattering length of a monomer of species $a = (a, b)$ is denoted by $\bar{a}_a$. For a binary homopolymer solution, $\bar{a}_c$ is the scattering length of the solvent. The number density $\rho_c(\vec{q})$ has been eliminated in case of a ternary incompressible mixture, by using the incompressibility constraint.

In the diffusion limit (i.e., $q \to 0$, $t \to \infty$ with $q^2 t$ fixed), the time evolution of $S(q, t)$ is governed by [3] $S(q, t) = \exp(-q^2 D t) S(q)$, where D is the diffusivity matrix and $S(q)$ is the matrix of static structure factors. D is, in general, not symmetric, but it can be diagonalized by a similarity transformation $Q D Q^{-1}$ $= d$, where $d = \mathrm{diag}[d_+, d_-]$. The transformation matrix Q is constructed in terms of the eigenvectors of D [3]. It can be shown that the eigenvalues are real. They are also positive, since the system is overdamped. Thus, $d_+ \geq d_- > 0$. The eigenvalues are given by $d_\pm = D_{av} \pm \sqrt{D_{av}^2 - |D|}$, where $D_{av} =$ $[D_{aa} + D_{bb}]/2$ and $|D| = D_{aa}D_{bb} - D_{ab}D_{ba}$.

The transformation matrix Q also diagonalizes the static structure matrix $S(q)$ by a congruent transformation, i.e., $Q S Q^T = s$, where $s =$ $\mathrm{diag}[s_+, s_-]$, with positive diagonal elements $s_\pm$. This is due to the fact that, whereas D is not symmetric, the mobility matrix μ, defined by $\mu =$

$DS/k_B T$, is symmetric and positive definite.

In the following, we will discuss the nature of the two exponential terms in Eq. (1). Therefore we perform the transformation $\vec{\xi}(t) = Q \vec{\rho}(t)$ on the number density vector $\vec{\rho}$. The components $\xi_+(t)$ and $\xi_-(t)$ of the vector $\vec{\xi}(t)$ are linear combinations of $\rho_a(t)$ and $\rho_b(t)$. The important property of $\vec{\xi}(t)$ is that its correlation $\langle \vec{\xi}(\vec{q}, t) \vec{\xi}^\dagger(\vec{q}, 0) \rangle$ is diagonal. Thus, $\langle \xi_+(t) \xi_-^*(0) \rangle = 0$ and

$$\langle \xi_\pm(t) \xi_\pm^*(0) \rangle = e^{-q^2 d_\pm t} s_\pm , \qquad (3)$$

and $\xi_\pm(\vec{q}, t)$ are seen to be the eigenmodes of the system. They are statistically uncoupled at all times.

3. Diffusion in polymer mixtures

In this section, general definitions are given of two diffusion mechanisms, which occur in polymer mixtures. These definitions are needed to address the question of whether the two normal modes observed in bimodal systems can be identified as the interdiffusion and cooperative diffusion modes.

The interdiffusion process mediates the relaxation of thermal fluctuations in the relative local concentration of a and b-type homopolymers towards their equilibrium values [3—5]. Species a and b may be dissolved into other species. The relative local concentration of species a relative to species b is defined by $c_a(\vec{r}, t) = [\rho_{a0} + \rho_a(\vec{r}, t)]/[\rho_{a0} + \rho_a(\vec{r}, t) + \rho_{b0} + \rho_b(\vec{r}, t)]$, where $\rho_{a0} = (N_a/V)$ is the equilibrium number density of the a-monomers, and $\rho_a(\vec{r}, t)$ is the local fluctuation about the mean. The concentration fluctuation δc_a about its mean value $x_a =$ $\rho_{a0}/(\rho_{a0} + \rho_{b0})$ is given, in the linear approximation, as $\delta c_a(\vec{q}, t) = x_a x_b \rho_-(\vec{q}, t)$, where $\rho_-(\vec{q}, t) = \vec{E}_-^T \vec{\rho}(\vec{q}, t)$ with $\vec{E}_-^T = [1/N_a, -1/N_b]$.

Thus, the interdiffusion process is related to the relaxation of $\rho_-(\vec{q}, t)$, which is a particular linear combination of the number densities $\rho_a(\vec{q}, t)$ and $\rho_b(\vec{q}, t)$. This linear combination is, in general, different from either of the two normal modes $\xi_+(t)$ and $\xi_-(t)$. Hence, its dynamic scattering function $S_{--}(\vec{q}, t) \equiv \langle \rho_-(\vec{q}, t) \rho_-(-\vec{q}, 0) \rangle$ is, for a bimodal system, a linear superposition of the two pure modes. In a multicomponent mixture, $S(q, t)$ contains as many exponential terms as given by the number of species, without counting the matrix component in case of an incompressible mixture.

The interdiffusion coefficient is related to the dynamic scattering function $S_{--}(q,t)$. If the interdiffusion process was a pure mode, $S_{--}(q,t)$ would relax with a single exponential as

$$S_{--}(q,t) = e^{-q^2 D_{in} t} S_{--}(q) ,\qquad (4)$$

which would provide the definition of the interdiffusion coefficient D_{in}. However, the interdiffusion process is not a pure mode, and its relaxation is bimodal.

An intuitive definition of the interdiffusion coefficient emerges if we try to approximate the bimodal relaxation of $S_{--}(q,t)$ by a single exponential. Hence, by matching the initial relaxation rates, we find with Eq. (4) that

$$D_{in} = k_B T \frac{\vec{E}_-^T \mu \vec{E}_-}{S_{--}}$$

$$= k_B T \frac{\dfrac{\mu_{aa}}{N_a^2} - 2 \dfrac{\mu_{ab}}{N_a N_b} + \dfrac{\mu_{bb}}{N_b^2}}{\dfrac{S_{aa}}{N_a^2} - 2 \dfrac{S_{ab}}{N_a N_b} + \dfrac{S_{bb}}{N_b^2}} ,\qquad (5)$$

where S_{--} is the long wavelength limit of $S_{--}(q)$.

This is a novel definition of the interdiffusion coefficient in terms of the partial mobilities $\mu_{\alpha\beta}$ and partial structure factors $S_{\alpha\beta}$ of the a and b components [3], valid both in the case of multicomponent polymer solutions and of multicomponent mixtures of polymers. We note that D_{in} is expressed as a product of a kinetic factor

$$\Lambda_{in} \equiv k_B T \vec{E}_-^T \mu \vec{E}_- ,\qquad (6)$$

and a thermodynamic factor $(1/S_{--})$, as conventionally done in the literature [4, 5].

The cooperative diffusion process mediates the relaxation of the total number density fluctuations of a pair of species a and b, viz. $\rho_+(\vec{q},t) = \rho_a(\vec{q},t) + \rho_b(\vec{q},t)$ towards the average total density $\rho_{a0} + \rho_{b0}$. This linear combination is also, in general, different from those corresponding to the normal modes.

The diffusion coefficient D_{coop} associated with this process is obtained as the initial slope of $S_{++}(q,t) \equiv \langle \rho_+(\vec{q},t)\rho_+(-\vec{q},t)\rangle$. Following the same arguments we used to define the interdiffusion coefficient, we obtain [3]

$$D_{coop} = k_B T \frac{\mu_{aa} + \mu_{bb} + 2\mu_{ab}}{S_{aa} + S_{bb} + 2S_{ab}} .\qquad (7)$$

4. Conditions for the identification of the normal modes as cooperative and interdiffusion modes

In this section, we briefly discuss the conditions under which the two normal modes, observed in dynamic scattering experiments, coincide with the interdiffusion and cooperative diffusion modes.

The normal modes are characterized by $\xi_+(\vec{q},t)$ and $\xi_-(\vec{q},t)$. The interdiffusion and cooperative diffusion modes, on the other hand, are characterized by $\rho_-(\vec{q},t)$ and $\rho_+(\vec{q},t)$, respectively. Hence, we want to find the conditions under which $\rho_+(\vec{q},t)$ and $\rho_-(\vec{q},t)$ become proportional to $\xi_+(\vec{q},t)$ and $\xi_-(\vec{q},t)$, respectively, for all t.

The desired conditions are obtained from the requirement that $\langle \rho_+(\vec{q},t)\rho_-(-\vec{q},0)\rangle = 0$ must hold at all times. This leads to two conditions [3] on the static and dynamic properties of the chains of components a and b, namely,

$$\frac{f_{aa}(q) + f_{ab}(q)}{N_a} = \frac{f_{bb}(q) + f_{ab}(q)}{N_b} ,\qquad (8)$$

where $f_{\alpha\beta} = S_{\alpha\beta}$ and $f_{\alpha\beta} = \mu_{\alpha\beta}$, respectively. Both conditions must hold for any ratio of N_a/N_b. When these two conditions are satisfied, we find $d_- = D_{in}$ and $d_+ = D_{coop}$.

It can be shown that the two conditions, Eq. (8), are satisfied only if all a-chains and b-chains are identical to each other, both statically and dynamically [3]. The chains differ from each other only in their labeling, so that they can be grouped as two different species. There is no restriction on the properties of the matrix component. Hence, this conclusion is valid both in binary polymer solutions and in ternary polymer mixtures. The interdiffusion coefficient in a symmetric bimodal system is identical with the tracer-diffusion coefficient [3, 5], i.e., $d_- = D_{in} = D$, where D is the tracer-diffusion coefficient of an a-chain or b-chain. That $D_{in} = D$ is to be expected, since the particle exchange between the two species is driven here only by tracer-diffusion.

Summarizing, we have shown that the two modes, observed in dynamic scattering from bimodal systems probed within the diffusive regime, cannot be identified, in general, as the interdiffusion and cooperative diffusion processes.

Let us now address the following question: Is it possible to extract D_{in} and D_{coop} from a single measurement of the dynamic-scattering function of a labeled component *a*? To answer this question, we consider the nodal expansion [3] of $S_{aa}(q,t)$, i.e.,

$$S_{aa}(q,t) = \frac{1}{d_+ - d_-} \{[k_B T\mu_{aa} - d_- S_{aa}(q)]^{-q^2_+ t}$$

$$+ [d_+ S_{aa}(q) - k_B T\mu_{aa}]e^{-q^2 d_- t}\} . \quad (9)$$

The measurement of $S_{aa}(q,t)$ yields d_+, d_- and both amplitudes. Equation (9) shows that $S_{aa}(q)$ and μ_{aa} can be determined from the amplitudes, in addition to d_+ and d_-. The general expressions of D_{in} and D_{coop} in Eqs. (5) and (7), respectively, indicate that μ_{bb}, μ_{ab}, $S_{bb}(q)$ and $S_{ab}(q)$ are also needed for the calculation of D_{in} and D_{coop}. The knowledge of d_+ and d_- is, therefore, not sufficient to determine these four unknowns, although it provides two relations among them, for d_+ and d_- are the eigenvalues of **D**. Hence, the interdiffusion and cooperative diffusion coefficients cannot be extracted, in general, from the measurement of only $S_{aa}(q,t)$. If, in addition, $S_{bb}(q,t)$ is also measured on the same system, one can then determine μ_{bb} and $S_{bb}(q)$ from the amplitudes of $S_{bb}(q,t)$, and it is possible to calculate μ_{ab} and $S_{ab}(q)$ from the knowledge of d_+ and d_- and infer D_{in} and D_{coop} from these two measurements.

5. Diffusion coefficients in RPA

In this section, we consider an incompressible ternary mixture of *a* and *b* homopolymers in a matrix of *c* homopolymers. The diffusion coefficients D_{in} and D_{coop} can then be calculated explicitly within the framework of the random phase approximation (RPA). Recently, Akcasu and Tombakoglu [6] have shown that, in RPA, the mobilities $\mu_{\alpha\beta}$ can be expressed in terms of the mobilities $\mu^0_{\alpha\beta}$ in the bare system. The "bare" system is defined [6] as the one in which the monomer-monomer interactions are not present, but the chain connectivity is maintained. The mobilities $\mu^0_{\alpha\beta}$ are related to the tracer diffusion coefficients in the bare system by $\mu^0_{\alpha\beta} = N_a D_a \delta_{\alpha\beta}/(k_B T)$, where $D_a = k_B T/\xi_a$ is the monomer tracer diffusion coefficient, and ξ_a is the monomer friction coefficient in the mixture. This result is obtained by adopting the Rouse model for the dynamics of a single Gaussian chain in the mixture.

The definition of Eq. (5) together with the RPA expressions for $\mu_{\alpha\beta}$ yields a novel expression for the kinetic factor [3]

$$\Lambda_{in} = \frac{D_a}{N_a} + \frac{D_b}{N_b} - \frac{(D_a - D_b)^2}{N_a D_a + N_b D_b + N_c D_c} . \quad (10)$$

The D_a in Eq. (10) are, in principle, the tracer diffusion coefficients in the bare system, in which the monomer-monomer interactions are not present. But, in Rouse dynamics, the mobilities are related to the friction coefficients by $\mu^0_{aa} = N_a/\xi_a$, regardless of whether the monomer interactions are present or not. The friction coefficients ξ_a enter the Rouse dynamics as parameters that must be specified as an input from elsewhere. Hence, ξ_a and D_a are interpreted as the friction coefficient and the tracer diffusion coefficient of a monomer of *a*-th kind in the actual interacting mixture. They, therefore, depend implicitly on the composition and the temperature of the mixture. Experimentally, one measures the c.o.m. tracer diffusion coefficient D^P_a of a polymer chain, rather than that of a monomer. They are related to each other by $D^P_a = D_a/p_a$, where p_a is the number of monomers in a chain of kind *a*. We mention that a closed expression [3] for the cooperative diffusion coefficient D_{coop} in the case of an incompressible ternary mixture can also be provided, based on the definition in Eq. (7) and the RPA.

An interesting feature of the expression Eq. (10) is that it contains, as two limiting cases, the results of both the slow mode and fast mode theories [5, 7] for the calculation of the kinetic factor.

The slow mode result follows when the matrix mobility μ^0_{cc} is much smaller than the sum of the mobilities of the other two components, or equivalently, when

$$N_c D_c \ll N_a D_a + N_b D_b . \quad (11)$$

Then, Eq. (10) yields

$$\frac{1}{\Lambda_{in}} = N x_a x_b \left[\frac{x_b}{D_a} + \frac{x_a}{D_b}\right] \quad (12)$$

for the kinetic factor, where $N = N_a + N_b$ and $x_a = N_a/N$. The above expression leads for $D_a \ll D_b$ to $\Lambda_{in} \propto D_a$, i.e., the kinetic factor is dominated by the slow species *a* (slow mode), as the name of the theory implies. The condition in Eq. (11) is automatically satisfied when the matrix is removed, i.e., $N_c \to 0$, so that one is left with an incompressible mixture of species a- and b (*bulk limit*). The slow mode result was obtained initially for a binary incompressible mixture using the RPA. However, Eq. (10) indicates that the slow mode result is still valid even in the presence of a third "matrix"-component, provided the unequality in Eq. (11) holds.

The fast mode result follows when the last term in Eq. (10) is negligible as compared to the sum of the first two. Explicitly,

$$\frac{1}{N_c D_c} \ll \frac{1}{N_a + N_b} \left(\frac{x_a}{D_a} + \frac{x_b}{D_b} \right) . \qquad (13)$$

Then, with this inequality assumed to be satisfied, the kinetic factor reduces to

$$\Lambda_{in} = \frac{1}{N x_a x_b} [x_b D_a + x_a D_b] . \qquad (14)$$

It is observed for $D_a \ll D_b$ that $\Lambda_{in} \propto D_b$, i.e., the kinetic factor is dominated by the fast component b (fast mode). The condition in Eq. (13) is satisfied when the number of matrix monomers is much larger than the total number of monomers of species a and b, or when the friction coefficient ξ_c of the matrix monomers is much smaller than the friction coefficients ξ_a and ξ_b, i.e., when the relaxation of the matrix is much faster than the relaxation of species a and b. This situation prevails in a binary solution of polymers, in which the solvent plays the role of the matrix. In these solutions, the solvent is usually described by the steady-state Navier Stokes equation with the implication that its response to volumetric point forces is instantaneous. Hence, when the condition in Eq. (13) is satisfied, the incompresible ternary mixture behaves like a polymer solution (*solution-like limit*), and the fast mode result emerges.

6. Conclusions

Summarizing, we have shown that the two exponential modes observed in dynamic scattering experiments in the diffusive regime on bimodal systems (i.e., binary solution of a and b homopolymers, or incompressible ternary mixtures of a and b homopolymers in a matrix of c homopolymers) cannot in general be identified as the interdiffusion and cooperative diffusion modes. We have arrived at this conclusion by first defining the interdiffusion and the cooperative diffusion coefficients for a given pair of species a and b, in the presence of a third "matrix" species (or solvent).

Conditions on $\mu_{\alpha\beta}$ and $S_{\alpha\beta}$ have been obtained under which the normal modes coincide with the interdiffusion and cooperative diffusion modes. The physical implication of these conditions is that the a and b chains must be identical, and are distinguished from each other only by their labeling. We have also shown that D_{in} and D_{coop} cannot be extracted, in general, from the measurement of a single dynamic scattering function $S_{aa}(q, t)$.

The above conclusions are quite general, and applicable to both solutions and incompressible ternary polymer mixtures. More specific results have been obtained within the framework of the RPA. Within RPA, a new expression for the interdiffusion coefficient of the a and b components, in the presence of a matrix of c homopolymers, has been obtained in terms of the tracer diffusion coefficients of all three species. This expression contains, as two limiting cases, the results of both the "slow mode" and "fast mode" theories, in which Λ_{in} is expressed in terms of the c.o.m. tracer diffusion coefficients of the a and b polymers only. The slow mode theory is recaptured when the matrix component is gradually removed, resulting in an incompressible binary mixture of a and b polymers. The fast mode theory, on the other hand, emerges in the opposite limit of large matrix concentration and/or in the limit of large tracer diffusivity of the matrix molecules.

This limit may be pictured physically either as a dilute solution of a and b polymers, in which the matrix corresponds to the solvent, or as a compressible mixture of a and b polymers, in which the matrix has a very high mobility, as would be the case, for example, with vacancies. If one is permitted to stretch the validity of the RPA by allowing the matrix to consist of vacancies instead of homopolymers, then Eq. (10) would also apply to binary compressible polymer mixtures. Then one would expect a gradual transition from the fast mode theory to the slow mode theory when the vacancy concentration, or the compressibility of the mixture is reduced. There seem to be some computer simulations supporting this conjecture [7]. However, the application of the RPA to polymer solutions and compressible polymer mixtures with vacancies may be questionable, for the RPA is found to be truly valid only for long chains. Nonetheless, the RPA may be indicative of trends in these systems.

Acknowledgements

This work has been supported (G.N. and R.K.) by the Deutsche Forschungsgemeinschaft (DFG) and the Petroleum Research Fund, administered by the American Chemical Society (A.Z.A.).

References

1. Zhou P, Brown W (1990) Macromolecules 23:1131
2. Benmouna M, Benoit H, Duval M, Akcasu AZ (1987) Macromolecules 20:1107
3. Akcasu AZ, Nägele G, Klein R (1991) Macromolecules 24:4408
4. Hess W, Akcasu AZ (1988) J Phys France 49:1251
5. Hess W, Nägele G, Akcasu AZ (1990) J Polym Sci B28:2233
6. Akcasu AZ, Tombakoglu M (1990) Macromolecules 23:607; Akcasu AZ (1991) Macromolecules 24:2109 (Addendum)
7. Jilge W, Carmesan I, Kremer K, Binder K: "A Monte Carlo Simulation of Polymer-Polymer Interdiffusion", preprint

Authors' address:

Gerhard Nägele
Universität Konstanz
Fakultät für Physik
Postfach 5560
D-7750 Konstanz
FRG

Progress in Colloid & Polymer Science   Progr Colloid Polym Sci 89:95—98 (1992)

Collective diffusion in colloidal suspensions:
A generalized Langevin equation approach

O. Alarcón-Waess*) and M. Medina-Noyola[1])

Departamento de Física y Matemáticas, Universidad de las Américas, Puebla, Mexico
[1]) Instituto de Física "Manuel Sandoval Vallarta", Universidad Autónoma de San Luis Potosí, Mexico

Abstract: The generalized Langevin equation and the procedure of contraction of the description is employed to study collective diffusion in colloidal suspensions without hydrodynamic interactions. We derive the time-evolution equation for the dynamic structure factor of the system. Our general results turn out to be fully equivalent to those derived on the basis of the many-particle Fokker-Planck equation. In particular, we show that, before taking the overdamped limit, the dynamic structure factor has the exact short-time dependence up to order t^4. We explain how our approach could be further extended to derive additional short-time exact conditions.

Key words: Colloidal suspensions; diffusion; Langevin equation

During the last decade, great progress has been achieved in the understanding of the dynamic properties of colloidal suspensions [1—4]. Much of this progress has been made possible by the fortunate analogy between a colloidal suspension and a molecular liquid, in which the colloidal particles play the role of molecules or atoms in an ordinary liquid, the solvent replaces the vacuum, and the effective forces between macroparticles replace the intermolecular forces. In this manner, the statistical mechanics of suspensions has consisted, to some extent, of the translation of theoretical approaches developed in the context of molecular fluids [5]. In a simple conception of this analogy, we could think of a suspension as a set of particles obeying Newton's laws of motion, interacting among themselves by means of direct (conservative) forces, just like the atoms of an ordinary liquid, but subjected, in addition, to the dissipative friction forces due to their interaction with the supporting solvent. Thus, the deterministic dynamics governing the motion of atoms in a liquid must be modified by adding these dissipative forces (and their corresponding fluc-

tuating forces), thus resulting in a set of coupled Langevin equations [1], one for each of the particles in the suspension.

This microscopic description of the dynamics of suspensions can be cast, alternatively, in terms of a many-body Fokker-Planck equation for the positions and momenta of the colloidal particles. In this manner, Hess and Klein [2] rooted in such a dynamic description their application to colloids of concepts such as generalized hydrodynamics, mode-mode coupling approximation, etc., previously developed for molecular liquids. Similarly, Ackerson's pioneering work [4], starting from the many-particle Smoluchowski equation, made use of protection operator techniques to derive the time evolution equations for the intermediate scattering function of a suspension, in close analogy with similar developments in liquids. On the other hand, Pusey and Tough [1], and Arauz-Lara and Medina-Noyola [4], derived short-time conditions for the self and collective dynamic properties of suspensions starting from the overdamped many-body Langevin, and Smoluchowski equations, respectively. Finally, Alarcón-Waess and García-Colín [6] approached the description of the collective dynamics of this type of systems using extended irreversible thermodynamics.

---

*) On leave of absence from Universidad Autónoma Metropolitana-Iztapalapa.

There are, of course, other theoretical approaches in the theory of molecular fluids which could also help in the development of the theory of the dynamics of suspensions. The main purpose of this paper is to derive a description of collective diffusion in colloidal suspensions using the Generalized Langevin equation (GLE) formalism [5], plus the procedure referred to as *contraction of the description* [7]. The idea of this approach is to expand the set of variables which describe the collective dynamics of the system. Although there is not a general prescription to select the set of state variables, the physics of the specific system generally indicates in terms of which variables one should describe its dynamics. Thus, we follow the previous application of GLE formalism to molecular fluids [5], in which one considers the mass, momentum, and energy densities, together with the corresponding fluxes, as the state variables of the fluid. In the case of colloidal suspensions, we shall consider the mass, momentum, and stress tensor as the necessary variables to describe its collective dynamics.

Let us consider a colloidal suspension as a mixture of two components, i.e., the colloidal particles and the supporting solvent. We are interested in describing only the dynamics of the particles. Thus, we consider the usual variables that describe the collective dynamics of simple fluids, which are the local concentration of macroparticles, $n(r, t)$, whose Fourier transform is given by

$$n(k, t) = \frac{1}{N^{1/2}} \sum_{l=1}^{N} e^{ik \cdot r_l(t)} , \qquad (1)$$

and the local particle current, defined as

$$j(k, t) = \frac{1}{N^{1/2}} \sum_{l=1}^{N} v_l(t) e^{ik \cdot r_l(t)} , \qquad (2)$$

where N is the total number of macroparticles, $r_l(t)$ the lth particle position at time t, and $v_l(t)$ its velocity. From Eq. (1), the continuity equation readily follows,

$$\frac{\partial \delta n(k, t)}{\partial t} = ik \cdot \delta j(k, t) , \qquad (3)$$

where "δ" means fluctuation around equilibrium, i.e., $\delta n(k, t) \equiv n(k, t) - \langle n(k, t) \rangle$, $\delta j(k, t) \equiv j(k, t)$,

where "$\langle \rangle$" means average over an equilibrium ensamble. Now, in order to obtain the time-evolution equation for the local particle-current, we derive Eq. (2) with respect to time, and employ the N-particle Langevin equation without hydrodynamic interactions [1]. In this manner, one is led to [8]

$$\frac{\partial \delta j(k, t)}{\partial t} = ik \cdot (\delta \sigma'(k, t) + \delta p I)$$
$$- \frac{\zeta}{m} \delta j(k, t) + \frac{1}{m} f(k, t) , \qquad (4)$$

where $\delta \sigma(k, t) = \delta \sigma'(k, t) + \delta p I$ is the usual definition of the stress tensor of a molecular fluid. It contains the information on the direct interactions between colloidal particles. ζ is the Stoke's friction coefficient of each colloidal particle of mass m, resulting from the friction due to the solvent, and $f(k, t)$ is a purely random Gaussian stochastic vector, given by

$$f(k, t) = \frac{1}{N^{1/2}} \sum_{l=1}^{N} f_l(t) e^{ik \cdot r_l(t)} , \qquad (5)$$

where $f_l(t)$ is the Brownian random force on particle l. It is easy to show that $f(k, t)$ satisfies a fluctuation-dissipation relation. Thus, we see that the difference between a molecular fluid and a colloidal suspension is the appearance of the last two terms of Eq. (4).

Recently, Medina-Noyola and del Rio-Correa [7] have shown that if a fluctuating process is described by an N-dimensional stochastic process $a(t)$ generated by a stochastic equation of the type

$$\frac{\partial a(t)}{\partial t} = -\int_0^t dt' \, G(t - t') a(t') + f(t) , \qquad (6)$$

where $f(t)$ is a stationary, generally non-white, random force, then the memory matrix $G(t)$ must be such that this GLE, must have the following structure

$$\frac{\partial a(t)}{\partial t} = -\omega \chi^{-1} a(t)$$
$$- \int_0^t L(t - t') \chi^{-1} a(t') + f(t) , \qquad (7)$$

where

$$\chi = \langle a(0) a^\dagger(0) \rangle \qquad (8)$$

is the static correlation matrix, ω is a time-independent antisymmetric matrix, $\omega = -\omega^T$, and $L(t)$ is given by $L(t) = L^\dagger(-t) = \langle f(t) f^\dagger(0) \rangle$. These conditions establish the minimum selection rules that the matrices ω and L must observe, if Eq. (6) is expected to describe fluctuations around the thermodynamic equilibrium state. These selection rules are a consequence of the stationarity condition of the system. In the event that the variables do possess additional symmetries themselves, then additional conditions on ω and $L(t)$ may be derived. For example, if the variables a_i have a well-defined time-reversal symmetry, that is, if a_i change to $\lambda_i a_i$, $\lambda_i = \pm 1$, when time is reversed, then one can show that $\omega = -\Lambda \omega \Lambda$ and $L(-t) = \Lambda L(t) \Lambda$, where the matrix Λ is defined as $\Lambda_{ij} = \lambda_i \delta_{ij}$.

To describe the collective dynamics of a colloidal suspension, we choose the vector $a(t)$ as follows

$$a^{\dagger}(t) = (\delta n(k,t), \ \delta j_l(k,t), \ \delta \sigma'(k,t)) , \quad (9)$$

where $\delta n(k,t)$ has just been defined (Eq. (1)), $\delta j_l(k,t) \equiv j_z(k,t)$, and $\delta \sigma_l'(k,t) \equiv \delta \sigma_{zz}'(k,t)$. For this vector, the static correlation matrix χ is diagonal, and its non-vanishing elements are

$$\chi_{nn} = \langle \delta n(-k,0)\delta n(k,0) \rangle$$
$$= 1 + n \int e^{ik \cdot r}[g(r) - 1]d^3r , \quad (10)$$

$$\chi_{jj} = \langle \delta j_l(-k,0)\delta j_l(k,0) \rangle = \frac{k_B T}{m} , \quad (11)$$

and

$$\chi_{\sigma\sigma} = \langle \delta \sigma_l(-k,0)\delta \sigma_l(k,0) \rangle$$

$$= \frac{n k_B T}{m^2} \left[3k_B T + n \int g(r) \frac{\partial^2 u(r)}{\partial z^2} \right.$$

$$\left. \cdot \frac{(1 - \cos kz)}{k^2} d^3r \right] - \frac{n}{\chi_{nn}} \left(\frac{k_B T}{m} \right)^2 , \quad (12)$$

where n is the bulk concentration, $g(r)$ the radial distribution function, k_B Boltzmann constant, T the temperature, and $u(r)$ the effective pair potential of the direct interactions between particles.

From the GLE formalism, one derives the time-evolution equation for the stress tensor, which turns out to be

$$\frac{\partial \delta \sigma_l'(k,t)}{\partial t} = ik \chi_{\sigma\sigma} \chi_{jj}^{-1} \delta j_l(k,t)$$

$$- \int_0^t dt' \chi_{\sigma\sigma}^{-1} L_{\sigma\sigma}(k,t-t')$$

$$\cdot \delta \sigma_l'(k,t) + f_\sigma(k,t) . \quad (13)$$

In deriving this equation [8], use has to be made of the selection rules referred to before.

Let us now employ the idea of contraction of the description, which in our case consists in reducing the set of Eqs. (3), (4), and (13) to a single equation for $\delta n(k,t)$. Thus, after some algebraic steps, one is led to

$$\frac{\partial \delta n(k,t)}{\partial t} = -\int_0^t \Gamma(k,t-t')\chi_{nn}^{-1}\delta n(k,t')dt'$$
$$+ \psi(k,t) , \quad (14)$$

where the Laplace transform of the memory function $\Gamma(k,t)$ is given by

$$\hat{\Gamma}(k,z) = \cfrac{k^2 k_B T}{z + \cfrac{\zeta}{m} + \cfrac{k^2 \chi_{\sigma\sigma}\chi_{jj}^{-1}}{z + \chi_{\sigma\sigma}^{-1}L_{\sigma\sigma}(k,z)}} . \quad (15)$$

In Eq. (14), $\psi(k,t)$ is a random term with zero mean and with its time-correlation function given by $\Gamma(k,t)$. In Eq. (15), the only unknown term is $L_{\sigma\sigma}(k,z)$, which is the memory function describing the relaxation of stress tensor fluctuations. From the definition of the dynamic structure factor $F(k,t) = \langle \delta n(-k,t)\delta n(k,0) \rangle$, and using Eq. (14), we obtain

$$\frac{\partial F(k,t)}{\partial t} = -\int_0^t \Gamma(k,t-t')\chi_{nn}^{-1}F(k,t')dt' . \quad (16)$$

Let us now write the Taylor expansion of $F(k,t)$ around $t = 0$,

$$F(k,t) = \sum_{i=0}^{\infty} \frac{m^{(m)}(k)t^i}{i!} , \quad (17)$$

where the ith coefficient (or moment) is defined by

$$m^{(i)}(k) = \lim_{t \to 0^+} \frac{\partial^i F(k,t)}{\partial t^i} , \quad (18)$$

From Eqs. (15), (17), and (18), one can calculate the first five moments, which turn out to be given by

$$m^{(0)}(k) = \chi_{nn} , \quad (19)$$

$$m^{(1)}(k) = 0 , \quad (20)$$

$$m^{(2)}(k) = -\frac{k^2}{m\beta} , \quad (21)$$

$$m^{(3)}(k) = \frac{k^2 \zeta}{m^2 \beta} , \quad (22)$$

and

$$m^{(4)}(k) = \frac{k^2}{m\beta} \left[k^2 C_T^2(k) - \frac{\zeta^2}{m^2} + m\beta k^2 \chi_{\sigma\sigma} \right], \quad (23)$$

where $\beta = (k_B T)^{-1}$ and $C_T^2(k) = k_B T/m\chi_{nn}$. Let us point out that these expressions for the moments correspond to the true $t \to 0$ limit, i.e., for "$t \to 0$", we mean $t \ll \tau_B(= m/\zeta)$. These results, which were obtained from the GLE formalism, turn out to be identical with those obtained by Hess and Klein [2]. In our method, higher-order moments could be evaluated directly in a similar manner, i.e., if we want to obtain the next moment, we need to expand the set of dynamic variables in a similar manner as we did here when we introduced the stress tensor as a state variable. Finally, if we compare our results in the appropriate limit (absence of interactions), with the corresponding exact results for an ideal suspension [9], we find full agreement. In summary, we have applied the GLE formalism to derive some properties of the dynamic structure factor of a colloidal suspension in the absence of hydrodynamic interactions. In particular, we found our results to be fully equivalent to those derived by Hess and Klein, starting from the many-particle Fokker-Planck equation [2].

Acknowledgements

This works was supported by the Consejo Nacional de Ciencia y Tecnología (CONACyT, México) and the Secretaría de Educación Pública (DGICSA-SEP, Convenio No. C90-01-0353, Anexo 5, México).

References

1. Pusey PN, Tough RJA (1985) In: Pecora R (ed) Dynamic Light Scattering and Velocimetry: Application of Photon Correlation Spectroscopy. Plenum Press, New York
2. Hess W, Klein R (1983) Adv Phys 32:173
3. Ackerson B (1976) J Chem Phys 64:242; (1978), ibid. 69:684
4. Arauz-Lara JL, Medina-Noyola M (1983) Physica A126:547
5. Boon JP, Yip S (1980) Molecular Hydrodynamics, McGraw Hill, New york
6. Alarcón-Waess O, García-Colín LS (1990) J Chem Phys 92:3086
7. Medina-Noyola M, del Rio-Correa JL (1987) Physica A146:483
8. Alarcón-Waess O, Medina-Noyola M (1991) Rev Mex Fis 37:Supl 1, p S38—S50
9. Chandrashekhar S (1943) Rev Mod Phys 15:1

Authors' address:

M. Medina-Noyola
Instituto de Física
Universidad Autónoma de San Luis Potosí
Apartado Postal 629
78000 San Luis Potosí, Mexico

Progress in Colloid & Polymer Science   Progr Colloid Polym Sci 89:99—102 (1992)

Influence of the preparation mode on the size of CdS particles synthesised "in situ" in reverse micelles

L. Motte[1], A. Lebrun[2], and M. P. Pileni[1,2]

[1]) C.E.N. Saclay, DRECAM.-S.C.M, Gif sur Yvette, France
[2]) Université P. et M. Curie, Laboratoire S.R.I. bâtiment de Chimie Physique Paris, France

Abstract: Functionalized reverse micelles are used to control the size of the CdS semiconductor particles. We show, in dilute solution, that the use of mixed sodium-cadmium AOT reverse micelles favors the formation of monodispersed particles. By increasing the water content, the size of the particle increases, and it maintains a high monodispersity. The increase of the water pool concentration and the amount of cadmium ions induces an increase in the size and in the polydispersity.

Key words: CdS; reverse micelles

Introduction

Surfactants dissolved in organic solvents form spheroidal aggregates called reverse micelles [1]. Water is readily solubilized in the polar core, forming a so-called water pool, characterized by w, $w = [H_2O]/[AOT]$. For AOT as a surfactant, the maximum amount of bound water in the micelle corresponds to a water-surfactant molar ratio $w = [H_2O]/[AOT]$ of about 10. Above $w = 15$, the water pool radius is found to depend linearly on the water content ($R_w(\text{Å}) = 1.5\ w$) [2]. Another property of reverse micelles is their dynamic character [1]: they can exchange the content of their water pools by a collision process.

In the present paper, we present quantitative data to show the change in the size of the CdS particles with the experimental conditions.

Experimental section

AOT was obtained from Sigma, isooctane from Fluka, sodium sulphide Na_2S by Janssen; they were gel used without further purification. The synthesis of functionalized surfactant has been previously described [5].

The water droplets concentration (RM) is the ratio of the AOT concentration over the aggregation number and, at a given water content w, is directly related to the volume fraction of water, ϕ_w, by the following expression:

$$[RM](\text{mol/l}) = 3 \cdot 10^{-3}\ \phi_w/[4 \cdot 10^{-30}\ N\pi(1,5w)^3]\ .$$

N is Avogadro's number.

The volume fraction of water ϕ_w is the ratio of the volume of water over the total volume.

Apparatus

The absorption spectra were obtained with a Perkin-Elmer lambda 5 and Hewlett Packard spectrophotometers. A Philips electron microscope (model CM 20, 200 kV) was used for electron microscopy.

Results and discussion

Cadmium sulphide suspensions are characterized by an absorption spectrum in the visible range. In the case of small particles, a quantum size effect [6—13, 17, 18] is observed due to the perturbation of the electronic structure of the semiconductor with the change in the particle size. For CdS semiconductor, as the diameter of the particles approaches the excitonic diameter, its electronic properties start to change [18]. This gives a widening of the forbidden band and therefore a blue shift in the absorption threshold as the size decreases. This phenomena occurs as the crystallite size is comparable to or below the excitonic diameter of 50—60 Å [18b]. In a first approximation a simple "electron-hole in a box" model can quantify this blue shift with the size variation [10, 18b, 17]. Thus, the absorption threshold is directly related to the average size of the particles in solution.

In the presence of an excess of cadmium ($x = 2$), Fig. 1 shows a red shift in the absorption spectra by increasing the water content. According to the data previously published using sodium di(ethyl-2-hexyl)sulfosuccinate (AOT) as a surfactant [3, 4] this can be attributed to an increase in the average size of particles with the water content.

Below the absorption onset several shoulders are observed (Fig. 1) and can be clearly recognized in the second derivative (inserts, Fig. 1). These weak absorption bands correspond to the excitonic transi-

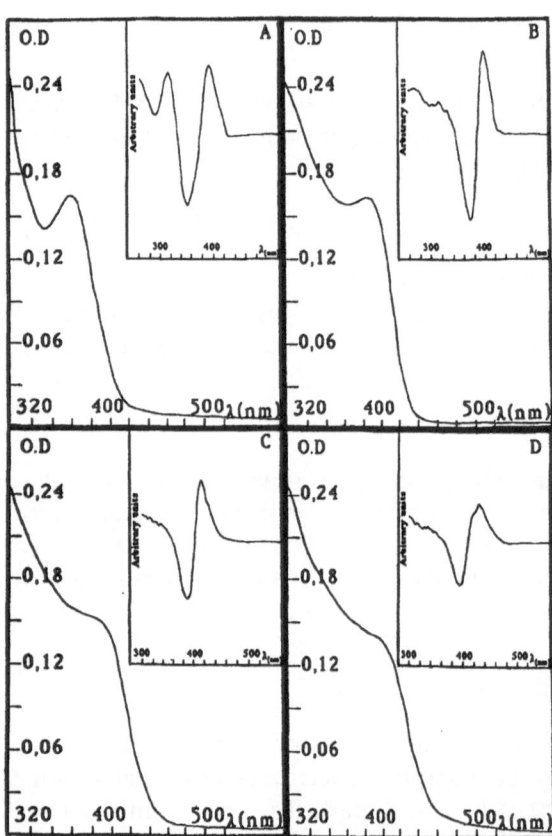

Fig. 1. Variation of the absorption spectrum of CdS in reverse micelles with the water content. [AOT-Na] = 0.1 M, [(AOT)$_2$Cd] = $2 \cdot 10^{-4}$ M, X = 2; A(w = 5); B(w = 10); C(w = 20); D(w = 40); insert: second derivative of the photoabsorption, the minimal indicates the position of the excitonic peak

tions. This clearly shows a narrow size distribution [14]. At low water content the first excitonic peak is well resolved and is followed by a shoulder. The second derivative shows a very high intensity of this bump (insert Fig. 1A). With a small crystallite, according to the data previously published [14], several bumps due to several excitonic peaks are expected. The insert Fig. 1A shows only one bump; this is due to the fact that the others are blue shifted and are not observable in our experimental conditions. By increasing the water content, that is to say, by increasing the size of the particles, several bumps are observed (insert Fig. 1B). The intensity of these bumps decreases with the water content, w (inserts, Fig. 1). This indicates a decrease in the number of excitonic transitions with the size of the particle. This is in agreement with the theoretical calculations previously published for the Q-particles [14].

From the relation between [6, 11—14] the absorption onset and the size of CdS particle, the average radius r is deduced. Figure 2A shows a large change in the size of the particle with the relative ratio of cadmium and sulphide ions ($x = [\mathrm{Cd}^{2+}]/[\mathrm{S}^{2-}]$). The largest sizes are obtained for $x = 1$, and the smallest for $x = 2$. It can be noticed that the size of CdS is always smaller when one of the two reactants are in excess ($x = 1/4, 1/2, 2$). This confirms that the crystallization process is faster when one of the species is in excess [15].

Electron microscopy has been performed using a sample synthesized at $w = 10$, $x = 2$, in mixed reverse micelles, characterized by 430 nm absorption onset, corresponding to a CdS diameter equal to 25 Å. The microanalysis study shows the characteristic lines of sulphide and cadmium ions, indicating that the observed particles are CdS semiconductor cristallites. The micrograph picture (Fig. 2B) shows spherical particles with an average size in the range of 40 ± 20 Å. The electron rays' diffraction pattern shows concentric circles, indicating a centered cubic face structure. The radius of the particle can be deduced from the line widening of the diffraction signals given by the following relationship [16]: $2r = \lambda L/\Delta R$, where λ is electron wavelength ($\lambda = 0.0251$ Å), L is camera length ($L = 4900$ mm), and ΔR is the widening line ($\Delta R = 3$ mm). The size determined from such a calculation is equal to 41 Å. The differences in the size of the particles are probably due to the extraction of CdS from the droplets with appearance of some aggregation processes.

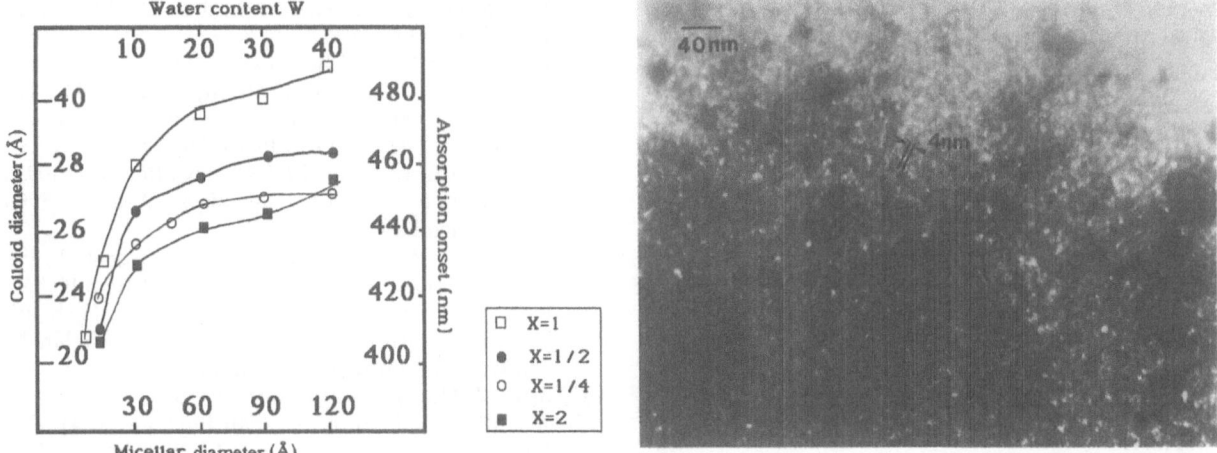

Fig. 2. A) Variation of the absorption onset and the CdS particle radius with the water content w at various X values. B) Electron microscopy of a sample after extraction from a micellar solution, $w = 10$, $x = 2$, 1 cm $\Leftrightarrow$ 40 nm

For a given polar volume fraction (ϕ_w) and w value ($w = 5$ and 10), at $X = 2$, the increase in the ratio of functionalized surfactant ([Cd(AOT)$_2$]/ [AOT] $= 0.1$ and 0.5%, respectively) induces a strong decrease in the relative intensity of the excitonic peak and a red shift in the absorption spectra (Fig. 3). This indicates an increase in the size and polydispersity of the particles with the number of cadmium ions per micelles.

Similarly, at fixed w value and for a given relative ratio of cadmium versus AOT, the increase in the polar volume fraction ϕ_w i.e., the increase in the water droplets concentration ([RM]) induces a decrease in the intensity and a broadening of the excitonic peak. The red shift of absorption onset is considerably reduced (Fig. 3). This indicates an unchanged average size of the particles with an increase in the polydispersity by increasing the number of droplets, i.e., by increasing the water pool exchange process.

Conclusion

We report the synthesis, in situ, of CdS in reverse AOT micelles using mixed micelles. The location of one of the reactants at the micelle interface makes it possible to favor the formation of monodispersed particles. The size of the particles depends on the mode of the synthesis: it changes with the ratio of cadmium versus sulphur ions, with the number of

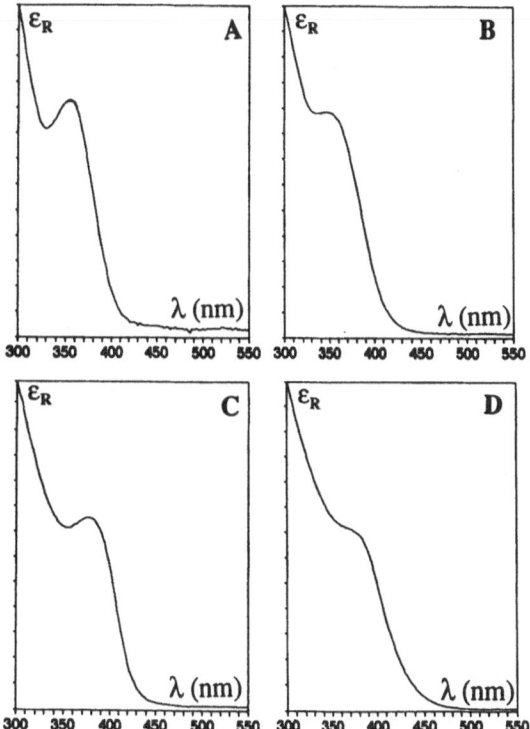

Fig. 3. Variation of the relative extinction coefficient (ε_r) of CdS particles with wavelength at $x = 2$, $w = 5$ and at various cadmium ions per micelles ([Cd(AOT)2]/[AOT] equal to 1‰ (A and B) and 5‰ (C and D), respectively) and for polar volume fraction ϕ_w equal to 0.9% (A and C) and 2.25% (B and D), respectively

droplets in the solution and the amount of cadmium ions presents in the solution. From these data it is reasonable to conclude that the smallest particles are obtained at very low water content, in very dilute water droplets and at low CdS concentration.

References

1. Structure and Reactivity in reverse micelles; Pileni MP (ed) (1989) Elsevier, Amsterdam
2. Pileni MP, Zemb T, Petit C (1985) Chem Phys Lett 118:414
3. Atkinson PJ, Grimson MJ, heenan RK, Howe AM, Robinson BH (1989) J Chem Soc, Chem Commun 1807
4. Petit C, Pileni MP (1988) J Phys Chem 92:2282
5. Petit C, Lixon P, Pileni MP (1990) J Phys Chem
6. Brus LE (1983) J Chem Phys 79:5566
7. Rossetti R, Ellison JL, Bigson JM, Brus LE (1984) J Chem Phys 80:4464
8. Nozik AJ, Williams F, Nenadocic MT, Rajh T, Micic OI (1985) J Phys Chem 89:397—399
9. Bawendi MG, Steigerwald ML, Brus LE (1990) Annu Rev Phys Chem 41:477
10. Henglein A (1989) Chem Rev 89:1861
11. 11a. Wang Y, herron N (1990) Phys Rev 41:6079; 11b. Wang Y, Herron N (1991) J Phys Chem 95:525
12. Kayanuma Y (1988) Phys Rev B 38:9797
13. Lippens PE, Lannoo M (1989) Phys Rev B 39:10935
14. Katsikas L, Eychmüller A, Giersig M, Weller H (1990) Chem Phys Lett 172:201
15. Fisher CH, Weller H, Lume-Periera C, Janata E, Heinglein A (1986) Ber Bunsenges Phys Chem 90:46
16. Electron microscopy of thin crystalite. Hish PB, Howie A, Nicholson RB, Pashley DW, Whelan MJ (1971) Butterworth press London
17. Brus LE (1983) J Chem Phys 79:5566
18. 18a. Wang Y, Herron N (1990) Phys Rev B 41:6079; 18b. Wang Y, Herron N (1991) J Phys Chem 95:525

Authors' address:

M. P. Pileni
C.E.N. Saclay
DRECAM-S.C.M.
91191 Gif sur Yvette, France

Progress in Colloid & Polymer Science   Progr Colloid Polym Sci 89:103—105 (1992)

Synthesis of copper metallic particles using functionalized surfactants in w/o and o/w microemulsions

I. Lisiecki[1]), L. Boulanger[2]), P. Lixon[3]), and M. P. Pileni[1,4])

[1]) Université P. et M. Curie, Laboratoire S.R.I. bâtiment de Chimie Physique, Paris, France
[2]) C.E.N. Saclay, CEREM-DTM-SRMP Gif sur Yvette, France
[3]) C.E.N. Saclay, D.R.E.C.A.M.-S.C.M Bât. 125, Gif sur Yvette, France
[4]) C.E.N. Sacley, D.R.E.C.A.M.-S.C.M Bât. 522, Gif sur Yvette, France

Abstract: Functionalized surfactant micelles are used to prepare small metallic copper particles. At low water content, small metallic particles are formed. By increasing the water content, passivation takes place.

Key words: Small particules; mixed micelles; electron micrographs; colloidal copper

Introduction

Metals constitute a wide class of catalysts, and because catalysis occurs on the surface, there is an economic incentive to obtain catalysts in the form of small metal particles. This, however, raises two main problems. One is fundamental in nature and addresses the question as to below which particle size the metallic properties are lost. The other is more practical and concerns the preparationn of very small particles. Surfactants dissolved in organic solvents, form spheroidal aggregates called reverse micelles [1]. Water is readily solubilized in the polar core, forming a so-called water pool, characterized by w, $w = [H_2O]/[AOT]$. For AOT as a surfactant, the maximum amount of bound water in the micelle corresponds to a water-surfactant molar ratio $w = [H_2O]/[AOT]$ of about 10. Above $w = 15$, the water pool radius is found to depend linearly on the water content ($R_w = 1.5\ w$) [2]. Another property of reverse micelles is their dynamic characters [1]. They can exchange the content of their water pools by a collision process.

In the present paper, we report the formation of colloidal copper metallic particles in reverse micelles. A change in the process of oxidation of the copper is observed by changing the experimental conditions.

Material and methods

AOT was obtained from Sigma, isooctane from Fluka, and sodium borohydride, $NaBH_4$, from Alfa; they were used without further purification. The synthesis of copper dioctylsulfosuccinate, $Cu(AOT)_2$, has been previously described [3].

A reverse micellar solution formed by solubilizing copper AOT, $\{[Cu(AOT)_2] = 10^{-3}\ M\}$ and AOT surfactant, $\{[AOT] = 0.25\ M\}$ is mixed with AOT reverse micelles containing $NaBH_4$, $[AOT] = 0.25\ M$, $[NaBH_4] = 2 \cdot 10^{-3}\ m$. The absorption spectra were recorded on Perkin-Elmer lambda 5 and Hewlett Packard HP 8452 A spectrophotometers. A drop of this solution is evaporated under vacuum and the electronm micrograph is obtained with a Philips electron microscope (model CM 20, 200 kV).

Results and discussion

The reduction of copper ions is followed by spectrophotometry.

In aqueous solution, in the absence of oxygen, the absorption spectrum is not well defined. As it is shown in Fig. 1A, a strong absorption is observed at 800 nm with a plateau at about 570 nm. In the presence of a protecting agent, such as sodium hexametaphosphate or HMP, the absorption spectrum is well resolved, with a peak centered at 560 nm and a plateau at about 250—300 nm (Fig. 1B). The comparison between the absorption spectra and EXAF experiments performed by Abe et al. [4] and Yanase and Komiyama [5], clearly indicates that the

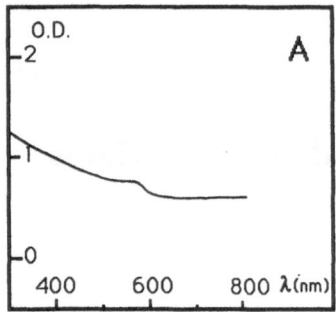

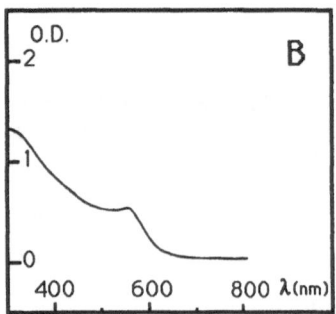

absorption at 800 nm is due to oxidized copper, and that at 560—570 nm is attributed to a plasmon spectrum due to formation of metallic clusters consisting of severals thousands of atoms.

In reverse mixed micelles, similar experiments have been performed at various water content. Figure 2 shows the change in the absorption spectra with time at two w values ($w = 5$ and $w = 9$), and the electron micrographs obtained in the same conditions. The change of the absorption spectrum with time can be attributed to the increase of the particle size. At low water content ($w = 5$), a very small absorption is observed compared to that obtained at higher water content ($w = 9$); Furthermore, the absorption spectrum obtained at infinite time at low water content is characterized by a

◆

Fig. 1. Absorption spectra of copper clusters obtained in aqueous solution in degassed solution; A) in the absence of HMP; $[CuSO_4] = 5 \cdot 10^{-4}$ M, $[NaBH_4] = 10^{-3}$ M. B) In the presence of HMP; $[CuSO_4] = 5 \cdot 10^{-4}$ M, $[NaBH_4] = 10^{-3}$ M. $[HMP] = 2.5 \cdot 10^{-4}$ M

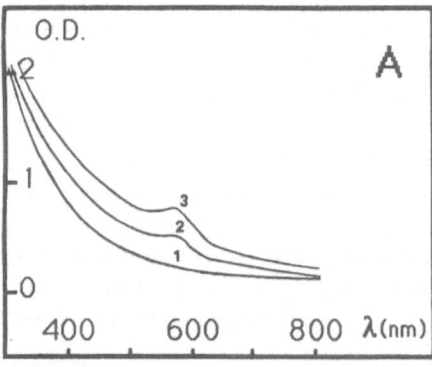

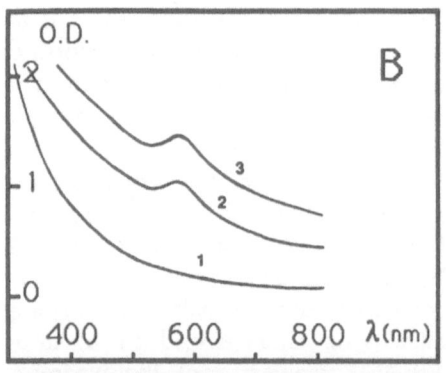

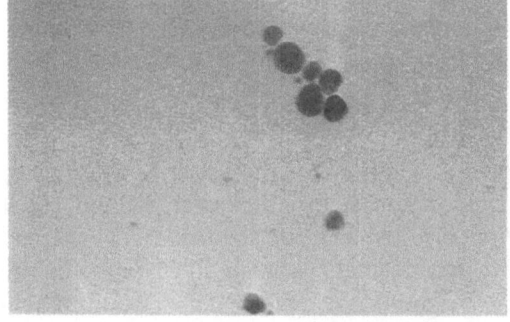

1 mm = 46 Å

1 mm = 46 Å

Fig. 2. Absorption spectra of copper clusters obtained in reverse micellar solution in degassed solution and electron microscopy pictures observed at various water content; (AOT) = 0.25 M.
A) $w = 5$, 1) $t = 0$, 2) $t = 2$ mn; 3) $t = 3$ mn
B) $w = 9$, 1) $t = 0$, 2) $t = 3$ mn; 3) $t = 5$ mn

plateau in the range of 560—570 nm with a very low absorption at 800 nm. At higher w values ($w = 9$), a plasmon peak is obtained at 570 nm with a residual absorption at 800 nm.

From the comparison between out data and those previously published [4, 5], it seems reasonable to conclude that the decrease in the water content favors the formation of pure metallic copper clusters, while at high w values the metallic copper clusters are surrounded by a surface oxide layer.

The electron micrographs shown in Fig. 2 change with the water content. As can be seen, the average size of the particles does not change greatly with the water content; it is found close to 150 Å. However, at low w value, the clusters are very well isolated, while by increasing w, the particles become more aggregated. This can be related to the structure of the clusters: at low water content the metallic particles do not strongly interact with the surfactant and are easily extracted when at high water content; interactions between the oxide surface and the surfactant creates an association among the particles.

Conclusion

We have reported, the synthesis, in situ, of $(Cu)_n$ in AOT reverse micelles. In reverse micelles, it is observed that functional surfactants favor the for-mation of copper clusters. At low water content, small metallic particles are formed. By increasing the w value, oxidation takes place. The size of the particles determined by electron microscopy is found to be 15 nm.

References

1. Chapter: "Structure of reversed micelles", D. Langevin. p 13
   Chapter: "Structural changes of reverse micelles and microemulsions by adding solutes or proteins", M. P. Pileni. p 44
   Structure and reactivity in reverse micelles (1989) Pileni MP (ed) Elsevier
2. Pileni MP, Zemb T, Petit C (1985) Chem Phys Lett 118:414
3. Petit C, Lixon P, Pileni MP (1991) Langmuir in press. Lisiecki I, Lixon P, Pileni MP (1991) Progr Colloid Polym Sci 84:342
4. Abe H, Charle KP, Tesche B, Schulze W (1982) Chem Phys 68:137
5. Akihisa Ayanase, Hiroshi Komiyama (1991) Surface science 248:11—19
6. Petit C, Pileni MP (1988) J Phys Chem 92:2282

Authors' address:

M. P. Pileni
Université P. et M. Curie
Laboratoire SRI
bâtiment de Chimie Physique
11 rue P. et M. Curie
75005 Paris, France

Progress in Colloid & Polymer Science

Progr Colloid Polym Sci 89:106—109 (1992)

Photochemical studies of nanosized CdS particles synthesized in micellar media

T. K. Jain[1]), F. Billoudet[1]), L. Motte[2]), I. Lisiecki[1,2]), and M. P. Pileni[1,2])

[1]) Université P. et M. Curie, Laboratoire S.R.I. bâtiment de Chimie Physique, Paris, France
[2]) C.E.N. Saclay, DRECAM.-S.C.M, Gif sur Yvette, France

Abstract: The photoinduced electron transfer reactions from nanosized CdS particles synthesized in micelles of cadmium lauryl sulphate and sodium lauryl sulphate to various dialkyl viologens is measured for a molar ratio of cadmium to sulphide ions equal to 2. — The yield of the reduced viologens obtained by CdS irradiations is maximum in the case of C_1 viologen, which decreases gradually with increase in the chain length of the viologens. Moeover, a back electron transfer reaction is observed in the case of C_1, C_3, and C_8 viologens. The addition of sodium lauryl sulphate (NaLS) inhibits the back electron transfer reaction. This is due to strong interactions between methyl viologen and lauryl sulphate, which gives the viologens a hydrophobic character and favors their entrance into the micellar core. This is confirmed by studying the photoelectron transfer from CdS particles to methylviologen laurylsulphate.

Key words: CdS particles; micellar solution; photoelectron transfer reaction

Introduction

Microemulsions are thermodynamically stable liquid phases of oil, water (or some other polar solvent), and surfactant. Microemulsion-based, "in situ" synthesis of microparticles has made considerable progress in the last few years and is found to be highly convenient for preparing particles in the nanometer size range [1—10].

Among various types of microparticles synthesized in microemulsions, synthesis of cadmium sulphide microparticles is of particular interest, as this provides a mean of storing solar energy and of mimicking photosynthesis systems [11]. Moreover, there is a direct relationship between the particle size and UV/VIS spectral properties of CdS [12].

In the present paper, we report the synthesis and photoelectron transfer reactions of CdS particles in oil-in-water-type microemulsion consisting of cadmium and sodium lauryl sulphates {Cd(LS)$_2$ and NaLS}.

Materials and methods

Sodium lauryl sulphate was purchased from BDH and Na$_2$S was procured from Janssen. Methylvio-logenlaurylsulfate, MV(LS)$_2$, was synthesized by mixing an aqueous solution of sodium lauryl sulphate with methylviologen chloride solution. A precipitate was obtained which was purified by several washings and crystallization in milipore distilled water. Cadmium lauryl sulphate, Cd(LS)$_2$, was synthesized as described in [8].

The surface-tension measurements of cadmium lauryl sulphate and methylviologen lauryl sulphate were performed on a Kruss equipment, and their CMDs were found to be 8×10^{-4} and 5×10^{-4} M, respectively.

A solution of sodium sulphide was added to cadmium lauryl sulphate micellar solution, keeping [CD(LS)$_2$] = 2 $\times 10^{-3}$ M at "x" = 2.0 (where x = [Cd^{2+}]/[S^{2-}]) to get a very fine dispersion of CdS particles in the system, which was stable, and no precipitation was observed within a few hours after preparation.

The absorption spectra were obtained on a Hewlett Packard HP8452A spectrophotometer. Electron microscopy was done with a Philips electron microscope (CM 20, 200 kv). Photoelectron transfer reactions were studied with the aid of an Applied Photophysics flash photolysis apparatus. To compare the efficiency of the photoelectron transfer with various viologens, we kept cadmium lauryl sulphate and viologen concentrations constant in all experiments and equal to 2×10^{-3} M and 10^{-4} M, respectively. In such experimental conditions, the concentrations of the various components in the system depends on the x value: In the presence of an excess of cadmium ions ($x = 2$), after formation of CdS particles, the solution contains 10^{-3} M of CdS particles, 10^{-3} M of Cd(LS)$_2$, and 10^{-3} M of NaLS.

Results and discussion

Cadmium sulphide suspensions are characterized by an absorption spectrum in the visible range. In the case of small particles, a quantum size effect [13—18] is observed, due to the perturbation of the electronic structure of the semiconductor with the change in the particle size [13, 18b, 19], and the absorption threshold is directly related to the average size of the particles in solution.

On adding sodium sulphide to the micellar solution of Cd(LS)$_2$, a fine dispersion of CdS particles if formed. In the presence of an excess of cadmium ions ($x = 2$), the absorption threshold is around 500 nm, which corresponds to a particle size of about 100 Å [19].

Electron microscopy was performed on a sample synthesized at $x = 2$ and characterized by 500 nm absorption onset, which corresponds to a CdS particle diameter equal to 100 Å. In order to extract CdS particles, the micellar solution containing CdS particles is treated with acetone. The electron micrograph and diffraction pattern of these particles is shown in Figs. 1A and B, respectively. The microanalysis study shows the characteristic lines of sulphide and cadmium ions, indicating that the observed particles are CdS semiconductor crystallites. Figure 1A shows spherical particles having average size in the range of 100 ± 20 Å. The diffraction pattern (Fig. 1B) depicts hexagonal compact structure for the CdS particles synthesized in normal micelles and extracted with the help of acetone.

The photoelectron transfer from CdS particles to various viologens has been studied by flash photolysis. The appearance of reduced viologens has been monitored spectrophotometrically at 610 nm.

The photoelectron transfer from CdS particles to methylviologen is characterized by a back electron

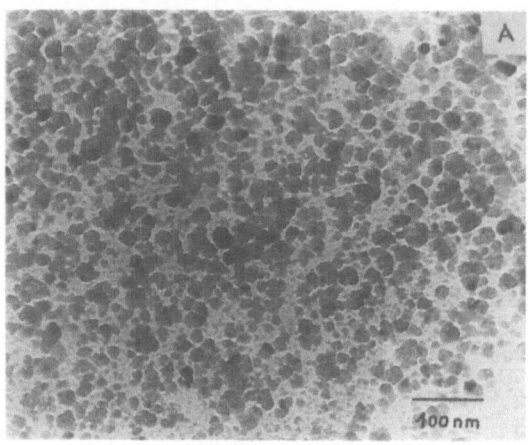

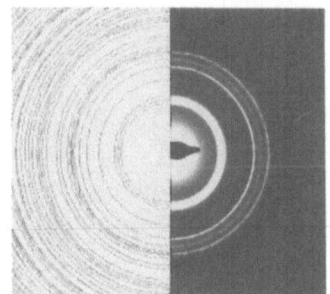

Fig. 1. A) Electron micrograph of CdS particles prepared for $x = 2.0$ and extracted with acetone; B) Diffraction pattern of the CdS particles prepared for $x = 2.0$ and extracted with acetone

transfer kinetic rate constant equal to 10^4 s^{-1}, and the yield of reduced viologen is found to be 3%. At [NaLS] $\leqslant 5 \times 10^{-3}$ M,a the back electron transfer kinetic rate constant and the yield of reduced methyl viologen remains unchanged.

By adding NaLS to the solution ([NaLS] $> 5 \times 10^{-3}$ M), the yield of reduced viologens decreases to reach a plateau at [NaLS] $= 10^{-2}$ M. The back reaction is totally prevented and the absorbance, measured at infinite time, is at maximum when NaLS concentration is equal to 5×10^{-3} M, is shown in Fig. 2A. This can be explained in the following way: At low NaLS concentration (below 5×10^{-3} M), methylviologen molecules are surrounded by CdS particles, Cd(LS)$_2$ micelles, and NaLS monomers. By increasing the NaLS con-

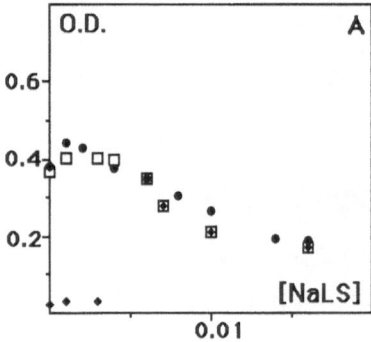

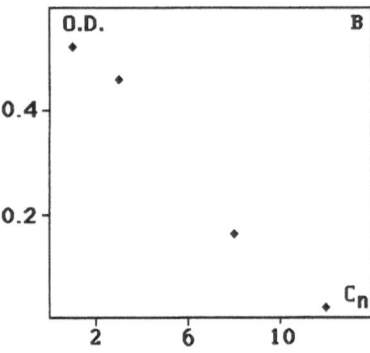

Fig. 2. A) Variation of the reduced viologen yield with sodium lauryl sulphate concentration for different micellar systems at $x = 2.0$: (□) immediately after the flash for methyl violotgen chloride; (♦) at infinite time for methyl viologen chloride; (●) in the presence of methyl viologen lauryl sulphate. B) Variation of the yield of various reduced dialkyl vologens with alkyl chain length of different viologens

centration, the methylviologen interacts with NaLS to form methylviologen lauryl sulphate, MV(LS)$_2$. The reduced methylviologen lauryl sulphate formed by photoelectron transfer (which has a strong hydrophobic character [20]) penetrates in the Cd(LS)$_2$ micelles. This process prevents the back electron transfer reaction. At higher NaLS concentration, mixed micelles are formed that contain less than one viologen molecular per micelle. The decrease in the yield of reduced viologen can be attributed to the decrease in the probability of finding CdS particles in the vicinity of methylviologen. This induces a decrease in the efficiency of the electron transfer reaction.

In order to understand this hypothesis, we synthesized MV(LS)$_2$ and the photoelectron transfer process from CdS particles to 10^{-4} M of MV(LS)$_2$ was studied. In the initial experimental conditions corresponding to a low NaLS concentration ([NALS] = 5×10^{-3} M), the back electron transfer reaction is totally prevented and the yield of reduced methylviologen lauryl sulphate is similar to that obtained with methylviologen chloride (Fig. 2A). It shows an unchanged behavior in the yield of reduced viologen lauryl sulphate compared to methylviologen chloride with NaLS concentration. This confirms the proposed mechanism, according to which the back electron transfer is prevented by the entrance of reduced viologen lauryl sulphate in the Cd(LS)$_2$ micelles.

The effect of the length of alkylchain of viologens on the electron transfer reactions is shown in Fig. 2B. The yield of reduced viologens decreases gradually on increasing the alkyl chain length of the viologens. Such decrease in the electron transfer efficiency could be interpreted in term of location of viologens: the hydrophobicity of the various viologens increases on going from C$_1$ to C$_{12}$ dialkyl viologen, which would make them participate in the formation of the mixed micelles with Cd(LS)$_2$. When sulphide ions are added to the mixed micellar solution, long-chain viologen can be trapped in the vacancies present in the CdS crystallite. By CdS excitation, because of the presence of viologen inside the particle, the forward electron transfer reaction is followed by a very fast reaction. The back electron transfer takes place before sulphide ions can give an electron to the hole.

Conclusions

CdS particles synthesized in o/w micellar solutions show a wide variation of CdS photophysical properties in aqueous medium.The size of the particles, as depicted by the absorption onset, is found to be around 100 Å. The electron transfer from CdS particles to various viologens reveals that the extent of the electron transfer depends on the alkyl chain length of the viologens. The back electron transfer reaction is strongly affected by the average location of viologen molecules before the synthesis of CdS particles and by the hydrophobic character of reduced viologens. The back reaction can be totally prevented by using MV(LS)$_2$.

Acknowledgement

One of the authors (TKJ) is grateful to C.N.R.S., France, for providing a post-doctoral fellowship.

References

1. Fendler JH (1987) Chem Rev 87:877
2. Structure and reactivity in reverse micelles (1989) Pileni MP (ed) Elsevier
3. Pileni MP, Zemb T, Petit C (1985) Chem Phys Lett 118:414
4. Lianos P, Thomas JK (1987) J Coll Inter Sci 117:50
5. Dannhauser T, O'Neil M, Johansson K, Witten D, McLendon G (1986) J Phys Chem 90:6074
6. Petit C, Pileni MP (1988) J Phys Chem 92:2282
7. Atkinson PJ, Grimson MJ, heenan RK, Howe AM, Robinson BH (1989) J Chem Soc, Chem Commun 1807
8. Petit C, Lixon P, Pileni MP (1990) J Phys Chem 94:1598
9. Mode S, Lianos P (1989) J Phys Chem 93:5854
10. Bloor DM, Wyn-Jones E (1990) (eds) The Structure, Dynamics and Equilibrium Properties of Colloidal Systems; NATO ASI Ser. Ser. C:324
11. Pelizzeti E, Schiavello M (1991) (eds) Photochemical Conversion and storage of solar energy; Kluwer Academic Publishers, The Netherlands
12. Watzke HJ, Fendler JH (1987) J Phys Chem 91:854
13. Brus LE (1983) J Chem Phys 79:5566
14. Rossetti R, Ellison JL, Bigson JM, Brus LE (1984) J Chem Phys 80:4464
15. Nozik AJ, Williams F, Nenadocic MT, Rajh T, Micic OI (1985) J Phys Chem 89:397—399
16. Bawendi MG, Steigerwald ML, Brus LE (1990) Annu Rev Phys Chem 41:477
17. a) Henglein A (1989) Chem Rev 89:1861. — b) Henglein A, Kurmer A, Jondre E, Weller H 61986) 132:133
18. a) Wang Y, Herron N (1990) Phys Rev B 41:6079. — b) Wang Y, Herron N (1991) J Phys Chem 95:525
19. Henglein A (1987) J Chim Phys 84:441
20. Lerebours B, Chevalier Y, Pileni MP (1985) Chem Phys Letters 117:89

Authors' address:

M. P. Pileni
Université P. et M. Curie
Laboratoire S.R.S.I
Bat 74 (F)
4 place Jussieu
75232 Paris Cedex 05

Progress in Colloid & Polymer Science   Progr Colloid Polym Sci 89:110—113 (1992)

Stability and rheological properties of gel emulsions

R. Pons[1]), C. Solans[1]), M. J. Stebé[2]), P. Erra[1]), and J. C. Ravey[2])

[1]) Instituto de Tecnología Química y Textil (C.S.I.C), Barcelona, Spain
[2]) Laboratoire de Physico-Chimie des Colloïdes, LESOC, URA CNRS 406, Université de Nancy I, Vandoevre-lès-Nancy, Cédex, France

Abstract: Stability and rheological studies on gel emulsions (w/o high internal phase ratio emulsions, HIPRE) both with pure and commercial polyethyleneglycol alkyl ether surfactants are reported. Stability of gel emulsions at room temperature is increased by addition of NaCl to the internal phase; in contrast, no appreciable effect is found at higher temperatures. Rheological properties are influenced by both NaCl concentration and temperature. Apparent yield stress and elastic modulus are increased either by NaCl addition or temperature increase. Shear modulus as a function of volume fraction is analyzed in terms of droplet size and interfacial tension.

Key words: Highly-concentrated emulsions; gel emulsion; stability; rheology; yield stress; shear modulus; storage modulus

Introduction

High internal phase-ratio emulsions (HIPRE) have been the object of numerous studies in the last 20 years [1—5]. Although the interest has mainly been focused on emulsions of the oil-in-water (o/w) type, in recent years studies on HIPRE of the water-in-oil (w/o) type (which form in either hydrogenated of fluorinated water/non-ionic surfactant/oil systems) have also been reported [6—15]. In this paper stability and rheological properties of w/o highly-concentrated emulsions or gelemulsions have been studied. The influence of volume fraction, temperature, and presence of salts were considered. Princen's equation for concentrated emulsions [4] has been used to analyze shear modulus dependence on volume fraction.

Materials and methods

Materials

Homogeneous polyethyleneglycol n-alkyl ethers (abbreviated as R_mEO_n) were obtained from Nikko Chemicals Co. Technical grade $C_{12}EO_4^-$ (Brij30) was from ICI. Pure n-decane and pro analysis NaCl were from Merck. Water was double distilled.

Methods

Gel-emulsion preparation: Increasing amounts of aqueous phase were added to a mixture of oil and surfactant with vigorous stirring by means of a vibromixer.

Visual assessment of stability: The occurrence of phase separation of thermostatized gelemulsions is recorded at different times.

Rheological measurements: A Ferranti-Shirley cone-plate rheometer was used in viscosity determinations of apparent yield stress. The samples were allowed to relax for 5 min before measurements. Shear rate was increased from 0 to 17 s^{-1} in 120 s in order to measure apparent yield stress. A stress-controlled rheometer, Carrimed, fitted with cone-plate geometry was used for oscillatory and creep measurements. The samples were allowed to relax for 5 min before measurement. Linear viscoelasticity was assessed by changing shear stress at a given frequency.

Microscopy: A Zeiss microscope coupled to a video camera was used. Digitalization and Video-

Table 1. Phase separation of gel emulsions with 99% w/w water or brine and a decane/$C_{12}EO_4^-$ weight ratio equal to 1

% NaCl in water	1 h		2 h		6 h		24 h		96 h	
	25°C	40°C	25°C	40°C	25°C	40°C	25°C	40°C	25°C	40°C
0	NO	NO	NO*)	NO	YES*)	YES	YES*)	YES	YES*)	YES
5	NO	NO	NO	NO	NO	YES	NO	YES	YES*)	YES
10	NO	NO	NO	NO	NO	YES	NO	YES	NO	YES

*) No gel appearance, but its stiffness is restored by shaking.

Table 2. Variation of droplet size with volume fraction for the system decane/$C_{16}EO_4$ w/w 1.5 and brine (5% NaCl) at 25°C

Volume fraction	0.752	0.872	0.935	0.987
R_{32}	0.5 μm	0.7 μm	0.8 μm	1.1 μm
$\sigma_{R_{32}}$	0.1 μm	0.2 μm	0.2 μm	0.4 μm

Table 3. Interfacial tension between dispersed and continuous phases from spinning drop measurements

Surfactant	s/o ratio*)	Salt	Interfacial tension		
			25°C	30°C	40°C
$C_{16}EO_4$	1.5	None	0.5 m Nm^{-1}	0.7 m Nm^{-1}	1.2 m Nm^{-1}
$C_{16}EO_4$	1.5	5% NaCl	0.6 m Nm^{-1}	0.8 m Nm^{-1}	1.8 m Nm^{-1}
$C_{12}EO_3$	1	None	1.0 m Nm^{-1}	1.7 m Nm^{-1}	2.3 m Nm^{-1}

*) Surfactant/decane w/w ratio.

Table 4. Fits of Princen's equation (equation (1)) for different systems

Surfactant	s/o*) ratio	T	Salt	Type of measurement	a	b
$C_{16}EO_4$	1.5	40°C	None	Creep	0.86 ± 0.25	0.73 ± 0.22
$C_{16}EO_4$	1.5	40°C	None	Oscillation	0.75 ± 0.18	0.73 ± 0.16
$C_{16}EO_4$	1.5	30°C	None	Oscillation	1.11 ± 0.29	0.73 ± 0.26
$C_{16}EO_4$	1.5	20°C	None	Oscillation	0.54 ± 0.35	0.68 ± 0.3
$C_{16}EO_4$	1.5	40°C	5% NaCl	Creep	1.41 ± 0.26	0.69 ± 0.23
$C_{16}EO_4$	1.5	30°C	5% NaCl	Creep	1.7 ± 0.7	0.6 ± 0.6
$C_{16}EO_4$	1.5	20°C	5% NaCl	Creep	2.3 ± 2.3	0.6 ± 0.6
$C_{12}EO_3$	1	40°C	None	Creep	0.62 ± 0.19	0.76 ± 0.16

*) Surfactant/decane w/w ratio.

Enhancement was performed by means of an Argus system. Phase contrast illumination is necessary to observe the samples due to the small refractive index difference between continuous and dispersed phases.

Interfacial tension: A Spinning Drop Texas University instrument was used. Aqueous phase and continuous phase were used as external and internal media, respectively.

Results and Discussion

The effect of NaCl concentration on emulsion stability, as a function of time and temperature was studied first. Stability results, assessed by visual determination of phase separation, are shown in Table 1 for emulsions with a 99% w/w water or brine and a decane/$C_{12}EO_4^-$ weight ratio equal to 1. With no salt present, stability is higher at the higher temperature. It can be observed that at the lower temperature (25°C) the increase in NaCl concentration correlates with the increase in stability, while at 40°C no correlation is observed.

In order to make a better interpretation of phase separation results, a study on rheological properties was undertaken. Apparent values of yield stress were obtained from continuous shear stress-shear rate measurements. Yield stress as a function of time at 25°C for different NaCl concentrations (Fig. 1) shows that for any of the gel emulsions studied, the shape of yield stress curves is similar; the higher the NaCl concentration, the higher the yield stress values. These results seemed to indicate that stability enhancement by salt addition could be due only to an increase of rigidity of the interfacial films, without affecting the rate of the processes responsible for reduction of yield stress (i.e., coalescence). This result agrees with results of stability obtained by visual inspection of phase separation, i.e., no sharp increase in stability is observed when NaCl is added.

Shear stress-shear rate measurements were also carried out as a function of NaCl concentration at different temperatures. Yield stress values are shown in Fig. 2. Gel emulsions from water (0% NaCl) shows correlation between yield stress and temperature, the higher the temperature, the higher the yield stress. When NaCl is added to the gelemulsion a maximum in yield stress is found at 40°C; these maxima are not correlated with the results of stability where at 40°C gelemulsions con-

taining salt are less stable than at 25°C. From Fig. 2 it is also observed that the actual concentration of NaCl seems to be an important factor only at room temperature; at 40°C and 60°C no major changes are visible after the addition of 1.5% NaCl.

Rheological measurements under small deformation, in the linear viscoelasticity region, were also carried out using pure nonionic surfactant systems. In this study, volume fraction, temperature, and addition of NaCl were considered. From dynamic measurements, storage (G') and loss (G'') moduli were obtained as a function of frequency. From transient measurements (creep measurements) instantaneous compliance J_0 and its reciprocal, shear modulus G_0 were obtained. In order to analyze the correlation between G_0 and volume fraction, Princen's equation [4] was used. This equation correlates the shear modulus with interfacial tension, droplet size, and volume fraction:

$$G_0 = a \, \frac{\gamma}{R_{32}} \, \phi^{1/3}(\phi - b) \,, \tag{1}$$

with γ being the interfacial tension and R_{32} the mean surface volume radius. The constants $a = 1.769$ and $b = 0.712$ are experimental values found by Princen. Constant b is supposed to be related to the maximum fraction packing of spheres (i.e., 0.74).

In order to fit this equation with our results, droplet size and interfacial tensions were measured (Tables 2 and 3). Although droplet size measurements are for one particular system, they are taken as being representative of droplet size for the other systems. It can be observed (Table 2) that droplet size and polydispersity increase with volume fraction. These results agree with observations made in other systems. The increase in size and polydispersity could be explained by considering the process of emulsion formation. We add the dispersed phase to the continuous phase little by little. At the beginning, small droplets can be formed. As we keep increasing the amount of dispersed phase, there is less continuous phase available to form the interfacial film; therefore, bigger droplets are formed. Interfacial tension values (Table 3) increase with temperature; they are low, but not so extremely low as at temperature close to HLB temperatures, where no gelemulsions can be formed.

The values of constants a and b from fits of Eq. (1) are shown in Table 4. Values of constant b are close to 0.74, the expected value. The values for constant

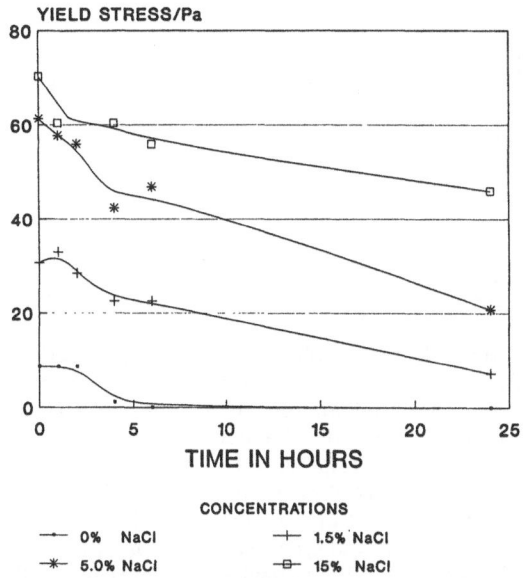

YIELD STRESS VERSUS TIME
T=25°C

CONCENTRATIONS

— 0% NaCl + 1.5% NaCl
—*— 5.0% NaCl —□— 15% NaCl

Fig. 1. Yield stress as a function of time for various NaCl concentrations in gel emulsions with 99% w/w water or brine and decane/$C_{12}EO_4^-$ weight ratio equal to 1 at 25 °C

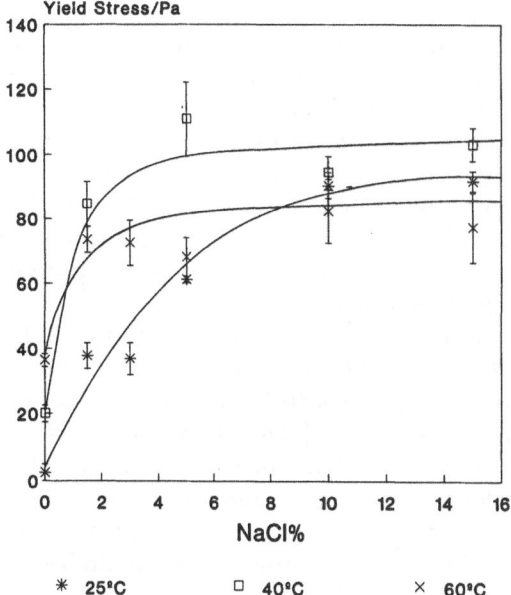

* 25°C □ 40°C × 60°C

Fig. 2. Yield stress as a function of NaCl concentration at different temperatures for gel emulsions with 99% w/w water or brine and a decane/$C_{12}EO_4^-$ weight ratio equal to 1

a are of the order of the constant found by Princen [4], but the differences, in particular for the systems where no salt is present, could be indicative of other factors affecting the elastic behavior and not considered in Princen's equation (i.e., continuous phase viscosity and film thickness).

Acknowledgements

Experimental assistance by Mrs. I. Carrera, and financial support from DGICYT PB 87/0210 Program and a CSIC/CNRS bilateral Cooperation Program are greatfully acknowledged.

References

1. Lissant KJ (1966) J Colloid Interface Sci 22:462—468
2. Lissant KJ, Mayhan KG (1973) J Colloid Interface Sci 42:201—208
3. Princen HM (1979) J Colloid Interface Sci 71:55—66
4. Princen HM, Kiss AD (1986) J Colloid Interface Sci 112:427—437
5. Princen HM (1988) Langmuir 4:486—487
6. Kunieda H, Solans C, Shida N, Parra JL (1987) Colloids and Surfaces 24:225—237
7. Solans C, Azemar N, Parra JL (1988) Progr Colloid Polym Sci 76:224—227
8. Solans C, Domínguez JG, Parra JL, Heuser J, Friberg SE (1988) Colloid Polym Sci 266:570—574
9. Kunieda H, Yano N, Solans C (1989) Colloids and Surfaces 36:313—322
10. Kunieda H, Evans DF, Solans C, Yoshida M (1990) Colloids and Surfaces 47:35—43
11. Ravey JC, Stébé MJ (1989) Physica B 394:156—157
12. Ravey JC, Stébé MJ (1990) Progr Colloid Polym Sci 82:218—288
13. Bampfield A, Cooper J (1988) In: Becher (ed) Encyclopedia of Emulsion Technology. Marcel Dekker, New York, Vol 3, pp 281—306
14. Kizling J, Kronberg B (1990) Colloids and Surfaces 50:131—140
15. Ruckenstein E, Ebert G, Platz G (1989) J Colloid Interface Sci 133:432—441
16. Courraze G, Grossiord JL (1983) In: Initiation à la Rheologie. Technique et Documentation (Lavoisier), Paris, pp 19—43

Authors' address:

Dr. Ramon Pons
Instituto de Tecnología Química y Textil (C.S.I.C.)
C/Jordi Girona 18—26
08034 Barcelona, Spain

Progress in Colloid & Polymer Science   Progr Colloid Polym Sci 89:114—117 (1992)

Coalescence lifetimes of oil and water drops at the planar oil-water interface and their relation to emulsion phase inversion

R. Aveyard, B. P. Binks, P. D. I. Fletcher, and X. Ye

School of Chemistry, University of Hull, Hull, U.K.

Abstract: We have measured the coalescence lifetimes of AOT-stabilised oil and water drops with the AOT monolayer present at the planar oil-water interface. Measurements were made as a function of aqueous phase NaCl concentration over the range over which phase inversion of the emulsions occurs and for heptane, dodecane and tetradecane as oil. The drop lifetimes decrease sharply with increasing [NaCl]. Below a crossover [NaCl], the oil drop lifetimes are longer than the water drops, whereas this behaviour is reversed at higher [NaCl]. The [NaCl] at the crossover points for the three oils increases with alkane chain length. This behaviour is compared with that of the corresponding emulsion systems which are water continuous at low [NaCl] but phase invert to give oil-continuous emulsions at high [NaCl]. The NaCl concentrations required for emulsion phase inversion increase with increasing oil chain length but are not identical to the [NaCl] values at the drop coalescence crossover points.

Key words: Coalescence; interface; emulsion; surfactant; microemulsion

Introduction

The addition of electrolyte to alkane + water mixtures containing the twintailed, anionic surfactant sodium bis(2-ethylhexyl)sulphosuccinate (AOT) leads to the Winsor I — Winsor III — Winsor II progression of equilibrium multi-phase systems [1]. The oil-water interfacial tension passes through a minimum at a salt concentration corresponding approximately to the mid-point of the Winsor III range. The macroemulsions produced by homogenising these multi-phase systems are generally of the same type (i.e. o/w or w/o) as the equilibrium microemulsion type which changes from o/w in Winsor I to w/o in Winsor II systems. Thus, for AOT as surfactant, the emulsion type can be phase inverted from o/w to w/o by the addition of NaCl to the aqueous phase [2]. The salt concentration required for emulsion inversion is generally similar to that required to achieve minimum interfacial tension. In addition to affecting emulsion type, the phase inversion variable ([NaCl] in this case) also changes the stability of the preferred emulsion type [3]. However, measurement of the emulsion stability only provides information concerning the preferred type. It is of interest to attempt to assess the effects of the phase inversion variable on the stabilities of oil and water drops separately. The stabilities of the two different drop types can be measured separately by determining the time taken for macroscopic sized drops to coalescence with the surfactant monolayer present at the planar oil-water interface [4—6].

We have measured the time taken for AOT-stabilised drops of the equilibrium aqueous and oil phases to coalescence with the planar oil-water interface as a function of [NaCl] covering the range over which the emulsions phase invert. Since the [NaCl] required to achieve minimum tension increases with the chain length of the alkane, we have investigated systems containing heptane, dodecane and tetradecane as oil.

Experimental

Water was distilled, deionised and passed through a Milli-Q reagent water system prior to use.

AOT (Sigma) was used without further purification. The heptane (Fisons, HPLC grade), dodecane and tetradecane (Aldrich, >99%) were passed over alumina prior to use to remove traces of polar impurities. NaCl was BDH AnalaR grade.

For the coalescence experiments, the multi-phase systems were equilibrated in a thermostatted vessel mounted on a vibration-free table. The equilibrium systems were prepared by shaking 20 mls of aqueous NaCl with 20 mls of 4 mM AOT in alkane. For the two-phase systems, the studied interface was that between the microemulsion and the excess phase. In the three-phase systems investigated here, the third, surfactant-rich phase was always the lowest (i.e. most dense) phase and was not present at the interface studied which was that between the excess oil and water phases. For water drop coalescence experiments, a drop of known volume of the water phase was drawn into an Agla syringe needle containing some oil phase. The needle tip was then moved into the upper oil phase and a further quantity of the oil phase was sucked into the needle. Thus the needle contained a water drop within oil. The needle tip was then positioned 5 mm from the interface and the needle contents gently expelled. This procedure released a drop of known volume, irrespective of the interfacial tension and needle tip diameter [7]. A curved needle was used similarly to release oil drops at the lower side of the interface. The drop volume was kept constant at 0.4 µl. The time taken for the drop to coalesce with the interface was then recorded and the measurement repeated for 40 drops, yielding a distribution of coalescence times.

Emulsions were prepared in a thermostatted vessel by homogenisation using an Ultra Turrax homogeniser fitted with a T25 shaft operating at 8000 rpm. Emulsion conductivities were measured using a platinum electrode dip cell within the emulsion vessel during homogenisation.

All measurements were made at 25°C.

Results and discussion

The distribution curves of coalescence times showed an induction period of time t_d followed by an exponential decay characterised by a first-order rate constant k. For the purposes of this paper, the distribution curves were simply characterised by the time taken for half the drops to coalesce ($t_{1/2}$), which is equal to $(t_d + \ln2/k)$. The detailed inter-

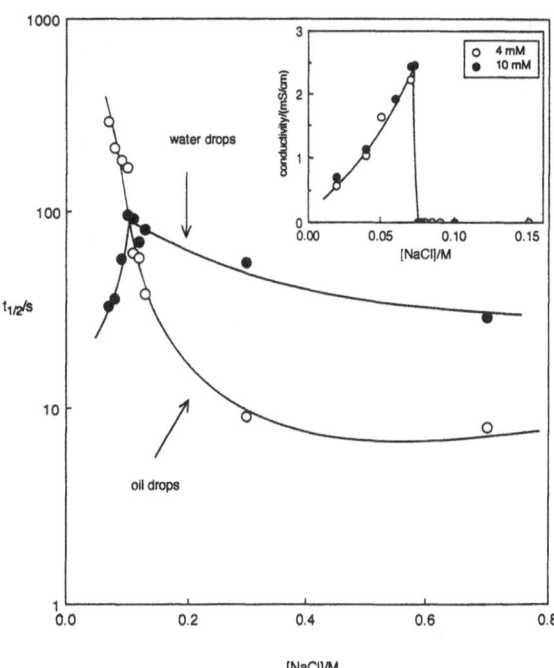

Fig. 1. Drop coalescence $t_{1/2}$ versus aqueous phase [NaCl] for heptane as oil. The inset shows the conductivity of the corresponding emulsions containing 4 mM or 10 mM AOT (initially in alkane) as indicated

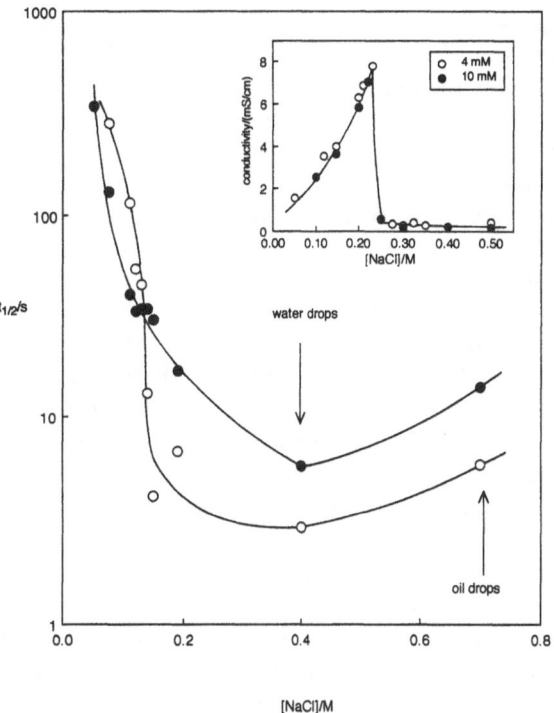

Fig. 2. As for Fig. 1 except the oil is dodecane

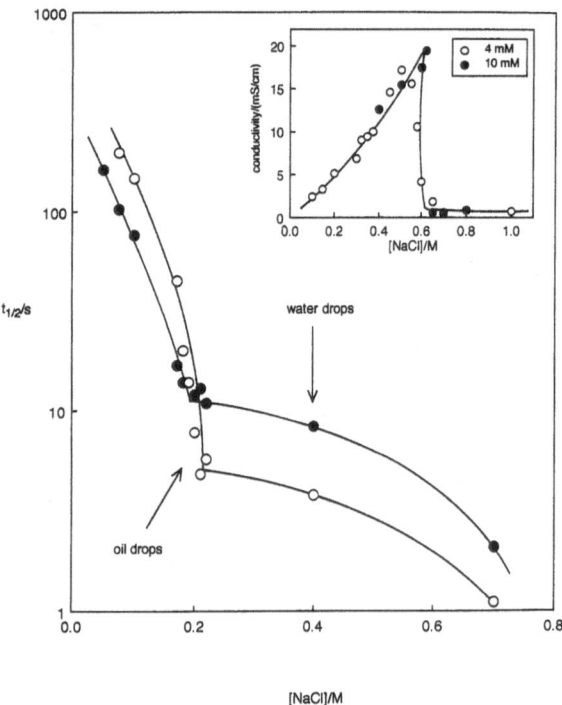

Fig. 3. As for Fig. 1 except the oil is tetradecane

pretation of such coalescence times is discussed fully in a recent review [8].

Figures 1—3 show the variation of $t_{1/2}$ for both water and oil drops with aqueous phase NaCl concentration for heptane, dodecane and tetradecane as the oil. For each oil, there is a "crossover" of the water and oil drop curves. At NaCl concentrations less than the crossover concentration, oil drops are more stable with respect to coalescence than water drops. The opposite behaviour is seen at salt concentrations higher than the crossover. The NaCl concentrations at the crossover points increase with alkane chain length and are approximately 0.10, 0.14 and 0.19 M for heptane, dodecane and tetradecane respectively. For comparision, the NaCl concentrations corresponding to the midpoints of the transitions from Winsor I to Winsor II microemulsion systems are 0.05, 0.15 and 0.3 M for the same oils.

The stability differences might be expected to relate to the tendency of the systems to form either o/w or w/o emulsions. The preferred emulsion types were investigated using conductivity measurements as shown in the insets to the figures. At low [NaCl], the emulsions show high conductivity, indicating the emulsions are o/w. At high [NaCl], the emulsions have phase inverted to give low conductivity, w/o emulsions. The NaCl concentrations required to phase invert the emulsions increase with increasing alkane chain length and are approximately 0.07, 0.23 and 0.6 M for heptane, dodecane and tetradecane respectively. Thus, although the NaCl concentrations at the crossover points and at phase inversion of the emulsions both increase with alkane chain length, the values are not coincident. The different values presumably reflect differences in the two types of experiment. Whereas the drop coalescence lifetime is primarily determined by the drainage and rupture characteristics of the thin film separating the drop and interface, emulsion formation and stability is also dependent on other factors, including Ostwald ripening and sedimentation/creaming of the droplets.

In addition to the crossover behaviour, it can be seen from the figures that the $t_{1/2}$ values of both the water and oil drops decrease by approximately two orders of magnitude with increasing [NaCl]. (We note the behaviour of the water drops in the heptane systems is anomalous in this respect.) In the case of the oil drops, the high $t_{1/2}$ values at low [NaCl] are presumably due to electrostatic repulsion between the negative charged drop and the AOT monolayer at the planar-oil-water interface. Addition of NaCl acts to screen the charges, thereby decreasing the repulsion and lowering $t_{1/2}$. We currently have no explanation for the case of the water drops.

The $t_{1/2}$ values at the crossover points decrease in the order heptane > dodecane > tetradecane. It is notworthy that the times taken for phase resolution of the emulsions are observed qualitatively to decrease in this same sequence.

In this preliminary study, we have shown that the coalescence lifetimes of both oil and water drops with the planar interfacial AOT monolayer are sensitive to both [NaCl] and alkane chain length under conditions in the vicinity of emulsion phase inversion. Further work, including measuring the effect of drop size on $t_{1/2}$, is required to test models of the coalescence process and extract parameters reflecting the intrinsic ability of monolayers to resist coalescence. Since many properties of AOT monolayers such as interfacial tension, composition and bending rigidity have been extensively studied [1, 9, 10], the eventual aim is to assess the roles of these properties in determining the resistance to coalescence.

Acknowledgement

We thank BP Research (Sunbury) for EMRA funding.

References

1. Aveyard R, Binks BP, Mead J (1986) J Chem Soc Fardaday Trans 1, 82:1755—1770
2. Aveyard R, Binks BP, Fletcher PDI, Ye X, Lu JR (1992) Proceedings of NATO ARW on "Emulsions — a fundamental and practical approach" Sjoblom J (Ed) Kluwer Amsterdam
3. Anton RE, Salager J-L (1986) J Colloid Interf Sci 111:54—59
4. Nielsen LE, Wall R, Adams G (1958) J Colloid Sci 13:441—458
5. Hodgson TD, Woods DR (1969) J Colloid Interf Sci 30:429—446
6. Davis SS, Smith A (1976) Colloid and Polym Sci 254:82—98
7. Liem AJS, Woods DR (1974) Am Inst Chem Eng Symp 70:8—23
8. Palermo T (1991) Revue de l'Institut Francais du Petrole 46:325—360
9. Aveyard R, Binks BP, Fletcher PDI in "The Structure, Dynamics and Equilibrium Properties of Colloidal Systems" (1990) Bloor DM, Wyn-Jones E (Eds) Kluwer Amsterdam pp 557—581
10. Binks BP, Kellay H, Meunier J (1991) Europhys Lett 16:53—58

Authors' address:

Dr. B. P. Binks
School of Chemistry
University of Hull
Hull, HU6 7RX, U.K.

Progress in Colloid & Polymer Science   Progr Colloid Polym Sci 89:118—121 (1992)

Interactions between hydrophobically modified polymers and surfactants

B. Magny[1]), I. Iliopoulos[1]), R. Audebert[1]), L. Piculell[2]), and B. Lindman[2])

[1]) Laboratoire de Physico-Chimie Macromoléculaire, Université P. et M. Curie, CNRS URA 278 — Paris, France
[2]) Physical Chemistry 1, Chemical Center, University of Lund, Sweden

Abstract: The rheological behavior of mixtures containing hydrophobically modified poly(sodium acrylate) (HMPAA) and surfactants is studied. The hydrophobic groups of the polymer (octadecyl or dodecyl chains) interact with the surfactant molecules to form mixed micellar-type aggregates. In semidilute solution these mixed aggregates act as cross-linkers between the polymer chains, inducing a viscosification or gelation of the system. The strongest effect was observed with a cationic surfactant, because in this case both electrostatic attraction and hydrophobic interactions contribute to the stabilization of the mixed aggregate. However, association occurs even between HMPAA and an anionic surfactant (SDS), despite the unfavorable electrostatic repulsions, indicating that the hydrophobic interactions are the driving force for the formation of the mixed aggregates.

Key words: Hydrophobically modified polymers; surfactants; interactions; rheology; gel

Introduction

Aqueous systems containing polymers and surfactants have received increasing attention over the last decades because of their importance in various industrial applications (paints, pharmaceuticals, oil recovery, etc.) and because of their similarity to biological systems (interactions between biopolymers and biomembranes or vesicles).

The behavior of polymer/surfactant mixtures is governed by a subtle balance between hydrophobic, hydrophilic, and ionic interactions. Nonionic polymers present rather weak interactions with (mainly anionic) surfactants [1] while polyelectrolytes present strong interactions with oppositely charged surfactants [2]. In the latter case, the electrostatic attractions play the main role for the stabilization of the polymer/surfactant complex. Decreasing the electrostatic attractions, for instance by increasing the ionic strength of the solution, can lead to the dissociation of the complex [3]. As a consequence, it is rather natural to expect that the association between polyelectrolytes and ionic surfactants of the same sign is very weak or absent because of the unfavorable electrostatic repulsions between these species in aqueous solution [4, 5]. On the other hand, the interactions between surfactants and water soluble polymers are increased with the hydrophobicity of the polymers [1, 6].

In a recent study Iliopoulos et al. have shown that sodium dodecyl sulfate (SDS) can bind to poly(sodium acrylate) if this polymer contains only 1 or 3 mol % of very hydrophobic alkyl groups (octadecylacrylamide units) [7]. In this case, mixed micelle-type aggregates seem to be formed, which contain both surfactant molecules and alkyl groups of the polymer.

In the present study, we compare the effects of anionic, non-ionic, and cationic surfactants on the rheological behavior of hydrophobically modified poly(sodium acrylate) (HMPAA). Our findings are interpreted in terms of the balance between electrostatic and hydrophobic interactions.

Experimental

Poly(acrylic acid) was purchased from Polysciences and its average molecular weight, given by the supplier, was 150000. The hydrophobically modified samples were prepared as described elsewhere [8]. They have the same polymerization degree as the precursor polymer and a random distribution of the alkyl groups along their chain [9]. All the polymers, including the precursor, were used in the fully neutralized sodium salt form. The typical structure of the modified samples is the following:

$$-(CH_2-CH)_{100-x}\!\!-\!\!-(CH_2-CH)_x$$
$$\begin{array}{ccc} | & & | \\ C=O & & C=O \\ | & & | \\ O^-Na^+ & & HN-(CH_2)_{n-1}-CH_3 \; . \end{array}$$

Where x is the modification degree in mol % and n the number of carbon atoms of the alkyl chain (in this study $n = 18$ or 12). 1-C18 denotes a modified poly(sodium acrylate) containing 1 mol % of octadecylacrylamide units.

Sodium dodecyl sulfate (SDS) was obtained from Kodak Lab. Chemicals as >99% pure, dodecyltrimethylammonium bromide (DTAB) from Tokyo Kasei as >99% pure and pentaethyleneglycol monododecyl ether from Fluka as >98% pure. All surfactants were used without further purification.

Polymer surfactant mixtures were prepared as described elsewhere [7]. Newtonian viscosities at 25 °C were obtained at low shear rates (between 0.06 and 1.28 s^{-1}) by using a Contraves LS-30 viscometer. The storage modulus of the gel-forming mixtures was measured with a Bohlin-Vor rheometer (cone and plate) at 25 °C and a frequency of 1 Hz.

Results and discussion

Figure 1 displays typical viscosity results versus the SDS concentration. The polymer concentration is kept constant at $C_p = 1\%$. The precursor, non-modified poly(sodium acrylate), exhibits a rather constant viscosity for SDS concentrations up to 10^{-2} mol · l^{-1} and then the viscosity decreases slightly. A similar behavior was observed also in mixtures of SDS and partially hydrolyzed polyacrylamide [5]. The slight decrease in viscosity is related to the increased ionic strength of the solution due to addition of SDS.

A very different behavior is found for the polymer bearing 3 mol % of octadecyl groups (3-C18). When the SDS concentration is higher than 10^{-3} mol · l^{-1}

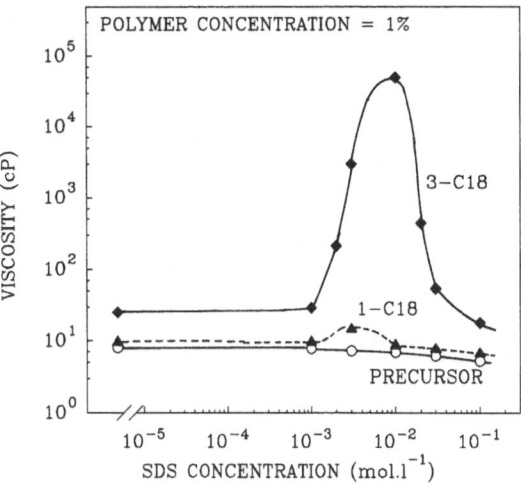

Fig. 1. Viscosity versus SDS concentration for mixtures of SDS with the precursor and two hydrophobically modified polymers: (o) precursor; (▲) 1-C18; (♦) 3-C18

the viscosity increases sharply and for surfactant concentrations between 3 10^{-3} and 10^{-2} mol · l^{-1} it becomes three orders of magnitude, or more, higher than that measured in pure water. At SDS concentrations higher than 10^{-2} mol · l^{-1} the viscosity decreases again and the system may exhibit viscosities even lower than that found in the absence of SDS ([SDS] = 10^{-1} mol · l^{-1}). The sample containing 1 mol % octadecyl groups (1-C18) exhibits only a very slight viscosity enhancement when the SDS concentration is between 10^{-3} and 10^{-2} mol · l^{-1}.

This peculiar behavior can be ascribed to the formation of mixed micellar-type aggregates containing alkyl groups belonging to the polymer and surfactant molecules. The structure of these aggregates will be discussed at the end of this paper.

The association between surfactants and HMPAA is expected to be stronger if the surfactant presents no repulsive electrostatic interactions with the polymer (non-ionic surfactant) and the strongest when electrostatic attractions are operative (cationic surfactants). This behavior is depicted in Fig. 2, where the viscosity of the polymer 1-C18 is plotted as a function of the surfactant concentration for three kinds of surfactant: anionic (SDS), non-ionic (C$_{12}$E$_5$), and cationic (DTAB). The cationic surfactant has the most pronounced effect on the rheology of the system. For DTAB concentrations between 10^{-3} and 10^{-2} mol · l^{-1} no viscosity measurements are possible, since a gel is formed.

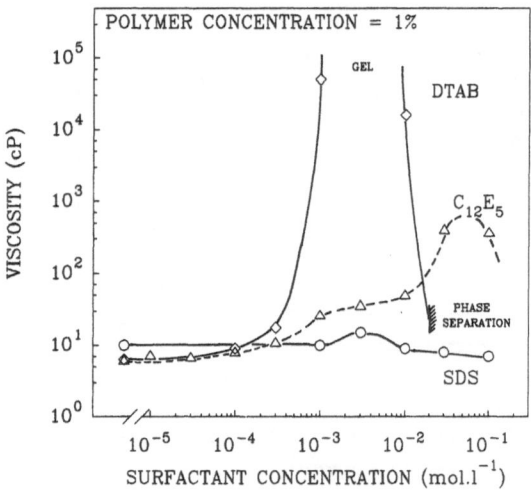

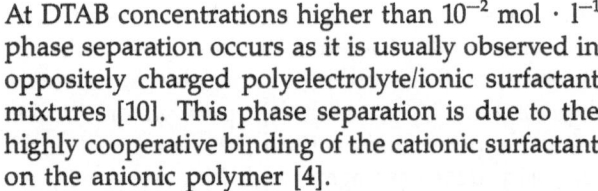

Fig. 2. Viscosity versus surfactant concentration for mixtures of a hydrophobically modified polymer (1-C18) with SDS (o), $C_{12}E_5$ (△) and DTAB (◇)

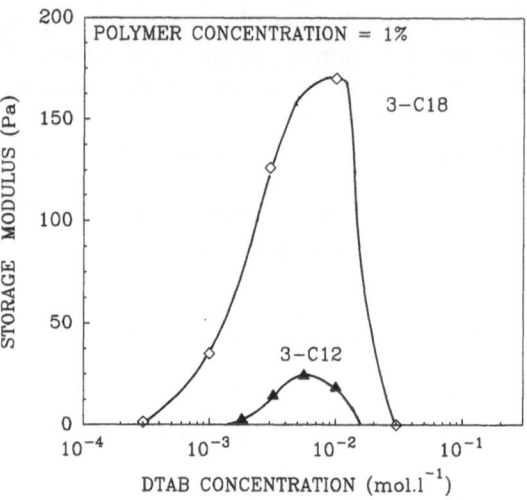

Fig. 3. The storage modulus of hydrophobically modified polymers as a function of DTAB concentration: (▲) 3-C12: (◇) 3-C18

At DTAB concentrations higher than 10^{-2} mol · l^{-1} phase separation occurs as it is usually observed in oppositely charged polyelectrolyte/ionic surfactant mixtures [10]. This phase separation is due to the highly cooperative binding of the cationic surfactant on the anionic polymer [4].

It is noteworthy that the non-modified PAA phase separates at a lower DTAB concentration than the sample 1-C18, while the sample 3-C18 gives a one-phase system even for a DTAB concentration of $3 \cdot 10^{-2}$ mol · l^{-1}. It is obvious that the HMPAA interacts with the cationic surfactant in a different way that the precursor PAA. Presumably, the binding of the DTAB molecules occurs, at first, close to the alkyl groups of the polymer chain forming mixed aggregates. As a consequence, the amount of free DTAB available for the cooperative electrostatic binding is decreased and the one-phase region is extended to higher surfactant concentrations. Because of the attractive electrostatic interactions the mixed aggregates are more stable than in the case of SDS/HMPAA mixtures, leading to a more pronounced effect on the rheological behavior of the system.

Similarly to the SDS/HMPAA system, the interactions between DTAB and HMPAA increase with the alkyl group content and length. Rather strong gels are formed in mixtures of DTAB and 3-C18 (Fig. 3). The strength of the gel, as expressed by the storage modulus, decreases when the polymer bears dodecyl groups (3-C12) instead of octadecyl (3-C18).

The viscosity enhancement obtained upon addition of a non-ionic surfactant ($C_{12}E_5$), in 1% solution of 1-C18, is in between that found with cationic and anionic surfactants (Fig. 2). Electrostatic interactions between surfactant and polymer are now absent, but some repulsive interactions could be operative between the sodium acrylate units of the polymer chain and the pentaethylene glycol polar heads of the surfactant micelle. These repulsive interactions should be similar to that between poly(sodium acrylate) and poly(ethylene glycol) [11].

A schematic illustration of the interactions between surfactant and HMPAA, which can explain our rheological results, is given in Fig. 4. In pure water or at very low surfactant concentrations few or no hydrophobic aggregates are formed and the viscosity of the system is very similar to that observed with the precursor non-modified poly(sodium acrylate) (scheme A). By increasing the surfactant concentration the surfactant molecules form micellar-type aggregates around the alkyl groups of the polymer chain. Depending on the molar ratio (alkyl group)/(surfactant) the mixed aggregates may contain two or more alkyl groups belonging to different polymer chains (scheme B). In semi-dilute solution these mixed hydrophobic aggregates induce a cross-linking of the polymer chains and, therefore, a viscosification or gelation of the system. At high surfactant concentrations ((alkyl group)/(surfactant)) ≪ 1, free surfactant micelles are in equilibrium with mixed aggregates containing only

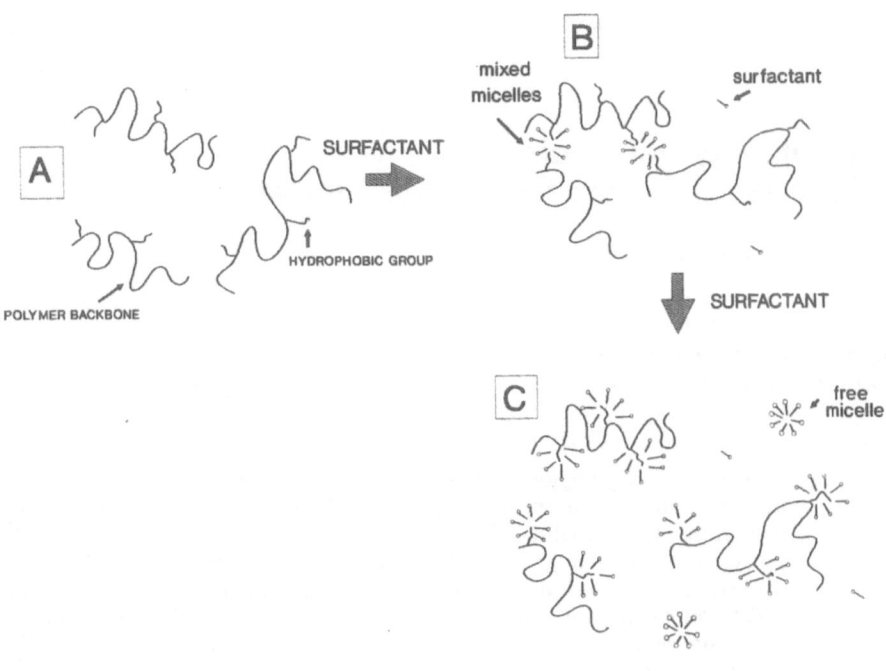

Fig. 4. Schematic illustration of the interactions between surfactant and hydrophobically modified polymer.
A) Polymer solution without surfactant.
B) The same solution after addition of a small amount of surfactant. Mixed aggregates are formed inducing the cross-linking of the polymer chains.
C) In the excess of surfactant free micelles are in equilibrium with mixed aggregates containing only one alkyl group from the polymer. No cross-linking occurs

one alkyl group (schema C). Thus, the effective cross-linking between polymer chains is prevented and the viscosity of the system decreases. Such an association model was also proposed for other similar systems [12, 13].

Acknowledgement

This work was partially supported by the "Société Française Hoechst". I.I.'s stay in Lund was supported by a grant from the Swedish Board of Technical Development. We thank Dr. K. Thalberg and Dr. U. Olsson for many helpful discusssions. Mr. V. Prunier and Miss S. Gasser are thanked for help in the viscosity measurements. Rheology measurements with the Bohlin-Vor rheometer were performed in the Food Technology Department, University of Lund.

References

1. Goddard ED (1986) Colloids and Surfaces 19:255
2. Goddard ED (1986) Colloids and Surfaces 19:301
3. Thalberg K, Lindman B, Karlström G (1991) J Phys Chem 95:6004
4. Hayakawa K, Kwak JCT (1991) In: Rubingh D, Holland PM (Eds) Cationic surfactants: Physical chemistry. Surfactants sciences series. Marcel Dekker, New York, p 189
5. Methemitis C, Morcellet M, Sabbadin J, François J (1986) Eur Polym J 22:619
6. McGlade MJ, Randall FJ, Tcheurekdjian N (1987) Macromolecules 20:1782
7. Iliopoulos I, Wang TK, Audebert R (1991) Langmuir 7:617
8. Wang TK, Iliopoulos I, Audebert R (1988) Polym Bull 20:577
9. Magny B, Lafuma F, Iliopoulos I (1992) Polymer (in press)
10. Thalberg K, Lindman B, Karlström G (1991) J Phys Chem 95:3370
11. Perrau MB, Iliopoulos I, Audebert R (1989) Polymer 30:2112
12. Gelman RA (1987) International Dissolving Pulps Conference, TAPPI Proceedings p 159
13. Tanaka R, Meadows J, Phillips GO, Williams PA Carbohydr Polym (1990) 12:443

Authors' address:

Dr. Ilias Iliopoulos
Laboratoire de Physico-Chimie Macromoléculaire
E.S.P.C.I.
10, rue Vauquelin
75231 Paris Cedex 05, France

Progress in Colloid & Polymer Science   Progr Colloid Polym Sci 89:122—124 (1992)

Aggregational behavior of polymeric micelles of methacrylate functionalized quaternary ammonium salts

G. Nika, C. M. Paleos[1]), P. Dais[2]), A. Xenakis[3]), and A. Malliaris[1])

[1]) N.R.C. "Demokritos", Agia Paraskevi, Athens, Greece
[2]) Dept. of Chemistry, University of Krete, Heraklion, Greece
[3]) National Hellenic Research Foundation, Athens, Greece

Abstract: The aggregational character of two methacrylate functionalized micelle-forming quaternary ammonium salts has been studied in aqueous environment. The methacrylate group was introduced, in the one case at the polar group (A, head), and in the other at the end of the long alkyl chain of a surfactant molecule (B, tail). The two monomers were irradiated in a Co-60 source and polymeric micelles were formed. Physicochemical studies, including molecular weight, electrical conductivity, fluorescence, and video-enhanced image processing were performed on these polymerized micelles.

Key words: Micellar polymerization; fluorescence quenching; functionalized micelles

Introduction

Extending our previous studies [1] on the polymerization and the aggregational behavior of micelle-forming quaternary ammonium surfactants, we present here our results on two such systems shown in Fig. 1. In these salts the polymerizable group is attached either on the quaternary ammonium head of the surfactant (A, head), or at the end of the long aliphatic chain (B, tail) [2]. Furthermore, in contrast to the previous work [1] where different polamerizable groups were introduced in the head and tail case, in the present study both monomers bear the same polymerizable unit, i.e., a methacrylate group. In this way any differences found in the two polymeric micelles can be attributed primarily to the effect of the position of the polymeric bond.

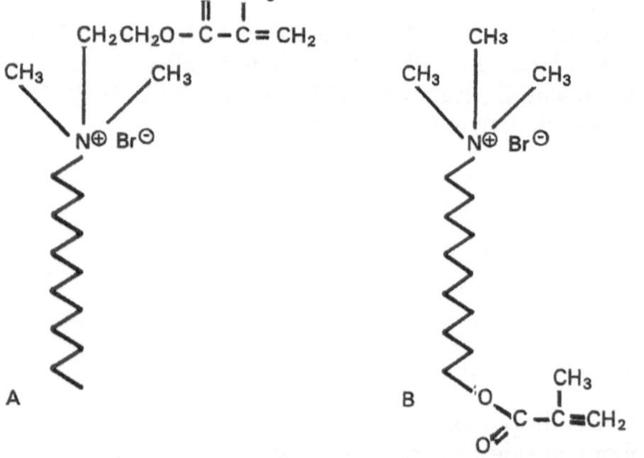

Experimental Part

Monomer A was synthesized by the quaternization of N,N-dimethyl aminoethyl methacrylate with bromododecane in acetone at room temperature and in the presence of inhibitor. The precipitated material was recrystallized from a 1:10 solution of ethanol:ethyl acetate. Monomer B was prepared by the reaction of ω-bromoundecyl methacrylate in ethyl acetate solution with trimethylamine gas at 0°C, also in the presence of inhibitor. The precipitate was filtered and dried. The critical micelle concentrations (CMC) of the above two

Fig. 1. Chemical formulas of head (A) and tail (B) monomers

surfactants A and B were determined by measuring the electrical conductivity and were found to be equal to 5.93×10^{-3} M [3] and 5.53×10^{-3} M [4], respectively. The two monomers were polymerized by γ-irradiation in a Cobalt-60 source at a temperature of 25 °C. They were both dissolved in water at concentrations above their corresponding CMC and, therefore, they were polymerized as monomeric micelles. Similar polymerizations of other monomeric micelles have been reported [2].

Number average molecular weights were determined with a Knauer vapor pressure osmometer. Electrical conductivity measurements were performed with the E518 conductometer of Metrohm in conjugation with a conductivity cell thermostated at 25 °C. For the video-enhanced image processing a previously described [5] set of S-VHS JVC video units (GF-S1000H, BR-S610E, BR-S811E, RM-G810U) was used. A Jenaval optical microscope equipped with differential interference contrast optics, coupled with a Philips solid-state CCD camera was also employed. For image-processing the FG100 card of Image Processing and the Imagepro software of Media Cybernetics were used. This technique allows direct visualization of particle sizes down to 500 Å [6]. Zone-refined pyrene (Aldrich, 99%) was used as the fluorophor, while recrystallized cetylpyridinium chloride (Kodak, 98%) was employed as the fluorescence quencher [7].

Results and Discussion

A first important differentiation between the two monomers was found in the rate of their polymerization. Thus, the polymerization of the head monomer A was very slow, while that of the tail monomer B was much faster, as determined by means of NMR spectra (shown in Fig. 2). Note that NMR studies have also indicated while polymer B is highly syndiotactic, polymer A is nearly purely heterotactic. The complete NMR study of these systems will be published separately. Another difference in the physicochemical behaviors of the two polymers involves their solubility. Thus, while the tail polymer is readily soluble in water, its head counterpart is practically insoluble. In ethanol both polymers are soluble.

The molecular weights of the two polymeric micelles, dissolved in ethanol and determined by

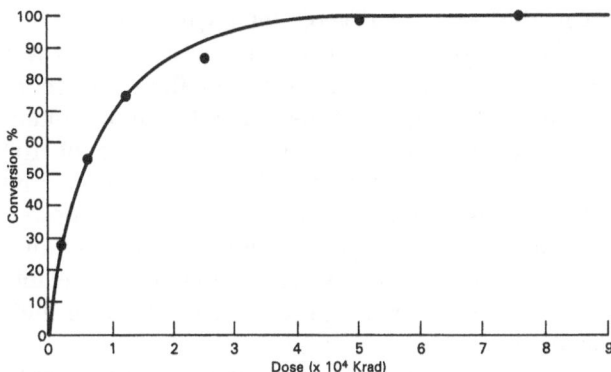

Fig. 2. Monomer to polymer % conversion vs dose of γ-irradiation for tail surfactant. The corresponding curve for the head surfactant was too fast to be followed

means of vapor pressure osmometry, were found equal to 20 000 and 3500 daltons for the head and tail polymers, respectively. These molecular weights correspond to a degree of polymerization (DP) equal to approximately 50 for the head and 10 for the tail polymer. Consequently, the size of the intramolecular polymeric micelles will correspond to an average of 10 monomeric units in the case of the tail and 50 in the case of the head surfactant.

Results apparently contrary to the above findings from molecular weight measurements were obtained from static fluorescence quenching [8—13]. Indeed, the monomeric micelles, were found to have an average aggregation number of 46 for the tail and 11 for the head surfactant, at total monomer surfactant concentration 2.5×10^{-2} M in both cases. A possible rationalization of this discrepancy is that polymerization in the larger tail micelle starts at several locations in the aggregate and results in the formation of a number of polymeric chains, viz. ca. five, from a single monomeric micelle. In the case of the head polymer, on the other hand, it appears that the polymerization and the formation of the final intramolecular polymeric micelles involves more than one — on average, five — monomeric micelles. The above explanation was supported by the number of tail monomers forming a tail polymeric micelle, also obtained by static fluorescence quenching. Thus, the tail polymer, which is soluble in water, was found to contain an average of 13 monomeric units, a number which is in excellent agreement with the DP of this polymer (10) determined from vapor pressure osmometry. Further information obtained from fluorescence was that the polarity of the environment at the palisade layer of the two monomeric micelles was very similar, as estimated from the fluorescence intensity ratio I_1/I_3 [14]. Where I_1 and I_3 are the intensities of the first and the third peaks of the fluorescence of pyrene. Moreover, from the fluorescence intensity ratio I_M/I_E of the monomer to the excimer pyrene emission it was concluded that the tail monomer micelles undergo a detectable size increase for up to approximately 12 h after their formation.

The failure of our image-processing equipment to optically detect either non-polymerized or polymerized micelles indicates that the average diameter of these aggregates is less than 500—700 Å [6]. This conclusion is in agreement with the aggregation numbers determined by fluorescence quenching for these micelles. Even the size of the largest of the aggregates studied here, i.e., the monomeric tail micelle, according to its mean aggregation number must correspond to a diameter less than ca. 15—20 Å.

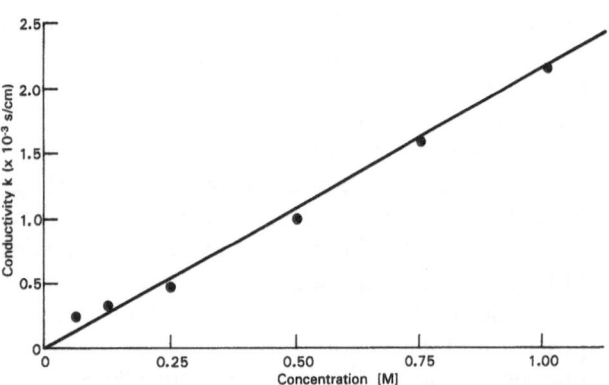

Fig. 3. Conductivity vs tail polymer concentration. Concentration is expressed with respect to the monomer

Interestingly enough, the electrical conductivity of the water soluble tail polymer measured up to 1 $\times 10^{-1}$ M, shown in Fig. 3, did not show any evidence of formation of intermolecular micelles. In other previously studied polymeric micelles [2, 15]

it was observed that above a certain concentration intermolecular aggregates were formed. These intermolecular micelles are known to be formed by the aggregation of several polymeric units [2].

Unfortunately, the nearly complete insolubility of the head polymer in water did not allow a thorough study of the aggregational properties of this polymeric surfactant.

References

1. Paleos CM, Margomenou G, Malliaris A (1988) Mol Cryst Liq Cryst 161:385
2. Mittal KL (ed) Plenum Press, New York 9:119 C28:403
3. Nagai K, Ohishi Y, Inaba H, Kudo S (1985) J Polym Sci Chem Ed 23:1221
4. Michas J, Paleos CM, Dais P (1989) Liq Cryst 5:1737
5. Malliaris A, Binana-Limbele W (1991) Prog Colloid Polym Sci 84:83
6. Miller DD, Bellare JR, Evans DF, Talmon Y, Ninham BW (1987) J Phys Chem 91:674
7. Malliaris A (1987) J Phys Chem 91:6511
8. Turro NJ, Yekta A (1978) J Am Chem Soc 100:5951
9. Infelta P (1979) Chem Phys Lett 70:179
10. Malliaris A (1987) J Photochem Photobiol A: Chem 40:79
11. Malliaris A (1987) Progr Colloid Polym Sci 73:161
12. Malliaris A (1987) Adv Colloid Interface Sci 27:153
13. Malliaris A (1988) Intern Rev Phys Chem 7:95
14. Kalyanasundaram K, Thomas JK (1977) J Am Chem Soc 99:2039
15. Paleos CM, Stassinopoulou C, Malliaris A (1983) J Phys Chem 87:251

Authors' address:

Dr. Angelos Malliaris
NRC "Demokritos"
Agia Paraskevi, Athens 15310, Greece

Progress in Colloid & Polymer Science   Progr Colloid Polym Sci 89:125—126 (1992)

Resonant phenomena in colloidal crystals

T. Palberg, M. Würth, P. König, E. Simnacher, and P. Leiderer

Universität Konstanz, Fakultät für Physik, Konstanz, FRG

Abstract: Colloidal crystals of completely deionized suspensions of latex speres are subjected to oscillatory and steady shear, as well as to homogeneous and inhomogeneous electric fields. Various resonant phenomena observed in such experiments are reported.

Key words: Colloidal crystals; shear resonances; elasticity; shear flow; electric fields

Colloidal crystals are prepared from highly charged monodisperse suspensions of polystyrene latex spheres in a continuous deionization circuit. Details of both materials used [1] and of the deionization setup [2] are given elsewhere. Resonant torsional and shear vibrations of crystals are detected by monitoring the amplitude of periodic shifts in the angle of Bragg scattered light as a function of frequency [3] (Lock-in technique or FFT-frequency analysis). From the resonance frequencies measured in a cylindrical geometry by periodic excitation the shear modulus G of the suspension is calculated [4]. Assuming a screened Coulomb potential for the interaction of particles a renormalized charge number [5] is derived and is found to increase with increasing particle concentration c_p.

Placed in a rectangular flow-through cell acting as resonator (depth:height = 1:10) with open ends, a crystalline suspension of $c_p = 2.5 \cdot 10^{18}$ m^{-3} shows a single resonance of frequency $v_{res} = 14$ Hz. The shear modulus derived via $G = 4\pi v_{res}^2/(1/k_x^2 + 1/k_y^2)$ agrees well with the values from the cylindrical geometry.

Under very slow *dc*-flow ($v_p \leqslant 7$ μm s^{-1}) the crystalline solid moves with a plug-like velocity profile, and it is excited to resonances of $v_{res} = 7$ Hz by alternating stick-slip at the walls of stick-slip frequency $v_{ss} = 3.5$ Hz. At slightly faster flow ($v_p = 7-12$ μm s^{-1}) there is depinning at all walls and the crystal moves without vibrational excitation, whereas increasing the flow further, results in, first, grinding of the solid into fine crystallites, and then, shear melting and a parabolic velocity profile [1, 6].

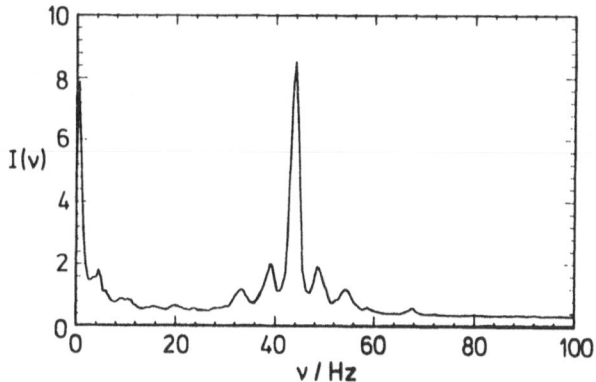

a

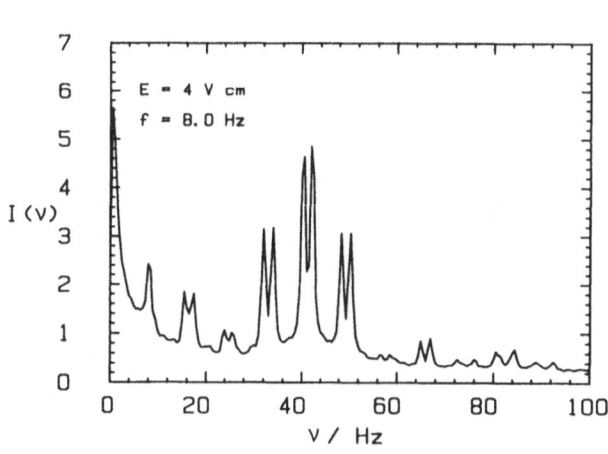

b

Fig. 1. Typical Doppler shift spectra (reference beam geometry) for a crystal in a weak homogeneous electric field of field strength $E = 4$ V cm^{-1}; a: *dc* field, b: *ac* field with $v_{ac} = 8$ Hz

When subjected to a weak homogeneous *dc* electric field the crystal also flows through the cell in a pluglike manner. Doppler shift spectroscopy [7] in the reference beam geometry shows a splitting of the velocity peak, as depicted in Fig. 1a. In an *ac* field two splittings may be observed, as shown in Fig. 1b: one being that of the frequency of the applied field v_{ac} [8], and the other being independent of both v_{ac} and field strength. An interpretation in terms of a resonant shear vibration seems tempting, but a rigorous light-scattering theory for this case still remains to be derived.

In the cylindrical resonator with quadrupolar electrode geometry inhomogeneous electric fields are generated. A crystalline material subjected to these shows electrophoretic behavior. In addition, resonant torsional vibrations are detected if the suspension is excited by a low-frequency *ac* field.

Acknowledgements

We sincerely thank the group of R. Weber for many helpful discussions, and gratefully acknowledge financial support of the DFG.

References

1. Deggelmann M, Palberg T, Hagenbüchle M, Maier EE, Krause R, Graf C, Weber R (1991) J Coll Interface Sci 143:318
2. Palberg T, Härtl W, Wittig U, Versmold H, Würth M, Simnacher E (1992) J Phys Chem (submitted)
3. Palberg T, Falcoz F, Hecht H, Simnacher E (1992) J Physique (submitted)
4. Dubois-Violette E, Pieranski P, Rothen F, Strzelecki L (1980) J Physique 41:369
5. Alexander S, Chaikin PM, Grant P, Morales GJ, Pincus P, Hone D (1984) J Chem Phys 80:5776
6. Palberg T, Streicher K, Würth M, König P (1992) Nature (submitted)
7. Palberg T, Versmold H (1989) J Phys Chem 93:5296
8. Bennet AJ, Uzgiris EE (1973) Phys Rev A 8(5):2667

Authors' address:

Dr. T. Palberg
Universität Konstanz
Fakultät f. Physik
Postfach 5560
D-7750 Konstanz, FRG

Progress in Colloid & Polymer Science   Progr Colloid Polym Sci 89:127—131 (1992)

Preparation of giant vesicles by external AC electric fields. Kinetics and applications

M. I. Angelova[1]), S. Soléau[2]), Ph. Méléard[2]), J. F. Faucon[2]), and P. Bothorel[2])

[1]) Central Laboratory of Biophysics, Bulgarian Academy of Sciences, Sofia, Bulgaria
[2]) Centre de Recherche Paul Pascal, C.N.R.S., Pessac, France

Abstract: We suggest a method for preparation of large ($\sim 15 \div 70\ \mu m$ in diameter), unilamellar, fluctuating, isolated vesicles, allowing us to perform routine measurements of membrane bending elasticity modulus k_c by image analysis of thermal fluctuations. The method is based on the effects of AC electric fields on the lipid swelling and liposome formation. We have observed a monotonous decrease of k_c for egg phosphatidylcholine (EPC) vesicles from $(0.66 \pm 0.06) \cdot 10^{-19}$ J for 1 day to $(0.45 \pm 0.05) \cdot 10^{-19}$ J for 13 days after vesicle formation. The lipid mixture Chol/PC 37:63 mol/mol led to $k_c = (1.06 \pm 0.15) \cdot 10^{-19}$ J, and PE/PC 25:75 mol/mol — to $(0.72 \pm 0.06) \cdot 10^{-19}$ J. — The mechanism of vesicle formation is discussed. It proved that the basic phenomenon is sequences of lateral fusions of smaller vesicles induced by the applied AC electric field.

Key words: Giant vesicles; liposomes; electroformation; bending elasticity

1. Introduction

Studies of the mechanical properties of the lipid bilayer are generally performed by optical microscopy, so they require the use of giant unilamellar vesicles of about 10 μm or more. Various methods have already been proposed to obtain giant vesicles, based on dialysis or dilution methods [1], on the use of chaotropic salts [2], or on vesicle fusion by freezing and thawing [3]. However, the vesicle preparations so obtained are generally very heterogeneous: giant vesicles are not well-isolated, and they often contain smaller particles. Furthermore, due to the formation process, which occurs under osmotic stress, the membrane tension is relatively high, and thermal fluctuations are not able to develop. So, all the researchers who are dealing with the mechanical properties of membranes use more gentle procedures, which are based on the early one proposed by Reeves and Dowben [4]: a dry lipid film is formed on a solid substrate (glass or teflon), then pure water is added, and giant vesicles are allowed to form spontaneously. Unfortunately, the yield of giant, unilamellar and fluctuating vesicles is usually exceedingly low and can even be zero, depending on the lipid used. This constitutes a major limitation in the study of the mechanical property of the lipid bilayer. Here, we propose a new method based on the previous works of Angelova and Dimitrov [5—8]. It uses the action of an AC electric field on the liposome formation. The usefulness of this method will be illustrated by some measurements of the bending modulus.

2. Materials and methods

Egg phosphatidylcholine (EPC) was prepared according to [9]. Egg phosphatidylethanolamine (EYPE) was Sigma (P-8798), egg lyso-phosphatidylethanolamine (lysoPE) was Sigma (L-4754), digalactosyl diacylglyceride (DGDG) was Sigma (D-4651), bovine phosphatidylserine (PS) was Lipid Products (No. 899), and cholesterol (Chol) was Serva (17090).

The cell for the vesicle preparation (Fig. 1) consisted of two ITO coated glasses separated by a silicone spacer of 0.3 mm. Lipids were dissolved in chloroform/methanol 9:1 at a concentration of

0.5 mg (total lipid)/ml. A drop of 2.5 µl was deposited on one of the conductive coated glasses, and dried using a nitrogen stream. Unless otherwise mentioned, and AC filed of 10 Hz, 1 V was applied to the cell, and milli-Q water ($R \sim 15$—18 MΩ/cm) was added. Usually, the time of swelling under AC field was 2 h.

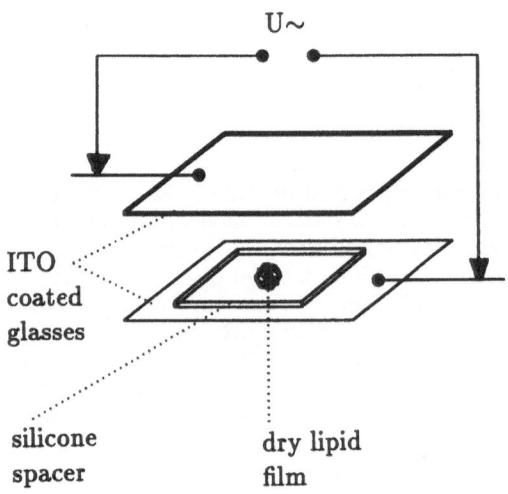

Fig. 1. Sketch of the cell for the vesicle preparation

Observation of the vesicle preparation was performed using a phase contrast inverted microscope (Zeiss IM35), with an objective Ph 40×, NA = 0.65 or Ph 100×, NA = 1.25 (in the case of bending elasticity measurements). Records were taken with a contrast-enhancing video camera (Hamamatsu C2400-07) and a U-matic video tape recorder (Sony, Japan). Images were digitized with a Pericolor 2001 image-processing system (Numelec) in the 512 × 512 8-bit pixels format. Image analysis and calculation of the bending modulus was then performed on a VAX-8600 computer, according to [10].

3. Results

3.1 Vesicle formation under an AC field

Angelova et al. have previously shown that the vesicle formation critically depends on the strength of the electric field, but also on the thickness of the dried lipid film [5, 7]. The thinner the lipid film, the higher the yield in thin-walled, probably unilamellar, vesicles. A very rough estimate of the thickness of the lipid film can be done from the area of the spot and the amount of deposited lipid,

assuming that the film is uniform (which is far from being the case). In our case, it corresponds to about 5—10 lipid bilayers. The spontaneous formation of vesicles is suppressed under such conditions. On the contrary, in the presence of an AC field, swelling occurs very fast and, within a few minutes, one can see in the microscope the circular contour of small structures which remain attached to the deposit. These structures grow progressively up to about 10 µm, until they come into contact with each other. Then, growing continues by a fusion mechanism between neighboring vesicles, as can be seen in Fig. 2. this process can lead to very large vesicles of several tenths of microns, which finally detach from the film. Then, this entire process starts again, which lead to a gradient of the vesicle size within the sample, with stacked layers of vesicles of increasing size from the glass to the inside of the cell (Fig. 3).

Three other points are worth mentioning about the vesicle formation. First, in the presence of the AC field, one can clearly see a motion of the vesicles as a whole, which are vibrating at the same frequency as the applied field. This is a direct evidence of the interaction between the field and vesicles. It shows that at least one of the effects of the electric field is to create a very gentle mechanical agitation, which certainly helps vesicles to form, fuse, and detach from the support (as in the case of sonication). Moreover, this effect can be easily observed and controlled by varying the intensity and frequency of the field. A second effect that can be noticed is that the thus obtained vesicles do not exhibit thermal fluctuations of their contour. This means that the electric-field-induced formation leads to a membrane tension for the new vesicles which must be of the order of at least 10^{-3}—10^{-4} mN/m. However, one has only to leave the cell at rest for a few hours in order to relax the vesicles and to observe thermal fluctuations. Third, it can be stressed that the effect of the electric field is completed within about 1 h, and practically no further evolution of the sample can be observed later on.

In order to check the efficiency of this method, it has been applied to various lipid and lipid mixtures. Of course, it works very well with pure phosphatidylcholines such as EPC, but also, as shown in Fig. 3, with mixtures Chol/EPC, for which the spontaneous formation is known to be very limited, and to lead to rather small and multilamellar vesicles. Under an AC electric field,

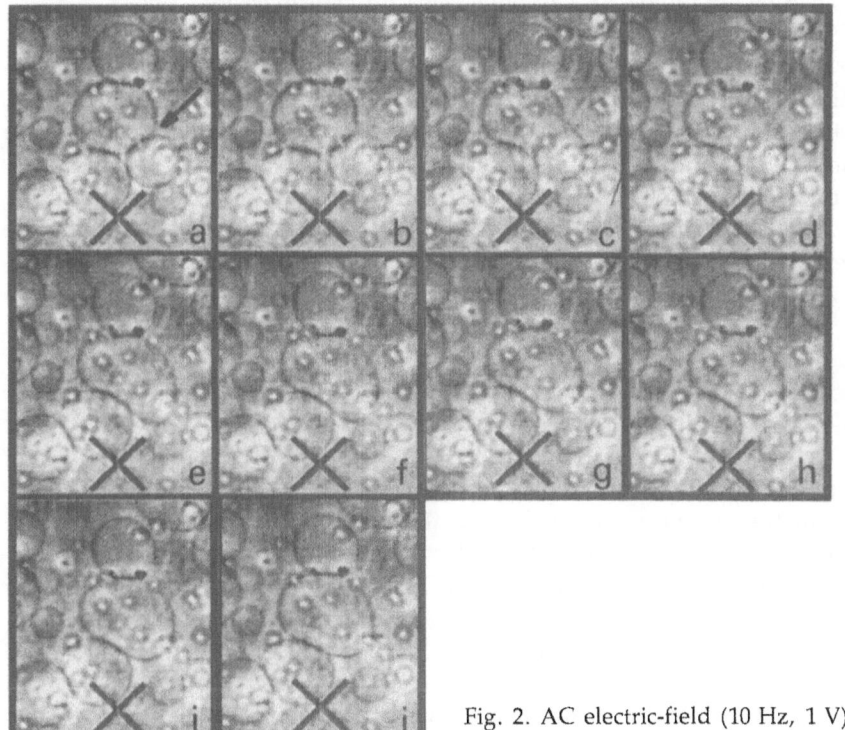

Fig. 2. AC electric-field (10 Hz, 1 V)-induced fusion of PC vesicles. Time intervals a—h) = 40 ms; i,j) = 320 ms. Cross = 25 µm

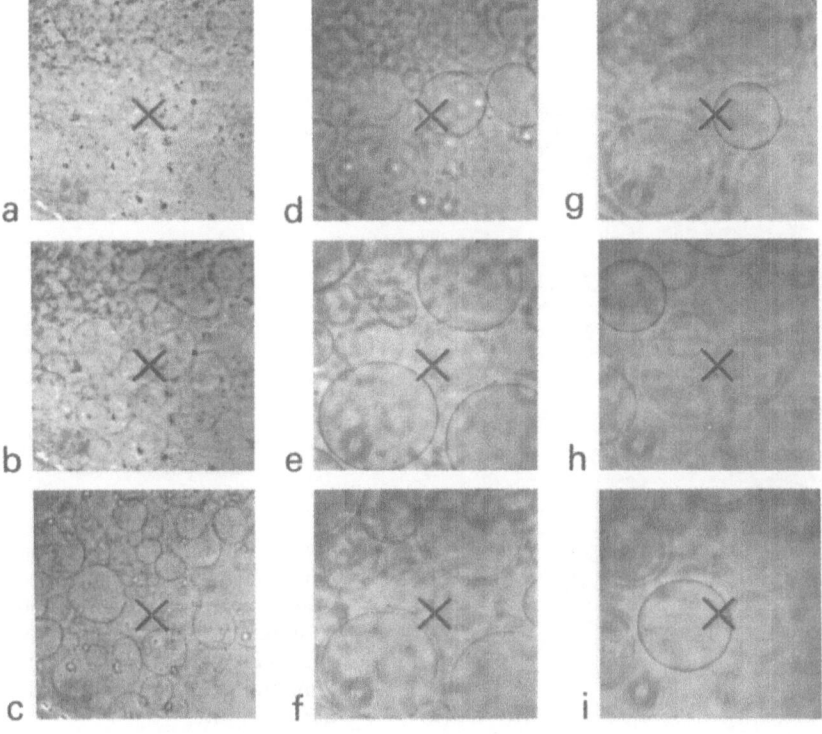

Fig. 3. Vesicle populations (Chol/PC 23:77 mol/mol, after 2 h 30 min swelling under AC field of 10 Hz, 10 V) at different distances from the electrode surface: a) at the very surface; b) at 5 µm; c) 10 µm; d) 20 µm; e) 30 µm; f) 40 µm; h) 60 µm; i) 70 µm. Cross = 25 µm

very large and thin-walled vesicles can be obtained up to 50% Chol. Mixtures of EPC with stearic acid also lead to the formation of giant vesicles, even in 100 mM sucrose, which shows that the electric field is able to overcome the osmotic stress.

Giant vesicles can also be obtained with PS and DGDG, although their behaviors differ significantly. In the case of PS, one can notice that giant vesicles often contain smaller ones. In the case of DGDG, numerous filaments and tubes are observed.

We were not able to obtain giant vesicles with pure EYPE, even on increasing the thickness of the lipid film and the intensity of the electric field up to 3 V. One must remind that the $L_a \rightarrow H_{II}$ transition temperature of EYPE is known to be about 25—30°C, so that it was normally in the lamellar phase under the experimental conditions used. The addition of lyso-PE up to 10% was ineffective for vesicle formation, while membranes formed rapidly at LysoPE/PE 50:50 mol/mol, but the vesicles were exceedingly fragile and broke down easily upon vibration. Similarly, the addition of EPC allowed the formation of vesicles, with a clear dependence on the composition of the mixture, as shown in Fig. 4. Very few and small vesicles are formed at 10% EPC. Small vesicles of about 10 µm start to appear at 15%, and their size progressively increases when further EPC is added.

3.2 Bending elasticity measurements

Some bending elasticity measurements were performed, following the method described in [10]. This method has been previously applied to a number of EPC vesicles obtained by spontaneous swelling of the lipid film, and led to a bending modulus $k_c = (0.45 \pm 0.11) \cdot 10^{-19}$ J. When the same method is applied to fresh EPC vesicles obtained by applying an electric field, one finds a much higher value of $k_c = (0.66 \pm 0.06) \cdot 10^{-19}$ J. However, if one measures the time dependence of k_c, one can see (Fig. 5) that a progressive decrease of k_c occurs, down to $k_c = (0.45 \pm 0.05) \cdot 10^{-19}$ J for 13 days after the vesicle preparation. The apparent discrepancy between the two measurements is certainly due to the fact that the spontaneous formation of vesicles require a very long time, so that bending elasticity measurements can be performed only several days after the beginning of the swelling. The observed decrease of k_c versus time is probably due to a chemical degradation of EPC, which leads to formation of lyso-derivatives, fatty acids, and oxidation products. Thus, k_c seems to be vary sensitive to the chemical composition of the bilayer.

It has also to be mentioned that the thermal fluctuations of all the studied vesicles were fitted with very low membrane tensions. Furthermore, the small dispersion obtained on the k_c values in-

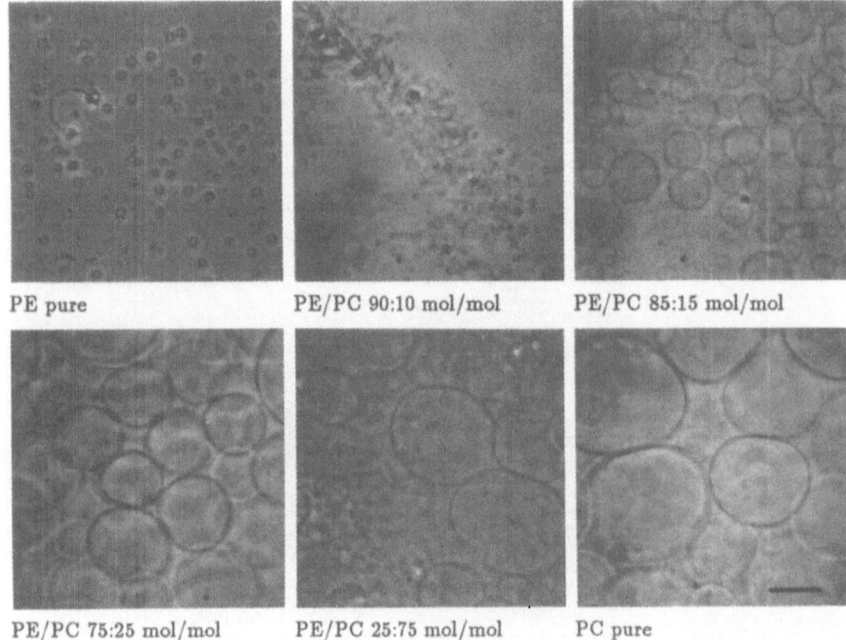

PE pure   PE/PC 90:10 mol/mol   PE/PC 85:15 mol/mol

PE/PC 75:25 mol/mol   PE/PC 25:75 mol/mol   PC pure

Fig. 4. The largest vesicles obtained from pure PE, mixtures PE/PC, and pure PC after 2 h 30 min swelling of lipid films in AC 10 Hz, 1 V. The temperature was 23°C. Bar = 25 µm

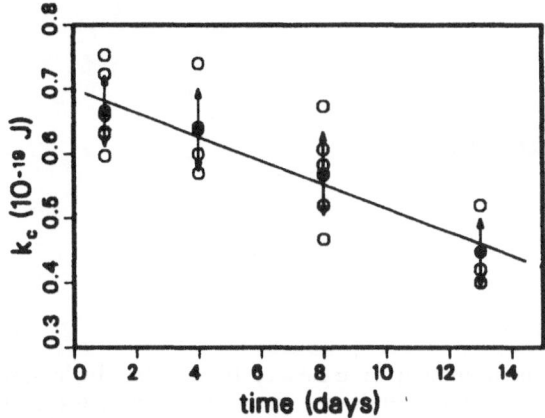

Fig. 5. The decrease of k_c upon the time for egg PC vesicles

dicates that the membranes of all the studied vesicles consist of the same number of bilayers, most probably one, and that they have the same composition.

The bending modulus has also been measured on some lipid mixtures. The EPC/PE 75:25 mol/mol mixture led to k_c (0.72 ± 0.06) · 10^{-19} J, very close to that of pure EPC. On the other hand, 37 mol% Chol in EPC led to an increase of k_c up to (1.06 ± 0.15) · 10^{-19} J. This corresponds to an important effect of about 60%. However, this is much smaller than the three- or four-fold increase previously reported [12, 11]. The main problem in this case lies in the number of bilayers constituting the contour, which is very difficult to assess unambiguously.

References

1. Oku N, Scheerer JF, MacDonald RC (1982) BBA 692:384
2. Oku N, MacDonald RC (1983) J Biol Chem 258:8733
3. Oku N, MacDonald RC (1983) Biochemistry 22:855
4. Reeves JP, Dowben RM (1969) J Cell Physiol 73:49
5. Angelova MI, Dimitrov DS (1986) Faraday Discuss Chem Soc No 81:303, 345
6. Dimitrov DS, Angelova MI (1987) Progr Coll Pol Sci 73:48
7. Angelova MI, Dimitrov DS (1987) Mol Cryst Liq Cryst 152:89
8. Angelova MI, Dimitrov DS (1988) Progr Coll Pol Sci 76:59
9. Singleton WS, Gray MS, Brown ML, White JL (1965) J A Oil Chem Soc 42:53
10. Faucon JF, Mitov MD, Méléard Ph, Bivas I, Bothorel P (1989) J Physique 50:2389
11. Evans E, Rawicz W (1990) Phys Rev Let 64:2094
12. Duwe HP, Kaes J, Sackmann E (1990) J Physique 51:945

Authors' address:

J. F. Faucon
Centre de Recherche Paul Pascal
C.N.R.S.
Avenue A. Schweitzer
33600 Pessac, France

Progress in Colloid & Polymer Science Progr Colloid Polym Sci 89:132—134 (1992)

Ordering phenomena in gyrotropic electrolytes

V. N. Bondarev

Physics Institute, I. I. Mechnikov University, Odessa, Ukraine

Abstract: The spontaneous formation of α-helices from chiral biological molecules in homogeneous aqueous solutions — gyrotropic electrolytes (GE) — is modeled as an instability caused by transverse fluctuations of the polarization due to the redistribution of mobile charges within the GE [3]. The chiral molecules condense in a cylindrical geometry, which is influenced by the chirality. Such an ordering is known to be characteristic for molecules like DNA, RNA or certain proteins, hence, the suggested theory may explain some basic processes occurring in the structuring of biological molecules.

Key words: Biological structures; spontaneous instability; Debye length; thermodynamic potential; helical modes

A characteristic feature of any living organism is the so-called chirality, which is displayed, in the visible activity of biological molecules. Up to now, it has been considered only an empirical fact [1]. In general, the conventional physical models describing macromolecules of DNA-type (e.g., [2] and references therein) do not take into account structural helicity. Therefore, the various macromolecular conformations are modeled by cylinders with different radii and linear charge densities.

Nevertheless, it is obvious that for biological structures the role of the chiral component (aminoacids, sugars, which are gyrotropic [1]) is as important as the role of electrostatics (salt and acid solutions). A way to investigate the ordered states in systems with such a combination of components, i.e., gyrotropic electrolytes (GE) has been outlined. In [3—5] it was shown that the homogeneous state of GE's becomes spontaneously unstable due to transverse fluctuations of polarization $\vec{P}$ arising from the redistribution of mobile charges in GE.

Due to the presence of a pseudoscalar γ typical for GE the transverse-type values formed by space derivatives of $\vec{P}$ may exist (unlike the usual, nongyrotropic electrolytes which can be described by a local longitudinal electric field $\vec{\nabla}\varphi$, expressed through the Coulomb potential φ, and the uncompensated charge density div $\vec{P}$, only). Thus, the dissipative current density must contain the term

$\sim \gamma$ curl $\vec{P}$ in the first order of the space derivatives. Just this term, in fact, causes the aforesaid instability of the homogeneous state of GE [3].

The concept of transverse polarization instability has been applied to plane-parallel layers of GE. These lead to [4] two-dimensional (2D) periodical $\vec{P}$-, φ superstructure, and models of real biological crystals [6]. In the present paper, we apply the ideas proposed in our previous articles [3, 4] to describe the physical properties of structures of polynucleotide-type. The latter we imagine as a cylinder of radius R and length L ($L \rightarrow \infty$), characterized by the dielectric constant ε, the Debye length R_D, and a pseudoscalar constant γ, which is proportional to the chiral component concentration. Such a cylinder is placed into a conventional electrolyte solution (nongyrotropic) with the Debye length R'_D and the dielectric constant ε'.

For the mathematical formulation of the problem, let us write the thermodynamic potential of the system as a sum of bulk terms, corresponding to GE and surrounding electrolyte (E), and the surface term

$$F = \int_{r<R} f_{GE} d^3r + \int_{r>R} f_E d^3r + \int_{r=R} f_S d^2r . \tag{1}$$

To second order approximation in electrostatic variables, the thermodynamic potential densities are (see [3—5])

$$f_{GE} = -\frac{\varepsilon}{8\pi}(\vec{\nabla}\varphi)^2 - \varphi\,\mathrm{div}\,\vec{P} + \frac{2\pi R_D^2}{\varepsilon}(\mathrm{div}\,\vec{P})^2$$

$$+ \frac{\gamma}{2}\vec{P}\cdot\mathrm{curl}\,\vec{P} + \frac{a_2}{2}(\mathrm{curl}\,\vec{P})^2, \qquad (2)$$

$$f_E = -\frac{\varepsilon'}{8\pi}(\vec{\nabla}\varphi)^2 - \varphi\,\mathrm{div}\,\vec{P} + \frac{2\pi R_D'^2}{\varepsilon'}(\mathrm{div}\,\vec{P})^2, \quad (3)$$

$$f_S = \rho_S\varphi + Ba_2\vec{P}\cdot\vec{\nabla}_2\rho_S. \qquad (4)$$

Here, $a_2 > 0$ is the transverse stiffness of GE, which in considered as a solid electrolyte- [3, 7]; the term containing the dimensionless constant B and the 2D gradient of the surface charge density ρ_S is added to the well-known surface contribution $\rho_S\varphi$ [8] to reflect the role of tangential $\vec{P}$-components in GE.

Varying Eqs. (1)—(4) with respect to φ and $\vec{P}$, we obtain the equations [3—5]

$$\mathrm{div}(-\varepsilon\vec{\nabla}\varphi + 4\pi\vec{P}) = 0, \quad r < R, \qquad (5)$$

$$\mathrm{div}(-\varepsilon'\vec{\nabla}\varphi + 4\pi\vec{P}) = 0, \quad r > R, \qquad (6)$$

$$-\vec{\nabla}\varphi - \gamma\,\mathrm{curl}\,\vec{P} + \frac{4\pi R_D^2}{\varepsilon}\,\mathrm{grad}\,\mathrm{div}\,\vec{P}$$

$$- a_2\,\mathrm{curl}\,\mathrm{curl}\,\vec{P} = 0, \quad r < R, \qquad (7)$$

$$-\vec{\nabla}\varphi + \frac{4\pi R_D'^2}{\varepsilon'}\,\mathrm{grad}\,\mathrm{div}\,\vec{P} = 0, \quad r > R. \qquad (8)$$

The boundary conditions in cylindrical geometry (r, χ, z) are

$$\varphi\,|_{R-0} = \varphi\,|_{R+0}, \qquad (9)$$

$$-\frac{\varepsilon'}{4\pi}\frac{\partial\varphi}{\partial r}\bigg|_{R+0} + \frac{\varepsilon}{4\pi}\frac{\partial\varphi}{\partial r}\bigg|_{R-0} - \rho_S = 0, \qquad (10)$$

$$\left[\frac{\gamma}{2}P_z + a_2(\mathrm{curl}\,\vec{P})_z\right]\bigg|_{R-0} + B\frac{a_2}{R}\frac{\partial\rho_S}{\partial\chi} = 0, \qquad (11)$$

$$\left[\frac{\gamma}{2}P_\chi + a_2(\mathrm{curl}\,\vec{P})_\chi\right]\bigg|_{R-0} - Ba_2\frac{\partial\rho_S}{\partial z} = 0, \qquad (12)$$

$$\left(\frac{4\pi R_D^2}{\varepsilon}\,\mathrm{div}\,\vec{P} - \varphi\right)\bigg|_{R-0} = 0, \qquad (13)$$

$$\left(\frac{4\pi R_D'^2}{\varepsilon'}\,\mathrm{div}\,\vec{P} - \varphi\right)\bigg|_{R+0} = 0. \qquad (14)$$

Assuming φ and $\vec{P}$ as oscillating in z and χ direction, i.e., $\exp[i(kz + m\chi)]$, with k as the wave number along the cylindrical axis z and $m = 0, \pm1, \pm2, \dots$ as the order of helicity, and solving the Eqs. (5)—(8) for radial parts, we can express the solutions in terms of usual (J_m) and modified (I_m, K_m) Bessel functions. Six arbitrary constants that appear in the solutions can be determined from the six boundary conditions (9)—(14). Then, we obtain an expression for the partial thermodynamic potential F_m by means of the procedure proposed in [4] (intermediate actions are omitted)

$$F_m = \frac{1}{2}\pi R^2 L\left[U_m(q) + \frac{A_m(q)}{W_m(q)}\right]\rho_{Sm}^2. \qquad (15)$$

Here, ρ_{Sm} is the partial amplitude of the surface charge density, also, the following definitions are used

$$U_m(q) = 4\pi/[m(\varepsilon' - \varepsilon) + \varepsilon S I_{m-1}(S)/$$

$$I_m(S) + \varepsilon' S' K_{m-1}(S')/K_m(S')], \qquad (16)$$

$$A_m(q) = 4B^2 a_2(m^2 + q^2)[(qt_0 - m^2 - q^2)tJ_m(t)$$

$$+ mt_0(t_0 - q)J_{m-1}(t)]/R^2 t_0^2, \qquad (17)$$

$$W_m(q) = mtJ_m(t) - q(t_0 - q)J_{m-1}(t) \qquad (18)$$

with

$$S = \sqrt{q^2 + (R/R_D)^2}, \quad S' = \sqrt{q^2 + (R/R_D')^2},$$

$$q = kR, \quad t_0 = R\gamma/a_2, \quad t = \sqrt{t_0^2 - q^2}. \qquad (19)$$

In the case $q > |t_0|$ only modified and no usual Bessel functions appear Eqs. (17), (18).

A thorough study of $W_m(q)$ shows that for every given m the function $W_m(q)$ passes through zero at a certain q, i.e., $F_m \to -\infty$. Such an instability occurs, if the t_0 value is large enough, i.e., $t_0 \geqslant t_0^{(-1)} = 3.112$. The upper index (-1) denotes that just F_{-1} will have a discontinuity occurring at $q = q^{(-1)} = 1.228$ for this threshold value $t_0^{(-1)}$ (we assume $\gamma > 0$; in the opposite case, we should discuss the F_1 function for $t_0^{(1)} = -3.112$).

The tendency of F_m towards $-\infty$ on one side from the discontinuity expresses the GE instability against the condensation of m-th harmonic with the period $\lambda_m \simeq 2\pi R/q^{(m)}$, where $q^{(m)}$ is the "dangerous" dimensionless wave number (compare with the results in [4] for the quasi-2D GE problem). Similar to GE in the plane geometry [4], such an instability is stopped by adding the fourth-order terms to the thermodynamic potential (the consideration of this question will be given elsewhere).

The obtained results may explain the mechanism of formation of primary biological structures, i.e., α-helices like DNA and RNA. Namely, helical harmonics with periods corresponding to the "dangerous" values $q^{(m)}$, the discontinuity points of F_m functions, just may be considered as analogous to the ordered states in real polynucleotides. At the threshold (see above), we have for the helix period along the z-axis $\lambda_{-1} = 2\pi R/q^{(-1)} \simeq 50$ Å, if $R \simeq 10$ Å, (which is the experimental value of the polynucleotide radius [6]).

When t_0 increases harmonics of a higher order than $m = -1$ appear. It can be shown that $m = -2$ (the analogue of the real righthand double helix [1]) is similar to the one indicated above and to $\lambda \simeq 35$ Å characterizing DNA [1].

In conclusion, we note that on the basis of our concept of the spontaneous ordering in GE, one is able to describe the variation of physical properties of polynucleotides as a function of external conditions. This problem will be the subject of our future investigation.

References

1. Cantor CR, Schimmel PR (1980) Biophysical chemistry vol III
2. Frank-Kamenetsky MD (1987) Uspekhi Fiz Nauk 151:595—618
3. Bondarev VN (1986) Pis'ma v ZhETF 43:200—202
4. Bondarev VN (1989) Physics Letters 136:139—142
5. Bondarev VN, Volyanskaya OO (1990) In: Proc 8th Symposium on Intermolecular Interactions and Molecular Conformations. Novosibirsk, pp 149—150 (in Russian)
6. Vainshtein BK (1979) In: Vainshtein BK, Chernov AA, Shuvalov LA (eds) Modern crystallography, vol 2. Nauka, Moscow, pp 193—245 (in Russian)
7. Boyce JB, Hayes TM (1979) In: Salamon MB (ed) Physics of superionic conductors. Springer-Verlag, Berlin Heidelberg New York, pp 20—60
8. Landau LD, Lifshitz EM (1982) Electrodynamics of continuous media. Nauka, Moscow, pp 50—52 (in Russian)

Author's address:

Dr. Victor N. Bondarev
Odessa University, Physics Institute
Ul. Pasteur'a, 27
270100 Odessa, Ukraine

Progress in Colloid & Polymer Science   Progr Colloid Polym Sci 89:135—139 (1992)

Structuration and elasticity of electrorheological fluids

G. Bossis[1]), E. Lemaire[1]), J. Persello[2]), and L. Petit[3])

[1]) Laboratoire de Physique de la Matiere Condensee, Universite de Nice, France
[2]) Rhône-Poulenc Recherches, Aubervilliers, France
[3]) Laboratoire de Physique, Ecole Normale Supérieure de Lyon, France

Abstract: Solid particles suspended in a liquid phase can build linear structures when a large enough electric field is applied to the suspension. These structures confer to the suspension a shear modulus G, which is measured with an oscillating plate device. This system allows to simultaneously visualize the change of structure of the suspension. We observe an important decrease of G with the amplitude of the strain, and explain it by the behavior of the forces between two conductive particles at short separations. Striking features relative to stripe formation upon large amplitude strain and subsequent considerable increase of elasticity are reported.

Key words: Electrorheological suspensions; shear modulus; rheology and structure

1. Introduction

Suspensions displaying a large increase in the transmission of stress upon application of an electric field are known as electrorheological fluids (ERF) [1—2]. These fluids, composed of solid particles suspended in a non polar fluid, solidify when submitted to an electric field of a few KV/mm. Numerous applications have been envisaged, such as, for instance, electrically triggered clutches or feedback controlled dampers, and many systems chosen with the rule of thumb have been tested for engineering applications [3]. Two fundamental quantities which characterize this transition from a fluid phase to a solid are the static yield stress τ_s and the static shear modulus G_0. The first one measures the force necessary to break fibers of particles in order to recover a fluid phase, and the second one measures the slope of the stress versus the strain before the fibers break. The first requirement for applications is to get the highest possible values for τ and G_0. Actually, for most of the applications, especially in the automotive industry, it still lacks one or two orders of magnitude to realize industrial devices with ER fluids [4]. An improvement of the performances of these fluids requires careful studies on well-defined systems in order to understand how the structure of the suspension and the dielectric properties of the particles (in-

cluding its ionic cloud) determine the resistance of these fibers. More fundamental studies of these suspensions are now focusing on the interpretation of viscoelasticity [5—7] or yield stress measurements [8—10] in terms of interparticle forces. In this work, we discuss first results of elasticity measurements of a silica based ER fluid made with an oscillating plate device which also allows to observe with a microscope, at the same time, the change of structure.

2. Experimental

Silica preparation

The solid particles are made of silica synthesized by the Stöber method [11]. This method involves the hydrolysis of an ethyl silicate by water using ethanol as solvent and ammonia as a catalyst. This synthesis is made in the presence of a blue dye, as described in [12]. The silica surface is modified by grafting a methyl-octyl-dimethoxysilane for the particles to be dispersible. The spherical silica particles are then transferred into a 500 cSt silicone oil to form a stable slurry. To keep the dispersion free from aggregates, we transferred in a first step the silica from ethanol to 2-butanone and in a second step from 2-butanone to silicone oil by azeotropic distillation.

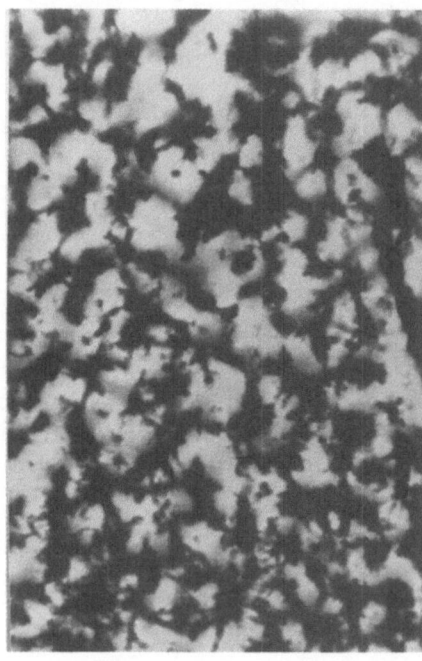

Fig. 1. Top view of the suspension structured by a field E = 4 KV/mm before applying a sinusoidal strain. The average diameter of the fibers is d = 10 μm

The blue dye allows to have a good contrast between the solid and the liquid phase, as can be seen in Fig. 1.

The suspension was placed between two glass plates with the field perpendicular to the plates. What we observe then is the projection of the fibers' cross-section. The typical diameter of these

fibers is 20 μm. The apparatus employed to measure the rigidity is shown in Fig. 2. The oscillating glass plate is mounted on an electromagnetic vibrator and the other plate is fixed on the stage of a micrometer. The two glass plates are coated with a transparent film of tin indium oxide which allows to apply the electric field and to maintain enough transparency to observe the structure with a microscope. The cell is illuminated with stroboscoped light, so we can visualize the periodic motion of the fibers. The parallelism of the two plates is adjusted with the help of a laser beam, and the gap h between the two electrodes was set either to 100 μm or 200 μm. For a sinusoidal applied force: $F(t) = Fo \cos \omega t$, which is proportional to the current in the coil of the vibrator, the acceleration is given by:

$$\frac{d^2x}{dt^2} = \frac{F(t)}{m} - \frac{1}{m}\left(k + \frac{GS}{h}\right)x$$

$$- \frac{1}{m}\left(a + \eta \frac{S}{d}\right)\frac{dx}{dt} ; \qquad (1)$$

k and a are, respectively, the elastic and the damping constants of the vibrator; G and η are the elasticity modulus and the viscosity of the suspension at the angular frequency ω, and S is the surface of the cell whose radius is 6 mm. This equation assumes that, in the presence of the electric field the mechanical response of the suspension can be characterized by the two quantities G and η. This is a quite plausible hypothesis as long as the strain

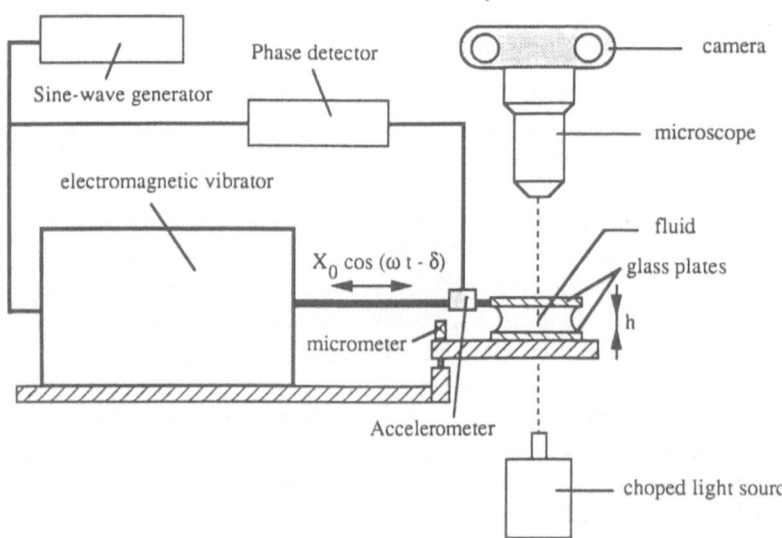

Fig. 2. Schematic description of the rheo-optical apparatus

remains smaller than 0.1. In this case, the fibers which are elongated by the motion of the upper plate do not break [13], and we measured with the help of an accelerometer fixed on the vibrating arm. The phase difference δ between the force and the amplitude is given by a phase detector. The solution of Eq. (1) yields:

$$tg\,\delta = \frac{2\beta\omega}{\omega_1^2 - \omega^2} \; ;$$

$$x_0 = \frac{F_0}{m}\,\frac{1}{\sqrt{(\omega_1^2 - \omega^2)^2 + 4\beta^2\omega^2}} \, , \qquad (2)$$

with

$$\beta = \frac{1}{2m}\left(a + \eta\,\frac{S}{h}\right) \; ; \quad \omega_1^2 = \frac{k}{m} + \frac{GS}{mh} \, . \quad (3)$$

When $\omega = \omega_1$, we have $\delta = \pi/2$ and we obtain directly the shear modulus corresponding to this frequency:

$$G(\omega_1) = \frac{mh}{S}\,4\pi^2(v_1^2 - v_0^2) \, , \qquad (4)$$

where $v_0 = 60$ Hz is the resonance frequency in the absence of the suspension.

We have plotted in Fig. 3 the variation of the rigidity modulus versus the strain amplitude for

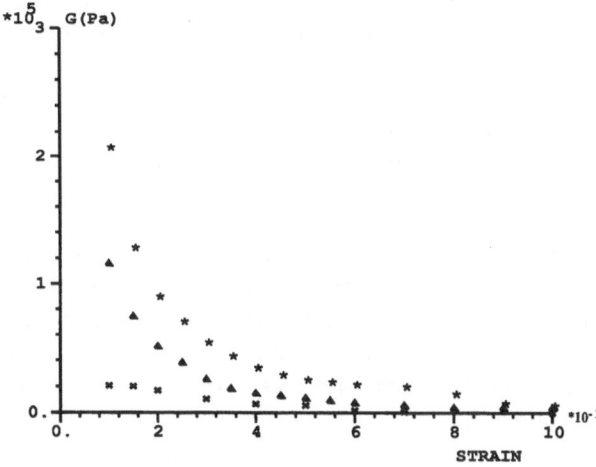

Fig. 3. Shear moduli at an average frequency $\omega_m = 63$ Hz versus the amplitude of the oscillating strain for different electric fields applied to the suspension:

△△△:	$E = 4$ KV/mm;	$v_E = 20$ Hz
***:	$E = 4\sqrt{2}$ KV/mm;	$v_E = 20$ Hz
×××:	$E = 4\sqrt{2}$ KV/mm;	$v_E = 10$ KHz

two values of the electric field. It appears that we have a large decrease of the shear modulus with the strain amplitude. Of course this variation is not consistent with the use of Eq. (4), which only holds for a constant value of G. Nevertheless, this quantity, determined from Eq. (4) can be considered as an average value of the rigidity modulus for strains lower than x/h. Its drop with the strain amplitude tells us that the elasticity of these fibers is strongly decreasing when they extend; we shall come back to this point in the discussion. The two curves A and B in Fig. 3 correspond respectively to a field of 4 KV/mm and $4 \times \sqrt{2}$ KV/mm. They have a homothetic ratio of 2 within the experimental error, which gives a square dependence in the electric field. This square dependence is expected as discussed in the next section. At last the lower curve in Fig. 3 is obtained for the same field amplitude as curve A, but at a higher frequency ($v_E = 10$ kHz) and the value of G is an order of magnitude lower. This is a proof that the polarization of the particles which gives rise to the fiber formation has a ionic origin since otherwise an electronic polarization would not change in this frequency range. Therefore, this ionic polarization is attributed to the presence of sodium ions surrounding the particles.

3. Discussion

At low field frequencies the motion of the ionic layer gives to the particles a very high apparent permittivity. So, as a first attempt, we can try to analyze these results by looking at the forces between two conductive particles. The restoring force between two conductive particles in a medium of permittivity ε, submitted to an electric field E, is given for small strain γ by [13]:

$$F_r^{12} \approx \varepsilon a^2 E^2(F_1 - 2F_8)\gamma \, . \qquad (7)$$

If we consider that we have N_c chains of particles per unit area, and introducing the volume fraction Φ of the particles, we get for the total restoring force per unit surface:

$$\frac{F_r}{S} = \frac{N_c}{S}\,F_r^{12} = \frac{3\Phi}{2\pi}\,\varepsilon E^2(F_1 - 2F_8)\gamma = G_0\gamma \, . (8)$$

The functions F_1 and F_8 are given for short separations ($\delta = r/a - 2 < 10^{-2}$) by Arp and Mason [14]:

$$F_1 \approx \frac{\pi^4}{18} \frac{1}{\delta(\ln\delta/12.689)^2} - 1.09 \tag{9}$$

$$F_8 \approx -0.7513 - \frac{\pi^4}{36} \ln^{-1}(\delta/12.689) + 0.28\,\delta \ .$$

At zero strain, the particles are prevented from coming into contact by the short-range repulsion of their ionic double layer. If we call δ_0 the minimum separation between their surfaces, then at a given strain $\gamma = x/h \ll 1$, and the separation between the surfaces of the spheres in the elongated chain will be $\delta \approx \delta_0 + \gamma^2$; the restoring force (8) will be expressed with the help of (9) by:

$$\frac{F_r}{S} \approx \frac{\Phi\varepsilon E^2 \gamma}{\delta_0 + \gamma^2} \ . \tag{10}$$

In our experiments γ extends from 10^{-2} to 10^{-1}, so δ remains lower than 10^{-2} which justifies the expansion leading to (10). This model can qualitatively explain the large decrease of the shear modulus with the strain which is experimentally observed. Of course, the expression (10) for the force between two particles is certainly a poor approximation of the real interaction between two polarized ionic layers. Furthermore, this model is oversimplified since we can observe thick fibers of particles instead of isolated chains of spheres and an important fraction of the particles could roll over each other inside a fiber, leading to a decrease of their section rather than to an increased separation between particles. Whatever the exact mechanism, it is clear that the elasticity of these ER fluids strongly decreases with the amplitude of the strain, and that more information about the dynamic response of this elasticity could be obtained by the application of a step-like strain.

Not only does the shear modulus of the ER fluids depend on the amplitude of the strain, but it also strongly depends on the past history of the applied strain. This can be easily understood if we compare Fig. 1 to Fig. 4. Contrary to the first picture obtained without strain, the second one refers to a suspension which was acted on by a periodic strain of amplitude $\gamma_0 \approx 0.1$ during about 15 min. We then observe that the rather loose structure in the beginning has been transformed in a well-defined one with isolated aggregates surrounded by a clear liquid. During this evolution the elasticity modulus

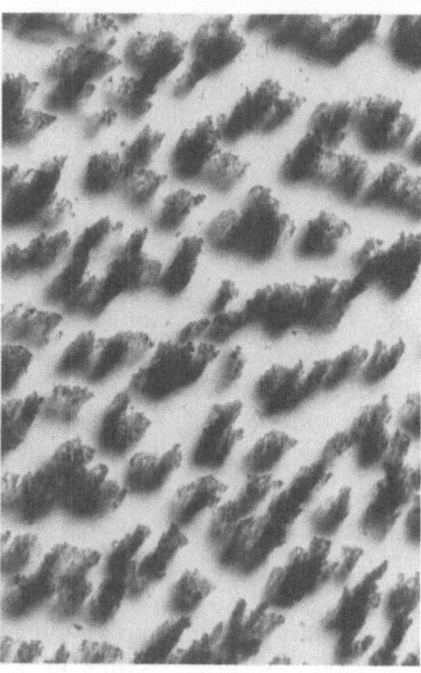

Fig. 4. Picture of the suspension after a strain of 15 min. Note the well-formed aggregates in comparison to the structure in Fig. 1

increases and the measurements in Fig. 3 correspond to the final stationary state of Fig. 4. For a strain amplitude below 0.1, this structure remains stable and we can see by stroboscopy that all the fibers follow the motion of the upper oscillating plate without any apparent slipping on the bottom plae. For larger strain amplitudes ($0.2 < \gamma \leqslant 1$) the fibers incline more and more in the flow and slip on the plates to form thin parallel stripes of particles aligned in the direction of the flow (which appear very clearly in Fig. 5A.

The interesting point is that, if we now measure the shear modulus at low strain amplitude ($\gamma = 0.02$), we find $G = 10^6$ instead of the previous value $G = 10^5$. The formation of stripes has considerably increased the elasticity modulus of the suspension, simply because a stripe is harder to strain in its own direction that a direction perpendicular to it. After a few minutes of application of this low-amplitude strain, fractures appear in the stripes, and even if the overall picture resembles Fig. 5A (compare Figs. 5A and 5B) these fractures and small disalignments relative to the flow lines are enough to recover a value $G = 2 \ 10^5$, which is closer to the original one. This effect is well

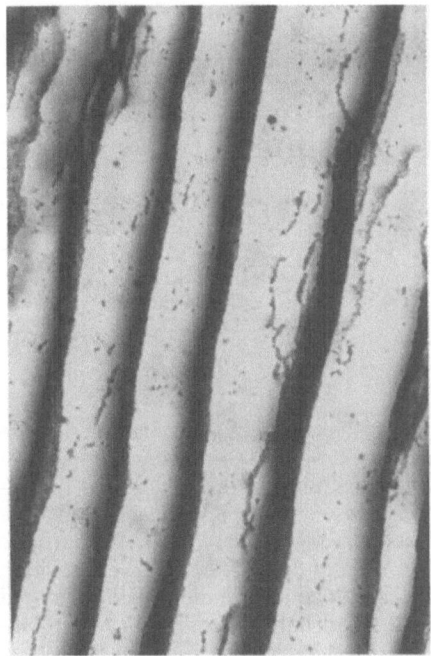

Fig. 5A. Structure obtained after applying a large amplitude strain ($\gamma = 1$). The particles are gathered in thin stripes aligned in the direction of the field

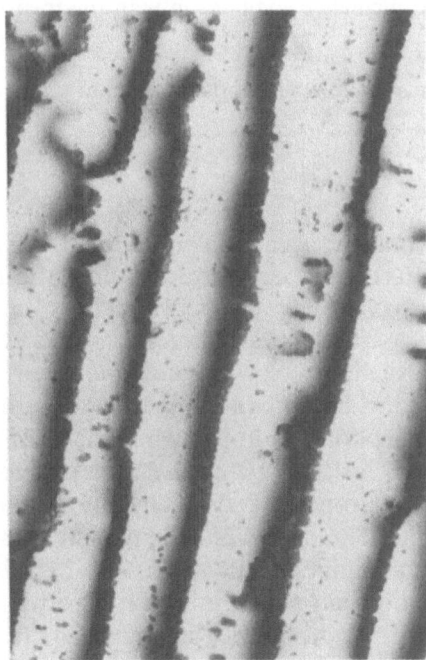

Fig. 5B. Same picture as Fig. 5A, but after 5 min of straining at amplitude $\gamma = 0.02$. Note the presence of small fractures which contribute very efficiently to lowering the shear modulus

reproducible. We can again apply a large strain; it will reform the stripes, and then at $\gamma = 0.02$, we again find $G = 10^6$. Finally, for very large strains we get a turbulent flow which allows to well redisperse the particles. A systematic study of these different structures in relation with the evolution of the relevant macroscopic properties (yield stress, shear modulus) is beyond the scope of this paper. Nevertheless, we hope that this preliminary work would have demonstrated that, even in the more simple *ER* fluids, we are faced with non-trivial situations: non-linearity, history dependency. A direct visualization of the structure evolution has been shown to be of great help for the interpretation of the experimental data.

References

1. Block H, Kelly JP (1988) J of Physics D; Appl Phys 21:1661
2. Jordan TC, Shaw NT (1989) IEEE Transactions on Electrical Insulation 24:849
3. Dienega YuF, Vinogradov GV (1984) Rheol Acta 23:636
4. Novak RF, Chaundy GJ (1990) 62nd Annual Meeting of the Society of Rheology, Santa Fe, Oct 1990
5. Brooks D, Goodwin J, Hjelm C, Marshall L, Zukoski C (1986) Colloids and Surfaces 18:293—312
6. Shulman ZP, Korobko EV, Yanoskü YG (1989) J of Non Newt Fluid Mechanics 33:181
7. Gassata DR, Filisko FE (1991) J Rheol 35:399
8. Sprecher AF, Carlson JD, Conrad H (1987) Material Science and Engineering 95:187
9. Gast AP, Zukoski CF (1989) Adv in Coll Sci 30:153
10. Lemaire E, Bossis G (1991) J Physics D Appl Phys 24:1473
11. Stober W, Fink A, Bohn E (1968) J Coll Interf Sci 26:62
12. Persello J (1988) Eur Pat EP 266.248, May 1988 (Rhône Poulenc Chimie)
13. Klingenberg DJ, Thesis PhD (1985) University of Missouri, Rolla 1985
14. Arp PA, Mason SG (1977) Colloid Polym Science 255:566

Authors' address:

G. Bossis
Laboratoire de Physique de la Matière Condensée
CNRS-URA No 190
Université de Nice-Sophia Antipolis
F-06034 Nice Cedex, France

Progress in Colloid & Polymer Science   Progr Colloid Polym Sci 89:140—144 (1992)

Electrophoretic mobility of aqueous colloidal suspensions in the gas and liquid-like phases

M. Deggelmann, H. Kramer, C. Martin, and R. Weber

Universität Konstanz, Fakultät für Physik, FRG

Abstract: New electrophoretic light scattering (ELS) measurements are reported on aqueous colloidal suspensions down to salt concentrations lower than $n_S = 10^{-6}$ M, where the polyions are liquid-like ordered due to screened Coulomb forces. Mobility data are presented for spherical polystyrene latices of different radius a and rodlike tobacco mosaic virus (TMV) particles (length l = 300 nm and radius a = 9 nm), and they are discussed in relation to the small ion concentration or κa. The salt concentration n_S is determined via electric conductivity measurements. Static light scattering (SLS) combined with computer calculations are used to characterize the structure; they also yield the salt concentration of the suspensions. The suspensions of latex spheres show, with decreasing salt concentration, a minimum in mobility. For lowering the salt content below 10^{-4} M an increase up to a factor of 3 is observed. The mobility of a dilute suspension of rodlike TMV is found to be about 5 $\frac{\mu m/s}{V/cm}$. Increasing the particle concentration leads to a decrease in mobility. In the limit of negligible salt concentration for a particle concentration below 0.1 $\frac{mg}{ml}$ the mobilities lie above 5 $\frac{\mu m/s}{V/cm}$. At higher concentrations they are smaller than this value. A qualitative comparison with the theory is made.

Key words: Latex spheres; tobacco mosaic virus; electrophoresis; light-scattering; ordered suspensions

Introduction

In recent years, polyelectrolyte suspensions have become popular model systems for studying phenomena such as diffusion, structural order, phase transition, etc. An appropriate method to investigate the properties of the double layer in detail, including adsorption and desorption processes, is to measure the velocity of the particles in an electric field, i.e., to determine the electrophoretic mobility.

The interpretation of mobility data yields the zeta-potential ζ at the plane of shear between the small ions fixed to the surface and the remaining diffuse part of the double layer moving freely. Theories for single spherical [1, 2] and rodlike [3] particles exist and relate the electrophoretic mobility to the zeta-potential as a function of κa, where κ is the Debye-Hückel screening parameter and a the particle radius. Unfortunately, structural or double layer overlapping effects are not included and these theories are only valid for relatively small ζ.

For spherical particles, Zukosky and Saville [4] use the standard theory of a dynamic model of a Stern layer to consider adsorption-desorption processes and transport within. For minimal ionic strength the particles carry their maximum charge. Increasing the salt concentration alters the charge by co- and counterion-adsorption. The mobility first decreases into a minimum at about 10^{-5} M salt concentration and then runs through a maximum to decrease at the end due to the compression of the diffuse double layer.

Experimentally, the maximum in mobility is well proved [5]. Data on lower salt content are rare, but the increase in mobility for vanishing salt concentration has been established [6].

The situation in the case of TMV as a rodlike suspended particle is more difficult. Nevertheless for low zeta-potentials ($\zeta < 100$ mV) the theoretical results differ only slightly from those of spherical particles [3].

A further feature of rods is the steric hindrance of rotational motion at concentrations above the overlap concentration c^*, defined as one rod per cubic particle length. We therefore expect an influence on the mobility and the viscosity of the suspension, but also, below c^* there might be an effect due to the long-range electrostatic interaction.

Interpretation of conductivity data

To discuss the mobility data in relation to the small ion concentration, we determine the salt concentration n_S from a measurement of the conductivity σ. Additionally, at low ionic strength when the polyions are liquid-like ordered n_S can be obtained by an independent experiment. Rescaled mean spherical approximation (RMSA) and/or Monte Carlo (MC) simulations can be performed to determine the salt concentration from measurements of the static structure factor $S(q)$ [6, 8].

From the salt concentration the Debye-Hückel parameter κ can be calculated by

$$\kappa^2 = \frac{e^2}{\varepsilon \cdot k_B \cdot T} \sum n_i \cdot Z_i^2 . \tag{1}$$

Here, ε is the dielectric constant of water, e the elementary charge, k_B the Boltzmann constant, and T the temperature. n_i denotes the number density of different small ions i with valency Z present in the suspension.

Assuming that all small ions outside the plane of shear move independently of each other and from the macroions, and taking the contribution of the macroions into account, we may approximately write [6]

$$\sigma = n_S \cdot e \cdot (\mu_{Na} + \mu_{Cl}) + n_P \cdot Z_{eff} \cdot e \cdot (\mu_H + \mu_P)$$

$$+ n_{OH} \cdot e \cdot \mu_{OH} . \tag{2}$$

Here, the mobilities of the small ions at infinite dilution μ_{Na}, μ_{Cl}, μ_H, μ_{OH} can be taken from the literature [7], while the mobility μ_P of the particles is known from electrophoresis. n_P is the particle

concentration and Z_{eff} is defined as the effective charge number transported in the electric field. In a first approach Z_{eff} is taken to be constant. For the case in which the sample is saltfree and the OH$^-$ concentration is small, i.e., no small ions except the H$^+$ counterions are present, Z_{eff} is given by

$$Z_{eff} = \frac{\sigma}{n_P \cdot e \cdot (\mu_H + \mu_P)} . \tag{3}$$

Using the dissociation product of water, the concentrations of OH$^-$ can be calculated from the H$^+$ concentration $n_H = Z_{eff} \cdot n_P$.

Experimental

For the measurements polystyrene latices with diameters $2a$ = 50 nm, 109 nm, and 233 nm (from Polysciences Inc., Warrington, USA, or Serva, Heidelberg, FRG) were used. The rodlike particles were tobacco mosaic viruses (TMV) with length l = 300 nm and radius a = 9 nm (kindly supplied by Prof. Wetter, Saarbrücken, FRG). The stock suspensions were diluted with doubly distilled water of conductivity σ = 0.05 μS/cm. In a closed circuit, including the electrophoretic and the conductivity cell, NaCl was added and the samples were filtered and continuously pumped through ion exchange resin (Serva) until the desired conductivity was reached. At the end of the procedure, when the conductivity became constant in time, the suspensions could be considered saltfree.

The electrophoretic mobility was measured with a conventional ELS apparatus (Zetasizer II, Malvern, England). Since TMV is a weak scatterer the original He-Ne laser (5 mW) was exchanged by a stronger one (35 mW). In all experiments the Doppler difference geometry was used. The crossing point of the beams was fixed on the stationary level of the electroosmotic flow profile. An alternating square electric field of magnitude up to E = ±25 V/cm was applied with a fixed frequency of 0.5 Hz. the conductivity was controlled in a WTW conductivity cell (WTW, Weilheim, FRG) with a Knick conductometer (Knick, Berlin, FRG). To determine the structure of the suspensions, a static light-scattering (SLS) apparatus (ALV, Langen, FRG) was used. All measurements were carried out at a temperature T = 21°C.

Results and discussion

In Fig. 1 the mobilities of aqueous suspensions of spheres of different diameters (Table 1) are displayed in a wide range of salt concentration as a function of the electric conductivity. The error bars were determined for all data points within 10%. At high salt concentration the mobilities of the spheres with diameters 109 nm and 250 nm are both found to be about 5 (μm/s)/(V/cm). For decreasing salt concentration the mobilities of these spheres first decrease and then show a minimum. Further lowering the conductivity of the suspension leads to a significant increase of the mobility of the 109 nm particles and a weak increase of mobility of the 250 nm particles. The spheres with 50 nm diameter show only an increase with decreasing conductivity. The results for the 100 nm spheres, which we published elsewhere [6], show the same behavior as that for 109 nm, but a stronger increase at minimal conductivity. No size dependence of mobility is found.

Table 1. Latex suspensions

Symbol	Diameter (nm)	Volume fraction (%)	Concentration n_p
□	50	0.018	$2.75 \cdot 10^{12}$
●	100	0.025	$4.77 \cdot 10^{11}$
○	109	0.025	$3.69 \cdot 10^{11}$
■	250	0.005	$6.11 \cdot 10^{9}$

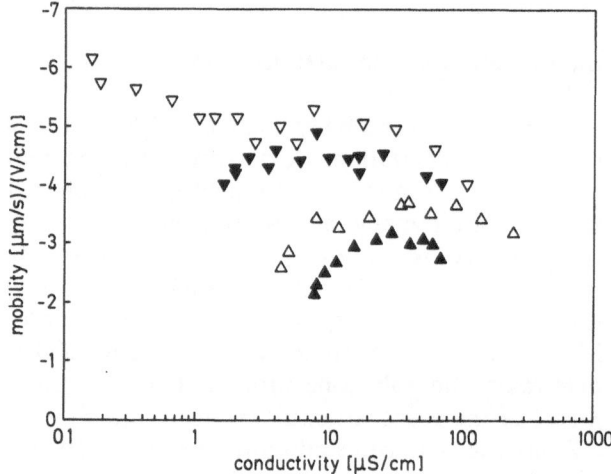

Fig. 2. Electrophoretic mobilities of aqueous suspensions of TMV versus electric conductivity. Symbols as in Table 2

Table 2. TMV suspensions

Symbol	Concentration n_W (mg/ml)	Concentration n_p (ml^{-1})
▽	0.037	$5.59 \cdot 10^{11}$
▼	1.1	$1.66 \cdot 10^{13}$
△	1.96	$2.96 \cdot 10^{13}$
▲	4.72	$7.13 \cdot 10^{13}$

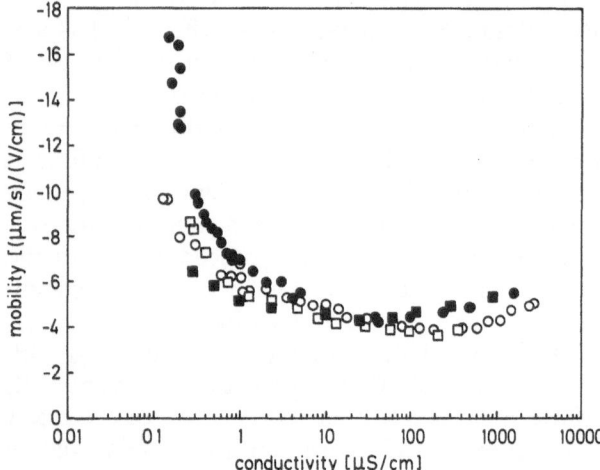

Fig. 1. Electrophoretic mobilities of aqueous suspensions of latex spheres versus electric conductivity. Symbols as in Table 1

Figure 2 shows the mobility of TMV versus the electric conductivity. For very dilute TMV particles the mobility is nearly 5 (μm/s)/(V/cm), with a weak rise at small conductivities (see Table 2). The behavior is qualitatively similar to that of the spheres. Increasing the particle concentration leads to a decrease in the mobility plateau, and for low salt content the mobility does not rise, but rather, it falls.

In Fig. 3 the values of Fig. 1 and Fig. 2 are plotted versus κa. The lines are from theory [1, 2] and show the relation between the ζ-potential (25 mV to 150 mV in 25 mV steps) and the mobility as a function of κa. Qualitatively, the spheres show the

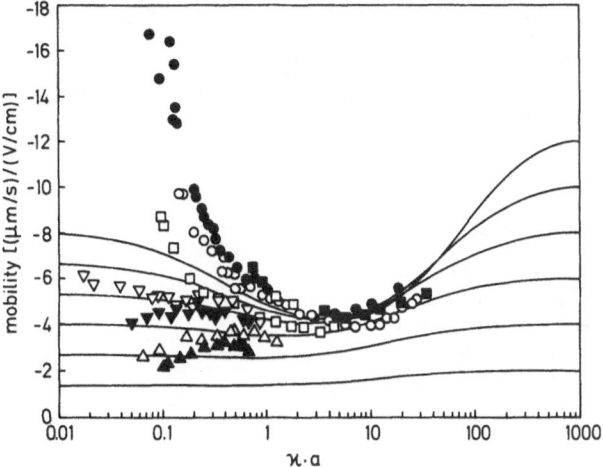

Fig. 3. Electrophoretic mobilities of aqueous suspensions of latex spheres and TMV versus κa. Lines are from theory [1, 2] for 25 mV to 150 mV steps. Symbols as in Table 1 and 2

predicted behavior [4]: The mobilities of the 100 nm, 109 nm, and 250 nm particles run with decreasing κa through a minimum and have their highest values at minimal κa. This minimum and the theoretical minimum caused by relaxation are found at the same κa-values. For most of the measured points a clear association of a mobility to a ζ-value is not possible. Several lines of the theory run through the minimum of the measured values and the high mobilities are too large compared to the theoretical values. The spheres with 50 nm diameter show, with decreasing κa, the increase to maximum mobility. Only for these spheres can an increase of the ζ-potential with decreasing κa be verified.

The mobilities of the TMV rods lie within the validity range of the theory, which, however, can only explain the behavior of the very diluted sample. Here, interaction between the rods can be excluded and the weak rise in mobility with decreasing κa is caused by counterion desorption. With raising particle concentration the interaction between the particles appears. The decrease in the mobility plateaus for increasing TMV concentration can be explained by the steric hindrance of the rodlike particles. This effect is strengthened by the Coulomb interaction, and is equivalent to an increase of viscosity. Thus, the curves of the higher concentrations additionally fall with decreasing κa.

To clarify this point, further viscosity measurements are in preparation.

In the last figure (Fig. 4) the mobility of TMV is displayed in a double logarithmic plot as a function of particle concentration. The mobility values of the suspensions without interaction are taken from the above discussed plateaus. In the limit of negligible particle concentration the mobility converges to about 5 $(\mu m/s)/(V/cm)$. Raising the particle concentration leads to a decrease in mobility at c^*, where the particle ends begin to overlap. The mobilities of the saltfree suspensions are measured at minimal κa. For a particle concentration below about 0.1 mg/ml these mobilities lie above 5 $(\mu m/s)/(V/cm)$,

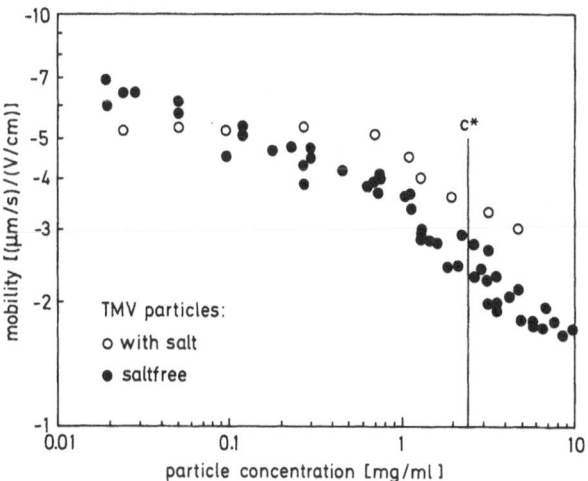

Fig. 4. Electrophoretic mobilities of aqueous suspensions of TMV versus particle concentration n_p

i.e., the values measured for the suspensions without interaction. Above this particle concentration the mobility values are smaller than those of the non-interacting suspensions. In the first case of negligible particle concentration charging mechanisms in the saltfree suspensions lead to an increase of mobility with decreasing κa. In the second case, the rise in viscosity due to the steric and electrostatic interaction of the particles predominates the charging mechanisms. This leads to a decrease of mobility with decreasing κa.

Acknowledgement

Financial support of the Deutsche Forschungsgemeinschaft (SFB 306) is gratefully acknowledged.

References

1. Wiersema PH, Loeb AL, Overbeek JThG (1966) J Colloid Interface Sci 22:78—99
2. O'Brien RW, White LR (1978) J Chem Soc Faraday Trans 2 47:1607—1626
3. Stigter DJ (1976) Phys Chem 82:1417—1423
4. Zukosky CF IV, Saville DA (1986) J Colloid Interface Sci 114:32—44, 45—53
5. Elimelech M, O'Melia CR (1990) Colloids and Surfaces 44:165—178
6. Deggelmann M, Palberg T, Hagenbüchle M, Maier EE, Krause R, Graf C, Weber R (1991) J Colloid Interface Sci 143:318—326
7. Flügge S (1955) In: Handbuch der Physik. Springer, Berlin
8. Maier EE, Fraden S, Deggelmann H, Hagenbüchle M, Krause R, Weber R (1991) Macromolecules 25:1125—1133

Authors' address:

Martin Deggelmann
Universität Konstanz
Fakultät für Physik
Postfach 5560
7750 Konstanz, Germany

Progress in Colloid & Polymer Science

Progr Colloid Polym Sci 89:145—148 (1992)

Reflectometry study of interbubble gas transfer in liquid foams

G. Marion, S. Sahnoun, B. Mendiboure, C.Dicharry, and J. Lachaise

L.T.E.M.P.M., Centre Universitaire de Recherche Scientifique, Pau, France

Abstract: We study interbubble gas transfer in liquid foams by measuring the relative decrease of the bubble area from reflectometry. Foams are obtained by independently fixing foaming solution and gas flows through a coarse porous structure. Precise measurements of these flows instantaneously give the gas volume fractions of the foams; real time diffractometry gives their initial bubble size distributions, which we found to be lognormal. — We calculate interbubble gas transfer within such foams on the basis of a theory developed by Lemlich for other bubble size distributions. For a given bubble mean size, we find that the experimentally measured width of the bubble size distribution provides the fastest transfer; then the calculated transfer is close to the one which corresponds to the Lemlich distribution. — Comparison between theoretical and experimental relative decreases of the bubble total area allows to obtain the initial effective permeability of the interbubble medium to gas transfer. We show that this permeability can be drastically lowered by associating an anionic surfactant with an amphoteric one in the foaming solution.

Key words: Foam; gas diffusion; anionic surfactant; amphoteric surfactant; lognormal distribution

Introduction

Four phenomena contribute to the degradation of a liquid foam: evaporation, interbubble gas transfer, drainage and breaking of its liquid films. Interbubble gas transfer can only be studied under good conditions if the three other phenomena are minimized, delayed or stopped. Evaporation of the liquid can be stopped by putting the foam in a closed vessel. Breaking of the liquid can be greatly delayed by using judicious associations of surfactants. The influence of the drainage can be minimized by studying the upper layers of the foam where it principally acts during the first instants of degradation. In a first approximation, we shall neglect this perturbation.

Interbubble gas transfer produces the decrease of the relative interfacial area of the bubbles. We have already shown that this decrease can be measured by reflectometry [1,2].

Recently, we have shown that reproducible foams can be obtained by independently fixing foaming solution and gas flows through a coarse porous structure and that the characterization of their initial structure can be obtained by real-time diffractometry [3]. The initial bubble size distributions of the foams that we obtained were found to be lognormal, which can be attributed to the random nature of the formation of the bubbles.

The theory of interbubble gas transfer has been developed by Clark and Blackman [4] and De Vries [5]; it has been recently improved by New [6] and Lemlich et al [7,8]. They considered distributions used up to date in foam technology as De Vries distribution, Bayens distribution, and Lemlich distribution. But they did not perform calculations on lognormal distributions.

In this paper, we extend the theory of interbubble gas transfer to foams' bubbles which are lognormally distributed. We use this extension to first calculate the influence of the width of the distribution on the speed of gas transfer, then to obtain an estimation of the initial effective permeability of the interbubble medium to gas transfer. The variation of this last parameter is studied as a function of the composition of an aqueous anionic/amphoteric surfactant mixture constituting the foaming solution; we discuss the contributions of volume and surface resistances to gas transfer as a function of the composition of the foaming mixture.

Interbubble gas transfer theory

The gas transfer mechanism comes from the bubble-size polydispersity. Consequently, the slightly higher pressure in the smaller bubbles forces gas to diffuse through the interbubble medium into the larger bubbles. Thus, the smaller bubbles shrink and the larger bubbles grow. This changes the distribution of bubble sizes and, eventually, the bubble number if the smaller bubbles shrink to the point of disappearance.

Instead of viewing the gas as diffusing directly from bubble to bubble, Lemlich [7] has suggested that it was more realistic to view the gas as first diffusing into the liquid region midway between the bubbles. The concentration of gas in this liquid can be considered as being equivalent (through Henry's law) to a gas pressure in the liquid. Then, by virtue of the classical law of Laplace and Young, this equivalent gas pressure can be viewed as that which would exist within a fictitious spherical bubble of radius ρ. Thus, the pressure difference between a bubble of radius r and the liquid is

$$\Delta P = 2\gamma \left(\frac{1}{\rho} - \frac{1}{r} \right) , \tag{1}$$

where γ is the interfacial tension. Of course, ΔP may be positive or negative.

The mass transfer rate:

$$\frac{dm}{dt} = JA\Delta P \tag{2}$$

is proportional to the interfacial area A, through which the transfer takes place, to pressure difference ΔP, and to the effective permeability J of the interbubble medium to gas transfer.

For a unit surface area, the effective permeability can be viewed as the reciprocal of the sum of two resistances, namely, the surface resistance $1/h$ of the membranes of the two bubbles, and the liquid resistance He/D. In this last expression, H is Henry's constant, D is the diffusion constant of the gas in the liquid, and e is the mean average liquid thickness between adjacent bubbles. Thus,

$$\frac{1}{J} = \frac{1}{h} + \frac{He}{D} . \tag{3}$$

Assuming conservation of gaseous moles throughout the foam, an average effective $\rho = r_{12}$, and considering the gas as perfect, it can be shown [7] that the variation with time of the size of a bubble of radius r is

$$\frac{dr}{dt} = \frac{2J\gamma \bar{R}T}{P_a} \left(\frac{1}{r_{12}} - \frac{1}{r} \right) , \tag{4}$$

where,

$$r_{12} = \int_0^\infty r^2 F(r,t)dr \Big/ \int_0^\infty rF(r,t)dr . \tag{5}$$

In these expressions, $\bar{R}$ is the ratio between the perfect gas constant and the gas molar mass, T is the absolute temperature, P_a the atmospheric pressure, and $F(r,t)$ the frequency distribution of r at time t.

We have used Eq. (4) to predict the changes in bubble size, assuming that the initial radius distribution is lognormal:

$$F(r,0) = \frac{1}{\sqrt{2\pi}} \frac{1}{\sigma_0} \frac{1}{r/r_0}$$

$$\cdot \exp \left[-\frac{1}{2} \left(\frac{\ln r/r_0}{\sigma_0} \right)^2 \right] , \tag{6}$$

where r_0 is the mean radius of the distribution and σ_0 its standard deviation.

Thus, if the gas volume fraction of the foam is φ, a good estimation of the initial value of e is

$$e = \frac{1}{3} r_0 \frac{1-\varphi}{\varphi} \exp \left(\frac{5}{4} \sigma_0 \right) . \tag{7}$$

Computations were performed from the radius evolution equation and from the radius distribution which was truncated and discretized; the volume was adjusted at every time step. Calculations were accelerated by using a method close to the fourth-order Runge-Kutta method.

Values of the effective permeability were actualized at each time step from the values of e modified by the changes in the radius distribution. For a given experiment, the $1/h$ value was determined after a series of tries on the initial ratio between the calculable volume resistance and the unknown surface resistance to gas transfer. The retained ratio is that which gives the better agreement between experiment and theory. When the volume resistance is negligible compared to the surface resistance, the effective permeability is constant for the whole range of the foam evolution.

Results and discussion

The simulated variations of the radius distribution of a foam, the bubbles of which are initially distributed according to a lognormal distribution, are drawn in Fig. 1. Calculations have been made with $\varphi = 0.83$, $r_0 = 25$ µm and $\sigma_0 = 0.5$, the standard deviation that we have practically always encountered for the foams that we produced by independently fixing foaming solution and gas flows through a coarse porous structure [3]. These variations have the same appearance as those reported in literature for the other distributions: when the parameter $\theta = \dfrac{2J\gamma\bar{R}T}{P_a r_0^2} t$ increases, mean radius is shifted towards higher values, width is increased, bubble total number is decreased.

Differences are more noticeable in Fig. 2 where are reported the relative variations with θ of the total surface of the bubbles. De Vries and Bayens distributions give faster variations than lognormal distribution, while differences between Lemlich and lognormal distributions are very small.

The study of the speed of the evolution in function of the standard deviation of the lognormal distribution shows that the fastest evolution is obtained when standard deviation is equal to 0.5, the common value of all the experimental distributions. This behavior could reflect the extreme disequilibrium of the initial pressure distribution in the foam, probably induced by the random nature of bubble formation.

Nitrogen foams were generated with 7.5 g/l aqueous foaming solutions mixing an anionic surfactant (Melanol LPI 45: monoisopropanolamine laurylethersulfate) with an amphoteric one (Amonyl 265 BA: cocobetaïne, sodium salt), both produced by the Société Seppic. We have studied the behavior of the foams in function of the proportions of the two surfactants in the mixture. The nitrogen volume fraction of the foams was fixed equal to 0.83. The initial radius distributions of the foams were carefully measured from real time diffractometry and the relative variations with time of

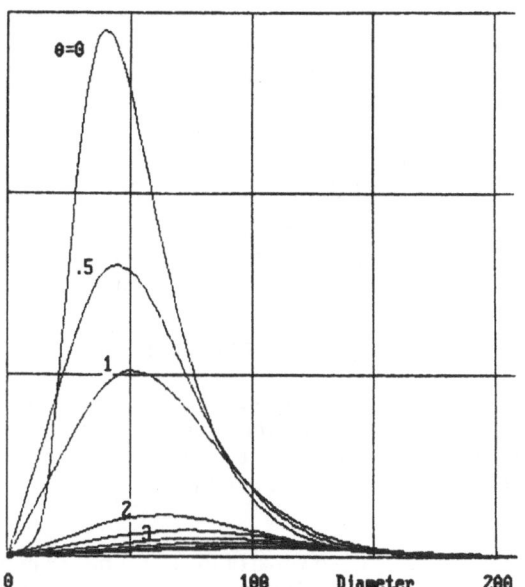

Fig. 1. Evolution of bubble size distribution induced by gas transfer within a foam ($\varphi = 0.83$), the bubble sizes of which were lognormally distributed ($r_0 = 25$ µm, $\sigma_0 = 0.5$)

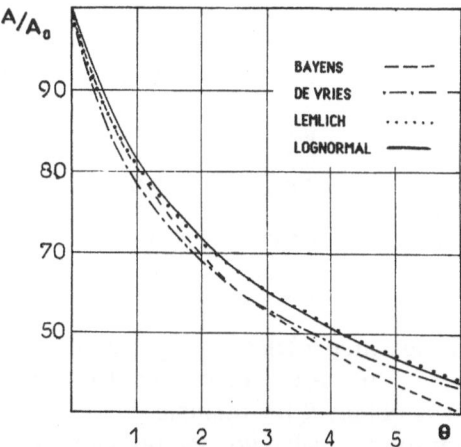

Fig. 2. Comparison between the relative decreases of bubble area for various bubble size distributions ($\varphi = 0.83$, $r_0 = 25$ µm). The standard deviation of the initial lognormal distribution is $\sigma_0 = 0.5$

the areas of their bubbles were determined from reflectometry.

We give in Fig. 3 the principle of the graphic determination of the initial effective permeability. After elimination of the area between calculated

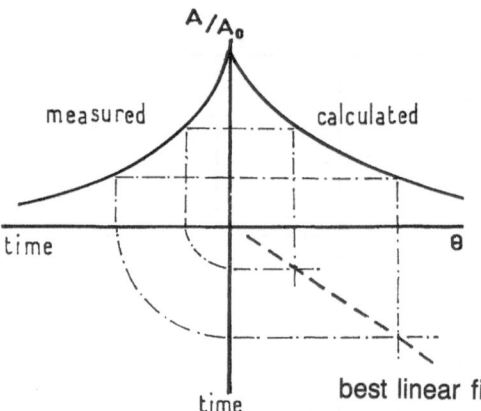

Fig. 3. Linearization of the calculated and measured bubble areas in a $t = f^{-1}(A/A_0)$ versus $\theta = f^{-1}(A/A_0)$ plot. The slope of the straight line is connected to the permeability J and to external parameters

Table 1. Variation of the inverse of the initial permeability, volume resistance (He/D) and surface resistance ($1/h$) to nitrogen transfer as a function of the mixture composition. Surface to total resistance ratio is given in the last column. Resistances are given in $10^{10}[\text{kg m}^{-2} \text{ s}^{-1} \text{ Pa}^{-1}]^{-1}$. Calculations were performed with $H/D = 2.4 \times 10^{+15}$ MKS

% Melanol	1/J	He/D	1/h	J/h
0	1.12	0.704	0.41	37%
12.5	4.12	0.713	3.40	82%
25.0	47.4	0.756	46.6	98%
50.0	36.4	0.798	35.6	98%
75.0	16.3	0.822	15.5	98%
80.0	4.31	0.752	3.56	82%
100	1.37	0.760	0.61	44%

$A(\theta)/A_0$ and measured $A(t)/A_0$, we get a line in a time versus θ plot. After successive tries, this linearization is obtained for a single value of the ratio between the calculable volume resistance and the unknown surface resistance to nitrogen transfer. The slope of the obtained straight line allows to deduce the initial value of the effective permeability; the value of the aforesaid ratio allows to estimate the surface resistance to nitrogen transfer.

For each of the studied foams, we report in Table 1 the inverse of the initial effective permeability, the volume resistance and the surface resistance to nitrogen transfer, and the ratio of the surface resistance to the total resistance. We see that the surface resistance to nitrogen transfer is greatly

favored by the mixture of the two surfactants. This effect could be attributed to the dipole-dipole interactions between the two surfactants.

Conclusions

We have calculated interbubble gas transfer within foams' sizes which are lognormally distributed. We have found that this transfer is fastest when the width of the distribution coincides with the experimental width, which expresses the high degree of disequilibrium of the generated foams.

Considering contributions of volume and surface resistances to gas transfer, we have indicated how the initial effective permeability of the interbubble medium can be deduced from comparison between theoretical and experimental relative decreases of the bubble total area with time. We have shown that the association of two surfactants, judiciously choosen to present dipole-dipole interactions, reinforces the surface resistance to nitrogen transfer and, in this way, delay foam degradation.

Acknowledgement

The authors are indebted to the Société Seppic for the gift of melanol and amonyl samples.

References

1. Lachaise J, Graciaa A, Marion G (1988) Second European Colloid and Interface Society Conference, Arcachon, France
2. Lachaise J, Graciaa A, Marion G, Salager JL (1990) J Dispersion Sci Tech 11:409—432
3. Lachaise J, Sahnoun S, Dicharry C, Mendiboure B, Salager JL (1991) Progr Colloid Polymer Sci 84:253—256
4. Clark NO, Blackman M (1948) Trans Farad Soc 44:1—7
5. De Vries AJ (1957) Foam Stability, Rubber-Stichling, Delft
6. New GE (1967) Proc Int Congress Surface Active Substances 1964, 2:1167
7. Lemlich R (1978) Ind & Eng Chemistry Fundam 17:89—93
8. Ranadive A, Lemlich R (1979) J Colloid Interf Sci 70:392

Author's address:

J. Lachaise
L.T.E.M.P.M.
Centre Universitaire de Recherche Scientifique
Avenue de l'Université
64000 Pau, France

Progress in Colloid & Polymer Science   Progr Colloid Polym Sci 89:149—155 (1992)

Perfluoroalkyl bilayer membranes prepared from saturated amphiphiles with fluorocarbon chains

S. Szönyi[1]), H. J. Watzke[2]), and A. Cambon[1])

[1]) Laboratoire de Chimie Organique du Fluor, Université de Nice-Sophia Antipolis, France
[2]) Institut für Polymere, ETH Zentrum, Zürich, Switzerland

Abstract: Ammonium-type F-alkylated surfactants characterized by the presence of two tails R_F/R_F' or R_F/R_F' have been prepared, and the vesicular behavior of some double-chain surfactants has been studied in ultrasound-treated aqueous dispersions. Stable vesicular aggregates were found which showed unilamellar membrane structures. Their shapes, hydrodynamic radii and size distribution were determined by dynamic light scattering and electron microscopy. The phase transition temperatures of aqueous dispersions of double-chain surfactants were measured by differential scanning calorimetry.

Key words: F-alkylated surfactants; vesicles; dynamic light scattering; electron microscopy; differential scanning calorimetry

Introduction

Vesicle formation can be considered as a general physicochemical phenomenon observed for a large variety of amphiphiles [1—4]. Liposomes from synthetic amphiphiles are of interest as models for biological membranes [5]; in addition, the introduction of long perfluoroalkyl chains in the hydrophobic portion of amphiphiles molecules leads to more stable model membranes [6]. Thus bilayer membranes of fluorocarbon amphiphiles show interesting characteristics that combine organized assemblage of the bilayer membrane and peculiar physicochemical properties of fluorocarbon compounds.

Nevertheless, recent investigations in the field of fluorocarbon vesicles are limited to small number of perfluoroalkyl surfactants, all derived from F-alkylated alcohols such as R_FCH_2OH and $R_FC_2H_4OH$, or from F-alkylated acids such as R_FCH_2COOH and $R_FC_2H_4COOH$ [7—9]. In the present paper, we describe the synthesis of a new class of vesicle-forming amphiphiles containing fluorocarbon chains and their aggregation behavior.

We prepared four series of ammonium amphiphiles. The molecular structures for each series,

A, B, C, and D, are summarized in Table 1. The results established that vesicles formed from amphiphiles of series A, B, C, and D.

Results and discussion

Synthesis

The different amphiphiles of series A, B, C, and D are obtained with good yields (70—90%). They are prepared from dimethyl fatty amines or tertiary amines deriving from the ring opening of F-alkyl oxiranes or from di-F-alkyl thiourea, according to the reaction sequences outlined in Schemes 1 and 2.

In series A the double-chain amphiphiles possess one fluorocarbon chain and one hydrocarbon chain.

In series B, C, and D the bitailed surfactants contain two fluorocarbon chains and an asymmetric carbon-bearing one. The presence of this center of chirality in the hydrophobic region of the vesicles might provide asymmetric recognition sites for various chiral guest molecules.

In series D, there is, in addition, a thiourea connector combining the double tail with the spacer portion, which is a simple methylene chain.

Table 1. Structure of amphiphiles containing fluorocarbon chains

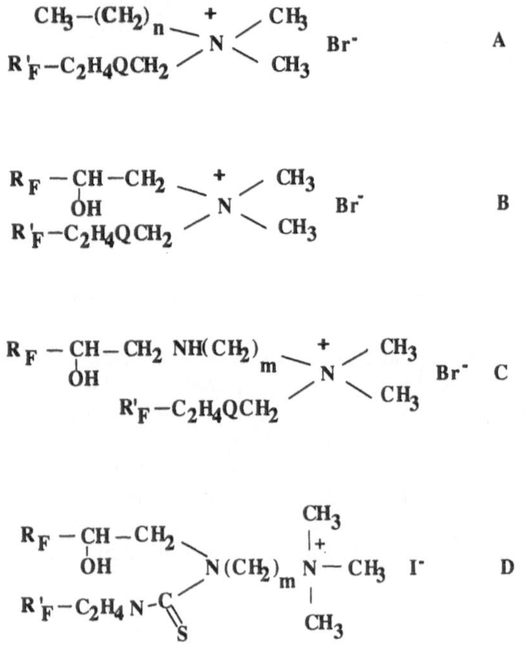

$$CH_3-(CH_2)_n \qquad CH_3$$
$$\qquad \qquad \overset{+}{N} \qquad \qquad Br^- \qquad \qquad A$$
$$R'_F-C_2H_4QCH_2 \qquad CH_3$$

$$R_F-CH-CH_2 \qquad CH_3$$
$$\quad | \qquad \qquad \overset{+}{N} \qquad \qquad Br^- \qquad \qquad B$$
$$OH$$
$$R'_F-C_2H_4QCH_2 \qquad CH_3$$

$$R_F-CH-CH_2\,NH(CH_2)_m \qquad CH_3$$
$$\quad | \qquad \qquad \qquad \overset{+}{N} \qquad \qquad Br^- \qquad C$$
$$OH$$
$$\qquad R'_F-C_2H_4QCH_2 \qquad CH_3$$

$$\qquad \qquad \qquad \qquad CH_3$$
$$R_F-CH-CH_2 \qquad \qquad |+$$
$$\quad | \qquad \qquad N(CH_2)_m\,N-CH_3 \quad I^- \qquad D$$
$$OH \qquad \qquad \qquad \quad |$$
$$R'_F-C_2H_4\,N-\overset{C}{\underset{S}{\|}} \qquad CH_3$$

$R_F, R'_F = C_4F_9, C_6F_{13}, C_8F_{17}$; $Q = S, SCO, OCO, NHCO$;
$n = 6,7,8,9,11$; $m = 2,3$.

Scheme 1

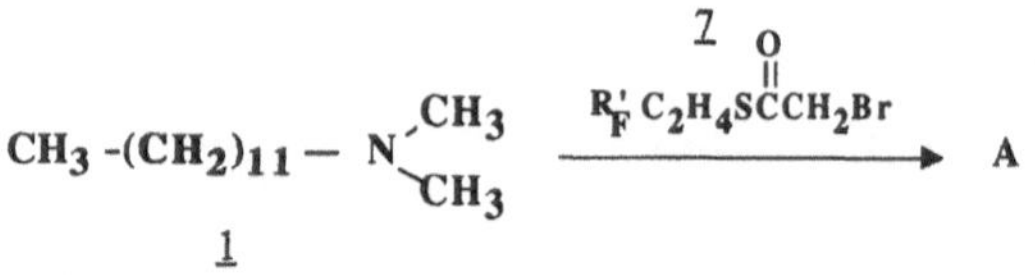

$$CH_3-(CH_2)_{11}-N\overset{CH_3}{\underset{CH_3}{\big<}} \quad \xrightarrow{\overset{\underset{\textstyle 7}{\quad} \;\; O}{R'_F C_2H_4 S\overset{\|}{C}CH_2Br}} \quad A$$
$$\underset{1}{}$$

Scheme 2

$$R_FCH-CH_2 \xrightarrow{(CH_3)_2NH} R_FCH-CH_2N\overset{CH_3}{\underset{CH_3}{\big<}} \xrightarrow{\overset{8}{R'_F C_2H_4SCH_2Br}} B$$
$$\underset{O}{\underset{2}{\diagdown \diagup}} \qquad \qquad OH \qquad \underset{3}{}$$

$$\downarrow {(CH_3)_2N(CH_2)_3NH_2}$$

$$R_FCH-CH_2\,NH(CH_2)_3\,N\overset{CH_3}{\underset{CH_3}{\big<}} \xrightarrow{\overset{\underset{\textstyle 7}{\quad}\;\; O}{R'_F C_2H_4 S\overset{\|}{C}CH_2Br}} C$$
$$\quad |$$
$$OH \qquad \qquad \underset{4}{}$$

$$\downarrow {R'_F C_2H_4NCS \; \underline{5}}$$

$$R_FCH-CH_2$$
$$\quad | \qquad \qquad N(CH_2)_3N\overset{CH_3}{\underset{CH_3}{\big<}} \xrightarrow{CH_3I} D$$
$$OH$$
$$R'_F\,C_2H_4N-\overset{C}{\underset{S}{\diagdown}}$$
$$\qquad \qquad \underset{6}{}$$

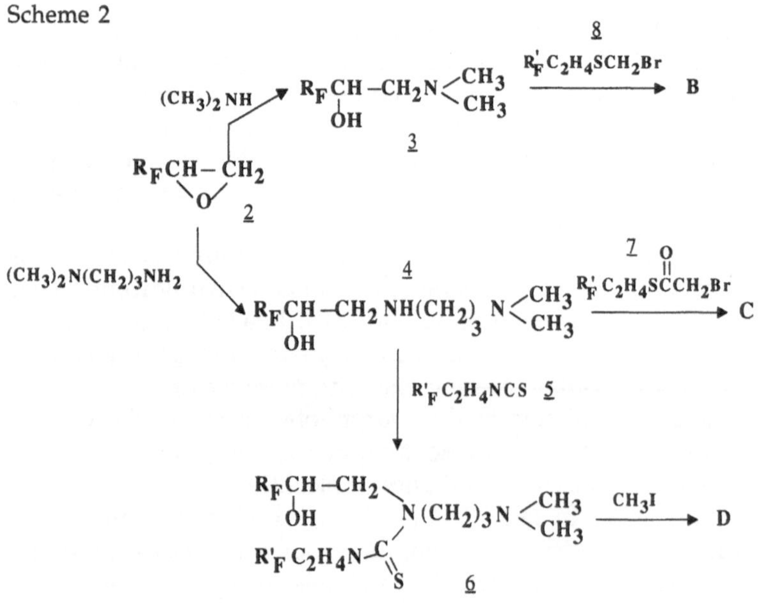

In all series, the tail and the spacer lengths can be varied independently. For example, it is possible to study changes in the rigidity of chains according to the asymmetry of the alkyl or perfluoroalkyl chains.

Aggregation behavior

Vesicle formation

We studied one compound of each series for their ability to form vesicular aggregates in aqueous solution. After dispersion in water under vigorously vortexing, compounds A4 ($R'_F = C_6F_{13}$, $Q = SCO$, $n = 11$), B1 ($R_F = C_8F_{17}$, $R'_F = C_6F_{13}$, $Q = S$), C3 ($R_F = R'_F = C_6F_{13}$, $Q = SCO$, $m = 3$) and D8 ($R_F = C_8F_{17}$, $R'_F = C_6F_{13}$, $m = 3$) were found to form liposomes upon sonication at 70°C during 30 min (compounds A4, B1, and C3), or at 40°C during 15 min (compound D8) (see experimental section). The sonicated solutions obtained are usually optically clear or translucent.

Electromicroscopy

The structure of molecular assemblies was elucidated by negative staining and freeze fracture electronmicroscopy (see Experimental section).

Figure 1a shows the vesicular aggregates of compound A4 in negative staining electronmicroscopy. Polydispersed unilamellar vesicles can be seen with spherical shapes. Their size distribution after the negative staining electronmicrographs is pictured in Fig. 2. It seems that compound A4 produces a

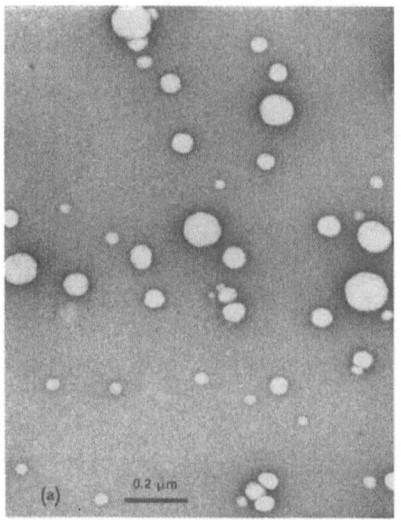

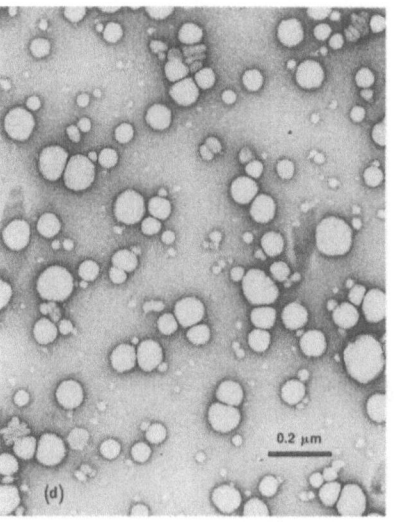

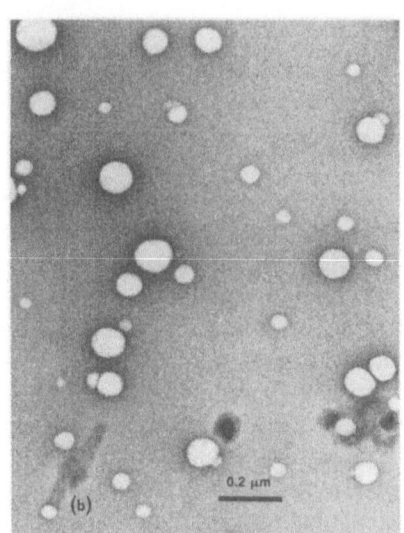

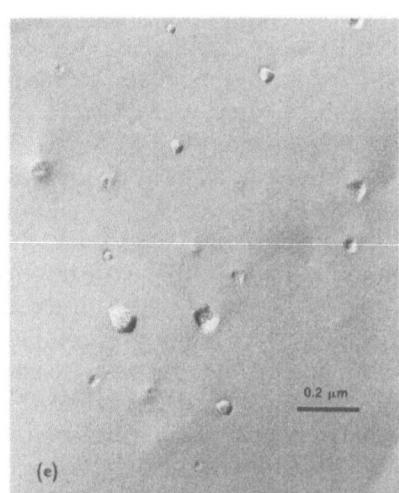

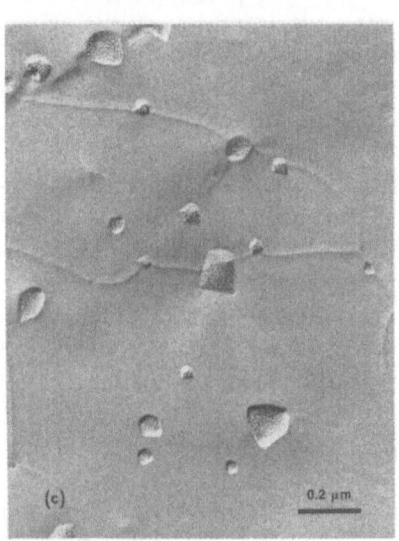

Fig. 1. Electron micrographs of vesicles; magnification ×
100000: a) amphiphile A4, negative staining; b) amphiphile A4 after 3.5 months, negative staining; c) amphiphile B1, freeze fracture; d) amphiphile C3, negative
staining + Cyt. C; e) amphiphile D8, freeze fracture

bimodal distribution of vesicles sizes with maxima
around 40 nm and 110 nm. The average diameter
appears to be ≈79.0 nm. To evaluate the stability of
the vesicular aggregate, vesicular samples of compound A4 were reinvestigated after 3.5 months by
negative staining electronmicroscopy. Figure 1b
again shows vesicular aggregates, but now the
larger vesicles have grown at the expense of the
small population. The preparation did not change
significantly in sizes or shapes, according to the

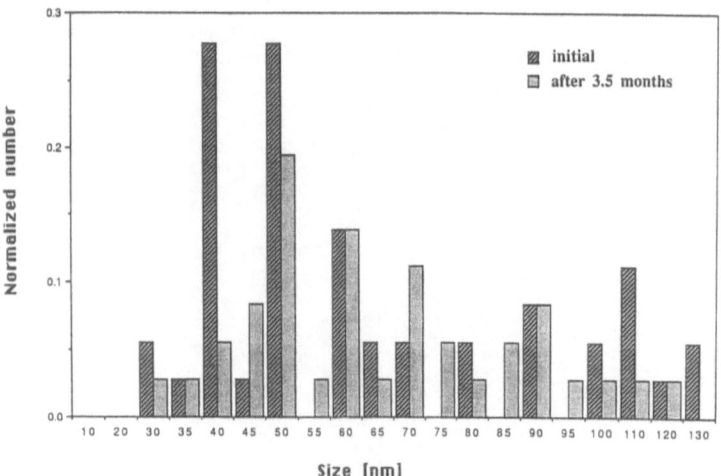

Fig. 2. Size distribution in negative staining (amphiphile A4)

freeze fracture electronmicrographs. The vesicles appear unilamellar and have a smooth surface, as in the case of normal phospholipid vesicles.

Negative staining technique applied to sonicated sample of compound B1 reveals a large polydispersed population of single-walled vesicles containing an inner water volume. Figure 1c shows the vesicular aggregates of compound B1 according to freeze fracture technique. The vesicles have unusual non-spherical shapes which are rather polygon-like with sharp edges. The polygon-like shapes are seen more frequently in the larger particles, while the smaller one resemble spherical shapes. Their size distribution after freeze fracture electronmicrographs is broad, with a maximum around 50.0 nm. The average diameter appears to be ≈ 65.0 nm. The diameter of non-spherical particles is represented by their large axis.

Figure 1d shows the closed-shell aggregates from compound C3 evidenced by negative staining electronmicroscopy. The preparation appears polydispersed, but spherical vesicles can be clearly seen. Because of an interaction between the positively charged surfactant and the negatively charged carbon films on the electronmicroscopy grids, the latter were unwettable by the uranyl acetate staining solution. Deposition of positively charge cytochrome C protein onto the carbon films suppressed this interaction, and the sample could be stained in a normal way. A fairly broad distribution can be observed from negative staining electronmicrographs with a broad maximum around 50—60 nm, which is also the average diameter size.

The vesicular feature is confirmed quite clearly from freeze fracture electronmicrographs. As in the case of compound B1, the larger vesicles resemble non-spherical shapes.

Figure 1e shows small aggregates of spherical and non-spherical shapes from compound D8 obtained with the freeze fracture technique. The surface features of the aggregates are the same as in the electronmicrographs of compounds B1 and C3. The size distribution of the diameters showing a maximum around 25.0 nm is pictured in Fig. 3. The distribution resembles rather unsymmetric form distribution of liposomes or vesicles, with a peaking maximum, combined with a tail, common to the larger vesicle sizes; the average diameter is around 50.0 nm. The extent of polydispersity can be seen in the negative staining electronmicrographs which show very large aggregates, but also unusual tubular structures. But, we cannot be sure that these structures are not artifacts of the negative staining procedure.

In contrast with compound A4, for all the compounds B1, C3, and D8 containing double fluorocarbon chains, freeze-fracture electronmicrographs show a remarkable decoration effect due to the different wetting behavior of platinum against vitrified water (smooth background) and the vesicle surface (leopard-like spots).

Dynamic light-scattering (DLS)

Table 2 collects all the DLS measurements on vesicular solutions prepared by sonication of

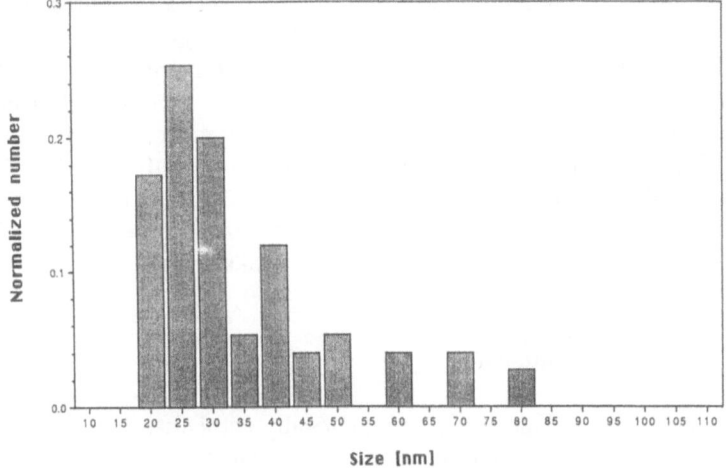

Fig. 3. Size distribution in freeze fracture (amphiphile D8)

aqueous dispersions of compounds A4, B1, C3, and D8 (see Experimental section). The DLS results confirm the polydispersity of vesicular solutions and the larger vesicles contribute more strongly to the scattered light intensity. As a result, the average diameter determined from the electronmicrographs does not always coincide with the average diameter measured with the DLS technique.

Table 2. Dynamic light scattering data

Amphiphile	Scattering angle [Degree]	Sample tin [μs]	Radii [nm]	Polydispersity
A4	90	5.0	~36.0	0.15
B1	90	5.0	~43.4	0.13
C3	90	6.0	~59.3	0.12
	60	12.0	~70.0	0.15
	45	20.0	~78.4	0.18
D8	90	3.2	~29.5	0.15
	60	6.4	~31.3	0.16
	45	11.0	~32.7	0.20

Differential Scanning Calorimetry (DSC)

Aqueous bilayer membranes usually show the crystal-to-liquid crystal phase transition. This phenomenon is readily observed by DSC. The phase transition parameters (enthalpy change ΔH, temperature at peak maximum Tc) of aqueous dispersions of double-chain amphiphiles A4, B1, C3, and D8 were measured by DSC as summarized in Table 3 (see Experimental section).

Table 3. Differential scanning calorimetry data: Phase transition temperature T_c of aqueous dispersions of double-chain amphiphiles bearing pure and mixed perfluorinated chains

Amphiphile[a]	T_c/°C heating	T_c/°C cooling	ΔH/kJ/mol heating	ΔH/kJ/mol cooling
A4	48.1[b]	—	0.15	—
B1	75.1	66.7	6.40	6.91
C3	69.5	49.7	16.55	16.40
D8	51.7[c]	27.6	2.23	1.16
	75.7[d]	—	1.88	—
	91.2[e]	80.9	14.48	11.21

[a] surfactant concentration was 10% w/w;
[b] a peak appears first at 43.5°C; after several runs this very weak and broad peak appears only in the heating mode;
[c] peak is very broad before equilibration;
[d] appears only at first heating (unequilibrated sample);
[e] reproducible peak difficult to assign.

Experimental section

Materials

N,N dimethyl undecylamine *1* was purchased from Fluka and used without further purification. Amphiphiles A4, B1, C3, and D8 were prepared by the reaction sequences outlined in schemes 1 and 2, respectively.

The synthetic procedures of intermediates 2 to 8 are reported elsewhere [10—14]. The structures of intermediates and final products were confirmed by elemental analysis, IR, ^{1}H NMR, ^{19}F NMR, and mass spectroscopies. The spectral data are in line with the proposed structures.

Preparation of amphiphiles A1

Stoichiometric amounts of N,N dimethyl undecylamine 2 and intermediate 7 are dissolved in 50 ml of diethyl ether, and the mixture was stirred for 8 h and refluxed for 2 h. The diethyl ether is evaporated off in vacuo to give white powders or waxy solids which are washed two or three times with 30 ml of diethyl ether, dried in vacuo, and recrystallized in ethyl acetate; m.p. 167°C.

Preparation of amphiphile D8

The intermediate 6 (4 mmol) and methyl iodide (20 mmol) are dissolved in chloroform (10 ml). The mixture is stirred vigorously and refluxed for 24 h. The precipitate is filtered, washed with chloroform, and dried in vacuo at room temperature; m.p. 217°C.

Methods

Generation of vesicles: 30 mg of amphiphiles was dispersed by vortexing in 10 ml of deionized and distilled water, and sonicated using a Bandelin Sonorex RK 100 H sonifier.

Electronmicroscopy: For negative staining procedure, the vesicular solutions obtained are applied to carbon grids which are prealably treated by glow discharge in vacuo or by cytochrome C. Uranyl acetate solutions (2% w/v) were applied immediately onto the wet sample grids, which were viewed in a Philips EM 301 electron microscope without further drying.

For freeze-fracture procedure the vesicle samples were propane jet vitrified following the procedure of Müller and coworker [15]; then replicas were studied in the electronmicroscope.

The dynamic light scattering mesurements were performed on a Malvern 4700 PS/MW Spectrometer using Argon laser light of 488 nm at a temperature of 25.0 ± 0.1°C and a scattering angle of 90°.

The differential scanning calorimetry measurements were made on a Perkin-Elmer 7 Series Thermal Analysis System with a heating rate of 10°C per min in the temperature range between 10° and 100°C; the results were taken after several heating and cooling runs, and the phase transition enthalpy ΔH was calculated from the peaks' areas.

Conclusion

This study unequivocally estblishes that double-chain cationic amphiphiles containing fluorocarbon moieties form closed-shell aggregates. The single-compartment vesicles obtained are structurally very stable and stay in solution over 3 months without meaningful morphological change. Such structural stability is explained by the presence of perfluoroalkyl tails in the molecular structure of amphiphiles which increases the hydrophobicity of the vesicle shell. Vesicles formed from mixed-chain surfactants show spherical-shaped aggregates of normal type, while those formed from the double fluorocarbon chain surfactants exhibit polygon-like shapes due to very high phase transition temperatures of the membranes. Fully F-alkylated cationic surfactants also revealed differences in the wetting behavior of platinum (decoration effect).

Acknowledgements

The authors are grateful to Dr. Ernst Wehrli for the electronmicroscopy and to Dr. Peter Schurtenberger for the dynamic light scattering measurements.

References

1. Kunitake T, Okahata Y (1977) J Am Chem Soc 99:3860
2. Fendler JH (1980) Acc Chem Res 13:7
3. Neumann R, Ringsdorf H (1986) J Am Chem Soc 108:487
4. Wagenaar A, Rupert LAM, Engberts JBFN (1989) J Org Chem 54:2638
5. Fendler JH (1982) Membrane Mimetic Chemistry, Wiley, New-York
6. Mukerjee P (1981) J Phys Chem 85:2298
7. Kunitake T, Okahata Y, Yasunami S (1982) J Am Chem Soc 104:5547
8. Kunitake T, Higashi N (1983) Makromol Chem Phys Suppl n° 14:81
9. Elbert R, Folda T, Ringsdorf H (1984) J Am Chem Soc 106:7687

10. Chaabouni MM, Szönyi S, Baklouti A, Cambon A (1990) J Fluorine Chem 46:307
11. Coudures C, Pastor R, Szönyi S, Cambon A (1984) J Fluorine Chem 24:105
12. Szönyi S, Vandamme R, Cambon A (1985) J Fluorine Chem 30:37
13. Bollens E, Szönyi F, Cambon A (1991) J Fluorine Chem 53:1
14. Trabelsi H, Szönyi S, Gaysinski M, Cambon A, Langmuir (in preparation)
15. Müller M, Meister N, Moor H (1980) Mikroscopie (Wien) 36:129

Authors' address:

Dr. S. Szönyi
Laboratoire de Chimie Organique du Fluor
Université de Nice-Sophia Antipolis
Parc Valrose, B.P. 71
06108 Nice Cedex 2, France

Progress in Colloid & Polymer Science

Progr Colloid Polym Sci 89:156—159 (1992)

Viscosity-percolation behavior of waterless microemulsions: a curious temperature effect

Z. Saïdi, C. Boned, and J. Peyrelasse

Laboratoire de Physique des Matériaux Industriels, Centre Universitaire de Recherche Scientifique, Pau, France

Abstract: The dynamic viscosity of the waterless microemulsion glycerol/AOT [sodium bis (2-ethylhexyl)sulfosuccinate]/isooctane has been studied at constant volume fraction ϕ as a function of temperature. The experimental curves are curious: depending on the value of ϕ, the viscosity either decrease or goes through a maximum when the temperature increases. The results are interpreted in the framework of the percolation theory. The percolation locus in the ϕ-T plane has been determined, as well as the cloud-point curve. An approximative model is used in order to fit the percolation line, taking into account the temperature dependence of the interaction strength.

Key words: Percolation; viscosity; waterless microemulsion

Introduction

In the past few years, we have seen a number of important results concerning fundamental phenomena in real systems or model systems, such as artificial structures of heterogeneous disordered systems. Behind the apparent multiplicity of these results there is, in fact, a profound unity reflected in the development of fundamental concepts, e.g., percolation, fractal, chaos, and change of scale. These concepts find their place in many models and, because of their profound and general scope, they bridge different disciplines [1]. In this paper, we propose to focus more particularly on how the concept of percolation has enabled an improved understanding of the variations of the dynamic viscosity of waterless microemulsions versus temperature.

Experimental techniques

1) Systems studied:

The samples were prepared with sodium bis (2-ethylhexyl) sulfosuccinate (AOT from SIGMA), isooctane (FLUKA AG puriss),and glycerol (PRO-LABO Rectapur). The molar ratio n = [glycerol]/[AOT] varies up to 4. As the microemulsions studied were of the glycerol in isooctane type, we define ϕ as the volume fraction of the dispersed matter (glycerol + AOT). All measurements were carried out between $5°C < T < 35°C$. The samples are characterized by n and ϕ (at T = 25°C). We should point out that the neighboring glycerol/AOT/heptane microemulsions have been studied by light scattering [2]. The resulting systems consist of discrete spherical droplets of glycerol stabilized by the surfactant. The droplet size is independent of temperature and depends primarily on the molar ratio n.

2) Experimental apparatus:

Kinematic viscosity measurements were made according to the capillary viscometer method (Viscotimer Lauda S/1). The density was determined using an automatic DMA 45 Anton Paar densitometer. Density measurements were carried out with an absolute uncertainty less than 10^{-3} g/cm³. Relative uncertainty of dynamic viscosity η can be estimated at less than 0.5%.

Experimental results

We have represented in Fig. 1 the variations of the dynamic viscosity versus ϕ at T = 10°C and T =

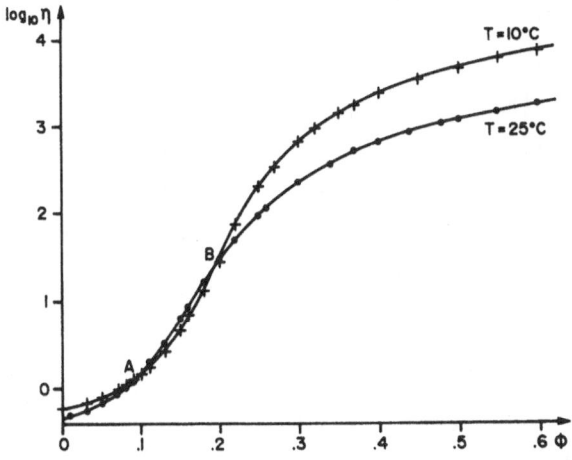

Fig. 1. Variations of $\log_{10}\eta$ (cP) versus ϕ ($n = 3.2$)

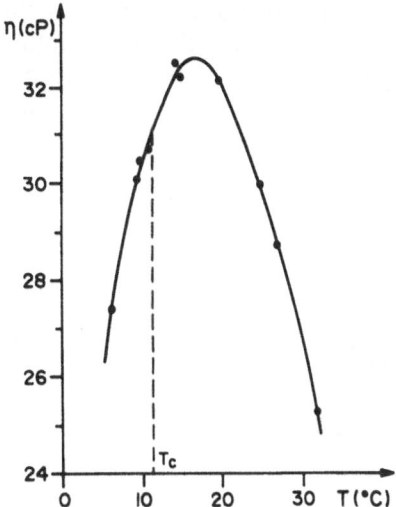

Fig. 3. Variations of η (cP) versus T (°C) at $\phi = 0.20$ ($n = 3.2$)

25°C ($n = 3.2$). When the value of ϕ is between ϕ_A and ϕ_B, the viscosity increases when temperature increases. However, when $\phi < \phi_A$ or $\phi > \phi_B$ there is a decrease in viscosity. Figure 2 represents, for the same microemulsions, the variations of $\log_{10}\eta$ versus T for several values of ϕ. Depending on the value of ϕ, the viscosity either decreases or goes through a maximum. Figure 3 represents the variations of the viscosity at $\phi = 0.20$; in this representation the maximum is very pronounced.

Discussion

1) Variations at constant T:

Analysis [3] of previous investigations carried out on percolation indicates that the following relationships (characteristic of a percolation phenomenon) should be adopted as the asymptotic behavior laws:

$$\phi > \phi_c + \delta \qquad \eta = C_1\eta_1(\phi - \phi_c)^\mu \, ;$$

$$\phi < \phi_c - \delta' \qquad \eta = C_2\eta_2(\phi_c - \phi)^{-s} \, , \qquad (1)$$

where $\mu = 2.00 \pm 0.25$ and $s = 1.2 \pm 0.2$. Here, ϕ is the volume fraction of the component 1. These values express the dynamic aspect of the phenomenon [3–5].

2) Variations at T variable:

By studying the $\eta(\phi)$ curves obtained for several temperatures, the variations of the percolation threshold ϕ_c as a function of the temperature can be determined. Figure 4 shows the location of the percolation line in the ϕ-T plane. In addition the figure also shows the cloud-point curve. As for water-based microemulsions [6, 7], the percolation threshold decreases when the temperature increases. Figure 4 shows that ϕ_c depends on T as well as η_1 and η_2 (viscosities of the component 1 and 2). If at variable T, $\phi_c(T)$ remains close to the value ϕ, one can write, developing to the first order,

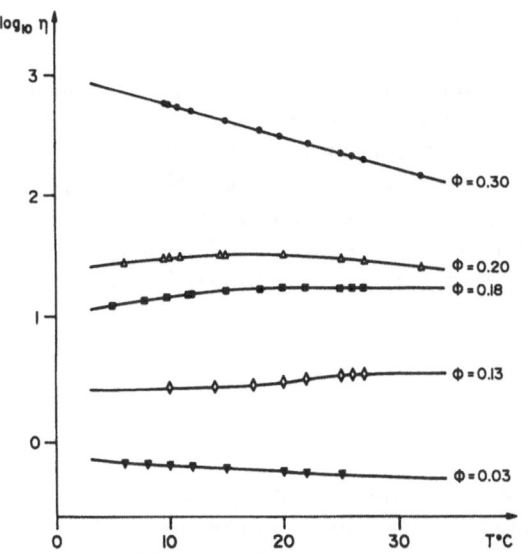

Fig. 2. Variations of $\log_{10}\eta$ (cP) versus T (°C) for different values of ϕ ($n = 3.2$)

$\phi_c(T) = \phi + K(T - T_c)$ in which $K = (d\phi_c/dT)_{T_c}$ and $\phi_c(T_c) = \phi$. This defines the temperature threshold T_c. If $K < 0$, which corresponds to our samples (Fig. 4), then $\phi > \phi_c$ if $T > T_c$. It follows that

$$\eta = C_1(T)\eta_1(T)[K(T_c)(T_c - T)]^\mu$$

$$\text{if} \quad T > T_c + \delta_T \tag{2}$$

$$\eta = C_2(T)\eta_2(T)[K(T_c)(T - T_c)]^{-s}$$

$$\text{if} \quad T < T_c - \delta'_T \tag{3}$$

These equations are not valid in the immediate vicinity of T_c (i.e., ϕ_c), where there is a continuous variation within a narrow interval around the percolation threshold. The crossover regime is $\Delta T = \delta_T + \delta'_T$. There are then scaling laws for the variations of η with T which have the same exponents μ and s as the variations of η with ϕ.

3) Difficulty of the analysis of the experimental curves:

It is necessary to emphasize that analysis of variations with T is only simple if K is independent of T_c, and if the prefactors η_1, η_2, C_1, C_2 are also independent of T. Moreover, there is a small variation of the volume fration ϕ of the sample when the temperature increases, due to the variations of the densities of the components 1 and 2. For instance, for the sample corresponding to $\phi = 0.180$ at $T = 25\,°C$, the volume fraction varies linearly between 0.1835 and 0.1792 when T increases from 5° to 30°C. It results that one has, in fact, $T_c(T)$, depending on the value $\phi(T)$. This variation could have some importance when the temperature is such that the sample is near the percolation threshold. If all these effects are negligible, $\log_{10}\eta = f(\log_{10} | T - T_c |)$ corresponds to two straight lines of slopes μ and $-s$. In these conditions η increases with increasing T. The percolation threshold is accompanied by a marked increase of the quantity measured with temperature. This has been observed for conductivity with water/AOT/oil systems [6—9]. In the present case, for viscosity and glycerol-based microemulsions, it is necessary to account for the very rapid decrease of $\eta_{glycerol}(T)$ with temperature (6260 cP at 6°C and 629 cP at 30°C). When $\phi_c(T)$ remains in the vicinity of $\phi(T)$, the increase due to the per-

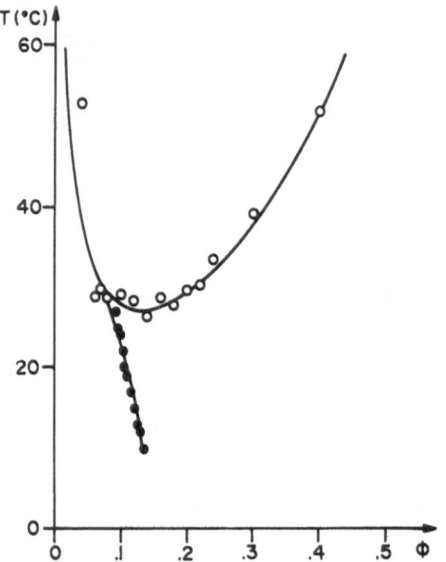

Fig. 4. The experimentally ($n = 4$) determined cloud-point curve (○) and percolation line (●) compared to the computed lines for the modified model (—) of Xu and Stell [10]

colation is greater than the decrease due to $\eta_1(T)$ and, then the viscosity increases. Conversely, far from $\phi(T)$ the decrease due to $\eta_1(T)$ dominates, and then the viscosity decreases. It results from these considerations that sometimes the viscosity can go to a maximum. This maximum position is at a higher temperature than the percolation temperature (Fig. 3). All these observations show, that it is difficult or even impossible to experimentally determine the exponents μ and s along this pathway, and that, in general, the correct analysis of the experimental curves is complicated. Let us recall here that for the more favorable water/AOT/oil decane system, a difference of evaluation for conductivity has been observed [8]: $\mu = 1.7$ and $s = 1.4$ with $| T - T_c |$ (pathway $\phi \sim$ Const.) and $\mu = 1.6$, $s = 1.2$ with $| \phi - \phi_c |$ (pathway $T =$ Const.).

4) Calculation of the percolation line:

In Fig. 4, we show in a ϕ-T plane the percolation threshold line and the cloud-point curve. For the water/AOT/decane system the percolation line has been calculated [7] using the second and third models of the theory of percolation introduced recently in simple fluids [10]. They explicitly take into account the temperature dependence of the interaction strength of the potential, using an empirical relation obtained by fitting small-angle neutron scattering data of the same system. For our system, we do not have such information. Therefore, we make the following crude approximation: we assume that the cloud-point curve is identical to the spinodal-line. A numerical fitting of this line, using the appropriate model [10], then gives the temperature dependence of the interaction strength, and it is possible to calculate the percolation line (Fig. 4). It is astonishing that this approximative method produces such a quantitative agreement with the experimental values. Furthermore, the numerical values we obtained [11] for the interactions in the glycerol/AOT/isooctane system are close to the ones obtained for a water/AOT/decane system.

Conclusion

In a previous article [12], we emphasized the importance of the experimental pathway. Here, we have chosen to vary temperature while keeping all other parameters constant. In fact, it is the distance $| \phi(T) - \phi_c(T) |$ which is important and here, as the percolation threshold varies at each point along the pathway (and also $\phi(T)$), it is possible to move through a percolation point. This defines a percolation temperature T_c. But as the prefactors in the asymptotic laws are also dependent on the temperature, this fact naturally makes correct analysis of the experimental curves more difficult. In that sense, the viscosity-percolation behavior of waterless microemulsions presents a curious temperature effect. The viscosity properties of the waterless glycerol/AOT/oil system give a good example of an unusual behavior which could look strange upon first analysis, but illustrate the difficulty in interpreting experimental data when a percolation phenomenon is involved.

References

1. Berroir A, Charpentier JC, Lehman P, Thoulouze D (1989) Images de la Physique, Ed CNRS, Suppl 74:1
2. Fletcher PDI, Galal MF, Robinson BH (1974) J Chem Soc Faraday Trans 1, 80:3307—3314
3. Saidi Z, Mathew C, Peyrelasse J, Boned C (1990) Phys Rev A 42:872—876
4. Lagues M (1979) J Phys Lett (Paris) 40:L331—L333
5. Grest GS, Webman J, Safran SA, Bug ALR (1986) Phys Rev A 33:2842—2845
6. Moha-Ouchane M, Peyrelasse J, Boned C (1987) Phys Rev A 35:3027—3032
7. Cametti C, Codastefano P, Tartaglia P, Rouch J, Chen SH (1990) Phys Rev Lett 64:1461—1464
8. Kim MW, Huang JS (1986) Phys Rev A 34:719—722
9. Eicke HF, Hilfiker R, Holz M (1984) Helv Chim Acta 67:361—372
10. Xu J, Stell G (1988) J Chem Phys 89:1101—1111
11. Peyrelasse J, Boned C, Saidi Z (unpublished results)
12. Peyrelasse J, Boned C (1990) Phys Rev A 41:938—953

Authors' address:

C. Boned and J. Peyrelasse
Laboratoire de Physique des Matériaux Industriels
Centre Universitaire de Recherche Scientifique
Avenue de l'Université
64000 Pau, France

Progress in Colloid & Polymer Science

Progr Colloid Polym Sci 89:160—164 (1992)

Fine-Tuning of the film thickness of ultrathin multilayer films composed of consecutively alternating layers of anionic and cationic polyelectrolytes

G. Decher and J. Schmitt

Institut für Physikalische Chemie; Johannes Gutenberg-Universität Mainz, FRG

Abstract: We have recently introduced a new method of creating ultrathin films [1—3] based on the electrostatic attraction between opposite charges. Consecutively, alternating adsorption of anionic and cationic polyelectrolytes leads to the formation of multilayer assemblies. Multilayer buildup is easily monitored by small angle x-ray scattering (SAXS). The total thickness of the multilayer assemblies increases linearly with the number of adsorbed layers, indicating a stepwise and regular deposition process. — Here, we report on the fine-tuning of the total film thickness by changing the ionic strength of the solvent from which the polyelectrolytes are adsorbed. When the anionic polyelectrolyte is adsorbed from solutions containing 1.0, 1.5, and 2.0 mol/l NaCl, the average thickness of each oppositely charged layer pair is precisely adjusted to 17.7 Å, 19.4 Å, and 22.6 Å, respectively.

Key words: Supramolecular architecture; ultrathin organic films; polyelectrolytes; adsorption; self assembly; x-ray scattering

1. Introduction

Ultrathin organic films are of interest in scientific as well as in applied research, e.g., in integrated optics, and as surface-coating or bio-sensors [4, 5]. A common method for the preparation of multilayered films is the now classic Langmuir-Blodgett technique, which is rather limited to planar substrates of a few square centimeters in area.

In contrast, adsorption processes from solution are, in principal, not restricted in substrate size or topology. Different self-assembly techniques, all based on chemisorption, have been developed (e.g., [6—8]). These techniques sometimes require several chemical activation steps, so that the multilayer buildup can become a time-consuming routine.

Our method is based on the electrostatic attraction between opposite charges and, thus, activation steps are not needed. A typical polyelectrolyte system for the multilayer buildup is polystyrene-sulfonate sodium salt (NaPSS) as the anionic polyelectrolyte, and polyallylamine hydrochloride (PAH) as the cationic polyelectrolyte, both of which

are commercially available. The molecules and their graphic representations are shown in Fig. 1. The process of the multilayer buildup is described in Fig. 2.

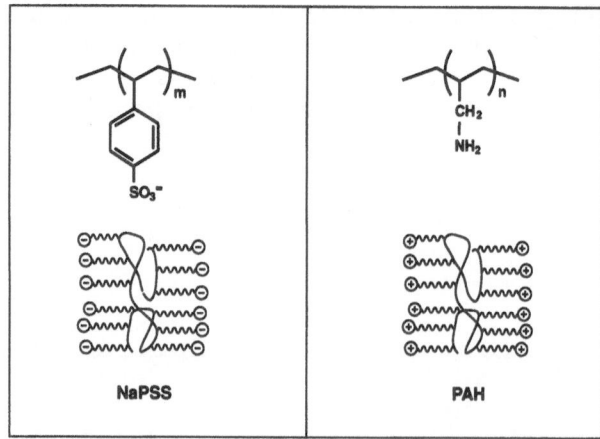

Fig. 1. Molecular structure of the polyelectrolytes and their graphic representations. The symbols are highly idealized and are not intended to represent the conformation of the polyelectrolyte chains in aqueous solution

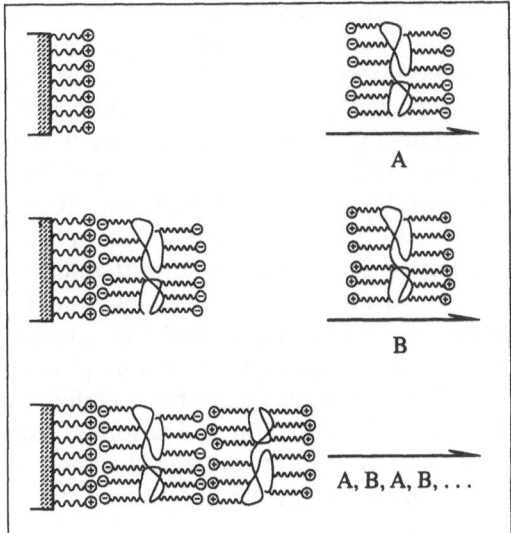

Fig. 2. Schematic for the buildup of multilayer assemblies: A solid substrate with a postively charged surface is immersed in a solution containing the anionic polyelectrolyte and a monolayer is adsorbed by reversing the surface charge (step A, first layer and all following odd numbered layers). After rinsing with water the substrate is immersed in the solution of the cationic polyelectrolyte and, again, a monolayer is adsorbed (step B, second layer and all following even-numberd layers). By repeating both steps in a cyclic fashion (A, B, A, B ...) alternating multilayer assemblies are obtained

We have previously reported on the buildup of multilayer assemblies by consecutively alternating adsorption of anionic and cationic polyelectrolytes. The buildup of the films was investigated by both UV/Vis spectroscopy and SAXS. A linear correlation of the amount of adsorbed material and of the total film thickness with the number of adsorbed layers was found [3].

Here, we present a method of fine-tuning the average thickness of each oppositely charged layer pair. The addition of sodium chloride in molar quantities to the solution of the anionic polyelectrolyte (NaPSS) leads to the possibility of very accurately adjusting the thickness of each individual layer pair.

2. Materials and methods

Polystyrenesulfonate (sodium salt, $Mr = 100\,000$) (NaPSS) was obtained from SERVA, polyallylamine (hydrochloride, $Mw = 50\,000—65\,000$) (PAH) was obtained from Aldrich. Both materials were used without further purification.

The ultrapure water used for all experiments and all cleaning steps was obtained by reversed osmosis (Milli-RO 35 TS; Millipore) followed by ion-exchange and filtration steps (Milli-Q; Millipore). The resistivity was better than 18.2 MΩ · cm, and the total organic content was less than 10 ppb (according to the manufacturer).

For all experiments aminobutyl-silanized fused quartz substrates were used. These substrates were prepared by first cleaning the surface ultrasonically in a hot H_2SO_4/H_2O_2 (7:3) mixture for 1 h and then washing with Milli-Q water. Further purification was carried out using the $H_2O_2/H_2O_2/NH_3$ (5:1:1) step of the RCA cleaning procedure [9]. After extensively washing with Milli-Q water, the substrates were immersed for 2 min in pure methanol, a methanol/toluene (1:1) mixture and, finally, in pure toluene. From the last solution the substrates were directly transferred to a 5% solution of 4-aminobutyldimethylmethoxysilane (Petrarch Systems) in dry toluene, and kept there for 15 h under an atmosphere of dry nitrogen. Afterwards, the substrates were immersed for 1 min each in toluene, toluene/methanol (1:1), methanol, and Milli-Q water. The freshly surface-modified substrates were stored in water and used within 2 h for the adsorption experiments.

Both polyelectrolytes were adsorbed from solutions containing $9.7 \cdot 10^{-3}$ monomol/l polymer and 0.01 mol/HCl (monomol refers to the molar concentration of monomer residues). The ionic strength of the NaPSS solution was adjusted by addition of different amounts of sodium chloride (Merck p.a.).

The adsorption was carried out as follows. A substrate (fused quartz) with a surface area of 8 cm^2 was immersed in 10 ml of the polyelectrolyte solution for 20 minutes at ambient temperature. After the deposition of each polyelectrolyte layer the substrate was rinsed three times for 1 min in pure water.

The multilayer assemblies were characterized by small angle x-ray scattering (Siemens D-500 powder diffractometer using Copper K_a-radiation with a wavelength of 1.541 Å, data acquisition via a DACO-MP interface connected to a personal computer). The dependence of the SAXS spectra on layer numbers was recorded from single multilayer specimen on fused quartz that was dried in a stream of nitrogen in between the deposition cycles.

3. Results and discussion

Multilayer assemblies consisting of 50 layers each were prepared by consecutive adsorption steps of NaPSS and PAH. Acidic solutions of both polyelectrolytes were used in order to assure quaternization of the aminogroups of either the polyallylamine hydrochloride or the aminobutyl-silanized quartz surface. Three sets of experiments were carried out in order to prepare multilayer assemblies with different thicknesses. The multilayer buildup was followed by small angle x-ray scattering, which is a convenient and direct method for the determination of the film thickness. SAXS-diffractograms were recorded after deposition of 6, 8, 10, 15, 21, and 25 oppositely charged layer pairs. The measurements were always performed during the multilayer buildup of a single specimen. Since the repeat unit is one oppositely charged layer pair, the film thickness was measured after the deposition of the PAH layers.

The recorded diffractograms show so-called Kiessig fringes [10], which result from the interference of x-ray beams reflected at the substrate/film and film/air interfaces. As an example,

Fig. 3 shows the dependence of the SAXS-diffractograms on the number of adsorbed layers for the system adsorbed from 1 molar NaCl solution.

From the periodicity of the Kiessig fringes the total film thickness is calculated. These values were plotted vs. the number of oppositely charged layer pairs, and from the slope the average thickness of the layer pairs was obtained (Fig. 4). The correlation coefficients for the linear fits are better than 0.9995 in all cases. The average thickness varies from 17.7 Å for the solution containing 1 mol/l NaCl, to 22.6 Å for the solution containing twice this amount (Table 1). This trend is also seen from earlier experiments where the polyanion was adsorbed from a NaCl free solution and an average thickness of each layer pair of approximately 10 Å was observed.

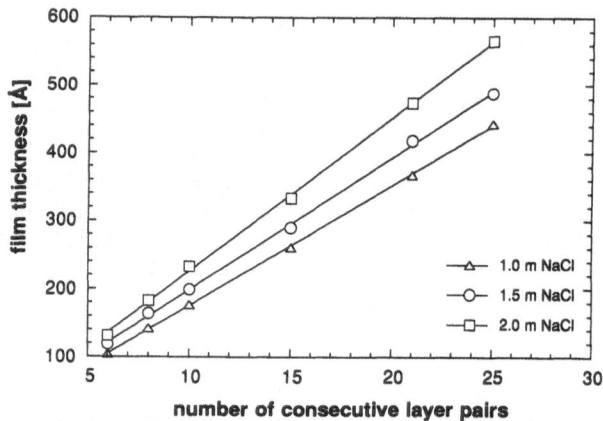

Fig. 4. Dependence of the total film thickness on the number of layers for three different NaCl concentrations of the NaPS solution. The triangles represent the values of the 1.0 molar, circles are the values of the 1.5 molar, and squares the values of the 2.0 molar NaCl solution. Errors are to small to be displayed.

Table 1 shows that the average thickness of the layer pairs can easily be fine-tuned by varying the quantities of added salt. In principle, it is possible to adjust the thickness of a single layer pair as well as the total film thickness.

Presently, we assume that the electrolyte added to the NaPSS solutions only affects the adsorbed NaPSS layers and not the structure of the PAH layers that were adsorbed from solutions of constant ionic strength.

It is known that the polyelectrolyte adsorption on different surfaces is influenced by the addition of

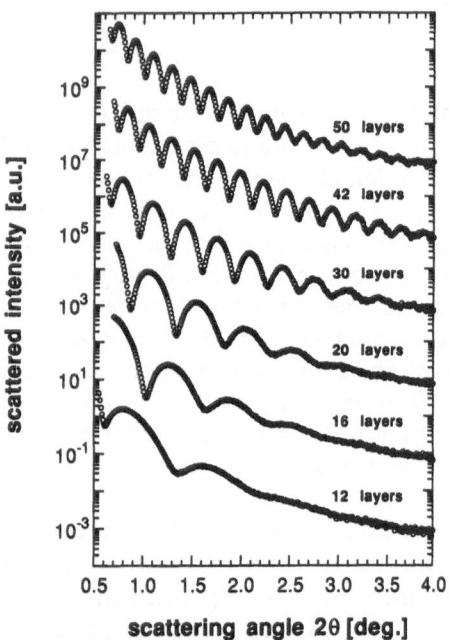

Fig. 3. Multilayer buildup as monitored by small angle x-ray scattering. The multilayer is composed of polyallylamine hydrochloride (PAH) and polystyrenesulfonate sodium salt (NaPS), the latter adsorbed from a solution containing 1 mol/l NaCl

salts [11-13], so that, in the absence of salts, the polyelectrolyte adsorbs in a more "flat" conformation, whereas in the presence of salts the adsorbed chains are in a more "loopy" conformation. This is in agreement with a theoretical approach on polyelectrolyte adsorption developed by van der Schee and Lyklema [14], and explained by the screening of the electrostatic repulsion of the ionic groups fixed on the chains of the polyelectrolyte.

The key to a regular multilayer buildup is the reversal of the surface charge in each adsorbed layer, as can easily be seen from Fig. 2. This makes it interesting to briefly discuss the molecular base for this charge reversal, especially the consequence of a "flat" or a "loopy" conformation. In Fig. 5 a schematic of possible conformations of adsorbed polymer chains is given. In the "loopy" conformation the reversal of the surface charge is easily conceived, since some segments of the polyelectrolyte chain are not attached to the surface. An excess of ionic side-groups is exposed to the solution interface and, thus, the surface charge is reversed (Fig. 5a). In a "flat" conformation this condition can also be realized as long as enough ionic groups on a polymer chain remain exposed to the solution interface (Fig. 5b).

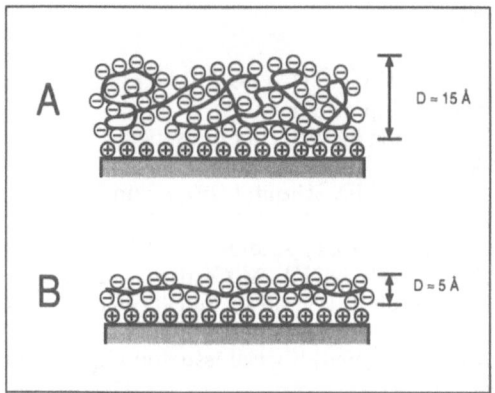

Fig. 5. Schematic of possible structures of the adsorbed layers (counterions omitted). The viewgraph is a two-dimensional projection of a three-dimensional assembly; it is not intended to represent the real conformation of the adsorbed layers. A) depicts the structure of an adsorbed chain in a "loopy" conformation. This is the case when the polyelectrolytes are adsorbed out of solutions containing additional electrolyte. B) depicts the structure of an adsorbed chain in a "flat" conformation. This is the case when the polyelectrolytes are adsorbed out of solutions containing no additional electrolyte

Table 1. Average thickness of oppositely charged layer pairs in dependence of the NaCl concentration. The average thicknesses and the errors are obtained from a linear least square fit of the total film thickness as a function of oppositely charged layer pairs (Fig. 4)

C_{NaCl} [mol/l]	Average thickness of oppositely charged layer pairs [Å]
1.0	17.7 ± 0.2
1.5	19.4 ± 0.3
2.0	22.6 ± 0.3

Marra and Hair observed a surface charge reversal by surface force measurements, when they adsorbed poly(2-vinylpyridine) at concentrations of approximately 10^{-4} monomol/l onto mica surfaces [15]. In this case, the thickness of the adsorbed monolayer was approximately 6 Å, which is roughly half the value we obtain for layer pairs also adsorbed in absence of added electrolyte.

4. Summary and conclusions

We have demonstrated that film thickness is strongly dependent on the ionic strength of the solution from which the polyelectrolyte was adsorbed. The recorded SAXS diffractograms showed well-developed interference fringes from both film interfaces. This allows the direct and unequivocal determination of the film thickness, which increased linearly with the number of adsorbed layers. By addition of molar quantities of electrolyte the average thickness of each oppositely charged layer pair can be adjusted with a precision of 0.5 Å. This opens the possibility of fine-tuning of the total thickness of a multilayer film with sublayer precision.

Acknowledgements

We thank the "Bundesministerium für Forschung und Technologie" for financial support and S. Münch for technical support. We are grateful to J. Maclennan and A. Leuthe for help with the x-ray measurements, and to H. Möhwald for stimulating discussions.

References

1. Decher G, Hong JD (1991) Makromol Chem, Macromol Symp 46:321
2. Decher G, Hong JD (1991) Ber Bunsenges Phys Chem 95:1430
3. Decher G, Hong JD, Schmitt J (1992) Thin Solid Films 207: in the press
4. Swalen JD, Allara DL, Andrade JD, Chandross EA, Garoff S, Israelachvili J, McCarthy TJ, Murray R, Pease RF, Rabolt JF, Wynne KJ, Yu H (1987) Langmuir 3:932
5. Adv Mater 3(1) (1991) Special Issue on Organic thin Films
6. Maoz R, Netzer L, Gun J, Sagiv J (1988) J de Chim Phys 85:1059
7. Lee H, Kepley LJ, Hong H, Akhter S, Mallouk TE (1988) J Phys Chem 88:2597
8. Tillman N, Ulman A, Penner T (1991) Langmuir 5:101
9. Kern W (1984) Semiconductor International 94
10. Kiessig H (1931) Ann d Phys 10:769
11. Cosgrove T, Obey TM, Vincent B (1986) J Colloid Interface Sci 111:409
12. Cafe MC, Robb ID (1982) J Colloid Interface Sci 86:411
13. Greene BW (1971) J Colloid Interface Sci 37:144
14. van der Schee HA, Lyklema J (1984) J Phys Chem 88:6661
15. Marra J, Hair ML (1988) J Phys Chem 92:6044

Authors' address:

Dr. Gero Decher
Institut für Physikalische Chemie
Johannes Gutenberg-Universität
Welder-Weg 11
D-6500 Mainz, F.R.G.

Progress in Colloid & Polymer Science   Progr Colloid Polym Sci 89:165—169 (1992)

Photoreactions in Langmuir-Blodgett-Kuhn multilayer assemblies of liquid crystalline azo-dye side-chain polymers

M. Sawodny[1]), A. Schmidt[1]), C. Urban[2]), H. Ringsdorf[2]), and W. Knoll[1])

[1]) MPI Polymerenforschung, Mainz
[2]) Institut für Organische Chemie, Johannes Gutenberg Universität, Mainz, FRG

Abstract: Reversible and persistent optical data storage in Langmuir-Blodgett-Kuhn liquid crystalline systems has been reported recently. Depending on the molecular structure, the trans-cis isomerization in these systems (induced by illumination with UV- and VIS-light) leads to different macroscopic behavior of the chromophore percentage on the photo-induced structural changes in a series of statistical copolymers.

Key words: Langmuir-Blodgett-Kuhn technique; azo-dye side-chain polymers; optical data storage

Introduction

Self-organizing liquid-crystalline materials are interesting from an applications point of view, because they allow for the preparation of novel supra-molecular structures not feasible in normal smectic LC-systems. An example is the control of the orientation of low-molecular mass nematic LC-crystals by a single monolayer [1]. In this paper, we deal with polyacrylate liquid crystalline side-chain polymers, containing different amounts of azo-dye. These materials form Langmuir-Blodgett-Kuhn (LBK) liquid crystalline systems, as described recently by Penner et al. [2]. They are photochromic, i.e., illumination with UV-light gives rise to a reversible change of the refractive index. This effect can be used for optical data storage, as shown for bulk liquid crystalline polymers [3, 4].

Experimental

Materials

The synthesis and the bulk liquid crystalline properties of the materials used in this study are described in detail elsewhere [5]. Structural formulae of the homopolymer and the copolymers containing azo-benzene side groups are shown in Fig. 1. We investigated the homopolymer (4L) and the

two copolymer systems with a mole fraction of the chromophore of 35% (1—40L) and 75% (1—80L), respectively.

Mono- and multilayer preparation

Stable monolayrs were prepared on a Langmuir trough (area = 1000 cm^2) in pure water (milli Q quality) by spreading a 10^{-3} M CH$_3$Cl solution at room temperature (4L, 1-80L) or at 5°C (1—40L). Compression speed was extremely slow (3 cm^2/min, total trough area: 350 cm^2). Sufficient care was taken, by working under "safe" red light, to keep the material in the trans-configuration. Using the LBK-technique, we transferred the monolayers onto various hydrophobic substrates at a surface pressure of π_0 = 25 mN/m (4L, 1—80L) and π_0 = 7 mN/m (1—40L) with a dipping speed of 3 mm/min.

Results

Structural characterization

X-ray reflectometry measurements combined with surface plasmon spectroscopy provided film-thickness, refractive indices, and structural data from the various multilayer systems deposited onto hydrophobic substrates. Figure 2 shows the X-ray pattern for a 24-layers system of the homopolymer

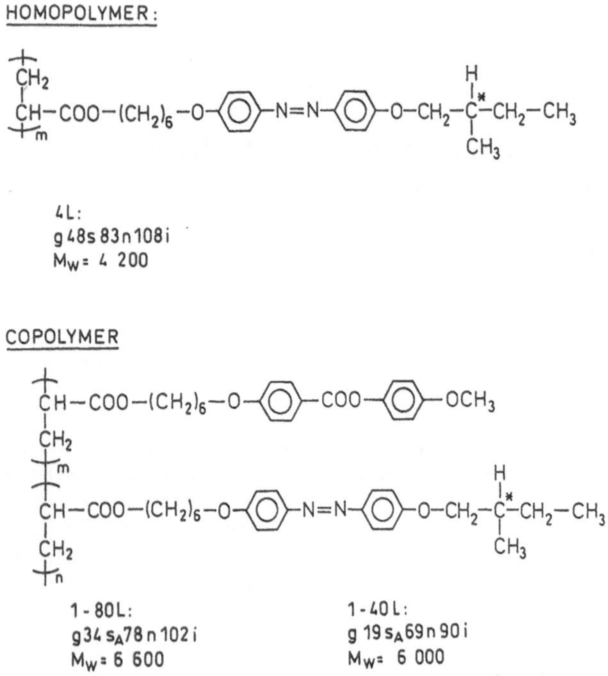

HOMOPOLYMER:

4L:
g 48 s 83 n 108 i
M_W = 4 200

COPOLYMER

1 - 80L: 1 - 40L:
g 34 s_A 78 n 102 i   g 19 s_A 69 n 90 i
M_W = 6 600 M_W = 6 000

Fig. 1. Structure formulae of the liquid crystalline side-chain polymers used in this study. 1—40L corresponds to a chromophore mole-fraction of 35% and 1-80L to a mole-fraction of 75%, respectively. Abbreviations of the systems used in the text, their bulk transition temperatures determined by DSC-measurements (g: glass transition, s_A: smectic A phase, s: smectic phase, n: nematic phase, i: isotropic phase), and their molecular weight (M_w) are shown

(4L) and the fit obtained from a calculation of the reflectivity curves b ased on the kinetmatic theory of diffraction [6]. The resulting electron-density profile corresponds to the double-layer periodicity given by a Y-type LBK-film and the chemical structure of the substance. In the amorphous area of the polymer backbone the electron density is signficantly decreased compared with the average value, whereas the mesogenic sidechains are more densely packed.

In contrast, the copolymer (1—40L), highly diluted with nonphotoreactive comesogenes, does not show such a "glassy" S_{C2}-phase induced by the LBK-process. The data can well be described by a molecular model, assuming interdigitating side-chains as in the bulk liquid crystalline phase. This implies thermodynamic equilibrium of the copolymer. This structural difference gives rise to a different behavior of the system 1—40L under il-

lumination with UV-light. This will be discussed in the forthcoming part of this paper.

Photoreactions in the homo- (4L)
and copolymer (1-80L) assemblies

It is well known that under illumination with UV-light, azobenzene molecules show a cis-trans photoisomerization in solution, as well as in solid films [8].

This effect can be used in liquid-crystalline polymers for persistent and reversible optical data storage. Figure 3 shows the UV-VIS spectra of 40 layers of the homopolymer transferred onto a quartz-slide during irradiation with 360 nm-light. The initial spectrum shows a broad $\pi \rightarrow \pi^*$ absorption between 300 and 400 nm resulting from different surroundings and aggregational states of the chromophores. From theoretical calculations, we derived four absorption bands at 325, 339, 358, and 381 nm, respectively. Following the assignments of Shimomura et al. [8], this spectral characteristic corresponds to the H-aggregate (325 nm), dimer (339 nm), monomer (358 nm), and J-like aggregate (381 nm).

Upon illumination with UV-light the spectra show decreasing absorption in the spectral range of 300—400 nm, corresponding to the trans-cis isomerisation of the abzobenzene molecules. This effect is pronounced for the H-aggregate and the chromophores in a liquid-crystalline environment, whereas dye molecules of J-aggregates seem to be less sensitive. Such a phenomenon is well known in cyanine-dye chromophores and is due to the desactivation of the dye molecules via fluorescence [9]. After thermal relaxation in the trans state, the UV-spectrum differs significantly from the initial one. This is due to the fact that the photoreaction causes a distortion of the well-ordered smectic S_{C2}-into a amorphous phase, as seen by X-ray measurements. Therefore, the environment of the chromophores induced by the LBK-process has been altered so that aggregation can no longer occur.

From this result, we conclude that the the data storage in these materials is possible because of an order-disorder transition.

Reversible photoreactions in the copolymer (1—40L)
highly diluted with non-photoreactive comesogens

Figure 4 shows UV-VIS-spectra of the copolymer 1—40L taken at normal incidence under illumina-

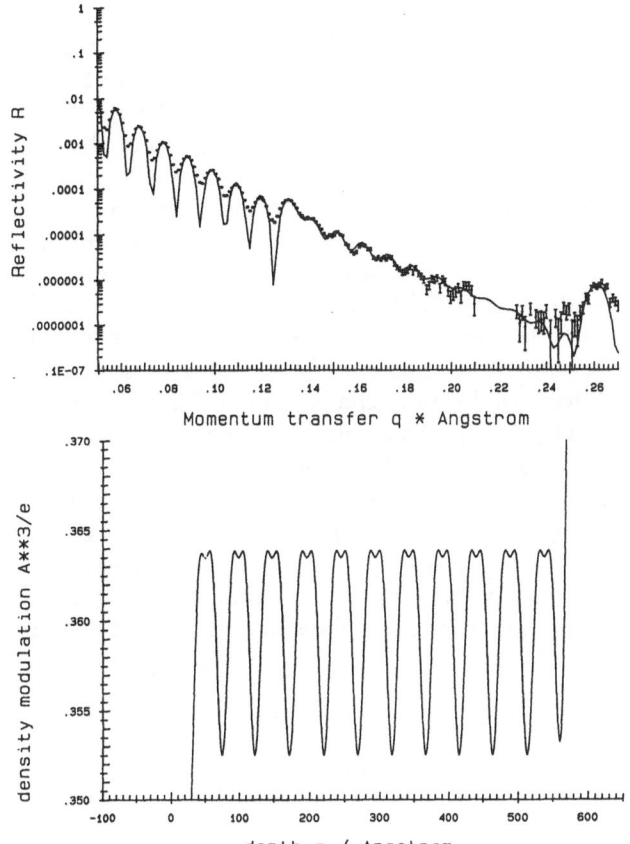

Fig. 2. X-ray reflectometry curve of 24 layers of the homopolymer (4L). The electron density profile resulting from fitting to the kinetmatic theory of X-ray refraction corresponds very well to the chemical structure of the molecules. The double-layer periodicity is determined to be 4.8 nm

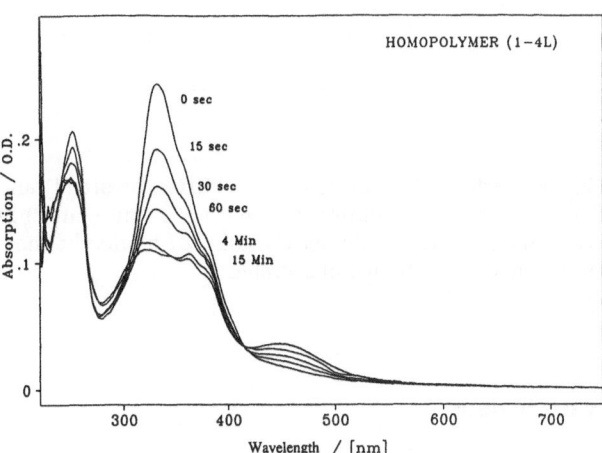

Fig. 3. Absorption spectra of the homopolymer sample consisting of 40 monolayers taken at various times as indicated after UV-irradiation with 360 nm-light

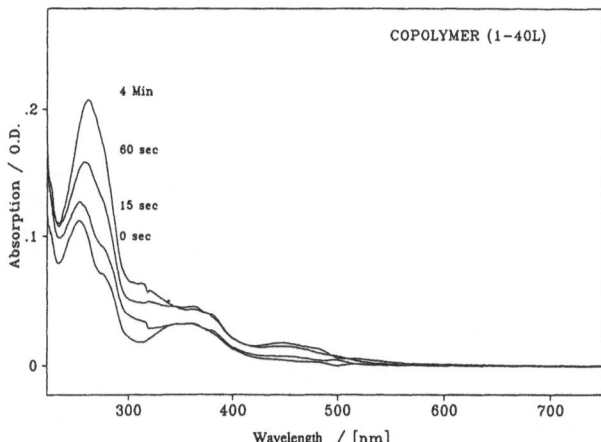

Fig. 4. Absorption spectra of the copolymer-system highly diluted with non-photoreactive chromophores taken at various times as indicating during UV-irradiation with 360 nm-light

tion with UV-light. Chromophore aggregation does not occur in LBK-films of this system because of spatial and, therefore, spectral decoupling of the dye-molecules by non-photoreactive comesogenes. Upon illumination with 360 nm light the absorption of the whole spectrum increases. This indicates that the complete liquid-crystalline matrix, chromophores as well as comesogenes, aligns itself parallel to the surface. Similar effects have been recently described by Seki et al. [1]; they found that one azo-command layer at the surface induces an initially homeotropic alignment in a whole-liquid crystalline system. Upon irradiation with 360 nm light the azo chromophores changed to the cis-conformation. The surface potential as measured by the contact angle method changed, which forced the system to align itself approximately parallel to the surface. With 450 nm light this process was reversed.

The UV spectra taken from our samples after one cycle of illumination (360 nm, then 450 nm) are identical to that taken prior to any irradiation. This means we have obtained a system in which the chromophores behave fully reversible in the thin LBK-film as in the solution.

To confirm the results derived from hte UV-VIS spectroscopy, we performed surface-plasmon measurements (PSPS) [10]. This method is sensitive to the refractive index n_z perpendicular to the surface. If the chromophores are aligned homeotropically, the revractive index is significantly higher than in the parallel orientation of the sample. In our experiments, we observed the expected refractive index change even in a two-layer system of the copolymer (1—40L). Figure 6 shows the PSPS curves before and under steady-state illumination with 360 nm light. The shift towards lower angles must be due to a decreasing refractive index, since X-ray measurements indicate that the geometrical thickness does not change under illumination with light [11, 12]. Observation of the reflected intensity at a fixed angle allows the calculation of the time-dependent refractive index change. This is depicted in Fig. 7 for several illumination cycles of two layers of the copolymer (1—40L). With surface-plasmon microscopy, we were able to read out the storage information with a lateral resolution better than 5 µm [13].

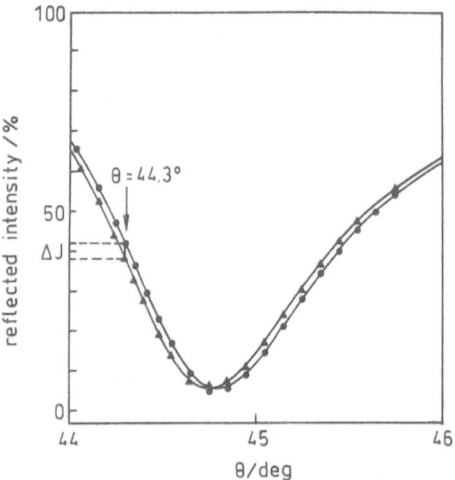

Fig. 6. Surface-plasmon resonance curves before (1) and under steady-state illumination with 360 nm-light (2) in two layers of the copolymer (1—40L). ΔI marks the maximum intensity-change observable

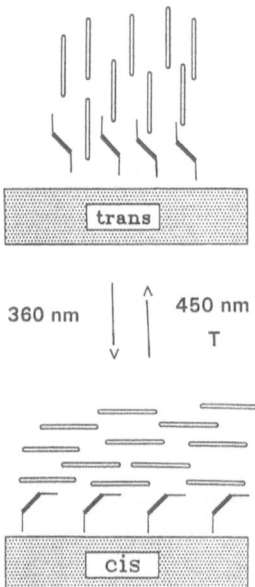

Fig. 5. Mechanism for reversible change in the alignment of the liquid-crystalline system, as proposed by Ichimura et al. [14]

Conclusions

Langmuir-Blodgett-Kuhn assemblies of liquid-crystalline azo-dye side-chain polymers can be used as persistent and reversible optical data storage media, because of the trans-cis isomerization of the

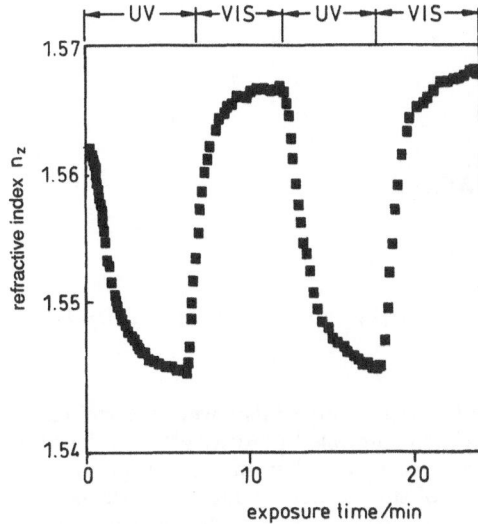

Fig. 7. Time-dependent refractive index change n_z calculated from PSPS measurements in a system of Cr/Au/eight layers polyglutamate/two layers of the copolymer 1-40L

chromophore molecules. Differences in the photochemical and photophysical behavior are results of the specific molecular architecture of the LBK-films. Systems with high amount of azo-dye form a bilayer structure. Illumination induces a order-disorder transition which can be used for persistent optical data storage. On the other hand, polymers which are highly diluted with non-photoreactive comesogens show a liquid-crystalline structure similar to the bulk-phase [12] and, therefore, allow for reversible photoreactions in the LBK-film as in solution.

Acknowledgements

We thank J. Stumpe, H. Möhwald, and G. Reiter for many helpful discussions. This work was supported by the Stiftung Volkswagenwerk (I/66 109).

References

1. Seki T, Tamaini T, Suzuki Y, Kawanishi Y, Ichimura K, Aoki K (1989) Macromolecules 22:3506
2. Penner TL, Schildkraut JS, Ringsdorf H, Schuster A (1991) Macromolecules 24:1041
3. Eich M, Wendorff JH, Reck B, Ringsdorf H (1987) Makromol Chem Rap Comm 8:59
4. Stumpe J, Müller L, Kreysig D, Hauk G, Koswig HD, Ruhmann R, Rübner J (1991) Makromol Chem Rap Comm 12:81
5. Ringsdorf H, Urban C, Knoll W, Sawodny M (1992) Makromol Chem (in press)
6. F. Rieutard, J. J. Benattar, L. Bosio (1986) J Physique 47:1249
7. Eisenbach CD (1978) Makromol Chem. 179:2489
8. Shimomura M, Ando R, Kunitake T (1983) Ber Bunsenges Phys Chem 87:1134
9. Kröhl T (1989) PhD-thesis, University Mainz
10. Wallis RF, Stegeman GI (1986) (eds) "Electromagnetic surface excitation", Springer Verlag, Berlin
11. Sawodny M, Schmidt A, Stamm M, Knoll W, Urban C, Ringsdorff (1991) Polym Adv Tech 2:127
12. Sawodny M, Schmidt A, Urban C, Ringsdorff H, Knoll W (1992) Thin Solid Films (in press)
13. Schmitt FJ, Weishorn AL, Hansma PK, Knoll W (1991) Makromol Chem Macromol Symp 46:133
14. Hickel W, Knoll W (1990) J Appl Phys 67:3572

Authors' address:

Michael Sawodny
Ackermannweg 10
D-6500 Mainz 1, FRG

Progress in Colloid & Polymer Science   Progr Colloid Polym Sci 89:170—172 (1992)

The surface dehydroxylation
and rehydroxylation of controlled porosity glasses

B. Biliński and W. Wójcik

Department of Physical Chemistry, Faculty of Chemistry, Maria Curie-Skłodowska University, Lublin, Poland

Abstract: Adsorption isotherms of n-octane and toluene were determined on thermally treated and rehydroxylated controlled porosity glasses. Based on the film pressure values which were obtained from these isotherms the dispersion and polar components of surface free energy were calculated. It was found that the dispersive interactions do not change significantly, however, they are always higher for the dehydroxylated surface. The rehydroxylation with water vapor seems to provide only the hydrolysis of surface B—O—B and B—O—Si groups, as well as those Si—O—Si bridges which are influenced by B atoms. The rehydroxylation with NaOH solution leads to values of the polar component comparable to that of bare glass. The possibility of surface rehydroxylation appeared affected by the presence of both boron and sodium atoms.

Key words: Surface free energy; controlled porosity glasses; thermal treatment; rehydroxylation

Introduction

Controlled porosity glasses (CPG) are very often used as adsorbents, as supports for chemically bonded or adhesively deposited stationary phases in chromatography, as carriers of biochemical ligands for affinity chromatography and biocatalytic transformations, and as molecular sieves, etc. The thermal treatment of such materials at a temperature range 400—600°C results in two fundamental surface processes [1]:

i) enrichment in boron and sodium; and
ii) condensation of hydroxyl groups.

This considerably changes the surface properties of the material.

The subsequent rehydroxylation of the surface of CPG provides a hydrolysis of Si—O—Si, B—O—B, and B—O—Si bridges, however, the results appear strongly dependent on the applied rehydroxylation procedure. Moreover, the presence of B atoms and Na ions on the surface also influences the process of rehydroxylation.

Surface free energy (γ_S) allows the anticipation of many physicochemical processes taking place at the interface. It results from the type and magnitude of intermolecular interactions, and may be expressed as a sum of two components [2, 3]:

$$\gamma_S = \gamma_S^d + \gamma_S^p \,. \tag{1}$$

The dispersion component (γ_S^d) relates to dispersive interactions. The polar one (γ_S^p) relates to other types of interactions such as dipole-dipole, π electrons, electrostatic or hydrogen bonds. It is almost impossible to distinguish various kinds of polar interactions, therefore, they used to be considered as one term (γ_S^p) [2, 3].

The components of surface free energy may be determined by various experimental methods such as contact angles, zeta potential or adsorption measurements. From the adsorption isotherm the film pressure π of an adsorbate may be calculated according to the well known Bangham-Razouk equation [4]:

$$\pi = \frac{RT}{A} \int_0^p a \, d(\ln p) \,, \tag{2}$$

where A is the specific surface area of the adsorbent, a is the adsorbed amount, p is the equilibrium vapor pressure of the adsorbate, and R, T are the gas constant and the temperature, respectively.

For thick adsorption films (at the saturation state) the maximal film pressure π_{max} (obtained by integration of Eq. (2) up to the saturated vapor pressure) may be equated to the work of spreading wetting of the solid surface with the liquid adsorbate [5]:

$$\pi_{max} = W_S = -2\gamma_L + 2\sqrt{\gamma_S^d \gamma_L^d} + 2\sqrt{\gamma_S^p \gamma_L^p}, \quad (3)$$

where γ_L is the surface tension of the liquid adsorbate and γ_L^d, γ_L^p are the dispersion and polar components of the surface tension, respectively.

Saturated hydrocarbons interact only by dispersive forces, therefore, the dispersion component may be calculated from the adsorption isotherm of a hydrocarbon. Knowing γ_S^d, one can determine the γ_S^p value from adsorption isotherm of a polar compound (water, alcohol, alkylbenzene).

Experimental

In this work the influence of dehydroxylation and rehydroxylation of CPG on the components of its surface free energy was investigated. The bare CPG was heated at 600 °C for 2, 20, and 120 h, and then rehydroxylated with either water vapor (20 h at 100 °C) or 0.1 N NaOH solution (30 min at room temperature). For each material the adsorption isotherms of n-octane and toluene were determined, applying the gas chromatographic method [6], and then the components of the surface free energy were calculated from Eq. (3).

Results and discussion

The numerical values of both components are collected in Table 1. As can be seen from this table the dehydroxylation-rehydroxylation procedure does not significantly affect the value of the dispersion component. However, the value for the dehydroxylated surface was always somewhat higher. As concerns the polar component, the influence of the rehydroxylation appears to be more complex. The rehydroxylation with water vapor leads to increasing γ_S^p values with increasing time of preliminary thermal treatment. As was mentioned above, the thermal treatment results in the enrichment of the

CPG surface in boron and sodium [1]. The previous experiments suggested that the rehydroxylation with water vapor results in hydrolysis of B—O—B and B—O—Si bridges, as well as those Si—O—Si ones which are influenced by boron atoms [7]. After the rehydroxylation with alkaline solution the γ_S^p values were higher than those after the water vapor treatment. This could be attributed to the hydrolysis of all kinds of surface bridges (Si—O—Si, B—O—B, B—O—Si).

The long-time heating of CPG provides an increase of sodium content in the boron-enriched layer. The γ_S^p value after the rehydroxylation with NaOH solution for CPG heated for 120 h appears lower than that for the glass heated for 20 h. This seems to result from the presence of Na and from the increase of the coordination number of B atoms from 3 to 4 [8].

Table 1. The values of dispersion (γ_S^d) and polar (γ_S^p) components of surface free energy of controlled porosity glasses

Material	γ_S^d [mJ/m²]	γ_S^p [mJ/m²]
Bare controlled porosity glass	37.34	144.74
Bare glass heated for 2 h	40.16	55.79
Glass rehydroxylated with vapor	38.27	79.80
Bare glass heated for 20 h	39.51	37.49
Glass rehydroxylated with vapor	38.35	73.49
Glass rehydroxylated with solution	34.94	138.24
Bare glass heated for 120 h	36.77	37.35
Glass rehydroxylated with vapor	34.43	100.78
Glass rehydroxylated with solution	33.03	130.35

The calculations of the adsorbed amount and the film pressure were conducted simultaneously. Therefore, it was possible to calculate the term $\delta\pi/\delta a$. It has been demonstrated that this term, as a function of either adsorbed amount or film pressure, may be very useful for the interpretation of the adsorption mechanism and the surface heterogeneity [9]. Figure 1 presents the dependences between $\delta\pi/\delta a$ and the film pressure of toluene for the CPG heated for 20 h, before and after the rehydroxylation. The maximum on all

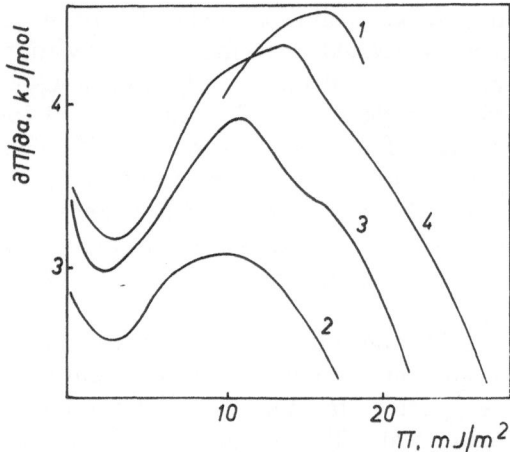

Fig. 1. The dependences between $\delta\pi/\delta a$ and the film pressure π of toluene for bare controlled porosity glass (1), the glass heated for 20 hours (2), the glass rehydroxylated with water vapour (3) and with NaOH solution (4)

curves corresponds to the BET monolayer coverage [5]. As can be seen, the height of this maximum is closely correlated with the magnitude of polar interactions (see Table 1). This was attributed to the interactions between surface hydroxyl groups and adsorbed toluene molecules.

Conclusions

Surface dehydroxylation and rehydroxylation of CPG does not affect the values of γ_S^d significantly, however, it is always slightly higher for the dehydroxylated surface. The value of γ_S^p increases considerably during the rehydroxylation, and this results from hydrolysis of surface Si—O—Si, B—O—B, and B—O—Si bridges. The rehydroxylation with water vapor provides the lower values of γ_S^p than those after the treatment with NaOH solution. It seems that during rehydroxylation with water vapor only those surface bridges hydrolyze which are influenced by B atoms. The presence of Na in

surface layer also affects the rehydroxylation process. Na$^+$ ions can change the coordination number of B atoms from 3 to 4, and formation of —OH groups bonded to 4-coordinated boron atoms may take place under different conditions. The knowledge of the surface free energy components of CPG and their changes during rehydroxylation seems very important for their practical applications as adsorbents, chromatographic supports, carriers of biochemical ligands, etc.

References

1. Low MJD, Ramasubramanian N (1967) J Phys Chem 71:730
2. Wu S (1978) In: Paul DR, Newman S (eds) Polymer Blends. Academic Press, New York, p 243
3. Jańczuk B, Białopiotrowicz T (1989) J Colloid Interf Sci 127:189
4. Adamson AW (1960) Physical Chemistry of Interfaces. Interscience, New York, p 263
5. Biliński B, Wójcik W, Dawidowicz AL (1991) Appl Surf Sci 41:99
6. Biliński B, Chibowski E (1983) Powder Technol 35:39
7. Biliński B (1991) J Mat Sci (submitted)
8. Kirutenko VM, Kiselev AV, Lygin VI, Scepalin KL (1974) Kinet Katal 15:1584
9. Biliński B (1988) Chromatographia 25:134

Authors' address:

B. Biliński
Dept. of Physical Chemistry
Marie Curie-Sklodowska Univ.
Marie Curie-Sklodowska Sq. 3
PL-20-031 Lublin, Poland

Progress in Colloid & Polymer Science

Progr Colloid Polym Sci 89:173—175 (1992)

Changes in barite surface free energy due to its surface precoverage with tetradecylamine chloride (TDACl) or sodium dodecylsulphate (DDSO$_4$Na)

E. Chibowski and L. Hołysz

Department of Physical Chemistry, Faculty of Chemistry, Marie Curie-Skłodowska University, Lublin, Poland

Abstract: Surface free energy components, apolar γ_s^{LW}, polar: electron donor γ_s^-, and electron acceptor γ_s^+ have been determined for the bare barite surface, as well as for those precovered with cationic collector: tetradecylamine chloride (TDACl) or anionic collector (DDSO$_4$Na). The surface coverages for TDACl were 0.25 and 1 statistical monolayers, and those for DDSO$_4$Na — 0.4 and 1 statistical monolayers. The γ_s^+ component was found to be meaningless. Therefore, polarity of the surface is due to the γ_s^- component. The adsorbed collector decreases the γ_s^- component resulting from hydrogen bonding. The decrease is larger in the case when DDSO$_4$Na was used as the collector. Simultaneously, the apolar component γ_s^{LW} increases for the surface precovered with TDCAl. It may be concluded that the adsorbing collector (or neutral TDA molecules) by partially remove water molecules from the barite surface. In the case of DDSO$_4$Na, γ_s^{LW} increased at 0.4 monolayer precoverage (probably as a result of the surface dehydration), and then at 1 monolayer precoverage it decreased to the value characteristic for long-chain hydrocarbon. These results are consistent with the flotation results. Straightline dependence for DDSO$_4$Na was found between free energy changes ΔG accompanying the flotation process and flotability changes.

Key words: Barite; surface free energy; collector; flotation

The concept of van Oss et al. [1—5] of the surface free energy components: apolar (Lifschitz — van der Waals, LW) and polar (electron donor, γ_s^-/electron acceptor, γ_s^+) is used in the model systems:

i) barite/tetradecylamine chloride (TDACl) — water;
ii) barite/sodium dodecylsulphate (DDSO$_4$Na) — water.

The components have been determined for bare barite surface, and the surface precovered with 0.25 and 1 statistical monolayers of TDACl or 0.4 and 1 satatistical monolayers of DDSO$_4$Na.

Surface free energy components, apolar γ_s^{LW} and polar γ_s^+, were determined by zeta potential measurements for a series of barite/collector/hexane or hexanol film samples in water. We described the details of the method and the theoretical back-

ground in [6, 7]. The polar component γ_s^- was determined by contact-angle measurements of glycerol or water droplets on barite pellets obtained from compressed barite powder [6, 7].

Correlations between changes of the experimentally determined components of the barite surface free energy and the barite samples flotability were investigated. The changes were caused by adsorption of TDACl and DDSO$_4$Na used as collectors and for the surface hyrophobization. It was found that using the determined values of the barite surface free energy components, flotability of the samples can be predicted in agreement with flotation tests. The prediction was based on the calculated values of ΔG attending the process of replacement of mineral/water by the mineral/gas interface. The thermodynamic condition obviously implies that for effective flotation change of the free energy, ΔG has to be negative. The changes of the components throw some light on the "energetic mechanism" of the collector adsorption.

In Fig. 1 are shown results of the surface free energy components and flotation recoveries of barite samples precovered with TDACl. The apolar component γ_s^{LW} increases from 26.2 mJ/m² for bare barite surface, to 44.1 mJ/m² for the surface covered with one TDACl monolayer. It seems that the bare surface is covered with two monolayers of physically adsorbed water [8, 9]. The adsorbing TDACl (or neutral molecules) may partially remove water molecules from the barite surface. This should lead to the increased dispersion interactions. The changes show that TDACl adsorption is not uniform, because, otherwise, the component should rather decrease to the value characteristic for the hydrocarbon chain, i.e., about 25 mJ/m², especially with one monolayer coverage.

The electron donor component γ_s^- is responsible for the hydrophobic properties of the barite surface (hydrogen bonding). The adsorbed ionic species of TDACl or its neutral molecules do not screen the hydrophilic centers, because for one statistical monolayer of the collector the value γ_s^- is still sufficiently high. This indicates that TDACl adsorption is a mosaic. The decrease in γ_s^- from 118.9 mJ/m² to 36.8 mJ/m² causes ▲ in the flotation from 10% to 38%, at $\Delta G = -16.5$ mJ/m² [6]. From these results, it can also be concluded that, even if the thermodynamic condition is fulfiled ($\Delta G \leqslant 0$), it is not always a sufficient condition for the fast flotation process, if its negative value is not large. A structured water film is probably present at the hydrophobic/hydrophilic barite-water interface. This may give rise to an energy barrier that has to be overcome by a gas bubble approaching the barite grain; this influences the kinetics of the process.

In Fig. 2 are shown the components of the barite surface free energy changes and the free energy changes accompanying the flotation process of the barite samples when DDSO₄Na was used.

At low coverage, DDSO₄Na probably causes dehydration of the barite surface. This appears as an increased value of the apolar component γ_s^{LW} (to 46 mJ/m²) [7]. At the same time, the electron donor component γ_s^- decreases from 119 to 23 mJ/m². The increase in the coverage to 1 statistical monolayer of DDSO₄Na is accompanied by the decrease in γ_s^{LW} to the value characteristic for long-chain hydrocarbons. This indicates that the coverage with DDSO₄Na is more uniform than that found for TDACl. This conclusion is also supported by the

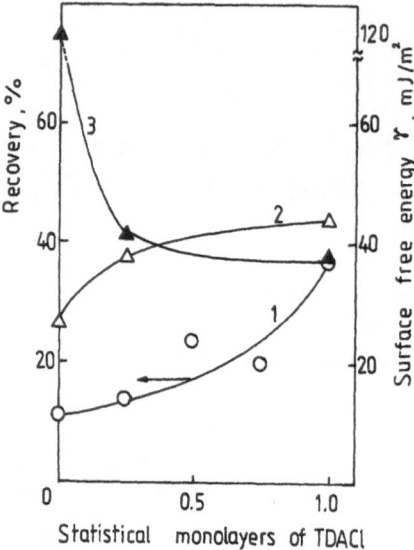

Fig. 1. Changes of flotability (curve 1), apolar component γ_s^{LW} (curve 2), and polar component γ_s^- (curve 3) of surface free energy as a function of the surface precoverage with TDACl of the barite samples

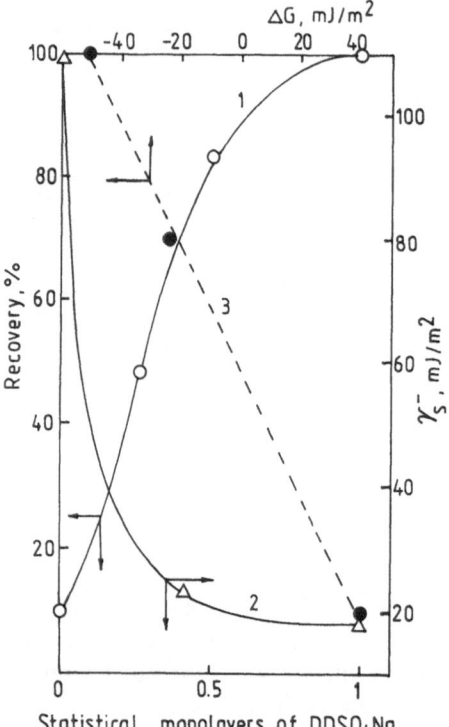

Fig. 2. Changes of flotability, electron-donor component γ_s^- (curve 2) as a function of the surface precoverage with DDSO₄Na, and the relationship between the flotability and ΔG changes of the barite samples

flotation results. At one monolayer precoverage the flotation already amounts to 100%. This increase in the flotation activity is mainly due to the sharp decrease (to 18 mJ/m^2) in the polar component γ_s^- of the surface free energy. As a result, $\Delta G = W_{spr}$ changes from +24.6 to —49.9 mJ/m^2. The consistency of the flotation results with the calculated ΔG values gives strong support for the determined surface free energy component values as meaningful quantities (Fig. 2). Straight-line dependence occurs between flotability and the free energy ΔG changes accompanying the flotation act, i.e., replacement of a mineral/water by a mineral/gas interface. However, this straight-line dependence probably takes place only in a definite range of ΔG changes around its zero value.

References

1. van Oss CJ, Good RJ, Chaudhury MK (1986) J Colloid Interface Sci 111:378—390
2. van Oss CJ, Good RJ, Chaudhury MK (1987) J Chromatog 391:53—65
3. van Oss CJ, Good RJ, Chaudhury MK (1988) Langmuir 4:884—891
4. van Oss CJ, Ju L, Chaudhury MK, Good RJ (1989) J Colloid Interface Sci 128:313—319
5. van Oss CJ, Good RJ (1989) J Macromol Sci Chem A26(8):1183—1203
6. Chibowski E, Hołysz L (1992) J Materials Sci Vol 2
7. Hołysz L, Chibowski E (1992) Langmuir 8:303—308
8. Staszczuk P, Jańczuk B, Chibowski E (1985) Materials Chem Phys 12:469—481
9. Jańczuk B, Chibowski E, Staszczuk P (1983) J Colloid Interface Sci 96:1—6

Authors' address:

L. Hołysz
Maria Curie-Skłodowska Univ.
Dept. of Physical Chemistry
20-031 Lublin, Poland

Progress in Colloid & Polymer Science

Progr Colloid Polym Sci 89:176—180 (1992)

Microcalorimetric study of cetylpyridinium-chloride adsorption onto different oxides

J. Seidel

Department of Interfacial Processes, Research Institute of Mineral Processing, Freiberg, FRG

Abstract: The adsorption behavior of cetylpyridinium-chloride (CPC) at different oxide surfaces (Al_2O_3, silicagel, SnO_2, TiO_2) was characterized by means of liquid-flow adsorption microcalorimetry. The corresponding adsorption isotherms were also measured by a flow method. It was found that all the experimental enthalpies of displacement were exothermic. The plot of the enthalpies of displacement against the amount of CPC adsorbed results in straight lines with approx. the same slope for all oxides, although the surface charge of the oxides is different. Electrostatic and lateral interactions seem to contribute only little to the measured enthalpies of displacement. In apparent contradiction to these results, the investigation of the pH and electrolyte concentration dependence of CPC adsorption onto silica gel suggests an electrostatic adsorption mechanism coupled with hemicellization.

Key words: Surfactant adsorption; flow microcalorimetry; enthalpy of displacement; adsorption isotherm; adsorption mechanism; oxide; pH effects; electrolyte effects; hemimicellization; specific interaction

Introduction

The modification of the surface properties of solids by surfactant adsorption is an essential step in many technological processes such as flotation, fine particle handling, etc. Because of the complex character of the adsorption process, our knowledge is limited or contradictory, especially concerning the adsorption mechanism.

Recently, new results on the adsorption mechanism of cationic alkylpyridinium salts at the surface of oxides have been published [1—4]. Unfortunately, there are some discrepancies in the postulated structures of the adsorption layer, as schematically shown in Fig. 1, and also about the role of chemisorption. During the last decade remarkable progress has been made in the application of adsorption calorimetry as an independent experimental method to study adsorption processes, especially to gain new insight into the mechanism of surfactant adsorption [5—8].

The aim of this work is to contribute to a better understanding of cationic cetylpyridinium-chloride (CPC) adsorption onto oxides by investigating the adsorption energetics by means of liquid-flow adsorption microcalorimetry. It is expected that the magnitude of the enthalpy of displacement and the dependence of the differential molar enthalpy of displacement from the adsorption density provide new information to discriminate between the different adsorption mechanisms mentioned above (Fig. 1).

Experimental

A scheme of the highly sensitive, home-made flow adsorption microcalorimeter is shown in Fig. 2. It is a Calvet-type microcalorimeter characterized by a minimal detectable heat effect of 1 μW and a good long time stability of the baseline. Further details concerning the properties and efficiency of the calorimeter and the measuring procedure were published in [8]. The corresponding adsorption isotherms were also measured by means of a flow method working under the same experimental conditions as in the flow calorimeter and using an UV-detector for recording the concentration of CPC

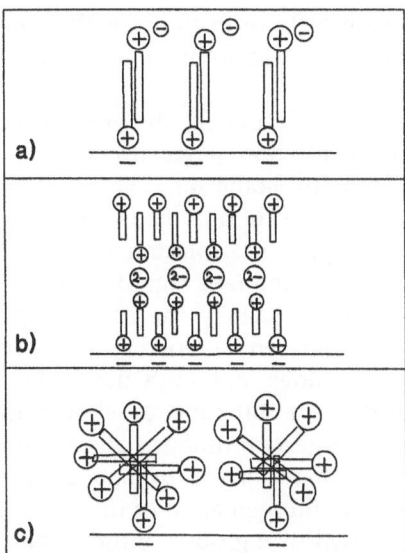

Fig. 1. Models of CPC adsorption: a) Model of Rupprecht [2]; b) Model of Schwarz et al. [3]; c) Model of Gu et al. [1]

Table 1. Selected properties of the CPC used

	CPC monohydrate (Merck, FRG)
Pretreatment	recrystallized twice from acetone/ethanol
Mol. weight	358 g/mol
Krafft point	11 °C
CMC	$8 \cdot 10^{-4}$ mol/l

Table 2. Selected properties of the oxides

Adsorbents	Spec. surface area	pH at point of zero charge
Alumina	4.0 m²/g	8.9
Silicagel	212 m²/g	2.2
Titania	6.7 m²/g	6.1
Tindioxide	6.3 m²/g	4.2

at the outlet of the adsorption column. The amount of CPC adsorbed was calculated from the retention volume as described in the literature [9, 10]. All adsorption and desorption measurements were carried out at 298 K, at a constant flow-rate of 20 cm³/h, and using a cumulative procedure, i.e., the concentration of the surfactant solutions was increased or decreased stepwise after reaching the equilibrium state.

Some important properties of the oxides and the surfactant used are shown in Tables 1 and 2.

Results and discussion

The experimentally determined adsorption and enthalpy of displacement isotherms at the uninfluenced pH-value of 5.6 are shown in Figs. 3 and 4, respectively. Assuming an area of 0.8 nm² occupied by one CPC molecule [11], monolayer adsorption would be reached at approximately

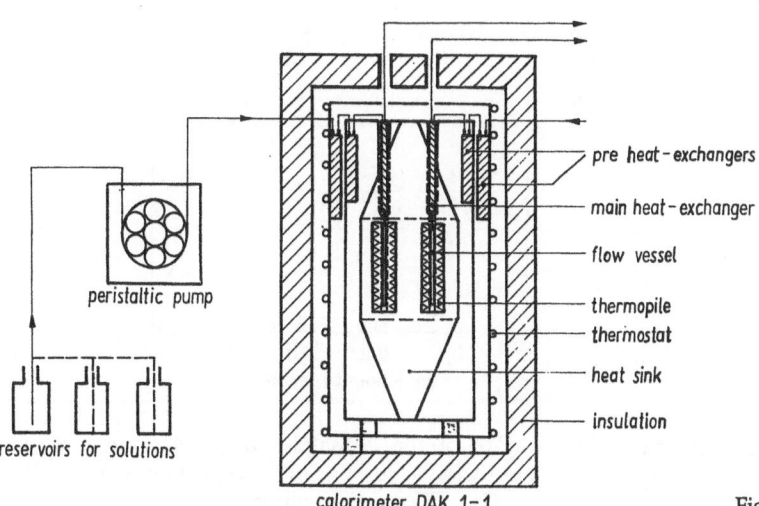

peristaltic pump

reservoirs for solutions

calorimeter DAK 1-1

pre heat-exchangers
main heat-exchanger
flow vessel
thermopile
thermostat
heat sink
insulation

Fig. 2. Scheme of the flow microcalorimeter

2 μmol/m² of adsorbed CPC. This suggests that under the given conditions only a sub-monolayer is formed (see Fig. 3).

Desorption measurements were carried out concerning the enthalpy effects, as well as the ad-sorbed amounts. Complete reversibility within the experimental uncertainty was found for both procedures.

The enthalpies of displacement are exothermic over the whole concentration range investigated. The shape of the enthalpy isotherms is very similar to that of the adsorption isotherms. The corresponding plot of the enthalpies of displacement against the amount of CPC adsorbed (see Fig. 5) results in straight lines with approximately the same slope (which represents the differential molar enthalpy of displacement) for all oxides, although the surface charges of the oxides are different, as shown in Table 2. This result suggests a uniform adsorption mechanism possibly characterized by specific interactions between the nitrogen atom of the surfactant and the surface hydroxyl groups of the oxides. Because of the relatively low value of the differential molar enthalpy of displacement (approx. −18 KJ/mol) and the reversibility of the CPC adsorption found by desorption measurements, a chemisorption mechanism is not probable. In addition, electrostatic and lateral (hemimicellization) interactions seem to contribute only little to the measured enthalpies of displacement. Referring to the formation of hemimicelles, this assumption is supported by the very small enthalpy of micellization of approx. −2 KJ/mol for CPC [12].

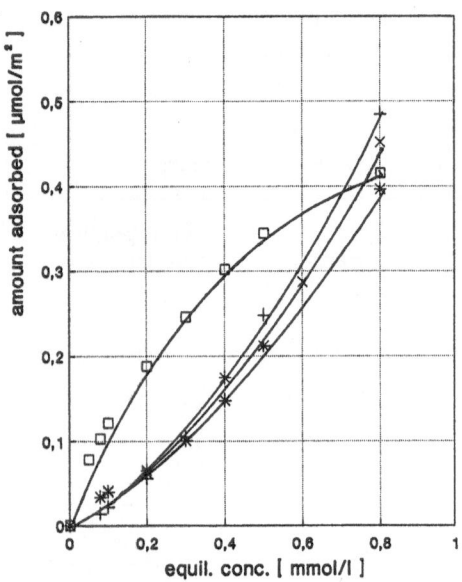

Fig. 3. Adsorption isotherms of CPC at different oxides pH = 5.6; —□— Al₂O₃; —+— SnO₂; —*— silica gel; —×— TiO₂

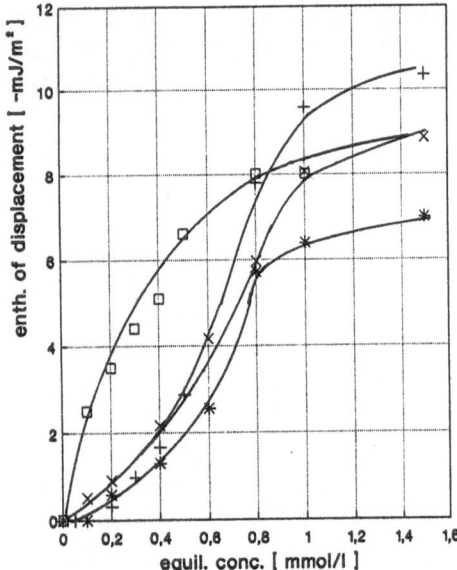

Fig. 4. Enthalpy isotherms for CPC adsorption onto different oxides; pH = 5.6; —□— Al₂O₃; —*— TiO₂; —+— SnO₂; —×— silicagel

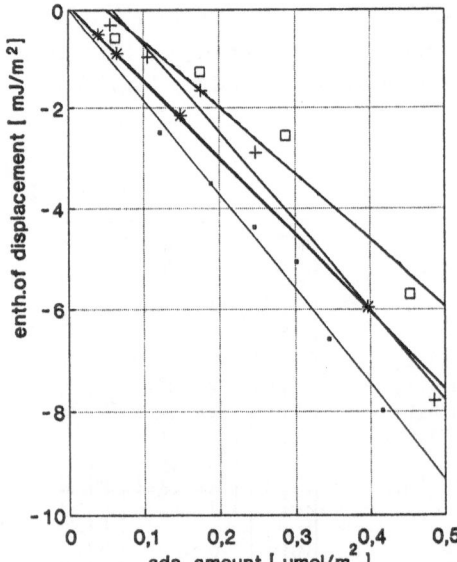

Fig. 5. Plot of the enthalpies of displacement against amount of CPC adsorbed; pH 3 5.6; —■— Al₂O₃; —+— SnO₂; —□— TiO₂; —*— silicagel

Furthermore, the influence of pH and electrolyte (KCl) concentration on the adsorption and enthalpy isotherms are shown in Figs. 6—9. It is obvious that both the amount adsorbed and the enthalpies of displacement are strongly influenced by these parameters, as expected for an electrostatic adsorption mechanism coupled with hemimicellization. In order to explain this apparent contradiction, further

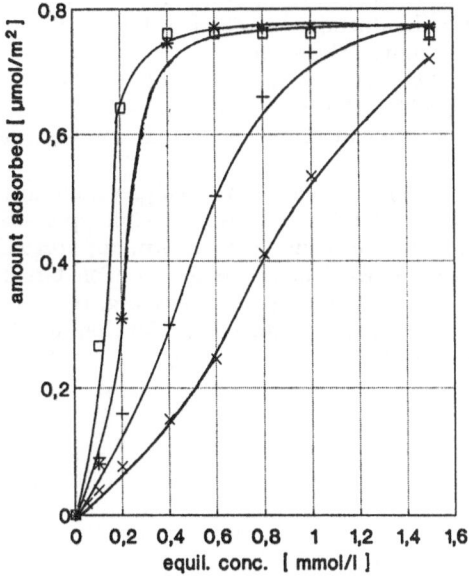

Fig. 6. Influence of KCl concentration on the adsorption of CPC onto silica gel; pH = 5.6; —×— pure water; —+— 1 mmol/l KCl; —*— 5 mmol/l KCl; —□— 10 mmol/l KCl

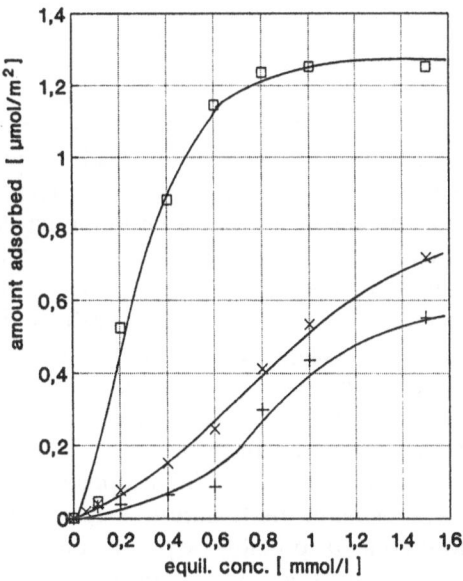

Fig. 8. Influence of pH on the adsorption of CPC onto silica gel —×— pH = 5.6; —+— pH = 4; —□— pH = 10

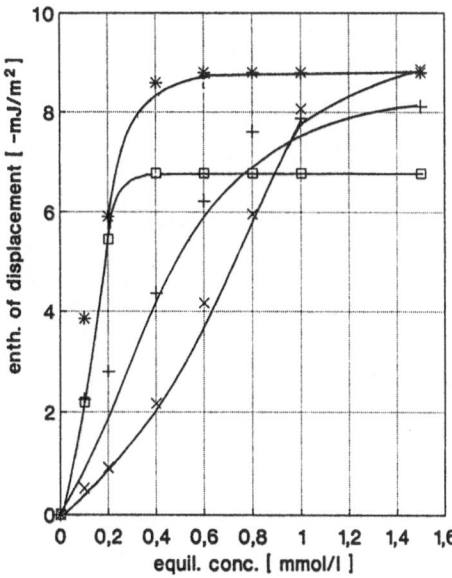

Fig. 7. Effect of KCl concentration on the enthalpy isotherms —×— pure water; —+— 1 mmol/l KCl; —*— 5 mmol/l KCl; —□— 10 mmol/l KCl

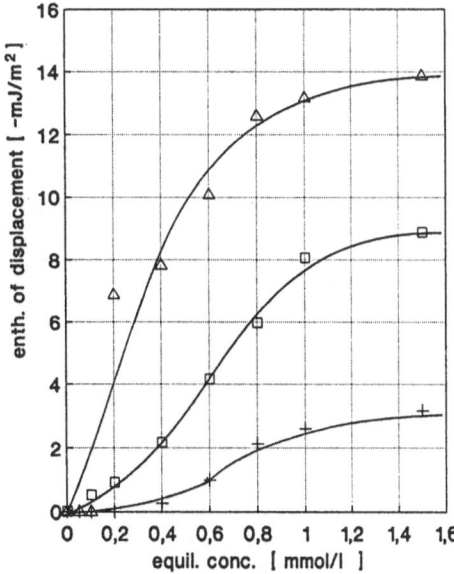

Fig. 9. Effect of pH on the enthalpy isotherms of CPC —□— pH = 5.6; —+— pH = 4; —△— pH = 10

systematic investigations will be carried out concerning the pH-dependence of adsorption onto the other oxides, especially alumina, and zetapotential measurements for the characterization of the appropriate surface charges.

Summary and conclusions

The adsorption and desorption behaviors of the cationic CPC surfactant onto different oxides have been studied by means of flow adsorption microcalorimetry. Generally, a reversible sub-monolayer adsorption with similar and constant differential molar enthalpies of displacement of approx. —18 KJ/mol was found for all oxides investigated, although the surface charges of the oxides were different. Because of the relatively low value of the differential molar enthalpy of displacement, a chemisorption mechanism is not probable.

The discrimination between the proposed adsorption mechanisms in the literature or new suggestions should only be possible by performing further systematic investigations concerning the pH-dependence of the adsorption and the behavior of the surface charges.

References

1. Gao Y, Du J, Gu T (1987) J Chem Soc, Faraday Trans I 83:2671—2679
2. Rupprecht H (1971) Kolloid Z Z Polym 249:1127
3. Schwarz R, Heckmann K, Strnad J (1988) J Colloid Interface Sci 124:50—56
4. Hough DB, Rendall HM (1983) In: Parfitt GD, Rochester CH (eds) Adsorption from solution at the solid/liquid interface. Academic Press, London, pp 247—319
5. Rouquerol J (1985) Pure Appl Chem 57:69—77
6. Partyka S, Lindheimer M, Zaini S, Keh E, Brun B (1986) Langmuir 2:101—105
7. Noll LA (1987) Colloids and Surfaces 26:43—54
8. Seidel J (1988) J Thermal Anal 33:317—322
9. Noll LA, Burchfield TE (1982) Colloids and Surfaces 5:33—42
10. Seidel J (1990) In: Thermische Analyseverfahren in Industrie und Forschung. University Jena, pp 49—54
11. Sigg J (1991) Untersuchungen zur Struktur von Adsorbaten ionischer Tenside an hydrophilen Grenzflächen. Thesis University of Regensburg, p 59
12. Fisicaro E, Pelizzetti E, Barbieri M (1990) Thermochim Acta 168:143—159

Authors' address:

Dr. Jürgen Seidel
MPI für Kolloid- und Grenzflächenforschung
Institutsteil Freiberg
Chemnitzer Str. 40
D(O)-9200 Freiberg, FRG

Progress in Colloid & Polymer Science   Progr Colloid Polym Sci 89:181—185 (1992)

Enhancing effects during the interaction of cationic surfactants and organic pollutants with clay minerals

E. Klumpp, H. Heitmann, H. Lewandowski, and M. J. Schwuger

Research Centre Jülich GmbH, FRG

Abstract: The adsorption of aromatic compounds (2-chlorophenol, 2-naphthol) in the presence of cationic C_{16}-surfactants by different clays (kaolin, illite, and bentonite (montmorillonite)) from water was studied. — The cationic surfactants alter the adsorption properties of clays by means of surface hydrophobing, which leads to a drastic increase and acceleration of the adsorption of these aromatic pollutants. If the surface is densely covered by surfactant ions, high exothermic heats of displacement for 2-naphthol are found by microcalorimetry. X-ray diffraction measurements of the basal spacing proved the intercalation of 2-chlorophenol within the hydrophobic interlayers of C_{16}-bentonite. At very high surfactant concentrations ($c \gg$ CMC) the decreased adsorption of 2-naphthol was explained by a competition for 2-naphthol molecules between micelles and hydrophobic adsorbed surfactant layer.

Key words: Adsorption; cationic surfactants; phenols; clay minerals; microcalorimetry; x-ray diffraction; kinetics

Introduction

The adsorption of organic pollutants in soils and sediments takes place primarily on colloidal clay minerals (kaolinite, montmorillonite, and illite). In the case of an annual input into the environment of several hundred thousand tons of synthetic and natural surfactants, large quantities of surface-active substances, in addition to pollutants, come into contact with these clay minerals, mainly via effluents and sewage sludge, but also as wetting and dispersing agents in pesticides and fertilizers.

The surface-active substances alter the adsorption properties of layer silicates by means of ion exchange processes, physisorption, and intercalation of molecules between the silicate layers [1]. They also affect the solubility and wetting ability of organic pollutants [2]. Thus, even small amounts of surface-active substances influence the transport and deposition of organic pollutants in soils and sediments.

There are certain technical applications which are based on these interactions between surfactants, clay minerals, and other organic substances, e.g.,

the additives in lubricants, paints, and oil drilling fluids, and those used in sealing landfills [3].

The adsorption of organic pollutants by clay minerals has been extensively studied [4]. The nature of clay-cationic surfactant complexes (organo-clays) has been also a subject of research [5]. In contrast, only few experiments on surfactant/pollutant/clay systems have been performed so far [6].

Materials and methods

The adsorbents, kaolin from Switzerland (supplied by FLUKA Chemie AG), Ca^{2+}-bentonite from Germany (supplied by Süd-Chemie AG), and illite from Hungary (supplied by Erbslöh AG) were purified and standardized according to [7]. Subsequent characterization yielded the following values. The cation exchange capacity (CEC) for kaolin was 7 meq/100 g, for Ca^{2+}-bentonite 90 meq/100 g, and for illite 27 meq/100 g [8]. The BET surface area (determined by N_2-adsorption) was 15 m^2/g for kaolin, 73.6 m^2/g for Ca^{2+}-bentonite and 36 m^2/g for illite.

Hexadecyl pyridinium chloride (CPC), hexadecyl trimethylammonium bromide (CTB) and hexadecyl benzyl dimethylammonium chloride (CBC), were chosen as cationic surfactants, and 2-naphthol and 2-chlorophenol as model-pollutants. All chemicals were delivered by FLUKA-Chemie AG in the purest form.

The adsorption isotherms were obtained by determining the equilibrium concentrations using conventional techniques such as UV derivative spectroscopy and two-phase titration. Before the analysis, the clay minerals were removed by centrifugation. More details of the experimental procedure and the nomenclature are given in [8—10].

If not otherwise specified, all experiments were performed at 298 K and the suspension concentration was 1 g clay/l.

Due to the very small particle size (0.1 μm < d < 2 μm) the centrifugation of samples was not suitable for kinetic investigations, since separation takes about the same time as the establishment of equilibria. For this reason, a continuous method was developed for separating the colloids. A double-piston pump was used to suck the clear solution from the suspension through stainless-steel frits and pump it through the high-pressure cells of a UV-VIS spectrophotometer. The solution was then passed back into the suspension through another filter frit. This arrangement made it possible to remove any filter cake that may have formed on the suction frit by changing the direction of flow at certain time intervals. The direction of flow was changed by means of a motor switch valve (Fig. 1). Kinetics within periods of minutes up to periods of several hundred hours were studied.

The heat of displacement ($\Delta_{21}H$) was isothermally determined using a Tronac titration calorimeter, model 1250.

The basal spacing of swellable clay minerals was measured with a theta-theta x-ray diffractometer XRD 3000 from Seifert. The wet samples covered by a Mylar film were measured directly after centrifugation [11].

Results

Figure 2 shows the adsorption isotherms of CPC on Ca^{2+}-bentonite and kaolin. The isotherms increase very steeply to a plateau at $n_1^{\sigma(v)}/m$, slightly above the CEC. The adsorbed amount of CPC on montmorillonite is higher by one order of magnitude as compared to kaolin [12]. In contrast, Ca^{2+}-bentonite and kaolin adsorbed the same amounts of 2-naphthol.

Kinetic measurements reveal a clearly faster adsorption of cation surfactants as compared to the adsorption of pollutants (Fig. 3). If cationic surfactants and pollutants are adsorbed simultaneously, a significant increase and acceleration in pollutant adsorption is observed. A maximum is passed during the first few minutes in certain surfactant/pollutants systems.

Results of the equilibrium investigations are summarized in Fig. 4. In contrast to adsorption isotherms, here the adsorbed amount of pollutant out of a 0.1 mmol/l 2-naphthol solution (1 g clay/l suspension) is plotted as a function of the total amount of surfactant in the system. At low surfactant concentrations higher amounts of naphthol are

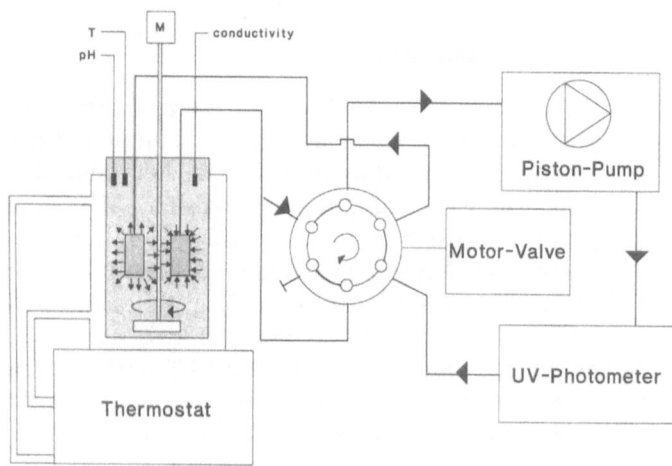

Fig. 1. A continuous determination of UV-active substances by alternating dead-end filtration of the suspension

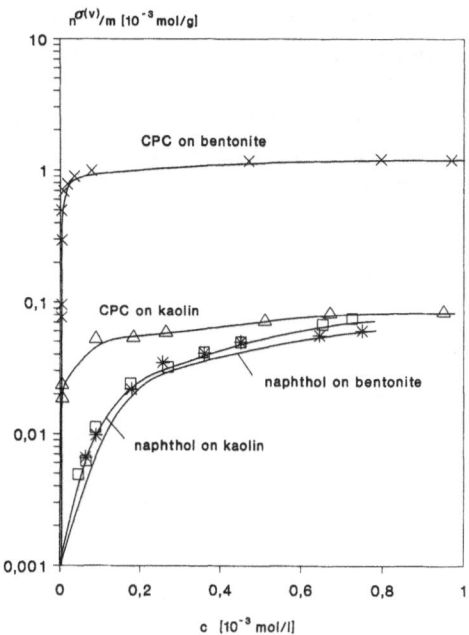

Fig. 2. Adsorption isotherms of cetylpyridinium chloride (CPC) and 2-naphthol on bentonite and kaolin

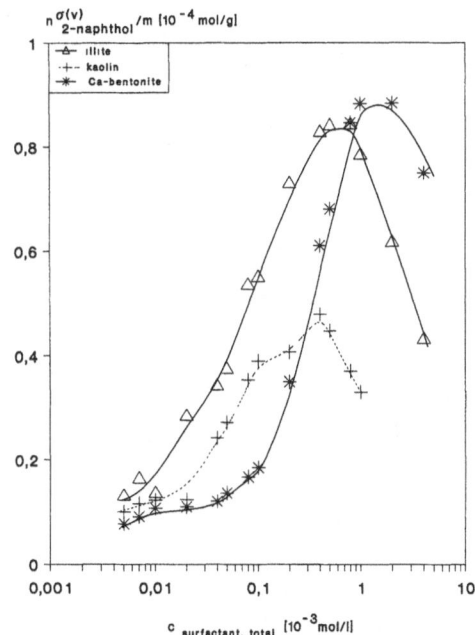

Fig. 4. The amount of 2-naphthol adsorbed from 0.1 mM 2-naphthol solution as a function of total surfactant quantities (CBC) in the system on different clays (1 g clay/l suspension)

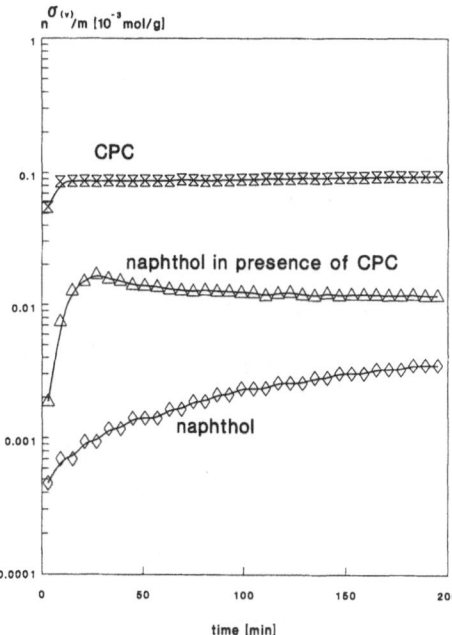

Fig. 3. Adsorption kinetics of 2-naphthol with and without the addition of 0.1 mM CPC and of CPC on bentonite (1 g clay/l suspension)

adsorbed on kaolin than on montmorillonite. This changes at higher surfactant concentrations [8]. It may be seen that naphthol adsorption for kaolin and montmorillonite has a maximum at a certain surfactant concentration.

Calorimetric measurements indicate that the adsorption of 2-naphthol on pure clay minerals is endothermic, while it is exothermic when the bentonite is covered by surfactant ions (Fig. 5, 6). For organo-kaolin (33% CEC) the enthalpy changes from exothermic to endothermic at increasing 2-naphthol concentration (Fig. 6).

The basal spacing of C_{16}-bentonite is not enlarged by the adsorption of small amounts of 2-naphthol ($>2 * 10^{-4}$ mol/g). Higher amounts of 2-naphthol adsorbed are hardly obtainable because of the low solubility of 2-naphthol in water. To prove the intercalation of pollutants, 2-chlorophenol with a higher solubility was chosen for x-ray diffraction experiments.

According to this, Fig. 7 shows the adsorption isotherm for 2-chlorophenol on 33%, 66%, and 100% CEC exhanged bentonite. It can be clearly seen that the 2-chlorophenol adsorption increases with increasing surfactant coverage [6]. The adsorb-

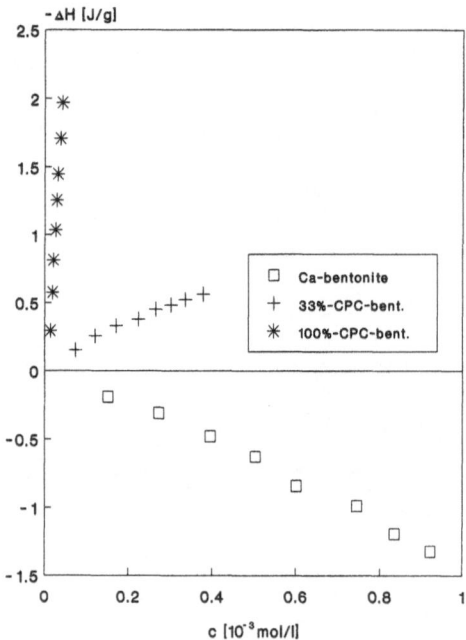

Fig. 5. Heats of displacement for 2-naphthol on Ca^{2+}-bentonite and on bentonite covered with CPC (33% and 100% CEC, 16 g clay/l suspension)

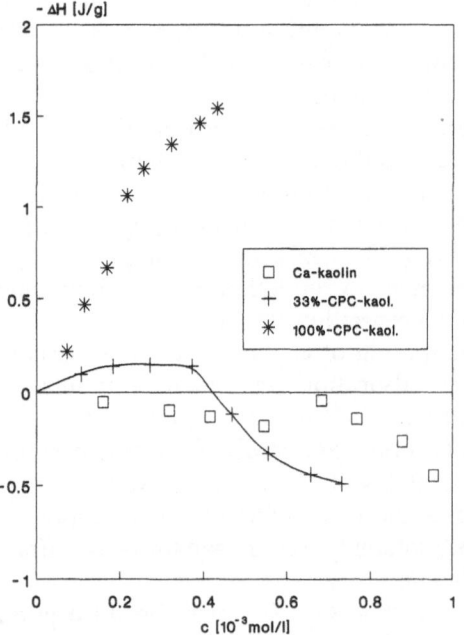

Fig. 6. Heats of displacement for 2-naphthol on kaolin and on kaolin covered with CPC (33% and 100% CEC, 16 g clay/l suspension)

ed amount of 2-chlorophenol is more than one order of magnitude higher than for 2-naphthol [8].

For 33% and 66% CEC organo-bentonites, the basal spacing is determined by the intercalated surfactant molecules (Fig. 8 lower curves). Only at very high amounts of adsorbed surfactant (100% CEC)

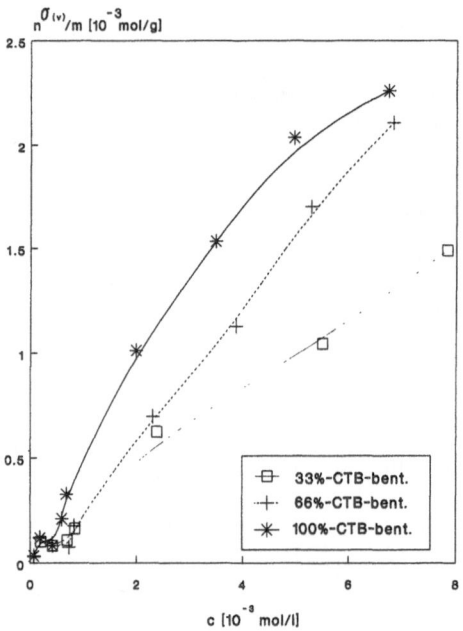

Fig. 7. Adsorption isotherms of 2-chlorophenol on CTB-bentonites

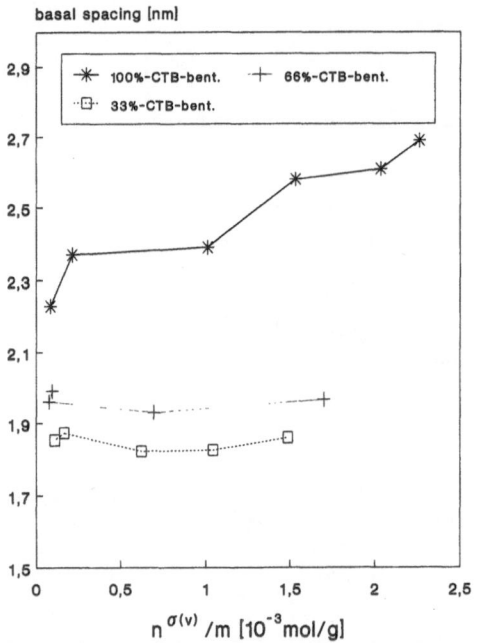

Fig. 8. Basal spacing of CTB-bentonites as a function of the adsorbed amount of 2-chlorophenol

an additional interlayer separation is observed (Fig. 8 upper curve) and it is proportional to the amount of adsorbed 2-chlorophenol.

Discussion

The cation exchange capacity and the accessible surface area of the adsorbent are of great importance for the adsorption of cationic surfactant, but less decisive for adsorption of nonionic organic pollutant, such as 2-naphthol.

Adsorption of cation surfactants is extremely fast, while the adsorption of naphthol needs several hours. Simultaneous adsorption of cation surfactant and pollutant leads to a drastic increase and acceleration of pollutant uptake.

In the case of simultaneous adsorption of pollutants and surfactants, a further variable must be included, i.e., charge density (CEC/specific surface area of the clay mineral). Although kaolin and illite have a smaller surface area, there are more exchangeable charges on these surfaces in relation to montmorillonite. In the case of ion-exchange by cationic surfactants, a higher charge density of the adsorbent results in the formation of a denser adsorbed layer. Dense and compact hydrophobic surfactant films have the power to store more pollutant molecules compared to the same number of surfactant molecules sporadically adsorbed on a greater surface area [13, 9]. This explains the course of the curves in Fig. 4. At low surfactant concentrations, illite and kaolin adsorb more pollutant compared to montmorillonite. Beyond a certain quantity of added surfactant, the larger montmorillonite surface is also covered with a dense surfactant film. Consequently, then montmorillonite adsorbs more pollutant.

At even higher surfactant concentrations (above CMC), micelles are formed in the solution, which compete with the adsorbed surfactant films for pollutant molecules. This leads to a decrease of the amount of adsorbed pollutant molecules. This phenomenon illustrates the soil-washing process [14] by means of cationic surfactants.

The endothermic heats of displacement of the pure clays can be explained by a dehydration of the clay surface as a consequence of the adsorption of 2-naphthol molecules. Exothermic enthalpies are measured when 2-naphthol adsorbs on a surface covered by higher amounts of surfactant molecules (0.92 mmol CPC/g bentonite); this means that 2-naphthol is preferably adsorbed within the domains of the surface covered by surfactant ions. In the case of CPC-kaolin (0.016 mmol CPC/g kaolin eqv. to 33% CEC), the adsorbed amount of surfactant is not sufficient to accomodate all 2-naphthol molecules. Thus, for smaller concentrations of 2-naphthol, adsorption at the hydrophobic patches is exothermic, while at higher amounts of adsorbed naphthol the enthalpy becomes positive.

The increased basal spacing at high 2-chlorophenol concentrations proves the intercalation of these organic pollutants into the interlayer space [15].

Acknowledgements

The experimental contributions of U. Paffen and B. Mainz are gratefully acknowledged. The kinetic studies and the novel kinetic method are part of the Ph. D. thesis of H. Heitmann at the University of Dortmund.

References

1. Lagaly G (1984) Phil Trans R Soc Lond A 311:315
2. Kile OE, Chiou CT (1989) Environ Sci Techn 23:832
3. Boyd SA, Lee JF, Mortland MM (1988) Nature 333:345
4. Hamaker JW, Thompson JM (1972) In: "Organic Chemicals in the Soil Environment", Vol 1, Ed.: Goring CAI, Hamaker JW, Marcel Dekker, New York
5. Lagaly G, Weiss A (1970) Kolloid Z Z Polymere 273:266—273, 364—368
6. Mortland MM, Shaobai S, Boyd SA (1986) Clays and Clay Minerals 34(5):581—585
7. Malberg R, Dekany I, Lagaly G (1989) Clay Minerals 24:631
8. Klumpp E, Heitmann H, Schwuger MJ (1991) Tenside, Surf Det 28:441
9. Dekany I, Szanto F, Weiss A, Lagaly G (1985) (1986) Ber Bunsenges Phys Chem 89:62—67, 90:422—427, 427—431
10. Rheinländer T, Klumpp E, Rossbach M, Schwuger MJ in this edition
11. Dekany I, Marosi T, Weiss A (1991) 3rd Int Conf on Fundamentals of Adsorption, Mersman AB (ed) New York, pp 221—228
12. Röhl W, von Rybinski W, Schwuger MJ (1991) Progress in Colloid and Polymer Sci 84:206
13. Somasundaran P, Fuerstenau DW (1966) J Phys Chem 70:90
14. Pramauro E, Pelizzetti E (1990) Colloids and Surfaces 48:193
15. Lagaly G, Witter R (1982) Ber Bunsenges Physik Chemie 86:74—80

Authors' address:

M. J. Schwuger
Institute for Applied Physical Chemistry
Research Centre Jülich
P.O. Box 1913
W-5170 Jülich, FRG

Progress in Colloid & Polymer Science   Progr Colloid Polym Sci 89:186—189 (1992)

Interfacial activity of acidic organophosphorus extractants and interfacial mechanism of metal extraction

J. Szymanowski and R. Cierpiszewski*)

Poznań Technical University, Institute of Chemical Technology and
*) Engineering and Academy of Economics, Poznań, Poland

Abstract: The interfacial tension isotherms for bis(di-2-ethylhexyl)phosphoric acid and 2-ethylhexyl phosphonic acid mono-2-ethylhexyl ester in aliphatic hydrocarbon-water systems were interpreted and used to propose an interfacial mechanism for the extraction of copper(II), zinc(II), and calcium(II). The surface excess isotherms were modeled using the Szyszkowski isotherm and spline function. The results suggest an interfacial mechanism of the metal extraction with the formation of the final product solvated by organic acid molecules as the rate-limiting step.

Key words: Acidic organophosphorus extractants; extraction mechanism; interfacial activity

Introduction

Acidic organoposphorus extractants have received much attention due to their potential in reprocessing nuclear fuel, in the selective recovery of metals, and in analytical chemistry [1]. Three different types of these acids are technologically important:

$$
\begin{array}{ccc}
\text{RO} \diagdown \quad \text{O} & \text{RO} \diagdown \quad \text{O} & \text{R} \diagdown \quad \text{O} \\
\text{P} & \text{P} & \text{P} \\
\text{RO} \diagup \quad \text{OH} & \text{R} \diagup \quad \text{OH} & \text{R} \diagup \quad \text{OH} \\
\text{I} & \text{II} & \text{III}
\end{array}
$$

Bis(2-ethylhexyl)phosphoric acid (I), abbreviated as DEHPA, has been studied for several years as an extractant of various metals from acidic sulphate solutions. 2-Ethylhexyl phosponic acid mono-2-ethylhexyl ester (II) is the active component of the commercial extractants PC-88A (Daihachi), SME 418 (Shell), and P-507 (China), whereas di(2-ethylhexyl)phosphonic acid (III) is the active substance of the extractant P-299 (China). Extractants of type II and III are used for cobalt(II) — nickel(II) separation.

Studies of extraction kinetics using hydrophobic hydroxy oximes have revealed that metal complexa-tion occurs at the interface after preadsorption of hydroxy oxime molecules [2—5]. An interfacial mechanism was also proposed for the extraction of metals with acidic organophosphorus extractants [6]. In this case, dimerization occurs and the association constants are high; values of $\log K_d$ near 4 are reported [7].

The aim of this work was to use the interfacial tension data to interpret the mechanism of metal extraction with acids containing phosphorus atoms.

Interfacial tension data

Interfacial tension data for bis(2-ethylhexyl)-phosphoric acid (I) and 2-ethylhexylphosphonic acid (II) were determined in the following systems: i) I — hexane-water (pH 1, 20°C) [8], ii) I — dodecane — 1 M HNO_3 (25°C) [9], iii) II — hep-tane-water (pH 1—3.5, 30°C) [10] and iv) II — dodecane — 1 M HNO_3 (25°C) [6]. The appropriate data for association, dissociation, partition and purity were given and discussed previously [7].

The computer program MINEX [11] was used for processing interfacial tension data and modeling the kinetics of the interface reactions, taking into account the effects of aqueous phase acidity and extractant concentration.

Model of interfacial reactions

The extraction kinetics of copper(II), zinc(II), and calcium(II) by compounds I and II were studied by several authors independently [1, 10, 12], and reaction schemes were proposed. They can be formulated as follows (reaction mechanism {A}):

$$\{A\}: \begin{cases} [(HL)_2]_o = [(HL)_2]_{ad} & (1) \\ M^{2+}_{int/w} + [(HL)_2]_{ad} = [ML_2]_{ad} + 2 H^+_{int/w} & (2) \\ [ML_2]_{ad} + n[(HL)_2]_{int/o} = [ML_2] + n[HL]_o, & (3) \end{cases}$$

where HL and ML_2 denote the acidic organophosphorus extractant and its complex with a metal, M^{2+}, respectively, subscripts o and w denote organic and aqueous phases, respectively, ad denotes the adsorption layer and int stands for the interface layer. Dehydration reactions were not considered to simplify the reaction scheme, i.e., it was assumed that dehydration does not affect the rate of extraction. The value of n depends upon the metal species extracted and is equal to 1 for copper(II) [10], equal to 0.5 for zinc(II) [12] and equal to 2 for calcium(II) [1]. Reaction 3 is assumed to be the rate limiting step of the process.

As an alternative mechanism, {B}, the dissociation of one proton from the dimer has been proposed, which is followed by the reaction with the metal in its ionic form [12]:

$$\{B\}: \begin{cases} [(HL)_2]_{ad} = [(HL_2^-)]_{ad} + [H^+_{int}]_w & (4) \\ [M^{2+}_{int}]_w + [(HL_2^-)]_{ad} = [ML_2] + [H^+_{int}]_w. & (5) \end{cases}$$

Although it is difficult to decide which mechanism is the more appropriate, both reaction schemes emphasize the role of the interface in the system and underline the importance of measurements of the interface tension and the interfacial concentration of organophosphorus extractants.

Interfacial concentration and reaction rate of metal extraction

Taking into account that reaction 3 of mechanism {A} is the rate limiting one and neglecting diffusion steps, the following kinetic equation is obtained

$$r = \frac{k[M^{2+}]_w [(HL)_2]_{ad} [(HL)_2]_o^n}{[H^+]_w^2} . \tag{6}$$

Only mechanism {A} yields quantitative agreement with the experimental data and describes successfully both the effect of reagent concentrations and aqueous phase acidity.

Thus, for $[M^{2+}]_w$ = const. and $[H^+]_w$ = const. the extraction rate should be proportional to the product of interfacial and bulk concentrations with various exponents:

$$r = a[(HL)_2]_{ad} [(HL)_2]_o^n , \tag{7}$$

where n = 1, 0.5 and 2 for Cu(II), Zn(II) and Ca(II), respectively. If one assumes that reaction 2 is the slowest one, the extraction rate should be proportional to the product of the metal concentration in the aqueous phase and interfacial concentration of the extractant dimer:

$$r = k[M^{2+}]_w [(HL)_2]_{ad} , \tag{8}$$

which is in disagreement with experimental kinetic data [1, 9, 12].

Results and discussion

Interfacial tension data (Fig. 1) can be well fitted by the Szyszkowski equation [9].

$$\gamma = \gamma_0[1 - B \ln(c/A + 1) , \tag{9}$$

where γ and γ_0 denote the interfacial tensions for extractant concentrations c and zero, respectively, and A and B are the adsorption coefficients.

The spline function, constructed from polynomials approximating small subranges [11], also fits very well to the experimental interfacial tension data. The average square error of such estimation, defined as $\Sigma(\gamma^{exp} - \gamma^{appr})^2/n$, where superscripts exp and appr denote experimental and approximated values of the interfacial tension in Nm^{-1} and n stands for number of experimental data, is about 10^{-7}. Extractant II is somewhat more surface active than I (Fig. 1). Both functions used as approximations give similar surface excess isotherms. The deviations between isotherms are small. The average difference between surface excess values of the extractant dimer at saturation is $0.15 \cdot 10^{-6}$ mol m^{-2}.

The interface becomes already saturated at a low extractant concentration of about 10^{-4}–10^{-3} mol dm^{-3} (Table 1). Thus, in the actual extraction system the interface is saturated.

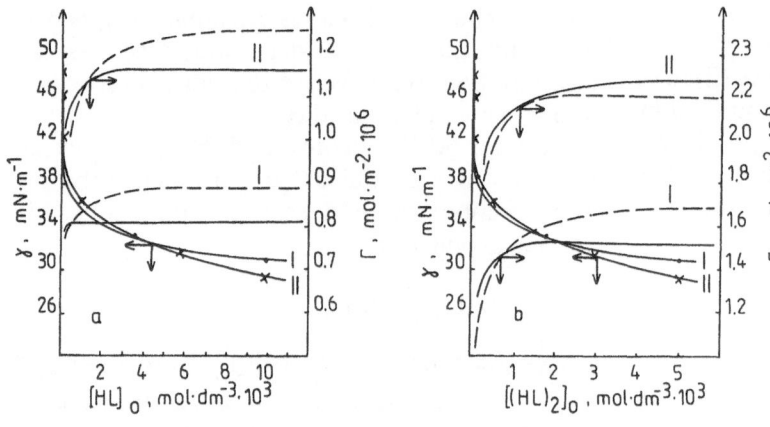

Fig. 1. Interfacial tension and surface excess isotherms for bis(2-ethylhexyl)phosphoric acid (I) (hexane-water pH 1) and 2-ethylhexyl phosphonic acid mono-2-ethylhexyl ester (II) (heptane-water, pH 1.1). a and b = functions of monomer and dimer concentration, respectively; o, × = experimental interfacial tension data for I and II, respectively; descending solid lines = approximated interfacial tension isotherms; ascending lines = approximated surface excess isotherms; solid lines = Szyszkowski equation; dahsed lines = spline function

Table 1. Surface excess at saturation for dimers

Extractant	System	Szyszkowski				Spline			
		C^{sat}		Γ		C^{sat}		Γ	
		mol	dm^{-3}	mol	$m^{-2}\ 10^6$	mol	dm^{-3}	mol	$m^{-2}\ 10^6$
I	hexane-water	2.3	10^{-4}		1.52	6	10^{-4}		1.69
	dodecane-1 M HNO_3	1.0	10^{-3}		1.43	6.3	10^{-5}		1.55
II	heptane-water pH 3.5	1.2	10^{-3}		1.87	1.2	10^{-3}		2.14
	heptane-water pH 1.1	4.0	10^{-3}		2.27	4.0	10^{-3}		2.21
	dodecane-1 M HNO_3	1.3	10^{-3}		2.00	1.3	10^{-4}		1.85

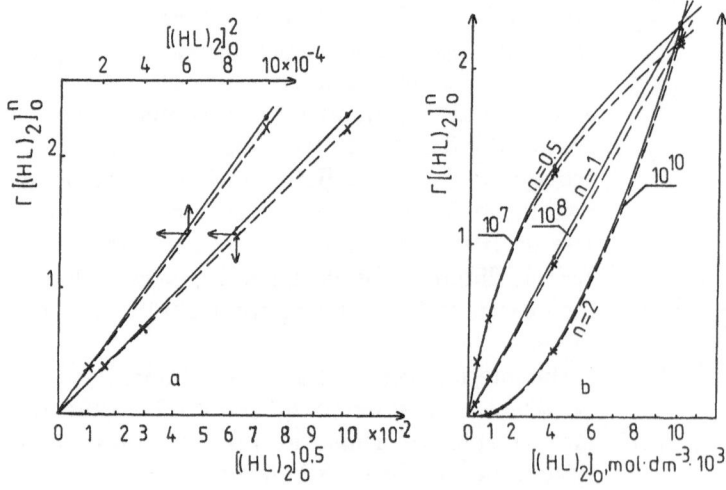

Fig. 2. Effect of extractant concentration upon the predicted relative reaction rates of metal extraction. a = functions of the dimer concentration with exponents of 0.5 or 2; b = functions of the dimer concentration; o, × = calculated values for I and II, respectively; solid lines = Szyszkowski equation; dashed lines = spline function

Both analytical descriptions of the data, i.e. the Szyszkowski isotherm and the Spline function, may be used to model the interfacial tension kinetics. Taking into account the high association constant of the extractant and the kinetic data, the dimer concentration was used for modeling.

Figure 2 shows the impact of the dimer concentration of bis(2-ethylhexyl)phosphoric acid and 2-ethylhexylphosphoric acid upon the product of surface excess and dimer concentration in the aqueous phase with various exponents, which are appropriate for copper (n = 1), zinc (n = 0.5), and calcium (n = 2). These graphs shows a typical behavior of the reaction kinetics (Fig. 2b) and with characteristic slopes can be easily converted into the corresponding straight lines (Fig. 2a). The slopes differences between the functions computed from the Szyszkowski isotherm or the spline function are negligible.

Thus, using scheme [A] of the interfacial reaction (equations 1—3) it is possible to derive kinetic equations containing the extractant interfacial concentration as estimated from the interfacial tension data. The kinetic equations are in agreement with the experimental data when it is assumed that the extractant is dimerized.

Conclusion

Both the Szyszkowski equation and the spline function may be used to model interfacial tension data in acidic organophosphorus extraction systems and kinetics of interfacial metal extraction.

Using interfacial tension and association data it is possible to describe the processes of metal extraction (Cu(II), Zn(II) and Ca(II)) in the framework of an interfacial reaction model.

Acknowledgement

This work was supported by the Polish grant KBN PB 843/3/91.

References

1. Danesi PR, Chiarizia R (1980) CRC Crit Rev Anal Chem 10:1
2. Harada M, Miyake Y (1989) In: Handbook of Heat and Mass Transfer. Gulf Publ 3:789—882
3. Szymanowski J, Prochaska K (1988) J Colloid Interface Sci 123:456—465
4. Szymanowski J, Prochaska K (1988) J Colloid Interface Sci 125:649—666
5. Szymanowski J, Prochaska K (1989) J Radioanal Nucl Chem 129:251—263
6. Vandegrift GF, Howritz EP (1980) J Inorg Nucl Chem 42:119-125
7. Cote G, Szymanowski J, J Chem Tech Biotechnol, submitted
8. Goankar AG, Neumann RD (1987) J Colloid Interface Sci 119:251—261
9. Nikitin S, Panteleewa, Szmidt W (1985) Radiochemia 2:179—183
10. Sato Y, Akiyoshi Y, Kondo K, Nakashio F (1989) J Chem Eng Japan 22:1182—1189
11. Prochaska K, Alejski K, Szymanowski J (1989) Proc 2nd International Conference of Separation Science and Technology, Hamilton, Ontario, Canada 181—182
12. Huang T, Juang R (1986) J Chem Eng Japan 19:379—386

Authors' address:

Prof. Dr. J. Szymanowski
Technical University of Poznań
Institute of Chemical Technology and Engineering
pl. Skłodowskiej-Curie 2
60-965 Poznań, Poland

Progress in Colloid & Polymer Science   Progr Colloid Polym Sci 89:190—193 (1992)

Adsorption studies on pesticide/cationic surfactant/bentonite systems

T. Rheinländer, E. Klumpp, M. Rossbach, and M. J. Schwuger

Institute for Applied Physical Chemistry, Research Centre Jülich, Germany

Abstract: Surfactant-bentonite complexes (organo-clays) with different amounts of adsorbed dodecyltrimethylammonium bromide (DTAB) were prepared. The adsorption of the pesticides paraquat and biphenyl from aqueous solution on clays was studied by the batch method with radiotracers. — On organo-clays the preferentially adsorbed di-cation paraquat only partially replaces the mono-cationic surfactant. This result was supported by investigations with microcalorimetry and x-ray diffraction. — On the contrary, the adsorption of the hydrophobic biphenyl is enhanced with an increased amount of adsorbed DTAB by hydrophobization of the bentonite.

Key words: Adsorption; pesticides; cationic surfactant; Ca^{2+}-bentonite; microcalorimetry; x-ray diffraction

Introduction

Pesticides are distributed worldwide on plants and soils by agricultural application. They are often used in combination with surfactants as dispersing or wetting agents, thus, the fate of both types of these substances within the soil is of great relevance to environmental science. In addition, pesticide formulations may include clay minerals as fillers, which sometimes are pretreated with cationic surfactants. The adsorption of individual organic substances on soils has already been investigated extensively [1, 2], whereas only a few experiments have been performed on the interactions between pesticides, surfactants, and soil components [3].

The study focuses on the physico-chemical interactions of pesticides and surfactants with clay minerals, i.e., the main inorganic component in the complex soil system.

Experimental

Materials

Ca^{2+}-bentonite from Bavaria (Süd-Chemie) with a content of 93% Ca^{2+}-montmorillonite was used

as the adsorbent [4]. The cation exchange capacity (CEC) of this bentonite is 0.9 meq/g and was determined according to Mehlich [5]. The BET surface area accessible to nitrogen molecules was 66 m^2/g. For the swelling layer silicate montmorillonite the theoretical specific surface area is approximately 750 m^2/g [6]. All data specified in this paper refer to air-dried Ca^{2+}-bentonite, which contained about 11 wt. % water according to thermogravimetric measurements (393 K).

The pesticides used were the post-harvest fungicide biphenyl, with a water solubility of only 45 μmol/l [7], and 1,1'-dimethyl-4,4'-bipyridinium dichloride, which is applied as herbicide under the common name paraquat. As cationic surfactant dodecyltrimethylammonium bromide (DTAB) is chosen. It has a critical micelle concentration (CMC) of 15 mmol/l [8].

Radioactively labeled organic substances were used. ^{14}C-labeled pesticides were supplied by Sigma. 1,1'-Dimethyl-4,4'-bipyridinium dichloride labeled in the methyl group had a specific activity of 450 MBq/mmol, whereas biphenyl had a specific activity of 280 MBq/mmol. Inactive DTAB was labeled by catalytic tritium exchange at the Zentralinstitut für Kernforschung Rossendorf. The resulting specific activity was 150 GBq/mmol.

If not specified otherwise, all data relate to 298 K, at which the experiments were performed.

Methods

The adsorption isotherms were investigated by the batch technique. The labeled substances were usually diluted with inactive material. Pesticide and/or surfactant solution were added to the bentonite swollen 24 h in water. The solid content was 1 g/l. The specimens were shaken for 2 (minimum) to 4 days, depending on the previously determined time, until the adsorption equilibrium was reached. After ultracentrifugation, the activity of the supernatant solution was measured by a LS 5000 TA scintillation counter from Beckman; it is capable of simultaneously determining activities of tritium and carbon-14 in one specimen.

The heat effects during adsorption were measured by a titration calorimeter from Tronac, model 1250, in the isothermal mode. The basal spacings of air-dried bentonite adsorbate complexes were determined by a XRD 3000 theta/theta diffractometer from Seifert.

The molecular exchange rate between the bulk phase and the adsorbed layer describes the dynamics of the adsorption equilibria. This was studied by the technique of isotopic molecular exchange [9]. An inactive solution was added to a suspension of bentonite. To the equilibrated solution an infinitesimal amount of the labeled solute was added and the decrease of activity in the solution was followed over time. For the cationic substances the exchange between free and adsorbed molecules is completed within 30 min. In this time, paraquat and DTAB also reach adsorption equilibrium. These results are valid for initial concentrations below and above the CEC [10].

For liquid mixtures, accumulation of one component is generally accompanied by displacement of the others (e.g., solvent) in the adsorbed layer. Therefore, the thermal effect during adsorption is the enthalpy of displacement $\Delta_{21}H$. According to the depletion method the difference of the concentration before the adsorption $c_{i,0}$ and at adsorption equilibrium c_i leads to the surface excess amount on volume basis $n_i^{\sigma(v)}$ ($i = 1, 2$) [11]. Since the interface of montmorillonite accessible in solution depends on many factors, the adsorbed amounts is related to the mass m. In diluted solutions with preferential adsorption of the solute, the surface excess equals the adsorbed amount of component i (n_i^s) in the adsorption layer [12].

$$n_i^{\sigma(v)}/m = (c_{i,0} - c_i)V/m \approx n_i^s/m ,\qquad (1)$$

where V is the total volume of solution.

Results and discussion

The cationic substances paraquat (PQ) and DTAB are strongly adsorbed on Ca^{2+}-bentonite (Fig. 1). Their isotherms are of the high-affinity type [13]. For PQ, saturation is reached close to the CEC of the Ca^{2+}-bentonite (0.41 mmol/g). This is the so-called strong adsorption capacity [14]. The PQ isotherm does not change at different temperatures or in the presence of cations like Ca^{2+}, which cannot compete with PQ [15—17].

DTAB is strongly adsorbed on bentonite up to a surface excess amount of 0.6 mmol/g (see Eq. 2), which corresponds to a monolayer of DTA^+-ions (1 nm^2) between the layers. The adsorption of DTAB above the range of strong adsoption is reversible and the increase over the CEC indicates that, apart from the Coulomb forces, the surfactant ions are adsorbed by other forces (van der Waals forces, hydrophobic bonding) [18]. For the nonionic biphenyl the adsorption isotherms are of the L-type (Fig. 3).

The adsorption of PQ was also investigated on clays which had been pretreated with DTAB solutions (organo-clays). For two organo-bentonites with different amounts of adsorbed DTAB, Fig. 2a illustrates the amounts of adsorbed PQ and DTAB as function of the equilibrium concentration of PQ. The preferentially adsorbed dication PQ only partially replaces the monovalent surfactant cation. The incomplete displacement of the adsorbed surfactant by PQ indicates that an exchange equilibrium is present (see Eq. (3)). For each equilibrium concentration the adsorbed amount of PQ and DTAB equals the CEC.

$$Ca^{2+}\text{-montm.} + 2\ DTA^+$$
$$\rightleftharpoons (DTA^+)_2\text{-montm.} + Ca^{2+}\qquad (2)$$
$$(DTA^+)_2\text{-montm.} + PQ^{2+}$$
$$\rightleftharpoons PQ^{2+}\text{-montm.} + 2\ DTA^+ .\qquad (3)$$

It can be seen from the calorimetric measurements that the interactions of the cationic PQ and

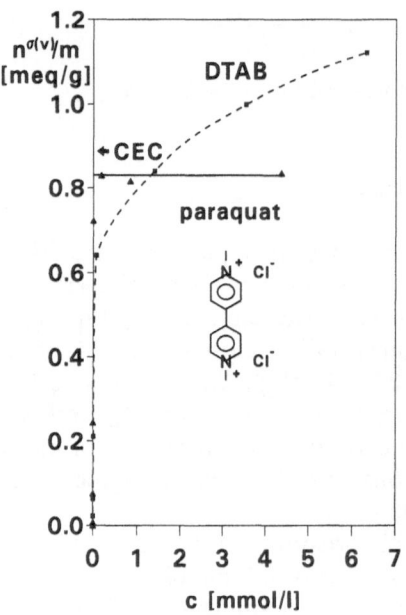

Fig. 1. Single-component adsorption of paraquat and DTAB on Ca²⁺-bentonite

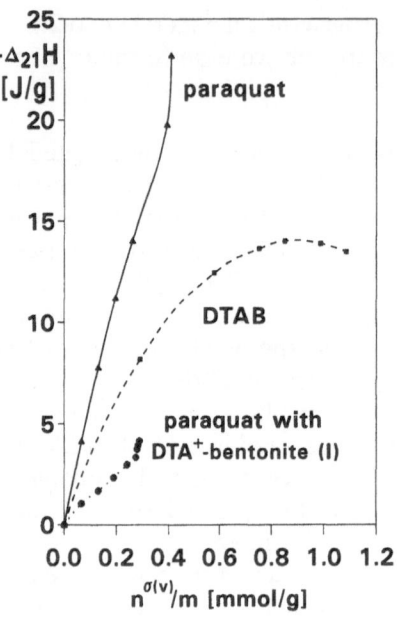

Fig. 2b. Enthalpies of displacement $\Delta_{21}H$ for paraquat, DTAB, and system (I) (in Fig. 2a) from aqueous solution on bentonite; $n_i^{\sigma(v)}/m$ = surface excess of paraquat or DTAB

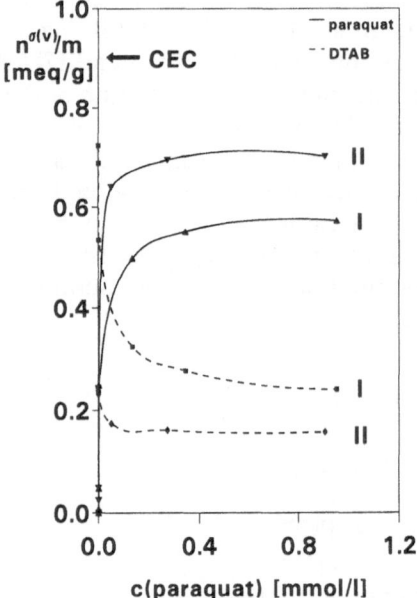

Fig. 2a. Adsorption of paraquat with partial displacement of DTAB from DTA⁺-bentonite with (I) 0.73 meq/g and (II) 0.25 meq/g DTA⁺ preadsorbed

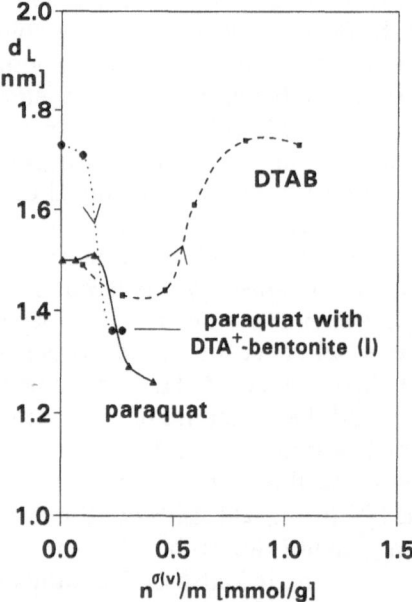

Fig. 2c. Basal spacings d_L of bentonite for system (I) (in Fig. 2a) and in the presence of paraquat and DTAB

DTAB with the negatively charged bentonite layers in aqueous solution are highly exothermic, especially for PQ (Fig. 2b). When the clay is pretreated with DTAB, the heat of displacement $\Delta_{21}H$ as function of the surface excess amount of PQ decreases considerably due to the energy required for the diplacement of DTAB.

Adsorption of PQ contracts the interlayer space because water molecules are removed from the interlayer space when the Ca^{2+}-ions are displaced by PQ ions (Fig. 2c). A basal spacing $d_L = 1.25$ nm indicates that the flat-lying PQ ions are strongly held between the layers [19]. DTAB with the voluminous trimethylammonium group expands the spacing from 1.45 nm (monolayer) to 1.75 nm, which indicates formation of bilayers of. flat-lying DTA^+-ions [20]. When PQ is added to a DTA^+-bentonite, the spacing at low amounts of PQ corresponds to that of DTAB, but is decreased at higher amounts of PQ.

Figure 3 illustrates the enhanced adsorption of the hydrophobic biphenyl with increased amount of adsorbed DTAB on bentonite [21]. Due to the cationic surfactant the surface of bentonite becomes hydrophobic and, thus, more attractive for biphenyl [22].

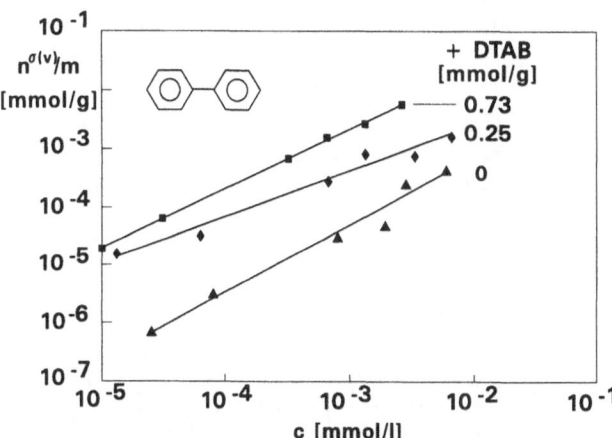

Fig. 3. Adsorption isotherms of biphenyl on Ca^{2+}-bentonite in the presence of different amounts of adsorbed DTAB

References

1. Hayes MHB, Pick ME, Toms BA (1978) J Colloid Interf Sci 65:254—265
2. Hower WF (1970) Clays Clay Min 18:97—105
3. Mortland MM, Shaobai S, Boyd SA (1986) Clays Clay Min 34:581—585
4. Röhl W, von Rybinski W, Schwuger MJ (1991) Progr Colloid Polym Sci 84:206—214
5. Mehlich A (1960) 7th Intern Congr of Soil Sci, Madison, Wisc, 292
6. van Olphen H (1977) An Introduction to Clay Colloid Chemistry, 2nd ed, Wiley, New York
7. Mackay D, Shiu WY (1977) J Chem Eng Data 22:399—402
8. Mukerjee P, Mysels KJ (1971) Critical Micelle Concentrations of Aqueous Systems, NSRDS-NBS 36, Washington
9. Nagy LG, Fóti G, Schay G (1980) J Colloid Interf Sci 75:338—345
10. Tomlinson TE, Knight BAG, Bastow AW, Heaver AA (1968) SCI Monogr 29:317—329
11. Everett DH (1986) Pure Appl Chem 58:967—984
12. Király Z, Dékány I (1988) Colloid Polym Sci 266:663—671
13. Giles CH, MacEwan TH, Nakhwa SN, Smith D (1960) J Chem Soc 3973—3993
14. de Keizer A (1990) Prog Colloid Polym Sci 83:118—126
15. Rheinländer T, Klumpp E, Rossbach M, Schwuger MJ (1991) unpublished results
16. Summers LA (1980) The Bipyridinium Herbicides, Academic Press, Sydney
17. Weed SB, Weber JB (1969) Soil Sci Soc Amer Proc 33:379—382
18. Rosen MJ (1989) Surfactants and Interfacial Phenomena, 2nd ed, Wiley, New York
19. Hayes MHB, Pick ME, Toms BA (1975) Res Rev 57:1—25
20. Dékány I, Szántó F, Weiss A, Lagaly G (1986) Ber Bunsenges Phys Chem 90:422—427, 427—431
21. Jaynes WF, Boyd SA (1991) Soil Sci Soc Am J 55:43—48
22. Schwuger MJ (1990) Tenside Surf Det 27:420—422

Acknowledgements

We thank Dr. Lewandowski, Mrs. Mainz, and Mrs. Paffen for helpful assistance. This article is part of the Ph. D. thesis of T. Rheinländer at the Heinrich-Heine-University Düsseldorf.

Authors' address:

M. J. Schwuger
Institute for Applied Physical Chemistry
Research Centre Jülich
P.O. Box 1913
W-5170 Jülich, FRG

Progress in Colloid & Polymer Science   Progr Colloid Polym Sci 89:194—196 (1992)

Modulation of reaction rate at inhomogeneous charged interfaces by electrohydrodynamic effects

A. Raudino

Dipartimento di Scienze Chimiche, Universitá di Catania, Italy

Abstract: We developed a theoretical model to investigate the effect of lateral inhomogeneities on the transport of ions across a charged membrane. The model focuses on the interfacial convective motions induced by non-uniforms stress inside the fluid. The formal solution has been obtained by combining the transport equations for the charged species with the Navier-Stokes and Poisson's equations. When the concentration of the reactive ions is such much smaller than that of inert electrolytes, a perturbation solution can be employed. The final formulas relate the flow enhancement to the spatial variation of the potential and to the properties of the medium. Numerical estimates show that convective effects may play a relevant role in modulating interfacial reactions.

Key words: Interfacial kinetics; inhomogeneous electric field; transport equations; perturbation methods

Introduction

Many surfaces of lipid assemblies containing two or more components exhibit an inhomogeneous lateral distribution forming micro-domains richer in one component [1—3]. Several properties of biological membranes, such as protein rectivity, permeability or electric potential, are affected by the patterning of the lipid-water interface. As far as the interfacial reactivity is concerned, it has been shown that clustering the reactive sites leads to rate variations [4]; conversely, uniformly reactive membranes experiencing an inhomogeneous potential show enhanced transport [5, 6]. In our opinion, a major effect of the lateral inhomogeneities could be the onset of convective motions caused by non-uniform stress inside the fluid. In this paper, we construct a model to describe the effect of convective motions on the transport of charged species across a membrane.

Theory

Let us consider a charged reactant (with charge $z_R e$ and bulk concentration $\bar{c}_R$) dissolved in a medium containing a large excess of inert electro-lytes. The solution is in contact with a charged membrane assumed to be uniformly reactive. Let Z and X, Y be the coordinates perpendicular and parallel to the membrane, and considering, for the sake of simplicity, a surface potential varying only along the X axis, the concentration of charged species in the presence of flow obeys the equation [7]:

$$v_x \frac{\partial c_j}{\partial X} + v_z \frac{\partial c_j}{\partial Z}$$

$$= D_j \left(\frac{\partial^2 c_j}{\partial X^2} + \frac{\partial^2 c_j}{\partial Z^2} \right) + z_j e \mu_j \left(\frac{\partial}{\partial X} \left(c_j \frac{\partial \phi}{\partial X} \right) \right.$$

$$\left. + \frac{\partial}{\partial Z} \left(c_j \frac{\partial \phi}{\partial Z} \right) \right), \tag{1}$$

where c_j is the concentration of c_+, c_- and c_R, $v_x (v_z)$ the flow velocity along the $X(Z)$ axis, D_j and μ_j the diffusion coefficient and mobility of the j-th species, and ϕ the electrostatic potential inside the solution. The boundary conditions for the reactant impose that its concentration vanishes at the interface, while outside the boundary layer (whose thickness

h is of order of 10^{-3}—10^{-4} cm), its concentration is constant:

$$c_R|_{z=0} = 0 \qquad c_R|_{z=h} = \bar{c}_R \ . \tag{2}$$

For $c_\pm$, we have $c_\pm|_{z=h} = \bar{c}_\pm$. Moreover, electroneutrality requires that $\sum_j z_j \bar{c}_j = 0$. The potential ϕ must satisfy Poisson's equation:

$$\frac{\partial^2 \phi}{\partial X^2} + \frac{\partial^2 \phi}{\partial Z^2} = -\frac{4\pi e}{\varepsilon} \sum_j z_j c_j \ , \tag{3}$$

where ε is the dielectric constant of the medium, while the flow velocities v_x and v_z obey the Navier-Stokes equation [7]:

$$\rho_0 \left(v_x \frac{\partial v_x}{\partial X} + v_z \frac{\partial v_x}{\partial Z} \right)$$

$$= \rho_0 \nu \left(\frac{\partial^2 v_x}{\partial X^2} + \frac{\partial^2 v_x}{\partial Z^2} \right) - e \sum_j z_j c_j \frac{\partial \phi}{\partial X} \tag{4a}$$

$$\rho_0 \left(v_z \frac{\partial v_z}{\partial X} + v_z \frac{\partial v_z}{\partial Z} \right)$$

$$= \rho_0 \nu \left(\frac{\partial^2 v_z}{\partial X^2} + \frac{\partial^2 v_z}{\partial Z^2} \right) - e \sum_j z_j c_j \frac{\partial \phi}{\partial Z} \ , \tag{4a}$$

ρ_0 and ν being the solvent density number and kinematic viscosity, respectively. The last term in Eqs. (4) describes the body forces acting on the unit volume of the fluid due to the interaction between charged species and surface potential. Finally, we take the fluid to be incompressible [7], thus,

$$\frac{\partial v_x}{\partial X} + \frac{\partial v_z}{\partial Z} = 0 \ . \tag{5}$$

The boundary conditions for the potential ϕ are

$$\lim_{z \to \infty} \phi = 0 \qquad \frac{\partial \phi}{\partial Z}\bigg|_{z=0} = -\frac{4\pi}{\varepsilon} \sigma(X) \ , \tag{6}$$

where $\sigma(X)$ is the laterally varying surface charge density of the membrane whose analytical profile is assumed to be

$$\sigma(X) = \sigma_0 \left(1 + H \cos\left(\frac{\pi X}{b}\right) \right) \quad |H| \leqslant 1 \ ; \tag{7}$$

H and b are two parameters measuring the magnitude and the period of the electrical inhomogeneity. The boundary conditions for the flow velcoties v_x and v_z are

$$v_x|_{z=0} = v_z|_{z=0} = 0 \quad \text{and} \quad \lim_{z \to \infty} v_x = \lim_{z \to \infty} v_z = 0 \ . \tag{8}$$

The assumption that inert electrolyte concentration is much larger than the reactant's one (typically $\bar{c}_\pm = 0.1$ M and $\bar{c}_R = 10^{-3}$—10^{-6} M in biological fluids) implies that the potential is very low, therefore, also the flow velocity induced by potential inhomogeneities is small. This suggests to seek for a perturbation solution where, for the above reasons, both the potential and the flow velocities have been assumed to be small, even at the zero-th order approximation:

$$\phi = \lambda \phi^{(0)} + \lambda^2 \phi^{(1)} + \dots \tag{9a}$$

$$v_i = \lambda v_i^{(0)} + \lambda^2 v_i^{(1)} + \dots \ (i = X \text{ or } Z) \tag{9b}$$

$$c_j = c_j^{(0)} + \lambda c_j^{(1)} + \lambda^2 c_j^{(2)} + \dots \ , \tag{9c}$$

where $\lambda = 1$ at the end of the calculations. It is easy to prove that a potential acting only along the Z-coordinate is unable to induce convective motions, therefore, in a first approximation we set $v_x^{(0)} = v_z^{(0)} = 0$. Inserting this result into the transport equations for c_+ and c_- (Eqn. (1)) and recalling the Einstein's relation $\mu_j = D_j/kT$, we obtain $c_\pm = \bar{c}_\pm \exp(-\phi^{(0)}/kT)$. Combining this result with Eq. (3) and neglecting the reactant concentration in respect to the inert electrolyte one, and retaining only linear terms in $\phi^{(0)}$, we find

$$\frac{\partial^2 \phi^{(0)}}{\partial X^2} + \frac{\partial^2 \phi^{(0)}}{\partial Z^2} - \kappa_0^2 \phi^{(0)} = 0, \tag{10}$$

where $\kappa_0^2 \equiv (z_+^2 \bar{c}_+ + z_-^2 \bar{c}_-) 4\pi e^2/\varepsilon kT$. Equation (10) with the boundary conditions (6) and (7) can be analytically solved. Expressing c_+ and c_- as a function of $\phi^{(0)}$, inserting them into the Navier-Stokes Eq. (4), and imposing the incompressibility condition (5), we have two coupled differential equations from which $v_x^{(1)}$ and $v_z^{(1)}$ can be calculated. We look for a Fourier series solution of Eq. (4), namely:

$$v_x^{(1)} = \sum_{n=1}^{\infty} \frac{\partial f_n(Z)}{\partial Z} \frac{b}{n\pi} \sin\frac{n\pi X}{b} \tag{11a}$$

$$v_x^{(1)} = \sum_{n=1}^{\infty} \frac{\partial f_n(Z)}{\partial Z} \frac{b}{n\pi} \sin\frac{n\pi X}{b} \tag{11a}$$

It is easy to see that Eqs. (11) satisfy condition (5). To obtain a close equation for the flow velocities, we selected a trial function for $f_n(Z)$ that satisfies the boundary conditions (8): $f_n(Z) = B_n Z^2 \exp(-\gamma_n Z)$, with B_n and γ_n to be determined. This function is not an exact solution of the Navier-Stokes equation, however. Since we are interested in the flow velocity behavior near the membrane-water interface ($Z \approx 0$), we may calculate B_n and γ_n by inserting Eqs. (11) into Eqs. (4), neglecting the quadratic terms in v_j and taking the limit for Z tending to zero. Once analytical expressions of $v_x^{(1)}$, $v_z^{(1)}$, and $\phi^{(0)}$ have been obtained, we inserted them into the transport Eq. (1) for c_R. Recalling the boundary conditions (2) and using the perturbation procedure outlined in Eqs. (9), we eventually calculated the reactant concentration profile. The flux J of the reactant across the interface is calculated by [7]

$$J = \int_S (D_R \vec{\nabla} v c_R|_{z=0} + \mu_R c_R \vec{\nabla}\phi|_{z=0} - \vec{v} c_R|_{z=0}) d^2S ,$$

(12)

where the three terms of the integrand describe the diffusion, migration, and convective contributions to the flux, and the integration is extended over the membrane surface. We introduce a dimensionless parameter a to measure the enhancement of the flux due to potential spatial inhomogeneities, i.e.,

$$a = \frac{\langle J(\phi) \rangle - J(\langle \phi \rangle)}{J(\langle \phi \rangle)} ;$$

(13)

$\langle J(\phi) \rangle$ and $J(\langle \phi \rangle)$ are the flux averaged over the membrane surface and the flux calculated using an averaged (constant) potential, respectively. Retaining the leading terms, we find after long algebra:

$a = a_{CONV} + a_{MIG}$, where

$$a_{MIG} \approx 2\pi^2 \frac{\sigma_0^2 z_R^2 e^2 H^2}{\varepsilon^2 (kT)^2 \kappa_0^3 h}$$

(14a)

$$a_{CONV} \approx 4\pi^2 \frac{\sigma_0^3 z_R e H^2}{\varepsilon^2 kT \kappa_0^2 \rho_0 \nu D_R} \left(\frac{b}{\pi}\right) ,$$

(14b)

which is valid for $b \gg \kappa_0^{-1}$.

Results and discussion

The final equations give some qualitative information on the effect of the potential-induced convective motions. Equation (14) shows that the enhancing parameter a_{CONV} can be either positive or negative depending on the sign of the reactant and charged surface. When both have the same sign a_{CONV} is positive, otherwise, it is negative. In contrast, the migration-enhancing factor a_{MIG} is always positive. Numerical estimates have been obtained using parameters typical of biological systems, i.e., the kinetmatic viscosity $\nu = 10^{-2}$ cm^2 sec^{-1}, the density $\rho_0 = 1$ gr cm^{-3}, the reactant diffusion coefficient $D_R = 10^{-5}$ cm^2 sec^{-1}, $z_R = 1$, the membrane charge density σ_0 = unit charge x fraction of charged lipids $\times$ (lipid area)$^{-1}$ = 4.8 $10^{-10} \times 10^{-1} \times$ (80 $10^{-16})^{-1}$ coul cm^{-2}, the Debye length $\kappa_0^{-1} = 10^{-7}$ cm, the boundary layer thickness $h = 10^{-3}$ cm, $T = 300$ K, the period b and the magnitude H of surface potential fluctuations were $b = 10^{-6}$ cm and $H = 0.5$. The results are a_{CONV} 10^{-5} and a_{MIG} 10^{-1}. It can be proved that when $b \to 0$ both a_{MIG} and a_{CONV} vanish.

References

1. Möhwald H (1990) Ann Rev Phys Chem 41:441—476
2. Larter R (1990) Chem Rev 90:355—381
3. Raudino A (1988) Liq Cryst 3:1055—1072
4. Baldo M, Grassi A, Raudino A (1991) J Phys Chem 95:6734—6740
5. Steinmetz CG, Larter R (1988) J Phys Chem 92:6113—6120
6. Kuehl SA, Sanderson RD (1988) J Phys Chem 92:517—525
7. Bird RB, Stewart WE, Lightfoot EN (1960) Transport Phenomena, Wiley, New York

Author's address:

A. Raudino
Dipartimento di Scienze Chimiche
Universitá di Catania
Viale A. Doria
6-95125 Catania, Italy

Progress in Colloid & Polymer Science   Progr Colloid Polym Sci 89:197—201 (1992)

Phase transitions and domain structures in ester and acid monolayers

X. Qiu, J. Ruiz-Garcia, and C. M. Knobler

Department of Chemistry and Biochemistry, University of California, Los Angeles, USA

Abstract: Fluorescence microscopic studies of ester and acid monolayers show the existence of several condensed phases, in accord with recent diffraction measurements. Studies with polarized excitation reveal "star defects," i.e., organized domains of uniform tilt similar to those seen in smectic liquid crystals. The observation of chiral patterns in monolayers composed of achiral molecules is in accord with the proposal that one of the phases is the analog of the smectic L liquid crystal. Monolayers of pentadecanoic acid develop large-scale chiral defect structures.

Key words: Monolayers; fluorescence microscopy; defects; phase transitions

1. Introduction

In the simplest picture of monolayers, there are only four phases: gaseous, liquid-expanded, liquid-condensed, and solid. But there has long been evidence of the existence of many more condensed phases. For example, extensive isotherm measurements of acids and esters by Stenhagen [1, 2] and Lundquist [3—5], some performed over 50 years ago, showed a number of features (some of them very subtle) that were interpreted as transitions between half a dozen condensed phases. Following Dervichian [6], Lundquist [4, 5] used thermodynamic and packing arguments to relate some of the monolayer phases to polymorphs of the three-dimensional crystals. The number of proposed monolayer phases exceeded the number of known crystalline phases, so it was necessary to postulate the existence of monolayer phases intermediate between solid and liquid. While such analogs of liquid crystalline phases were plausible, no direct evidence of their existence was presented.

Doubts about the existence and nature of the monolayer phases could have been removed by experiments that probed the microscopic character of the phases directly, but there were no methods by which this could be done. Such experiments have now been made possible by the availability of synchrotron sources, which have allowed diffraction and reflection studies to be performed on floating monolayers (see e.g., [7]). The results of these experiments are in accord with the complicated polymorphism deduced by Stenhagen from isotherm studies. The phase diagram of no one amphiphile has been verified in its entirety, but if one makes the reasonable (and usual) assumption that the phase boundaries in the fatty acids shift smoothly with chain length, it is now possible to associate a structure with each of the phases.

This identification is not unequivocal, however. Because of the width of the beam, the diffraction patterns are powder-averaged, and there are too few reflections to allow a unique determination of the structure. The assignments can be made only by bringing in additional information, such as the known structures of the three-dimensional crystals. Thus, it is desirable to find other experiments that provide structural information. We will describe here observations by polarized fluorescence microscopy that can confirm structural assignments. These studies also provide information about monolayer structure on an intermediate scale that shed light on defect structures and the approach to equilibrium.

2. Fluorescence microscopic studies of LC phases

We will focus here on liquid-condensed (LC) phases, by which we mean those condensed phases that can coexist with the liquid-expanded (LE) phase. The fact that there was more than one LC phase, which was evident from the early phase diagrams, was rediscovered in fluorescence microscopy studies with unpolarized sources [8—10]. When a monolayer doped with a small amount of a fluorescent probe is examined by fluorescence microscopy, the LE-LC coexistence region is clearly distinguishable [11]. The LE phase appears bright because it has a high solubility for the probe; the LC phase, which has a lower concentration of probe, appears dark. The lever rule applies in the two-phase region, so the fraction of LC phase increases as the monolayer is compressed.

Isolated domains of the LC phase of the ester methyl hexadecanoate are circular at high temperature (Fig. 1(a)). The domains become hexagonal when the monolayer is cooled below 27.0°C at constant area (Fig. 1(b)). This change in morphology is completely reversible [9] and is unaffected by the concentration of the probe. A similar transition can be observed in all of the methyl ester homologues from the tetradecanoate to the nonadecanoate (Fig. 2).

We interpreted [9] these circle-to-hexagon transitions as phase transitions between two LC phases, a view strengthened by the realization that they exactly parallel a series of triple points found by Lundquist [4] in the ethyl esters (Fig. 2). These points occur when there is an equilibrium between the LE phase and two condensed phases that she labeled LS (superliquid) and L_2'.

Lundquist argued on the grounds of the area per molecule that the hydrocarbon chains are normal to the surface in the LS phase and tilted in the L_2' phase. This identification of the LC phase is supported by the recent analyses by Bibo et al. [12]. They related the phase diagrams of acids determined from precise isotherm measurements to the structures obtained by diffraction measurements. In addition, they performed a classical miscibility study that allowed them to connect the phases in the acids to those in the esters.

Bibo et al. propose that there may be four LC phases in the esters and that they are all hexatics, i.e., phases with short-range positional order but quasi-long-range bond orientational order [13]. (The term quasi-long-range means that correlations fall off with distance according to a power law; short-range order decays exponentially.) The essential distinction between the phases is in the tilt order. In one of the phases the chains are untilted;

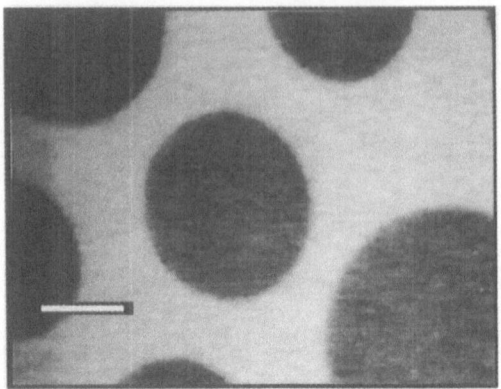

a

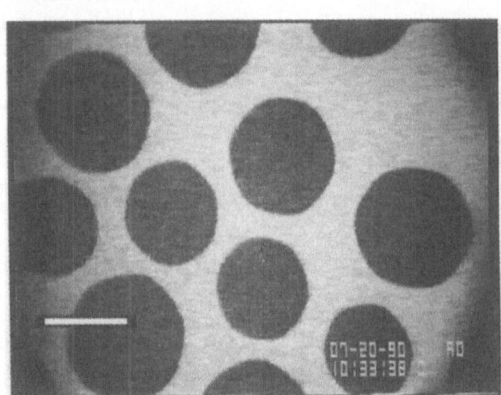

b

Fig. 1. Fluorescence microscopy images of methyl hexadecanoate in the LE-LC two-phase region. The probe in this and all other experiments in this paper is NBD-hexadecylamine at a concentration of 1%; the bar represents 100 μm. a) Unfacetted LC domains at 28°C; b) hexagonal domains below 19°C

this is Lundquist's LS phase and it has the character of a B_H smectic phase. The chains in the other three phases are tilted and the tilt azimuths are locked in a fixed direction with respect to the locally hexagonal bond orientational order. One of these tilted phases, which they call L_2^*, has the character of the smectic F phase in which the tilt is in the direction of the next-nearest neighbor. In phase L_2, the tilt azimuth is locked in the nearest-neighbor direction as in smectic I. The third tilted phase, L_1', which is like smectic L, is one in which the tilt is locked in some intermediate direction between next-nearest and nearest neighbors.

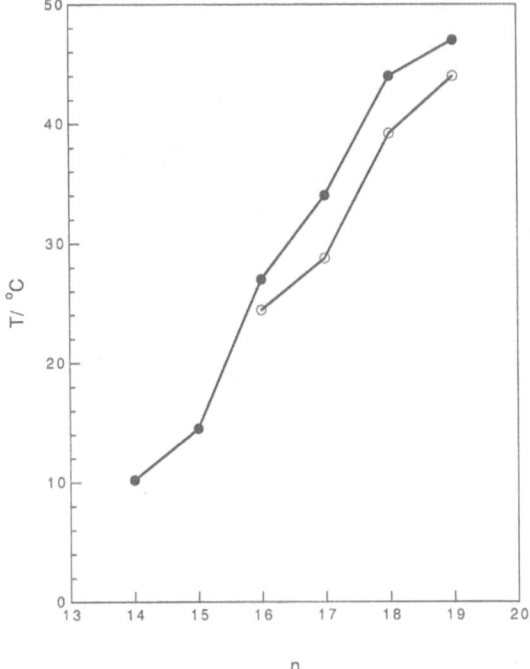

Fig. 2. Dependence of the circle-to-hexagon transition temperature on the chain length n in methyl and ethyl esters. ● methyl esters; ○ LS-L_2'-LE triple point in ethyl esters [4]

3. Polarized excitation studies of esters

McConnell and his coworkers demonstrated that regions of different tilt in monolayers can be distinguished by fluorescence microscopy if the exciting radiation is polarized and strikes the monolayer obliquely. In order for the method to be successful, the fluorophore on the probe molecule must remain in some fixed orientation with respect to the chain axes of the amphiphiles that form the monolayer. If this is the case, then the fluorescence intensity, which depends on the degree of overlap between the transition moment of the fluorophore and the electric field of the exciting radiation, will vary with chain orientation.

Figure 3 shows fluorescence images of methyl octadecanoate and methyl heptadecanoate obtained with polarized radiation [9]. When the LC domains are circular (Fig. 3(a)) their intensity of fluorescence is uniform and shows no change with the direction of the exciting beam, a result consistent with the identification of the high-temperature LC phase as an untilted phase. When the domains become hexa-

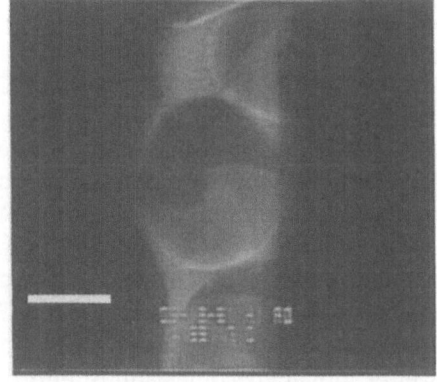

Fig. 3. Fluorescence microscopy images with polarized excitation (bar: 100 μm). a) Untilted phase in methyl octadecanoate at 46 °C; b) star defect with straight arms in methyl octadecanoate at 24 °C; c) star defect with curved arms in methyl heptadecanoate at 10 °C

gonal, however (Fig. 3(b)), they show a remarkable six-fold variation in intensity. Each of the pie-shaped segments represents a region in which the tilt azimuth is constant. Similar five-arm patterns that have been called "star defects" are observed [15] in freely-suspended films of smectic *I* liquid crystals; six-arm stars have been found [16] in domains of smectic *F* liquid crystals surrounded by an isotropic phase. Selinger and Nelson [17] have developed a theory that accounts in detail for the domain structure in freely suspended films; it has been shown [9] to account for the six-arm stars as well. While the polarized fluorescence images do not allow us to distinguish between nearest-neighbor and next-nearest neighbor tilted phases, a phase with intermediate tilt can be identified because it is the only one that is inherently chiral; it will have chiral features even when it is composed of achiral molecules. Thus, a phase in which the star defects have right- and left-handed curved arms (Fig. 3(c)) can only be L'_1, the equivalent of smectic L. For the C_{18} and shorter chain length esters, chiral domains appear at the circle-to-hexagon transition. We have observed only straight-armed star defects in C_{19} to C_{21} esters.

4. Polarized excitation studies of pentadecanoic acid

Organized regions of uniform tilt can also be observed in monolayers of fatty acids. We have begun detailed investigations of the evolution of defect structures in pentadecanoic acid at temperatures below the LE-LC-G triple point, 17°C [11]. Studies carried out for a variety of initial conditions yield the same long-term behavior. In one experiment the monolayer was deposited at a temperature of about 8°C and an area of about 24 $Å^2$/molecule. The monolayer was initially quite heterogeneous and contained regions of LE phase that slowly decrease with time, leaving the LC phase and a small amount of the gas.

The monolayer eventually becomes almost entirely LC phase and is uniformly fluorescent with unpolarized excitation. Examination with polarized excitation provides an entirely different picture, however (Fig. 4). The LC phase consists of striped regions interrupted by defects of several kinds. The scale of the patterns is comparable to that of the trough, i.e., several centimeters.

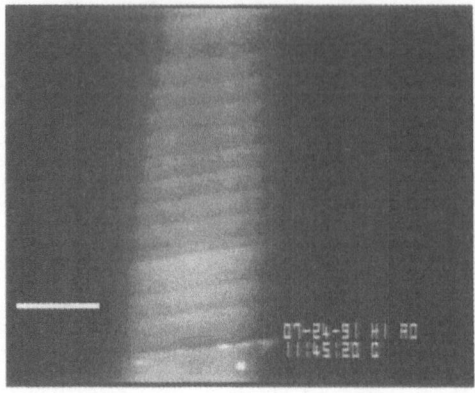

a

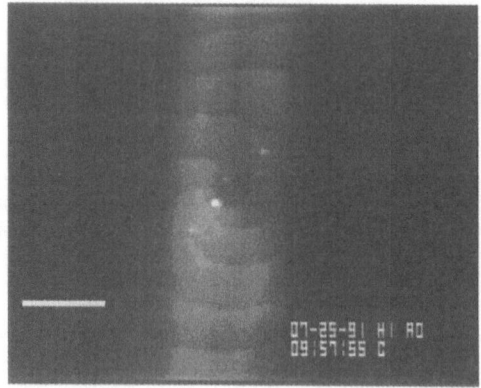

b

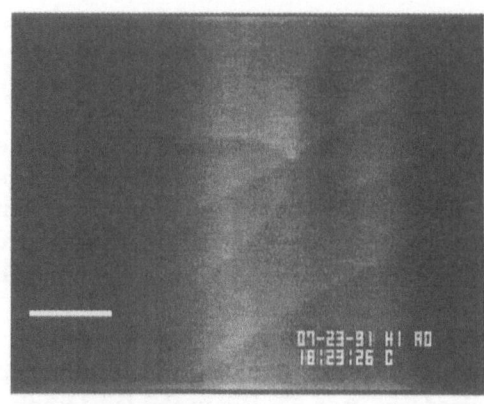

c

Fig. 4. Fluorescence microscopy images with polarized excitation in pentadecanoic acid at 8°C (bar: 100 μm). a) Stripe pattern; b) spiral defect; c) hybrid defect

The stripes appear in groups of uniform width, but the widths vary from 10 to 140 µm. The variation of fluorescence intensity as a function of the fractional distance across a stripe is essentially independent of the stripe width. By comparison with the intensity variations between the segments of a star defect, it appears that there is a 120° rotation of the tilt azimuth within each stripe.

A close examination [18] of the point defects shows that they are of two types: a) Spirals with a definite chirality similar to those observed in star defects but of much greater size (Fig. 4(b)), and b) "Hybrids" that have the appearance of being derived from a combination of two halves of star defects of different chirality (Fig. 4(c)). The presence of the chiral defects identifies the LC phase as L_1'; Bibo et al. [12] had indications that this phase existed in fatty acids, which they were unable to prove.

By following the stripes that originate at the hybrid defects, it can be seen that adjacent stripes are of opposite chirality. Langer and Sethna [19] have discussed the origin of striped textures observed in freely suspended films of the smectic I phase of a chiral liquid crystal. We believe that the stripes found in the monolayer are of a similar nature. The Langer-Sethna theory is not immediately applicable, however, because it does not apply to systems in which the chirality is not a molecular property.

Acknowledgement

This work was supported by the U. S. National Science Foundation. We have benefited from discussions with Robijn Bruinsma and Jonathan Selinger.

References

1. Stenhagen E (1955) In: Braude EA, Nachod FC (eds) Determination of organic structures by physical methods. Academic Press, New York, pp 325—371
2. Ställberg-Stenhagen S, Stenhagen E (1945) Nature 156:239—240
3. Lundquist M (1971) Chem Scr 1:5—20
4. Lundquist M (1971) Chem Scr 1:197—209
5. Lundquist M (1978) Fats other lipids. Prog Chem 16:101—124
6. Dervichian DG (1939) J Chem Phys 7:931—948
7. Kenn RM, Böhm C, Bibo AM, Peterson IR, Möhwald H (1991) J Phys Chem 95:2092—2097
8. Knobler CM (1990) Science 249:870—874
9. Qui X, Ruiz-Garcia J, Stine KJ, Knobler CM, Selinger J (1991) Phys Rev Lett 67:703—706
10. Akamatsu S, Rondelez F (1991) J Phys II (Paris)
11. Moore BG, Knobler CM, Akamatsu S, Rondelez F (1990) J Phys Chem 94:4588—4595
12. Bibo AM, Knobler CM, Peterson IR (1991) J Phys Chem 95:5591—5599
13. Brock JD, Birgeneau RJ, Litster JD, Aharony A (1989) Contemp Phys 30:321—335
14. Moy VT, Keller DJ, Gaub HE, McConnell HM (1986) J Phys Chem 90:3198—3202
15. Dierker SB, Pindak R, Meyer RB (1986) Phys Rev Lett 56:1819—1822
16. Walton CR, Goodby JW (1984) Mol Cryst Liq Cryst 92:263—269
17. Selinger JV, Nelson DR (1989) Phys Rev A 39:3135—3147
18. Qiu X, Ruiz-Garcia J, Knobler CM (to be published)
19. Langer SA, Sethna JP (1986) Phys Rev A 34:5035—5045

Authors' address:

Prof. Charles M. Knobler
Department of Chemistry and Biochemistry
University of California
Los Angeles, California 90024, USA

Progress in Colloid & Polymer Science

Progr Colloid Polym Sci 89:202—208 (1992)

Lateral diffusion of macromolecules in monolayers at the air/water interface

S. Kim and H. Yu

Department of Chemistry, University of Wisconsin, Madison, Wisconsin, USA

Abstract: We report a lateral diffusion study of a surface active protein, bacterial lipase from pseudomonas fluorescens, and a vinyl polymer, poly(t-butyl methacrylate) on the air/water interface by the technique of fluorescence recovery after photobleaching. For the validation of our technique and the calibration of the instrument, we relied on a phospholipid system that Peters and Beck used earlier, and found that our results were in accord with theirs within 20% in absolute magnitude. For both the phospholipid and lipase, we analyzed the lateral diffusion data in terms of the free area model of Sackmann and Träuble. We conclude that the results of lipase could be interpreted by invoking a conformational change induced by lateral compression in the monolayer state, and those of the polymer by postulating the quenching of the diffusion process in the dilute region when its surface concentration enters into the semidilute region.

Key words: Lateral diffusion, air/water interface; lipase, surface pressure; poly(t-butyl methacrylate); DLPC

Introduction

This is to report a study of lateral diffusion of macromolecules in dilute monolayer state at the air/water interface (A/W). The molecular species studied here are a surface active protein, lipase, and a vinyl polymer, poly(t-butyl methacrylate). The objective of the study was to characterize hydrodynamic behaviors of constituent species of monolayers at A/W, and to develop a framework for understanding the collective monolayer dynamics. In addition, there exists a substantial excitement associated with the monolayer studies per se since they afford us an emerging research area of liquid state in terms of reduced dimensionality [1, 2] and interfacial perspectives [3, 4], even if one does not consider the technological undertakings of current research activities directed to organic thin films and interfaces [5]. Thus, our interest in this area has been on the collective dynamics in monolayers and thin films of small molecule amphiphiles and amphiphilic polymers probed by the techniques of surface light scattering of spontaneous capillary waves [6—8] and electrocapillary wave diffraction [9].

These studies have been focused on the macroscopic characterization of the monolayers and thin films in the context of the lateral viscoelastic parameters, which are extracted from the two experimental techniques. Our interest is thus naturally extended to lateral diffusion of monolayer constituents at different monolayer states, quite analogous to the interest in the chain self-diffusion of polymer sample in bulk state after having examined its steady shear viscosity of the sample.

Here, we report our first study on the subject; Peters and Beck [10] pioneered the same type of measurements some years ago with a phospholipid system. Our principal claim here is the scope of systems that we have examined. Although we have made several innovations for the technique, we essentially followed the trailblazing work of Peters and Beck.

Experimentally, the monolayers pose an immense challenge because of the surface flows coupled with, inherently, a small number of available photoprobes for detection in a range of femtomole to picomole. Since a detailed report will appear elsewhere [11], we will only summarize our findings here.

Experimental section

Materials

Phospholipids: L-α-dilauroylphosphatdiylcholine (DLPC) was purchased from Sigma and used without further purification. N-4-nitrobenzo-2-oxa-1,3 diazole phosphatidylcholine (NBD-PC), which was used as photolabeled phospholipid molecules in the monolayer matrix of DLPC, was purchased from Avanti Polar Lipids (Birmingham, Alabama, USA); it was also used without further purification.

Lipase: A bacterial lipase from pseudomonas fluorescens, Amano LPL-200S, was a gift from the Corporate Research Division of Proctor & Gamble Company. The molar mass of the lipase was determined to be 33 089 Daltons (M. P. Lacey, private communication). For FRAP measurements, the lipase was labeled with fluorescein-isothiocyanate (FITC) (Aldrich), and the labeled protein was purified by eluting through a Sephadex (G-25 fine; Pharmacia) column. The eluent component of labeled lipase was well separated from the free dye, and the labeling yield of was determined to be approximately 0.3 dyes per lipase molecule. For the quasi-elastic light scattering experiment (see below), the lipase was purified by eluting through the same Sephadex column using a small sample of the labeled lipase as the marker.

Vinyl polymer: Poly(t-butyl methacrylate) sample we used here was purchased from Polysciences (Warrington, Pennsylvania, USA), whose molar mass is estimated as about 100 000 g/mol, and it was labeled according to the following chemistry:

$$RCOOtBu + Me_3SiI \rightarrow RCOOSiMe_3 + tBuI$$

$$2\ RCOOSiMe_3 + H_2O$$

$$\rightarrow\ 2\ RCOOH + (Me_3Si)_2O\ .$$

For the coupling reaction of the dye to a partially dealkylated polymer sample, N-hydroxysuccinimide (NHS) was employed to make an activated intermediate [12], which would then react with an amino group functionalized dye molecule. Here, one uses an equimolar mixture NHS and dicyclohexyl carbodiimide (DCC) as the catalyst.

Bovine serum albumin (BSA): BSA was purchased from ICN Biomedical, Inc. (Costa Mesa, California, USA). The labeling procedure was the same as in the case of lipase.

Methods

Fluorescence recovery after photobleaching (FRAP)

Optics: A schematic diagram of the optical system used for our FRAP instrument is presented in Fig. 1.

Electronics: A block diagram for the electronics is presented in Fig. 2. The output of PM is processed by the lock-in amplifier (Ithaco, Model 391 A) to shape the signal to a simple recovery curve, and the resulting signals are amplified to a maximum of 10 V for the A/D conversion. The signal acquisition software is designed to take every data point within 1 ms by making use of a machine code subroutine and directly storing the data in the random access memory (RAM). One set of data is composed of 256 data points, and it is sent to a VAX computer for further analyses. Whenever the S/N ratio of the acquired data is below a certain threshold leve, several sets of the data are accumulated and averaged before being sent to the VAX.

Signal processing: Since our instrument employs a pattern bleaching method, the signal processing procedure follows a published scheme [13, 14]

$$V(t) = V(0)\exp(-t/\tau)\ , \tag{1}$$

with

$$1/\tau = Dk^2\ , \tag{2}$$

where $V(t)$ is the output voltage, $V(0)$ is the output immediately following the bleaching period, which is the point of $t = 0$, and D is a diffusion coefficient of labeled species in the sample, k is the spatial wavevector of the pattern (equal in magnitude to

Schematic Diagram of FRAP Instrument

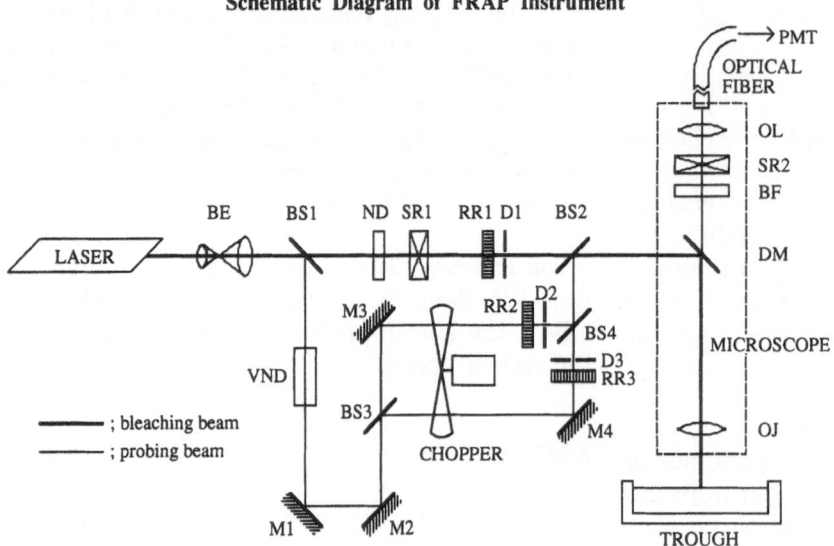

BE; beam expander   BF; barrier filter   BS; beam splitter
D; diaphragm   DM; dichroic mirror   OL; ocular
M; mirror   ND; neutral density filter   OJ; objective
PMT; photomultiplier tube   RR; Ronchi ruling   SR; shutter
VND; variable neutral density filter

Fig. 1. Schematic diagram of optical configuration of the surface FRAP

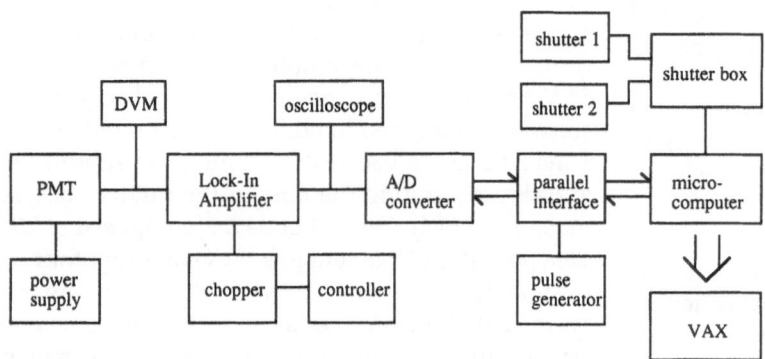

Fig. 2. Block diagram of the electronics of the surface FRAP instrument

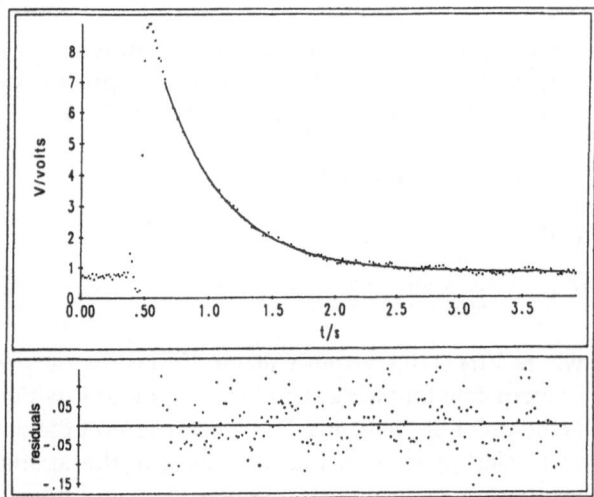

Fig. 3. FRAP profile of labeled BSA in bulk solution at a concentration of 0.1% by weight

$2\pi/p$ with p being the spacing of the Ronchi ruling). Thus, the diffusion coefficient D is deduced from the measured time constant, τ, but only at a single value of k instead of various values of k and fitting to Eq. (2).

FRAP sample preparations: Our FRAP instrument is designed to determine the diffusion coefficients for both bulk samples and on monolayers. For convenience, we refer to the bulk sample measurements as three-dimensional (3D) type, and the monolayer measurements as two-dimensional (2D) type. For 3D type measurements, one drop of the sample solution (10—30 µL) is placed between a microscope slide glass and a cover glass with a 1 mil (25.4 µm) thick Teflon sheet as the spacer. The four sides of the cover glass are then sealed with a vacuum grease. Since BSA and lipase both adsorb strongly to glass surfaces, we saturate the surfaces with the unlabeled sample of the protein to be studied beforehand by soaking slide glass and cover glass in the protein solution prior to use.

For the lateral diffusion studies, a small Langmuir trough (5 cm × 12.5 cm × 1.5 cm) was milled out from a preannealed teflon block. An inverted-cone shaped teflon barrier is devised to confine a small area of the surfaces in order to minimize the surface flow problem.

FRAP instrument calibration: Before we could begin the lateral diffusion experiment, we had to confirm that our instrument worked correctly. For this purpose, the diffusion coefficient of labeled BSA in dilute bulk solution at 0.1% by weight was examined. In Fig. 3 are presented the raw data of the fluorescence recovery profile and a fitted curve to a single exponential function Eq. (1), over the data points (upper plot) together with the normalized residuals plot (lower). The relaxation time thus deduced by the fitting gives rise to $D = (5.9 \pm 0.1) \times 10^{-11}$ m²/s, which agrees with that obtained by our separate QELS measurements (see below) and the published values. We thus have established that the FRAP instrument is calibrated in giving rise to the correct signal time profile relative to an input fluorescence signal.

Quasi-elastic light scattering (QELS)

QELS was used to determine the 3D diffusion constants of both lipase and BSA in dilute concen-

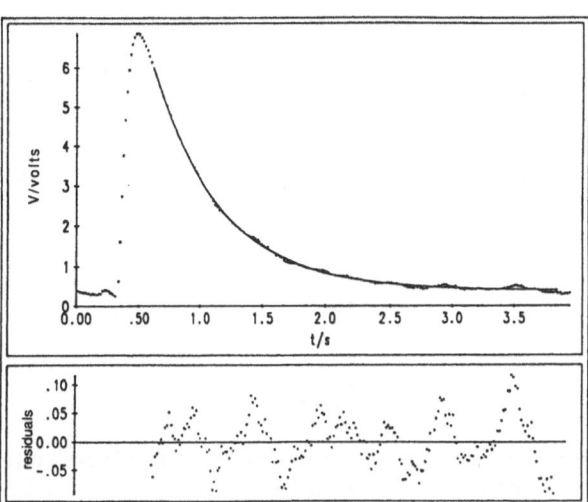

Fig. 4. Surface FRAP profile after 12 accumulations of NBD-PC as the photoprobe in DLPC monolayer at a concentration of 1% at the air/water interface at $\Pi = 20$ mN/m

solutions. The instrumental details and signal analysis method have appeared elsewhere [15].

Results and discussion

We first turn to the results of our experiment with a phospholipid, DLPC, which was chosen to compare against the earlier published results by Peters and Beck [10]. The surface pressure isotherm (Π-A) was obtained by the compression method. The spreading solvent for monolayer was chloroform, and any DLPC solution in chloroform older than 1 week was not used for the experiment because of degradation problem in the solvent. Our isotherms were in complete accord with within experimental error.

For the lateral diffusion measurement, 1% of NBD-PC as the photoprobe was incorporated to the DLPC solution. Π-A isotherm was checked to find no appreciable change (as was noticed by Peters and Beck). The results of the lateral diffusion measurements of NBD-PC as the probe in DLPC monolayer are displayed in Fig. 5 together with Π-A isotherm against the surface concentration as the common independent variable. It is to be noted that the range of reciprocal surface concentration A for both Π and D overlaps substantially, although the range for Π is greater.

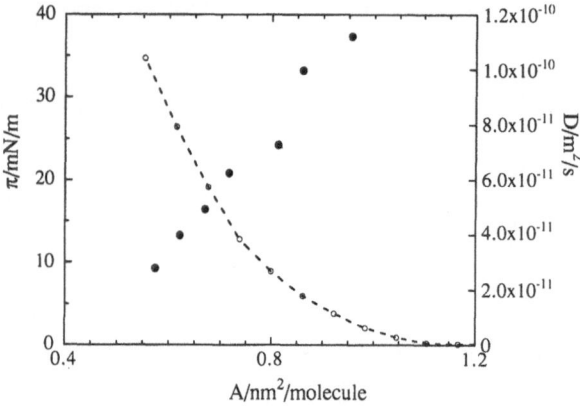

Fig. 5. Surface pressure Π (open circles), and lateral diffusion coefficient D (filled circles) against reciprocal surface concentration A of DLPC. The dashed curve over the open circles is drawn merely to indicate a trend and is not a result of fitting to a specific model

Comparing our results to those of Peters and Beck, all of our values for the lateral diffusion coefficient are about 20% greater than theirs. On the other hand, the surface concentration dependence of ours parallels theirs quite closely (a finding which we will discuss shortly). As for the discrepancy, there may be two possible sources. First, we used hydrocarbon tail labeled phosphatidylcholine (PC), whereas they used polar head labeled phosphatidylethanolamine (PE), whereby NBD-PC could experience a less hydrodynamic friction from the subphase in addition to having more defects in lateral packing of hydrophobic tails attributable to the presence of the labeled photoprobes. Secondly, their FRAP technique employed the circular spot bleaching method, which is known to impart about 20% uncertainty in the diffusion coefficients, mainly due to errors in the determination of the size of the effective bleached area. Regardless of the validity of these speculations, we find that the agreement within 20% in absolute magnitude is encouraging.

We now try to present our results in a little more quantitative terms. As with the analysis by Peters and Beck, we use a semi-empirical free area model of Sackman and Träuble, and Galla et al. [16, 17], which is the 2D analog of the free volume model of Cohen and Turnbull [18]. The diffusion coefficient of a rigid cylinder with the cross-sectional area of a_0 confined to a layer of hydrodynamic continuum with local viscosity η_{loc} is derived as

$$D = P\exp(-Q/a_f) , \tag{3}$$

where $P = da\eta_{loc}$ (with d standing for a linear dimension of the diffusant and a standing for a geometric factor of the diffusant close to unity), $Q = \gamma a^*$, (with γ being a numerical factor between 1/2 and unity, representing 2D packing fraction of the cylinder and a^* being a threshold free area per cylinder when the displacement probability becomes finite), and a_f is the free area per cylinder, i.e., $a_f = A - a_0$. Thus, Eq. (3) invites plotting of D vs. a_f semi-logarithmically if the model is to be tested. We show such a plot in Fig. 6, which the model is well borne out at least for one logarithmic decade in D. An a_0 value of 0.40 nm², which is rather close to the value of 0.425 nm² chosen by Peters and Beck, is also obtained from the plot.

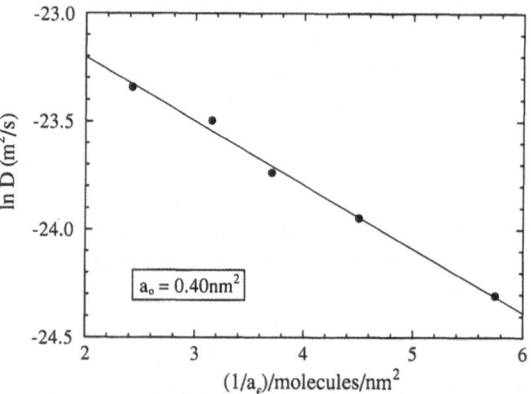

Fig. 6. Free area model plot of the lateral diffusion coefficient of DLPC, $\ln D$ vs. $1/a_f$, where the free area $a_f = A - a_0$, with $a_0 = 0.40$ nm² as an adjustable parameter for fitting to Eq. (3), is to be compared to $a_0 = 0.425$ nm² as used by Peters and Beck [10]

We now turn to the bacterial lipase from pseudomonas fluorescens. The surface pressure isotherm was obtained with 10 mM phosphate buffer (pH 7 with 0.15 M NaCl) as the subphase. This is presented in Fig. 7. The lipase was spread from aqueous solution, using Trunit's method [19], with some minor modifications. The spreading method was validated by obtaining the same Π-A isotherm of BSA repeatedly, and the results agreed well with the literature [20].

The lateral diffusion study was performed with the same subphase, and repeated measurements with freshly spread lipase monolayers agreed within a 5% variability, which is about the ex-

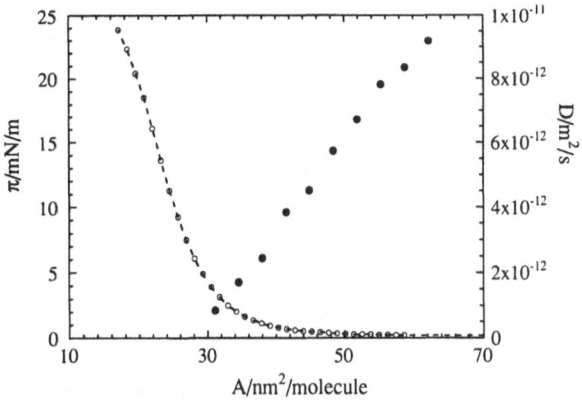

Fig. 7. Surface pressure Π (open circles), and lateral diffusion coefficient D (filled circles) against reciprocal surface concentration A of bacterial lipase from Pseudomonas fluorescens. The dashed curve over the open circles is drawn merely to indicate a trend and is not a result of fitting to a specific model

perimental error in a given measurement. The results are also plotted in Fig. 7. Unlike the case of DLPC, we have here a limited range of overlap in the surface concentration for both Π and D, 30—62 nm²/molecule. Analyzing the lateral diffusion data only for the dilute region according to the free area model, Eq. (3), we obtain $a_0 = 28.5$ nm²/molecule, and this is presented in Fig. 8. During the measurement of Π-A isotherm, it is observed that the instability begins at around $A = 28.0$ nm²/molecule. It is interesting to note that the extrapolated diffusion plot to $D = 0$ occurs at the same position. If we

assume that this is not fortuitous, then we might conclude that the cross-sectional area of a lipase molecule ranges from 25.4 to 28.9 nm²/molecule, depending on its conformation at the interface.

In Fig. 9 are displayed plots of the lateral diffusion coefficient vs. A and Π-A isotherm of PtBMA. This polymer is chosen as a model for a flexible polymer, and it has also been studied by surface light scattering in our laboratory. It is noteworthy that the lateral diffusion seems to quench once the monolayer attains a surface concentration corresponding to a semidilute solution. Thus, our goal is to study the collective viscoelastic behaviors of monolayers of such vinyl polymers vis-a-vis more complex block and star copolymers in terms of molecular architecture. At this stage, we have only presented a preliminary result.

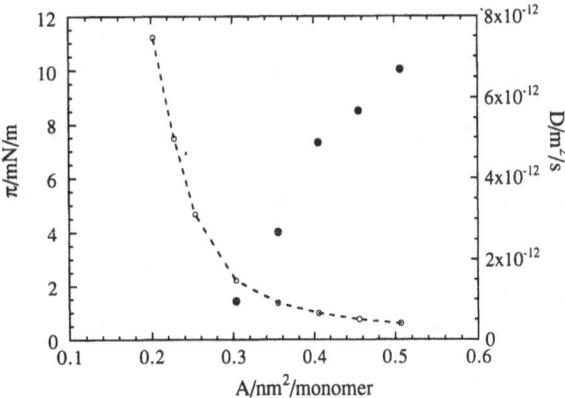

Fig. 9. Surface pressure Π (open circles), and lateral diffusion coefficient D (filled circles) against reciprocal surface concentration A of the poly(t-butyl methacrylate) sample. The dashed curve over the open circles is drawn merely to indicate a trend and is not a result of fitting to a specific model

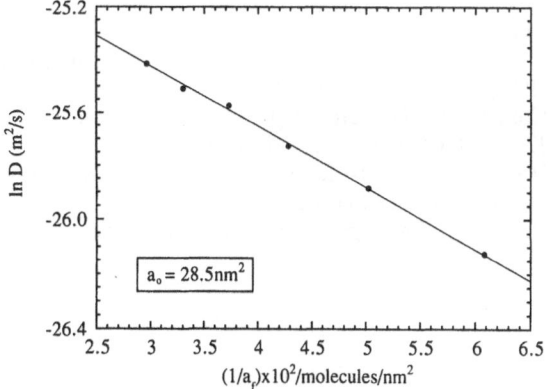

Fig. 8. Free area model plot of the lateral diffusion coefficient of the bacterial lipase ln D vs. $1/a_f$, where the free area $a_f = A - a_0$, with $a_0 = 28.5$ nm² as an adjustable parameter for fitting to Eq. (3)

A major difficulty in studying this polymer lies in the labeling step. After partial dealkylation of the t-butyl group and subsequent hydrolysis to recover partial methacrylic acid moety, the polymer seems to change its conformation in hydrophobic media. The bulkiness of the t-butyl group might be blocking the access of dye molecules to the carboxylic acid groups. As seen in Fig. 9, Π-A isotherm is more expanded than the published isotherm, which is perhaps indicative of a conformational change from pure poly(t-butyl methacrylate). Nevertheless, it is obvious that the diffusion is quenched or slowed down quickly at the beginning of the semidilute region and the conventional plot for obtaining a_0 seems entirely normal.

Finally, we come to a probing study of subphase viscosity effect on lateral diffusion, with DLPC as the test monolayer material. Sucrose was used to vary the viscosity of subphase. The results are displaced in Fig. 10 in terms of the free area model plot. While the viscosity of 40% sucrose solution is 5.9 ~ 6.0 times greater than that of pure water, the lateral diffusion is slowed down by a factor of 4. There are two factors to be considered in the interpretation of this result. First, there should be a sucrose depletion layer at A/W, giving rise to a concentration gradient in the vicinity of interface, whereby the effective local viscosity amounts to only about a factor of 4, whereas the bulk viscosity is increased by a factor of 6. Alternatively, sucrose binds with water through hydrogen bonding and this bonding is stronger or more abundant than that between water molecule themselves [21]. Thus, these bindings would prevent water molecules from participating in hydration of PC polar heads of DLPC molecules, resulting in a change in the frictional characteristics. Whatever would be the final outcome of this line of investigation, it is certainly intriguing if the lateral hydrodynamics scale differently than in the bulk liquid state. For this probing of subphase depth from the interface, we plan to use diblock, multiarm star polymers with varying length of hydrophilic polyelectrolyte blocks which will be made to penetrate into the subphase by taking advantage of electrostatic repulsions.

Acknowledgement

This work is in part supported by the Polymers Program of NSF(DMR-8903943) and the Corporate Research Division of the Procter and Gamble Company. H.Y. gratefully acknowledges a fruitful conversation with Dr. Reiner Peters several years ago.

References

1. des Cloiseaux J (1975) J Phys (LeoUlis Fr) 36:1199
2. de Gennes P-G (1981) Macromolecules 14:1637
3. Shah DO (1972) In: Progress in Surface Science Davidson SG (ed) (Pergamon Press), Vol 3, Part 3, p 221
4. v. Tscharner V, McConnell HM (1981) Biophys J 36:409
5. Swalen JD, Allara DL, Andrade JD, Chandross EA, Garoff S, Israelachvili J, McCarthy TJ, Murray R, Pease RF, Rabolt JF, Wynne KJ, Yu H (1987) Langmuir 3:932
6. Sano M, Kawaguchi M, Chen Y-L, Skarlupka RJ, Chang T, Zografi G, Yu H (1986) Rev Sci Instrum 57:1158
7. Sauer BB, Yu H, Tien C-F, Hager DF (1987) Macromolecules 20:393
8. Kawaguchi M, Sauer BB, Yu H (1989) Macromolecules 22:1735
9. Ito K, Sauer BB, Skarlupka RJ, Sano M, Yu H (1990) Langmuir 6:1379
10. Peters R, Beck K (1983) Proc Natl Acad Sci USA 80:7183
11. Kim S, Yu H (1992) J Phys Chem 96:4034
12. Petersen NO (1983) Spectroscopy (Ottawa) 2:408
13. Smith BA, McConnell HM (1978) Proc Natl Acad Sci USA 75:2759
14. Lanni F, Ware BR (1982) Rev Sci Instrum 53:905
15. Amis EJ, Janmey PA, Ferry JD, Yu H (1983) Macromolecules 16:441
16. Träuble H, Sackmann E (1972) J Am Chem Soc 94:4499
17. Galla HJ, Hartmann W, Theilen U, Sackmann E (1979) J Membr Biol 48:215
18. Cohen MH, Turnbull D (1959) J Chem Phys 31:1164
19. Trurnit HJ (1960) J Colloid Sci 15:1
20. MacRitchie F, Ter-Minassian-Saraga L (1983) Progr Colloid Polymer Sci 68:14
21. Taylor JB, Rowlingson JS (1955) Trans Faraday Soc 51:1183

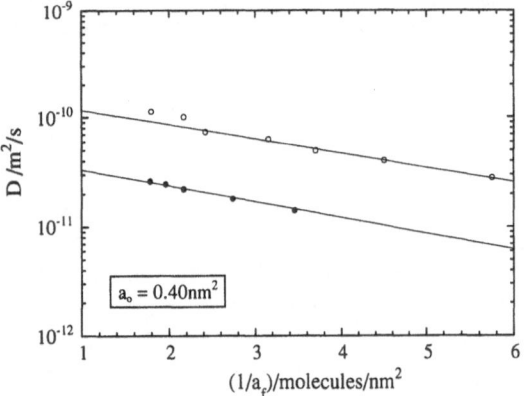

Fig. 10. Comparison of the free area model plots of the lateral diffusion coefficient of DLPC on 40% sucrose solution and deionized water, where both plots are drawn with the limiting area/molecule, $a_0 = 0.40$ nm².

Authors' address:

H. Yu
Dept. of Chemistry
University of Wisconsin
Madison, Wisconsin 53706, USA

Progress in Colloid & Polymer Science   Progr Colloid Polym Sci 89:209—213 (1992)

Two-dimensional pattern formation in Langmuir monolayers

S. Akamatsu and F. Rondelez

Institut Curie, Laboratoire de Physico-Chimie des Surfaces et Interfaces[a]), Paris, France

Abstract: Two dimensional pattern formation during the phase transitions between the liquid expanded phase and various condensed phases in Langmuir monolayers is observed by fluorescence microscopy. For fatty acids, circular or fractal domains are generated, depending on temperature. The analysis of their geometrical properties yields information on characteristic lengths involved in this diffusion-controlled growth process and on line tension. For acyl-amino-acids, spectacular dendritic growth is observed. We have checked the predicted theoretical laws of shape selection ($\rho^2 V = constant$, V: tip growth velocity, ρ: tip curvature radius). From this, an estimation of the line tension anisotropy can be obtained.

Key words: Langmuir-monolayers; two-dimensional growth; fluorescence microscopy; dendrites; fractals

Introduction

Pattern formation is of great interest in such fields as crystal growth, electrochemistry, colloidal aggregation, and viscous flow [1]. Most experiments have been performed on three-dimensional systems where the shape of the advancing front is controlled by a continuously adjustable parameter, i.e., the electric field for electrolytic aggregates, pressure for viscous fingering, or supersaturation for crystal growth [2]. In addition, it has been shown that microstructural properties and anisotropy also play an important role [2—4]. Theoretically, however, the most recent models for unstable growth have been developed in two dimensions [4]. In this respect, Langmuir monolayers, in which purely two-dimensional nucleation and growth phenomena occur in a horizontal plane [5], are particularly suitable systems for a direct comparison with the theoretical predictions. We present here three examples of two-dimensional pattern formation during the first-order transitions between the liquid expanded (LE) phase and various liquid condensed (LC) phases, observed by fluorescence microscopy [5]. In one case, one is dealing with an equilibrium structure (circles). In two other cases, one is dealing with metastable patterns (fractals and dendrites), typical of diffusion-controlled growth. Characteristic lengths and information on line tension are derived from their geometrical properties.

Experimental

Our experimental setup has been fully described in a previous paper [6]. Fatty acids were purchased from Fluka and used either as received (purity > 99%) or after recrystallization. Myristoyl alanine, which is an acyl-amino-acid, was synthesized in our laboratory. The fluorescent dye NBD-HDA (Molecular Probes, USA) was used as received, in concentration of 0.1—1 mol% within the monolayer.

1. Equilibrium structure

We have recently shown, in good agreement with Π-A studies by Bibo et al. [7], that, for myristic (MYA) and pentadecanoic (PDA) acid monolayers, the LE phase can coexist with two different LC phases, according to temperature T [6]. Below a well-defined temperature T^* (24.0°C for MYA and 35.6°C for PDA), the condensed phase, called LC_1,

[a]) Laboratoire associé au CNRS (URA 1379).

a

b

c

Fig. 1. Fluorescence microscopy images (the horizontal dimensions is 600 µm) of various Langmuir monolayers (condensed phase: dark, LE phase: bright): a) Pentadecanoic acid at 20 °C: equilibrium circular domains of the fluid-like LC_1 phase; b) Myristic acid at 25 °C: metastable fractal domain of the viscoelastic-like LC_2 phase; c) D-Myristoyl-alanine: steady-state dendrite of an anisotropic solid-like phase

grows in the form of circular islands. Figure 1a shows a fluorescence microscopy image of LC_1 domains (dark) in the LE continuum phase (bright), during PDA monolayer compression at 20 °C, i.e., for $T < T^*$. This circular geometry corresponds to equilibrium, since it minimizes the line energy. The domains are fluid-like: submitted to shear stresses (e.g., air convections above the Langmuir trough) they deform into ellipses, but return to circles in a matter of seconds when the stress is removed.

2. Non-equilibrium structures

a) Fractal domains: In fatty acid monolayers, and for temperatures above T^*, the LE phase transforms into another condensed phase called LC_2. The LC_2 domains are ramified, even for the lowest compression speeds, as shown in Fig. 1b for a MYA monolayer at 25 °C ($T > T^*$). They are much more rigid than LC_1 domains, and hardly deform even when they bump into each other. This is characteristic of a h ighly viscous or viscoelatic behavior, which explains why line tension is unable to restore the equilibrium circular structure during the time scale of the experiment. However, these metastable shapes evolve in time scales of the order of 1 h towards a quasi-circular geometry in order to minimize the line energy.

The growth initially proceeds in the form of small circular islands, and the first (four to six) bumps

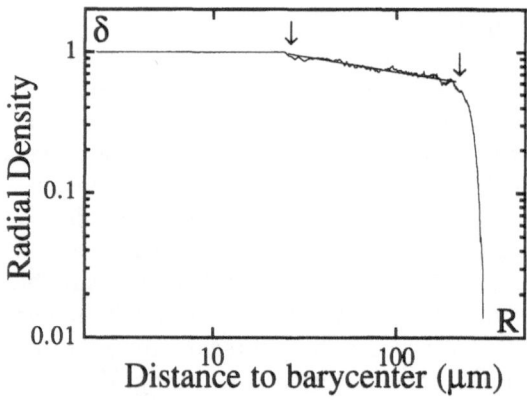

Fig. 2. Radial density δ averaged on all angular orientations as a function of the distance R to the barycenter, calculated for the ramified domain of Figure 1b. The straight line results from a best fit calculation for a power law, in the region delimited by small arrows

appear only after a finite radius R_0. In a well-developed pattern ($R \gg R_0$), three internal parts with different geometrical properties can be easily distinguished by studying the radial density δ of the domain, averaged over all angles, as a function of the distance R to the center of mass [1] (Fig. 2). There are i) a central dense core ($\delta = 1$) which corresponds exactly to the initial non-distorted nucleus, with a constant radius $R_0 = 27.7 \pm 0.4 \, \mu m$, ii) a large intermediate region, where tip-splitting is dominant, where δ exhibits a power-law (fractal) behavior, and where further aggregation is screened, and iii) a distal active zone, where growth takes place.

In general, unstable growth in Langmuir monolayers is shown to be controlled by the diffusion of impurities (e.g., the fluorescent dye) expelled from the condensed phase [8]. We propose here that density gradients and the diffusion of the matrix molecules within the LE phase (the monolayer being then considered as a two-dimensional solution) can themselves destabilize the growth [9]. This hypothesis is supported by our observations of ramified domains, even at the lowest dye concentrations experimentally accessible (0.005 mol%).

The existence of a critical radius of instability, R_{MS}^j, above which thermal fluctuations at the boundary of a circular germ are enhanced by diffusion phenomena is predicted by the Mullins-Sekerka (M-S) thoery [10]. This radius depends on the mode j of the fluctuation, decomposed in spherical harmonics. As five primary bumps are observed in mean around the initial circular nucleus, one can write $R_{MS}^5 = R_0$. After calculating (from the phase diagram) the supersaturation $\Delta = 8.8 \times 10^{-3}$, the critical radius of nucleation, R_c, above which a germ will grow rather than vanish, can be derived from the general relationship between R_{MS}^j and R_c for a two-dimensional system [11]: $R_{MS}^j = R_c (1 + j(j - 1)a(\Delta))$, where $a(\Delta)$ is a constant which depends only on Δ. This yields $R_c = 0.3 \pm 0.1 \mu m$, which is consistent with our observations that any visible seed (the resolution of our optical setup is 3—4 μm) is stable. One can also calculate the capillary length, d_0, which estimates the width of the interline [4], by knowing that $d_0 = R_c\Delta$. We find a value of 34 Å, i.e., of a few molecular lengths, in good agreement with literature results. Finally, one can derive an approximate value of the line tension σ [12], since $d_0 \sim \sigma/L$ (the latent heat L is deduced from the phase

diagram [6]). This yields a low σ value of roughly 10^{-11} J $\cdot$ m^{-1}, which explains that the tip-splitting instabilities are observed even for the slowest compression speeds, and, thus, the driving force towards equilibrium is very weak. Moreover, σ is certainly not (or extremely weakly) anisotropic, since the ramifications are randomly distributed. In comparison, the high σ value ($5 \cdot 10^{-11}$ J $\cdot$ m^{-1}) and the high line tension anisotropy (≈ 0.9) found for NBD-stearic acid monolayers explains why, in this case, the needle-like equilibrium shape is stabilized [13].

The fractal region is a classical morphology for diffusion-controlled growth [1]. The associated Hausdorf exponent is 1.78 $\pm$ 0.07 for the displayed image. When averaged over different objects and by using several methods, one obtains a most probable value of 1.75 $\pm$ 0.02. This non-linear tip-splitting regime is hardly addressed analytically. On the contrary, numerical models like DLA [1] have been successfully used to simulate such growth phenomena. In these studies, tenuous structures, with long-range spatial fractal behavior are generated, with a Hausdorff exponent ranging from 1.67 to 1.75. Our experimental value is close to predictions for off-lattice aggregation, with finite curvature-dependent sticking probabilities used to model line tension [14].

b) Dendrites: The above paragraphs show that the growth morphology of condensed phases in Langmuir monolayers can be related, for a single molecule, to their temperature-dependent physical structure. It is also possible to obtain radically different growth morphologies by changing the nature of the amphiphile [15]. We have used the D-myristoyl alanine (D-MAla), a chiral enantiomer species of an acyl-amino-acid which differs from MYA by only its alanine head group. The LE-LC transition in D-MAla monolayers has been studied by Π-A measurements by Bouloussa et al. [16]. Fluorescence microscopy observations have revealed dendritic growth during this transition [17]. Since dendritic growth requires, as a necessary condition, line tension anisotropy, the condensed phase cannot be an isotropic liquid [3, 4].

The dendrite shown in Fig. 1c corresponds to a two-step experiment, where a small nucleus, first selected by a compression/expansion method, is then forced to grow by again compressing the monolayer at a given compression rate $R = A^{-1} dA/dt$. Controlling R is roughly equivalent to clamping the supersaturation at a constant value. The bright halo surrounding the pattern shows that dye diffusion probably dominates the unstable growth process. Contrary to fatty acids, thin straight branches grow with stable tips. Their width (5—25 µm) and velocity V (1—25 µm · s^{-1}) do not vary over times of several tens of seconds, and the growth is steady-state.

In good agreement with the theoretical predictions of Ivantsov [4], the tips are parabolic over distances of at least 10 µm from their top. This has been checked indirectly by looking at the shape of the lines of iso dye concentration in the LE phase, which is less affected by the limited resolution than the small tip.

Since Langmuir monolayers are purely two-dimensional systems, they are well suited to check the theories of shape selection, which have been developed only for planar geometries. In particular, the product $\rho^2 V$ has to be a constant independent of the supersaturation, and therefore of V [4]. We have deduced the radius of curvature at the tip ρ by measuring the dendrite width at a given distance from the tip, which univocally determines the equation of the parabola. The results, plotted in Fig. 3 as $\rho^2 V$ against V, show that $\rho^2 V$ is indeed constant around a mean value of 25.9 ± 1.9 µm^3 · s^{-1}.

one. That they grow at a 90° angle suggests an underlying fourfold symmetry. They are apparently regularly distributed, with periods of 10 to 20 µm, especially for the first five near the tip. At long times, they undergo a screening effect, and the longest ones grow at the expense of the shortest. There are recent theories which interpret this side-branching as a selective amplification of noise [3, 4]. Mechanical vibrations of our liquid surface, which annot be totally suppressed, are probably the source of this secondary instability.

Conclusion

Langmuir monolayers are optimum two-dimensional systems to study nucleation and growth phenomena. Depending on the nature of the growing condensed phase, different growth morphologies have been observed, corresponding to equilibrium, as well as metastable structures. By using recent theoretical results for two-dimensional growth, characteristic lengths and line tension values have been estimated.

Acknowledgements

We are grateful to O. Bouloussa for the synthesis of the D-MAla. We gratefully acknowledge M. Ben Amar, C. Caroli, B. Caroli, B. Roulet, P. Tabeling, and K. To for fruitful discussions. This study was partly supported by a grant from the Direction of Research, Development and Innovation of Elf-Aquitaine and by a BDI Doctorate position for one of us (S.A.).

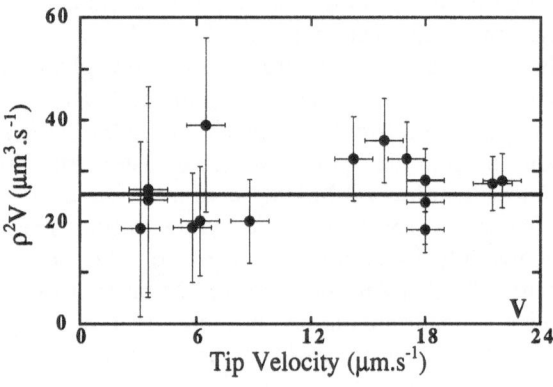

Fig. 3. $\rho^2 V$ vs. V (ρ: radius of curvature at the tip; V: growth velocity) for D-MAla dendrites (see text for details)

Our dendrites are never perfect parabolic needles. As in Fig. 1c, for a fully developed dendrite (which can be 1 cm long) several tens of secondary branches developed on both sides of the main

References

1. Vicsek T (1989) Fractal Growth Phenomena, World Scientific
2. Sawada Y, Dougherty A, Gollub JP (1986) Phys Rev Lett 56:1260. Bensimon D, Kadanoff LP, Liang S, Shraiman BI, Tang C (1986) Rev Mod Phys 58:977. Honjo H, Ohta S, Matsushita M (1987) Phys Rev A 36:4555
3. Huang SC, Glicksman ME (1981) Acta Metall 29:701—715 and 717—734. Janiaud B, Bouissou P, Perrin B, Tabeling P (1990) Phys Rev A 41:7059
4. Ivantsov GP (1947) Doklady Akad Nauk SSR 58:567. Langer JS (1980) Rev Mod Phys 52:1. Ben Aman M, Pomeau Y (1986) Europhys Lett 2:307. Kessler DA, Koplik J, Levire H (1988) Adv Phys 37:255

5. Gaines GL Jr (1966) Insoluble Monolayers at Liquid-Gas Interfaces, Wiley, New York. Knobler CM, Stine K, Moore BG (1990) In: Dynamics and Patterns in Complex Fluids, Springer Proceedings in Physics, Springer Verlag, Berlin. Möhwald H (1990) Ann Rev Phys Chem 41:441

6. Akamatsu S, Rondelez F (1991) J Physique II (Paris) 1:1309

7. Bibo A-M, Knobler CM, Peterson IR (1991) J Phys Chem 95:5591

8. Miller A, Möhwald H (1987) J Chem Phys 86:4258

9. Suresh KA, Nittmann J, Rondelez F (1988) Europhys Lett 6:437. Akamatsu S, Rondelez F, to be published

10. Mullins WW, Sekerka RF (1963) J Appl Phys 34:323 and (1964) 35:444

11. Oswald P (1988) J Physique (Paris) 49:2119

12. Weis RM, McConnell HM (1985) J Phys Chem 89:4453

13. Muller P, Gallet F (1991) Phys Rev Lett 67:1106

14. Meakin P, Family F, Vicsek T (1987) J Coll Interface Sci 117:394

15. Göbel HD, Gaub HD, Möhwald H (1987) Chem Phys Lett 138:441. Knobler CM (1990) Science 249:870

16. Bouloussa O, Dupeyrat M (1988) Biochim Biophys Acta 938:395

17. Akamatsu S, Bouloussa O, To K, Rondelez F (1991) submitted to Phys Rev Lett

Authors' address:

S. Akamatsu
Laboratoire Léon Brillouin
CEN Sacley
91191 Gif sur Yvette Cedex, France

Progress in Colloid & Polymer Science   Progr Colloid Polym Sci 89:214—0217 (1992)

Electronic speckle pattern interferometry used to characterize monolayer films

G. Aschero, E. Piano, C. Pontiggia, and R. Rolandi

Department of Physics, University of Genoa, Italy

Abstract: Electronic speckle pattern interferometry (ESPI) can detect micro displacements and deformation of optically rough surfaces. — Experiments have been performed on pure liquid surfaces and Langmuir films at the air-water interface to determine whether ESPI can be implemented to study liquid surface properties. — Mechanically induced resonant modes of liquid surfaces have been visualized as interference patterns similar to Chladni figures. Oscillation amplitudes on the order of a few microns have also been measured. — The resonance frequency depends only slightly on surface tension and surface pressure, while there is greater correlation between vibration amplitude and the characteristics of the liquid surface. — On the basis of these preliminary results it may be possible to use this technique to detect short wavelength oscillations (<1 mm) and to measure the damping of capillary ripples to obtain the viscoelastic parameters of liquid surfaces.

Key words: Electronic speckel pattern interferometry; liquid surfaces

Introduction

Laser light scattered by a rough surface has a grainy aspect, due to the random intensity distribution caused by the interference of coherent waves coming from a large number of scatterers on the surface. The speckle pattern changes when the surface is translated or deformed. Speckle interferometry is designed to obtain information about surface movements from speckle pattern changes. A plane reference wave is combined with the speckle field to convert phase variations into irradiance variations. Hence, any change in the optical path length by just a fraction of the laser wavelength can be recorded as a variation of the irradiance of the speckle interference pattern.

Subtracting the intensities of two of these patterns, recorded before and after a surface displacement, produces a system of fringes, whose separation contains information about such a displacement. In electronic speckle pattern interferometry (ESPI), speckle patterns are recorded via a television (TV) camera and electronically subtracted, and the result is displayed as a set of fringes on a TV monitor.

This kind of interferometry is often used in quality control and nondestructive testing, since it can be used to detect cracks, stress lines, weak points, and resonant modes of mechanical structures and devices. The amplitude of displacements and deformations can also be measured. A sensitivity of $\lambda/100$ has been attained for oscillation amplitude measurements [1].

The detection and analysis of liquid surface micromovements can be implemented to investigate evaporation rate, surfactant deposition and adsorption, surface tension, and visco-elasticity properties.

To determine whether ESPI is suitable to study liquid surface properties, we have carried out a series of measurements on pure liquid surfaces (water and ethanol) and Langmuir films prepared with phosphatidyl-serine (PS) and a polymerizable styrene functionalized surfactant, the bis[2-(n-hexadecanoyloxy)ethyl]methyl(p-vinylbenzyl) ammonium chloride (BHO VAC).

Experimental details

The experiment was conducted using a modified commercial speckle interferometer, Vidispec (Ealing Electro-Optics p.l.c., G.B.). The laser beam (He-Ne, λ = 632.8 nm) was deviated by a mirror to illuminate the liquid surface perpendicularly, from above. The liquid was contained in a stainless steel cylindrical trough with an internal diameter of 40 mm and depth of 30 mm. The internal surfaces of the trough were sanded to eliminate reflections. The object lens was focused on the liquid surface.

Spectroscopic grade ethanol and water purified by a Millipore Milli-Q filter system were used. The specific resistivity of the water was $\geqslant$18.2 MΩ × cm). The PS was generously provided by Fidia Research Laboratories (Padova, Italy). The BHOVAC was synthesized by Prof. P. Tundo (University of Venice) according to the method previously described [2].

To prepare the monolayers the phospholipid and the monomeric surfactant were dissolved in spectroscopic grade chloroform at concentrations of 1—2 mg/ml. These solutions were spread at room temperature (22°—23°C) with a microsyringe on purified water. The solution amounts were adjusted to obtain PS films with a density of 48 $\mathring{A}^2$/mol and monomeric BHOVAC films with a density of 38 $\mathring{A}^2$/mol.

The BHOVAC monolayers were polymerized by an ultra-violet (UV) lamp (Mineral Light UVG-11, $\lambda \cong$ 254 nm) suspended 10 cm over the liquid surface. The interferometer and trough were mounted on an optical table. The interference figures, readable in real time on a TV monitor, were usually stored by a Sony video recorder. Images were digitized by a Matrox board to be filtered and analyzed utilizing an image analysis software EIDOIPS, (Eidosoft, Milan, Italy) running on IBM personal computer. Surface oscillations were mechanically induced by dipping a 1-mm diameter stainless stell needle into the surface and connecting it to a small "woofer" loudspeaker. Temperature, pressure, and humidity were not controlled. During an experiment, the temperature generally remained stable within 0.4°C, owing to the high thermal capacity of the steel vessel and the optical bench. Humidity ranged from 40—50%.

Results and discussion

Different resonant modes of a pure water surface are shown in Fig. 1. The surface oscillations were

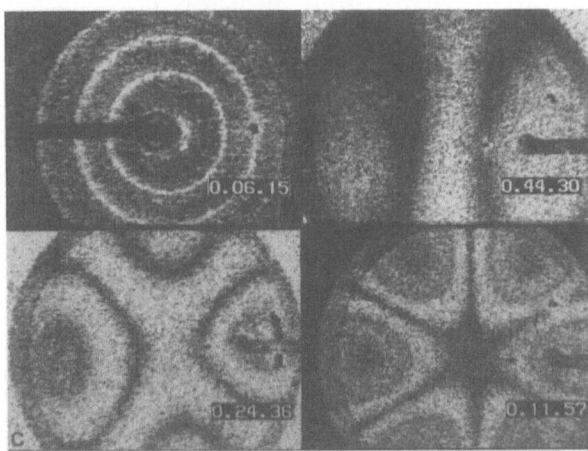

Fig. 1. Surface resonant modes of the water contained in the circular trough. The needle used to generate the oscillation is visible. a) The needle dips into the center of the circular trough. Interference fringes have circular symmetry for all resonant frequencies. Resonance frequency = (21 ± 1) Hz; b) The needle dips into the water surface at about one-fourth of the diameter. A diametral node appears at the resonant frequency of (6 ± 1) Hz; c) Similar to (b), but the frequency is (8 ± 1) Hz; two diametral nodes appear; d) Similar to (b), but the frequency is (11 ± 1) Hz; three diametral nodes appear. The fringe color is inverted by the software in an attempt to improve the image definition

induced by the needle when it was placed in different positions. Intense white fringes are nodal lines; alternate gray and black fringes are zones ocillating with different amplitudes. These figures are analogous to Chladni figures, obtained by spreading thin dyed powders on membranes and solid surfaces. When the needle was dipped in the liquid surface at one-fourth of the diameter of the trough, different resonance frequencies generated fringe patterns with different symmetries which could be identified by the number of diametral nodal lines.

In Fig. 2 resonant frequencies of water and ethanol surfaces, PS and BHOVAC films are plotted versus the number of diametral nodal lines. High resonance frequencies of the different surfaces show better separation. The resonance frequencies of the water surface with a surface tension $\simeq$71 mN/m are higher than those of the ethanol surface ($\gamma \simeq$ 22 mN/m).

The wavelength must be at least 1 mm; below this value fringe figures are confused. This limit is indicative: it is influenced by the statistical nature of

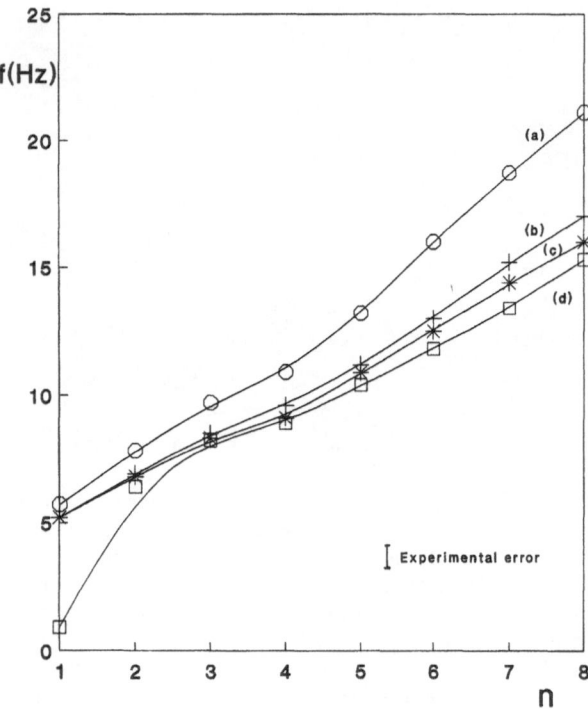

Fig. 2. Resonance frequencies (f) are plotted versus the number of diametral nodes (n) for different surfaces. The needle was dipped at one-fourth of the trough diameter. The continuous lines have been drawn as visual guides: a) water; b) BHOVAC film; c) PS film; d) ethanol

speckle patterns, the optical magnifying characteristics of our interferometer, and the resolution of the Vidicon camera mounted on it. Since the experimental error of the resonant frequency is ~1 Hz, the difference between the resonant frequencies of films prepared with BHOVAC (surface pressure, $\pi \simeq 65$ mN/m) and PS ($\pi \simeq 43$ mN/m) cannot be resolved. Similar measurements on other liquid surfaces and films at different surface pressures pointed out the poor correlation of resonant frequency with surface pressure.

There is greater correlation between the vibration amplitude and the characteristics of the liquid surface. Figure 3 illustrates the maxima amplitudes of the oscillations of the resonant mode with one diametral nodal line versus the maxima of the sinusoidal voltages applied to the loudspeaker. The amplitudes were measured by using the relation between amplitude an contrast of fringes (3). In these cases the mean experimental error for the amplitude was 0.08 μm and the lowest detectable value was 0.12 μm.

The curves belonging to different surfaces are well separated. It is interesting to note that amplitude of the polymerized BHOVAC film are much lower than those of the monomeric film, which shows how the sensitivity of the method

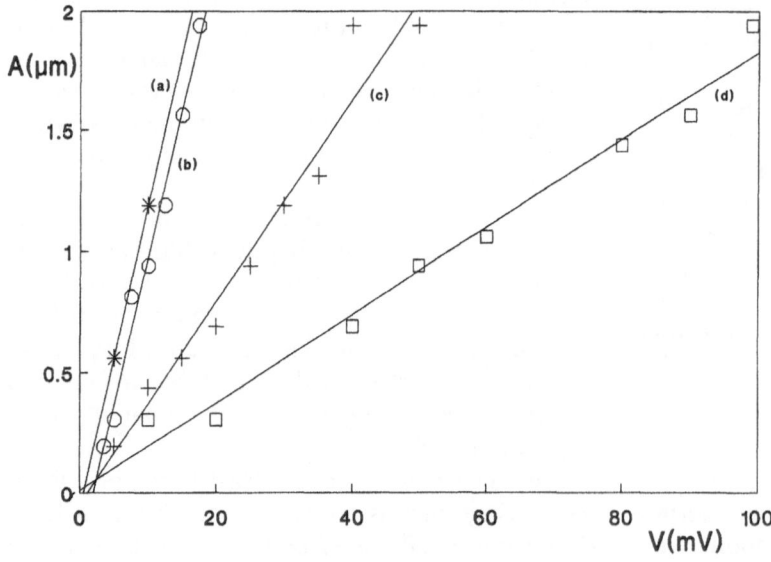

Fig. 3. Maxima amplitudes (A) of the induced oscillation as a function of the maxima sinusoidal voltages (V) applied to the loudspeaker. The experimental values have been fitted by straight lines of equation $A = aV + b$; r^2 is the regression coefficient.
a) Monomeric BHOVAC film: $a = (119 \pm 4) \cdot 10^{-6}$ m/V; b) $(-1.2 \pm 3) \cdot 10^{-2}$ μm, $r^2 = 0.99$;
b) PS film: $a = (113 \pm 7) \cdot 10^{-6}$ m/V, $b = (-14 \pm 7) \cdot 10^{-2}$ μm, $r^2 = 0.97$; c) water: $a = (41 \pm 3) \cdot 10^{-6}$ m/V; $b = (-3 \pm 8) \cdot 10^{-2}$ μm, $r^2 = 0.96$; d) BHOVAC film, partially polymerized by an irradiation of 1.5 min: $a = (18.1 \pm 0.8)$ m/V, $b = (0.9 \pm 5) \cdot 10^{-2}$ μm, $r^2 = 0.98$

varies with the physical properties of the films. When the BHOVAC films are polymerized by UV radiation over a constant area, the surface pressure drops exponentially as a function of the irradiation time [4]. The surface pressure was not continuously monitored during the experiment, but from the irradiation time (1.5 min) and other controlling experiments, it can be estimated to be about 30 mN/m. The viscosity of the polymerized film should be higher than that of the monomeric film [5].

Conclusion

ESPI techniques proved to be a simple and easily reproducible way to study the resonant waves induced on a liquid surface. Waves are detectable with amplitude on the order of micrometers with an average uncertainty of 0.08 μm.

By utilizing an optical system with higher magnification and laser beam modulation techniques [6, 7] it should be possible to detect oscillations with short wavelengths (less than 1 mm) and small amplitudes, to measure the damping characteristics of capillary ripples and to obtain the viscoelastic parameters of the liquid surface.

Possible applications include the study of the dynamic rheological properties of liquid surfaces, as well as the detection and visualization of resonance modes, in addition to a vibration amplitude map.

Acknowledgements

This work has been supported by: EEC within ESPRIT II Basic Research Action 3200-OLDS; MURST, with 40% and 60% grants; CNR within "Progetto Finalizzato Chimica Fine II".

References

1. Moran SE, Luganmani R, Craig PN, Law RL (1989) J Opt Soc Am A 6:252—269
2. Reed W, Guterman L, Tundo P, Fendler JH (1984) J Am Chem Soc 106:1897—1907
3. Ek L, Molin SE (1970) Opt Comm 2:419—424
4. Rolandi R, Paradiso R, Xu SQ, Palmer C, Fendler JH (1989) J Am Chem Soc 111:5233—5239
5. Rolandi R, Gussoni A, Maga L, Robello M (1992) Thin Solid Films 207
6. Hogmoen K, Pederson HM (1977) J Opt Soc Am 67:1578—1583
7. Nakadate S (1989) Appl Opt 25:4162—4167

Authors' address:

Dr. R. Rolandi
University of Genoa
Department of Physics
Via Dodecaneso 33
16146 Genoa, Italy

Progress in Colloid & Polymer Science   Progr Colloid Polym Sci 89:218—222 (1992)

Photo-induced electron transfer in monolayers: effect of acceptor location at the interface

G. Caminati*, R. C. Ahuja, and D. Möbius

Max-Planck-Institut für Biophysikalische Chemie, Göttingen, FRG
*) Dipartimento di Chimica, Università di Firenze, Italy

Abstract: We have investigated photo-induced electron transfer (PET) in monolayers at the water-air interface. The donor was a cyanine dye (di-octadecyl-oxacyanine perchlorate) incorporated in a matrix monolayer of dipalmitoyl-phosphatidic acid. Three different viologens derivatives were used as acceptors, namely methyl-, tetradecyl-, and benzylviologen. The acceptors were dissolved in the aqueous subphase and the process of adsorption at the interface was studied by measuring surface pressure-area isotherms, as well as UV-Vis reflection spectra. In the case of methylviologen, we did not find any evidence for penetration of the acceptor in the monolayer, whereas adsorption of benzyl- and tetradecylviologen led to changes in the arrangement of the molecules in the monolayer. — Photo-induced electron transfer was then monitored by measuring the change in fluorescence of the donor in the monolayer as the quencher was adsorbed at the monolayer/subphase interface. Fluorescence intensity-area isotherms were measured simultaneously with surface pressure-area isotherms. Fluorescence quenching due to the adsorbed acceptor was found to increase with increasing hydrophobicity of the quencher molecule. These results are correlated to the different modes of adsorption of the acceptors.

Key words: Photoinduced electron transfer; monolayers; water-air interface; acceptor adsorption

Introduction

Electron transfer from the excited state of the donor molecule to the ground-state acceptor (PET) has been the subject of many theoretical and experimental studies [1] related to the importance of this process in many areas of chemistry and physics. Recently, photo-induced electron transfer (PET) phenomena have been extensively investigated in microheterogeneous environments such as micelles, vesicles or microemulsions [2]. These studies have revealed the important role of the interface in enhancing the efficiency of photochemical processes. Investigation of PET in monolayer assemblies at air/water interface is especially attractive since these systems may be considered as rigid structures of specifically designed architecture [3]. One important advantage of monolayer organizes over disperse systems in solution is the possibility to control experimentally the position and the orientation of the redox couple [3—5]. The present work examines the conditions and the mechanism of the electron transfer between an excited cyanine dye diluted in a matrix monolayer, and an appropriate acceptor solubilized in the aqueous subphase. Three different viologen derivatives, namely, methyl-, tetradecyl- and benzyl-viologen, are used as acceptors. The viologen molecules thus differ only in hydrophobicity, whereas their redox properties are almost identical [1]. The adsorption process from the subphase to the monolayer/water interface was then preliminarily studied in order to determine the distribution and density of the acceptors at the interface. Finally, the fluorescence quenching, resulting from the electron transfer process was studied at the air/water interface as a function of acceptor structure, acceptor surface density, and of the compression of the monolayer.

Experimental

The N,N'-dioctedecyloxacyanine perchlorate (OC) was synthesized by Sondermann [6]. Dipalmitoylphosphatidic acid (DPPA) was obtained from Sigma Chemical Co., and used without further purification. Methylviologen dichloride (MV) and benzylviologen dichloride (BV) were supplied by Aldrich, whereas 1-methyl-1-tetradecyl-4,4'-bipyridinium dichloride (TDV) was obtained from Fluka. Schematic structures of the redox couples are shown in Fig. 1. Monolayers were prepared by spreading a chloroform solution of the donor OC and the matrix (DPPA) on the aqueous subphase (Millipore water, pH: 5.6), a 1:10 molar ratio between OC and DPPA was used in all experiments, unless otherwise specified. The viologens were solubilized in the subphase (concentration range 1×10^{-8} to 1×10^{-5} M). Surface pressure and surface potential were measured simultaneously on a round through at 20°C. Reflection measurements (spectra at constant π and isotherms at 280 and 380 nm) were obtained using an apparatus previously described [7]. Fluorescence measurements were done with the excitation beam at 30°, whereas the emitted light was collected normal to the water interface using fiber optics. Fluorescence

was measured at 425 nm with excitation at 365 nm, both in the presence and in the absence of acceptor.

Results and discussion

Figure 2 shows the reflection spectra obtained for mixed monolayers of DPPA/OC with a 10:1 molar ratio for different subphase compositions. All spectra were recorded at $\pi = 35$ mN/m. Curve (a) is the reflection spectrum for mixed monolayers spread on pure water subphase. A band at 380 nm and a shoulder at 365 nm are observed, and they may be attributed to the cyanine dye [8, 9]. The shape and

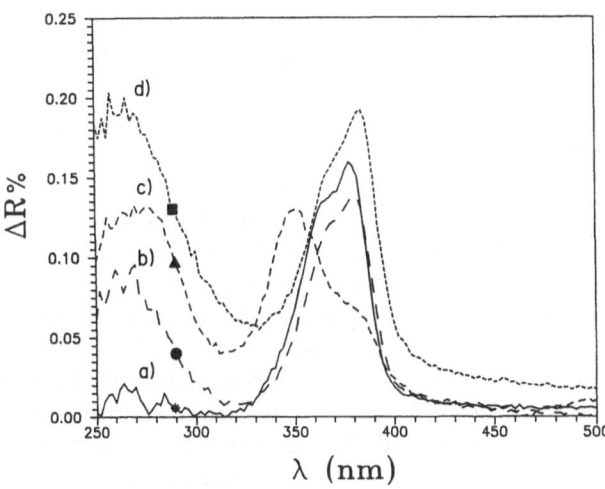

Fig. 2. Reflection spectra for the 1:10 OC/DPPA mixed monolayer on different subphases. Surface pressure = 35 mN/M, acceptor concentration 1×10^{-6} M. Curve a: * water subphase; curve b: ● MV; curve c: ▲ TDV; curve d: ■ BV

the position of the band indicates that the donor is present mainly in its monomeric form at this monolayer composition. Higher aggregates of cyanine dyes were previously found in monolayers assemblies [9], these aggregates are characterized by a narrow absorption band red shifted with respect to the monomer absorption. Curves (b), (c), and (d) are the reflection spectra obtained for subphases containing MV, TDV, and BV, respectively. The acceptor concentration in the subphase was 1×10^{-6} M. The appearance of a new band around 275 nm in the reflection spectra reveals the presence of the viologens at the monolayer/subphase interface. The intensity of the reflection signal is maximum for BV and minimum for MV, this may be due

DONOR:

OC

ACCEPTORS:

R = R' = —CH₃	MV
R = —CH₃ ; R' = —C₁₄H₂₄	TDV
R = R' = —CH₂—C₆H₅	BV

Fig. 1. Schematic structures of the donor and acceptor molecules

either to a different orientation of the adsorbed chromophore or to a difference in surface densities. Since the viologen chromophore will orient its long axis (also the optical transition moment axis) parallel to the interface in order to maximize electrostatic interactions with the negatively charged DPPA, we may reasonably assume that this would hold for all three viologens derivatives. Although the electrostatic contribution to adsorption is the same for the three viologens, a higher interfacial concentration of BV and TDV may be expected since these substances are slightly surface active. Reflection spectra from the acceptor solution subphase without DPPA monolayer were also recorded, but no reflection signal was detected in the range of concentration studied. The position of the band for TDV is shifted towards higher wavelengths, thus implying additional interactions with the monolayer molecules or a different moiety of the chromophore.

The adsorption process was then studied for the pure DPPA monolayer. Surface pressure was recorded simultaneously with reflection at 280 nm as a function of compression and the results are illustrated in Fig. 3. The π-A isotherm in the presence of MV (curve b) is virtually identical to the isotherm obtained on pure water subphase (curve a), but the reflection signal at 280 nm increases sharply at ca. 0.65 nm^2/mole. This indicates that MV is adsorbed at the monolayer/subphase interface due to electrostatic attraction, but it does not penetrate in the monolayer. Penetration of the acceptor molecule in the monolayer should in fact be reflected in either a shift or a change of shape of the isotherms [10]. Moreover, the reflection signal, when normalized to the change in surface density, remains almost constant up to π = 35 mN/m, indicating that the orientation of the chromophore does not change as the monolayer is compressed. The reflection signal at 280 nm increases with increasing acceptor concentration in the subphase, reaching saturation for concentrations larger than 2×10^{-6} M (results reported in [11]). The surface-pressure-area isotherm show shifts towards higher areas when TDV is present in the subphase. The limiting area per molecule is around 0.58 nm^2, a value much larger than that on pure water (0.40 nm^2/mole). The additional area of 0.18 nm^2 is close to that of one hydrocarbon chain. These data support the hypothesis of the insertion of the hydrocarbon tail of the asymmetric TDV in DPPA monolayer. Reflection values as a function of surface area indicate

that the orientation of TDV chromophore does not change significantly upon compression of the monolayer.

Surface pressure-area isotherms for BV (curve d, Fig. 3) exhibit a plateau for molecular areas between 0.9 and 0.6 nm^2, but this feature is found only for BV concentrations larger than 1×10^{-6} M. This behavior may correspond to a change in the arrangement of the acceptor molecule at the interface. The change in molecular distribution in the monolayer is followed by the squetzing out of the acceptor into the water layer underneath the polar groups. The expulsion of BV from the interfacial layer is supported by the decrease in molecular area after the plateau and by the value of the limiting area for DPPA monolayer on BV subphase which is 0.44 nm^2, a value similar to the MV case.

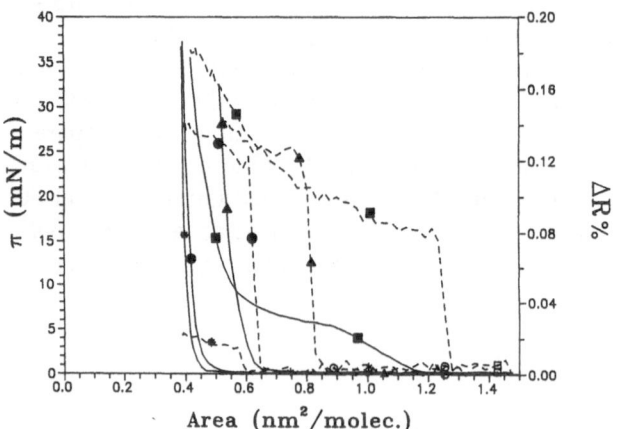

Fig. 3. Reflection-area and surface pressure-area isotherms for mixed monolayers of OC/DPPA (1:10) on different subphases. Solid lines refer to surface pressure values; dashed lines refer to reflection values. Curve a: * water subphase; curve b: ● MV; curve c: ▲ TDV; curve d: ■ BV

Acceptor adsorption is observed also when the donor is added to the matrix monolayer (see Fig. 2). In the case of MV and BV the presence of the positively charged oxacyanine only slightly reduces the amount of acceptor adsorbed. The decrease in reflection signal is due to the presence of an additional positive charge of the cyanine dye which repels acceptor molecules. The reflection signal due to OC in the monolayer is not affected by the presence of the acceptors which, at this surface pressure (35 mN/m), are located underneath the monolayer polar groups. However, in the case of

TDV, at a concentration higher than 1×10^{-6} M, a plateau is observed in the π-A isotherms (results not shown here). The analysis of the reflection spectra at different surface pressures allows us to ascribe this transition to the formation of higher aggregates of cyanine dyes, as can also be seen from the corresponding spectrum (Fig. 2, curve c).

Steady-state fluorescence intensity-area isotherms for the OC/DPPA monolayer were measured both in the absence and in the presence of quencher in the subphase. In Fig. 4 the results obtained for an acceptor concentration of 5×10^{-8} M are reported. Addition of the viologens in the subphase leads to a decrease in fluorescence intensity and this is attributed to the electron transfer between the excited

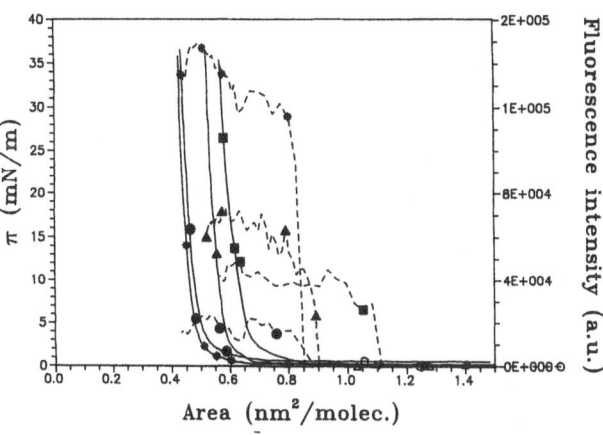

Fig. 4. Fluorescence intensity-area and surface pressure-area isotherms for mixed monolayers of OC/DPPA (1:10) on different subphases. Solid lines refer to surface pressure values; dashed lines refer to fluorescence intensity values. Curve a: * water subphase; curve b: ● MV; curve c: ▲ TDV; curve d: ■ BV

donor located in the monolyer and the acceptor present at the monolayer water/interface. Maximum quenching efficiency is obtained for methylviologen, and minimum quenching for TDV. In the case of MV, fluorescence intensity decreases rapidly as the concentration is changed from 10^{-8} M to 10^{-7} M, and fluorescence is completely quenched at MV $= 5 \times 10^{-7}$ M. These concentration values are much lower than the values obtained from reflection data. The detailed investigation of the PET process as a function of MV concentration are reported in [11]. For TDV and BV the fluorescence

intensity is almost totally quenched at concentrations higher than 5×10^{-6} M. In this case, additional processes (beside PET) should be taken into consideration, e.g., squeezing out of benzylviologen from the interfacial layer at high surface pressures and induced phase segregation in the mixed monolayer in the presence of TDV in the subphase. The presence of these phenomena, which superimpose to the electron transfer process, prevents a detailed analysis of the quenching mechanism in the high concentration regime. For quencher concentrations lower than 5×10^{-6} M some general deductions may be drawn. The efficiency of quenching is directly related to the position of the quencher at the interface. When the acceptor is inserted and anchored with its hydrophobic chain in the monomolecular film, as for TDV, quenching efficiency is greatly reduced compared to the case where the acceptor is adsorbed at monolayer-water inteface (MV). Since the driving force for electron transfer is nearly the same for all three quenchers, the difference in electron transfer yield has to be related to their different mobilities and distributions at the interface. Analogous experiments on subphases with higher viscosities or on transferred layer systems should be done to gain further insight into the process.

Conclusions

The results presented show that photo-induced electron transfer between the excited donor and the acceptor takes place at the monolayer/subphase interface. The quenching efficiencies decrease in the order MV > BV > TDV, and this is probably correlated to the position and mobility of the acceptor in the interfacial layer. Adsorption studies have, in fact, shown that the three viologens are localized in different environments at the interface. MV is only electrostatically attracted towards the negatively charged monolayer, whereas TDV partially penetrates the monolayer with its hydrophobic tail. BV is located in the monolayer at low surface density and it is successively squeezed out into the aqueous subphase as the monolayer is progressively compressed. The orientation of the chromophore may be assumed to be similar for the three molecules, i.e., the long axis lies parallel to the interface, in order to maximize the electrostatic interaction with the negatively charged DPPA. Nevertheless, BV and TDV mobility is restricted compared to the free-moving MV. The results obtained have also shown that high concentrations of TDV ($>5 \times 10^{-6}$ M) induce a change in the aggregation state of the dye in the mixed monolayer. These observations may have consequences also for PET studies in self-aggregating systems (such as micelles, vesicles, etc.) where similar acceptors are often used without taking into account that they may perturb the system and induce donor aggregation at the interface.

References

1. Fox MA, Chanon M (eds) (1988) Photoinduced electron transfer. Elsevier, Amsterdam parts A—D
2. Kalyanasundaram K (ed) (1988) Photochemistry in microhetereogeneous systems, Academic Press, New York
3. Möbius D (1981) Acc Chem Res 14:63
4. Kuhn H (1979) In Gerisher H, Katz JJ (eds) Light-induced Charge Separation in Biology and Chemistry. Verlag Chemie
5. Ahuja RC, Möbius D (1989) Thin Solid Films 179:475
6. Sondermann J (1971) Liebig Ann Chem 183:749
7. Gruniger H, Möbius D, Meyer H (1983) J Chem Phys 79:3701
8. Ahuja RC, Möbius D (1991) Thin Solid Films (submitted)
9. Möbius D, Kuhn H (1988) J Appl Phys 64:5138
10. Kozarac Z, Dhathatreyan A, Möbius D (1989) Colloid Polym Sci 267:722
11. Caminati G, Ahuja RC, Möbius D (1991) J Phys Chem (submitted)

Authors' address:

Gabriella Caminati
Dipartimento di Chimica
Via Gino Capponi, 9
50121 Firenze, Italy

Progress in Colloid & Polymer Science   Progr Colloid Polym Sci 89:223—226 (1992)

Dye/dihexadecylphosphate monolayers: a spectroscopic and thermodynamic study

G. Caminati, G. Gabrielli, E. Barni[1]), P. Savarino[1]), and D. Möbius[2])

Dipartimento di Chimica, Università di Firenze, Italy
[1]) Dipartimento di Chimica Generale ed Organica Applicata, Università di Torino, Italy
[2]) Max-Planck-Institut für Biophysikalische Chemie, Göttingen, FRG

Abstract: We have studied monolayers of a dye containing hydrophobic moieties at the air/water interface and in LB films. In particular, the aims of the present study were to characterize the interfacial behavior of the alkylsubstituted cyanine dye, (N,N'-tetrahexadecylcyanine iodide) to investigate the surface miscibility of the cyanine dye with the matrix molecule (dihexadecylphosphate), and to correlate the results obtained at the air/water interface with the data collected on transferred layers of the same system. For these purposes, mixtures with different molar ratios of matrix and dye were prepared and their interfacial distribution and orientation were studied, both at the air/water interface, and in multilayers transferred on solid substrates. The monolayers at the air/water interface were studied by measuring surface pressure-area and reflection at constant area or at constant wavelength, whereas absorption spectra were performed on the transferred layer system.

Key words: Monolayers; water-air interface; cyanine dyes; Langmuir-Blodgett films

Introduction

Dye molecules are often used in industrial applications, as well as in basic research in chemistry and physics. Cyanine dyes in particular have been widely employed in such various fields as textiles, photographic sensitizers, and liquid crystal technology [1]. In all these processes it is important to define and control the state of aggregation of the dye. Previous studies have shown that incorporation of the cyanine molecules in self-aggregating systems such as microemulsions or micelles may prevent the dye from aggregating [2]. Moreover, cyanine dyes have been successfully used as donors in the study of photo-induced electron transfer [3, 4] in monolayer, where the question of distribution of the dye in the matrix monolayer is always first addressed. For this reason, we started a systematic study of the interfacial behavior of an alkyl-substituted cyanine dye at the water-air interface and in transferred layers. The cyanine dye chosen for the present study contains four hexadecyl chains and it was studied both alone and in mixed monolayers with a matrix molecule at the water-air interface. As a matrix molecule, we have chosen dihexadecylphosphate; this two-chains molecule with a rather bulky head group has been recently employed in Langmuir-Blodgett studies [5]. The chemical structure of the molecule makes dihexadecylphosphate a perfect candidate as a matrix for a dye molecule. The double chains ensure high hydrophobic interactions with the dye, compared with conventional fatty acids, whereas the larger fluidity of the monolayer allows investigation over a large range of surface phases. Mixtures of dihexadecylphosphate-cyanine dye with different molar ratios were studied at the air/water interface by measuring surface pressure-area isotherms. Reflection spectra were recorded at constant surface pressure and the signal, at constant wavelength, was collected as a function of surface area. The results obtained at the air/water interface are correlated to the data collected for multilayers of the same mixtures transferred on quartz slides.

Experimental

The N,N'-tetrahexadecylcyanine iodide (B9) was synthesized as reported in [2], the chemical structure of the molecule is reported in Fig. 1. Dihexadecylphosphate (DHP) was obtained from Sigma Chemical Co., and used without further purification. Monolayers were obtained by spreading a chloroform solution of the cyanine dye and the matrix molecule (DHP) on the aqueous subphase (Millipore water, pH: 5.6). Molar ratios ranging from 1:10 to 5:1 between DHP and B9 were used. Buffer solution at pH: 9 were obtained using KCl, H_3BO_3, and NaOH (all supplied by Merck). Surface pressure was measured on a round trough at 20°C. Reflection measurements (isotherms and spectra at constant π) were obtained using an apparatus previously described [6]. Fluorescence measurements were obtained by exciting at 30° and collecting the emission normal to the water interface using fiber optics. All measurements were carried out both under nitrogen and oxygen atmosphere. The mixed monolayers were transferred on quartz slides at 30 mN/m using KSV (Finland) instrumentation. Absorption spectra of transferred films were taken with a Lambda 5 Perkin Elmer spectrophotometer.

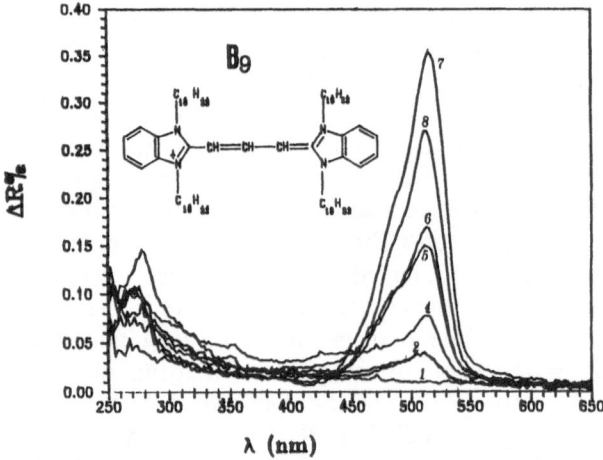

Fig. 1. Reflection spectra for the B9/DHP mixed monolayer on water subphase with different molar ratios. Surface pressure = 35 mN/M. Curve 1: DHP; curve 2: DHP/B9 10:1; curve 3: 5/1; curve 4: DHP/B9 2:1; curve 5: DHP/B9 1:1; curve 6: BHP/B9 1:2; curve 7: DHP/B9 1:5; curve 8: B9

Results and discussion

Mixed B9/DHP monolayers were spread on water subphase and compressed at $\pi = 35$ mN/m under nitrogen atmosphere. Reflection spectra were then recorded and the results are reported in Fig. 1. Curve (8) is the spectrum obtained for the pure B9 monolayer; the band centered around 510 nm is specific for the cyanine dye (1). It is evident from the shape and the position of the band in the spectrum that no aggregation of cyanine dye takes place in the monolayer. The fact that the dye is present mainly in its monomeric form is easily understood, considering its chemical structure. We expect the cyanine molecule to be flat at the interface at low surface pressures; with compression the four hydrophobic chains are pushed to an upright position almost normal to the interface. In this way, the molecule adopts an "inverse table" configuration that prevents formation of cyanine aggregates in the monolayer. We have found that the magnitude of the 510 nm band is progressively reduced in the presence of oxygen, while a new band at 440 nm appears in the spectrum. Similar effects were also observed in preliminary studies on the pH dependence in B9 monolayer. The band at 510 nm was found to increase with increasing pH. This was ascribed to surface-enhanced protonation of the molecule in the monolayer. The intensity of the cyanine dye in the mixed monolayer decreases with decreasing content of the cyanine dye in the mixture, except for the 5:1 molar ratio where reflection signal reaches a maximum value (2). Surface pressure-area isotherms for the different molar ratios were recorded simultaneously with the reflection signal at 510 nm. Figure 2 shows the results obtained: the π-A isotherm for the pure B9 monolayer has a limiting area value of 1.7 nm²/molec; this number is larger than the one found for cyanines with only two alkyl chains (1.1 nm²/molecule at maximum packing) [4]. In that case, the plane of the chromophore was oriented perpendicularly to the interface with the alkyl chains anchored in the monolayer. Pure DHP isotherm is rather steep and it shows a limiting area of 0.4 nm² per molecule, roughly corresponding to the area of two aliphatic chains. The DHP/B9 mixtures exhibit isotherms very similar to the pure DHP up to B9 molar fractions of 0.85, above this value the isotherms resembles B9 behavior and show a change in slope in the low surface pressure region. This discontinuity in the π-area plots corresponds to similar

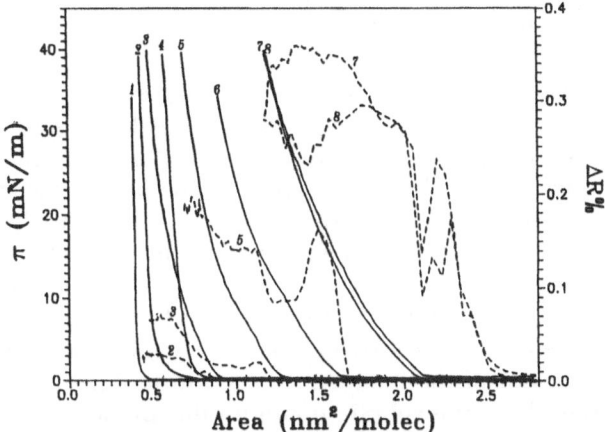

Fig. 2. Reflection-area and surface-pressure-area isotherms for mixed monolayers of DHP/B9 system on water subphase. Solid lines refer to surface-pressure values; dashed lines refer to reflection values. Curve 1: DHP; curve 2: DHP/B9 10:1; curve 3: 5/1; curve 4: DHP/B9 2:1; curve 5: DHP/B9 1:1; curve 6: DHP/B9 1:2; curve 7: DHP/B9 1:5; curve 8: B9

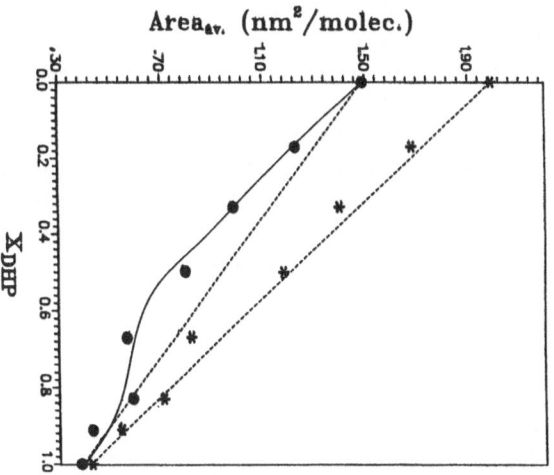

Fig. 3. Average molecular areas plotted against DHP molar fraction. * $\pi = 2$ mN/m; ● $\pi = 30$ mN/m

behavior in the reflection (see dashed and solid lines (5) in Fig. 2) and fluorescence measurements that will be reported elsewhere. Due to the chemical structure of the B9 molecule, changes in the orientation of the chromophore are not expected. The variation in reflection signal at higher surface pressure may then be ascribed to a difference in cyanine surface density. This may be a consequence of a change in the interactions between the two components and in the packing of the monolayer. Further information may be gained from the average molecular areas at constant surface pres-

sure determined from the π-area isotherms. Typical values in the low surface-pressure regime and in the condensed phase are plotted in Fig. 3 as a function of DHP molar fraction. In the expanded phase (up to $\pi = 5$ mN/m) the areas follow ideal additivity in the whole concentration range. At high surface pressures (π above 10 mN/m), average molecular areas are lower than those predicted from additivity up to DHP/B9 molar ratios 5:1; minimum values are found for molar ratios around 2:1. Deviations from linearity are taken to indicate miscibility and non-ideal mixing [7], following a conventional thermodynamic treatment, these results suggest that presence of attractive interaction between the components in the more condensed phases and a maximum thermodynamic stability for mixtures with 1:1 and 2:1 molar ratios. On the other hand, this also agrees with the considerations derived from the evaluation of the difference in areas at different molar ratios. In fact, considering the geometrical requirements of the molecules, we can accordingly hpyothesize that the DHP molecules will insert on top of the cyanine dye once the 1:2 molar ratio is overcome. This feature is strictly dependent on the character of the polar head group. In fact, when a buffer solution (pH: 9) is used as subphase, different results are obtained. At high pH DHP gives more expanded isotherms, whereas only minor changes are found for pure B9 monolayers. At this pH the shape of the isotherms does not change with molar ratios, and the average molecular areas always show positive deviations from ideality when plotted against DHP molar fractions. This result excludes insertion of DHP molecules on top of B9 at higher pH. The full analysis of the two systems will be given in a forthcoming paper.

DHP/B9 were transferred on solid substrate from water subphase (pH: 5.6); Table 1 reports the transfer ratios, both for the down- and for the upstroke for the mixtures: transferability was found to increase with DHP content in the monolayer, and even more so for the upstrokes. Absorption spectra of the transferred films were taken (Fig. 3) under oxygen. A peak at 510 nm is present only for mixtures with high B9 content, and since the transfer was not complete, the magnitude of the 510 nm band is not expected to be proportional to the content of the cyanine in the monolayer. For mixtures containing high DHP molar fraction the band at 510 nm disappears, whereas a broad band centered around 440 nm may be found. A similar band was also present in the reflection spectra when the

Table 1

DHP/B9 molar ratio	Transfer ratio		
	1° layer	2° layer	3° layer
DHP	1.1	0.7	1.0
10:1	0.9	0.7	0.9
5:1	1.1	0.4	0.9
1:1	1.2	0.0	0.6
1:5	1.1	−0.1	0.6
B9	1.2	−0.6	1.0

monolayer was kept ageing under oxygen. Similar experiments on subphase at pH: 9 gave unsatisfactory results for the transfer process and, in this case, we could not detect any absorption in the 300—600 nm range. Further experiments on the characterization of B9 in solution, to clarify the dependence of the absorption spectra on oxygen and pH, and in LB films on different substrates are in progress.

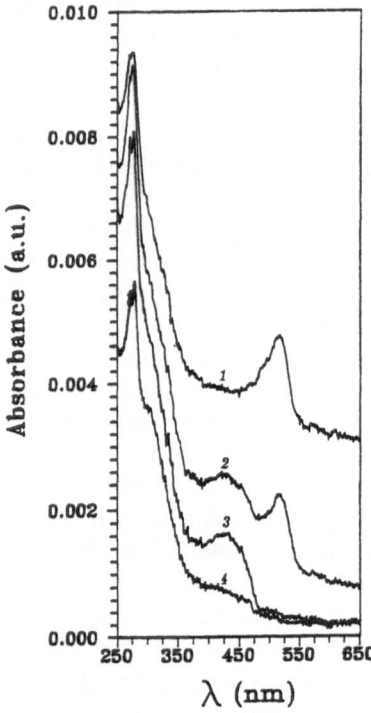

Fig. 4. Absorption spectra of three layers of DHP/B9 mixtures with different molar fractions on quarz substrate. Curve 1: DHP/B9 1:5; curve 2: DHP/B9 1:1; curve 3: DHP/B9 5:1; curve 4: DHP/B9 10:1

Conclusions

The results presented so far show that spectroscopic techniques in conjunction with pressure-area isotherms may give complementary information on the microenvironment of the interface and the thermodynamics of the system. We have also shown that the cyanine dye used in the present study does not form aggregates in spreading monolayers due to its distribution at the interface. When the dye is mixed with a matrix substance, the behavior of the mixtures was found to depend on the pH of the subphase and on the surface phase examined. At high pH, the mixtures always show positive deviations from ideality, suggesting miscibility with prevailing repulsion energies. At lower pH the molecular distribution in the monolayer changes with compression: for DHP/B9 molar ratios above 2:1, we observed that DHP is probably intercalating on top of B9 molecules when surface pressure is raised above a critical value. The results of reflection, fluorescence, and surface-pressure measurements support this hypothesis. The transfer of the mixed monolayers onto solid substrate was found to be highly dependent on the DHP content of the mixture and on the pH of the aqueous subphase. The above results are particularly valuable in those processes where prevention of dye aggregation is of high priority.

References

1. a) Sturmer DM, Heseltine DW In: James TH (ed) (1977) Theory of the Photographic Process, Mac Millian, New York. b) Zollinger H (1987) Color Chemistry, VCH, Weinheim
2. Savarino P, Viscardi G, Barni E (1990) Colloids and Surfaces 48:47
3. Möbius D (1981) Acc Chem Res 14:63
4. Caminati G, Ahuja RC, Möbius D (1991) Thin Solid Film (accepted for publication)
5. Bettarini S, Bonosi F, Gabrielli G, Martini G, Puggelli M (1991) Thin Solid Film (submitted for publication)
6. Gruniger H, Möbius D, Meyer H (1983) J Chem Phys 79:3701
7. Möbius D, Kuhn H (1988) J Appl Phys 64:5138
8. Barnes GT (1991) J Coll Interf Sci 144:229

Authors' address:

Gabriella Caminati
Dipartimento di Chimica
Via Gino Capponi, 9
50121 Firenze, Italy

Progress in Colloid & Polymer Science   Progr Colloid Polym Sci 89:227—232 (1992)

Mixed monolyers:
Support acidity, two-dimensional phases and compatibility

G. Gabrielli, M. Puggelli, and A. Gilardoni

Department of Chemistry, University of Florence, Italy

Abstract: The following two-dimensional binary systems at the W/A interface are studied: stearic acid (SA)-polymethylmethacrylate (PMMA), stearyl-amine (SAm)-PMMA, SA-valinomycin (VAl), SAm-VAL, methylstearate (MS)-VAL. — The behavior of area, surface compressional moduli, and overall collapse pressure as a function of molar ratios of components shows: i) SA and SAm are compatible with PMMA if the pH of support allows their ionization; ii) SA and SAm are incompatible with VAL at any pH value of the support studied. — The different behavior is attributed to the impossibility of interactions between polar groups of VAL and ionizable groups of the other component. The bidimensional distribution of the VAL is confirmed by transfer of K^+ from subphase to monolayers, proved by ATR-FTIR experiments on mono and plurilayers of VAL. — The polar groups are therefore oriented toward the inner molecular cavity of VAL. — The study extended to the MS-VAL system shows compatibility between components with prevalent repulsive interaction between the two components.

Key words: Monolayers; plurilayers; bidimensional system compatibility; bidimensional orientation

Introduction

In the last few years many studies on mixed monolayers have appeared which have investigated both the compatibility of substances of high and low molecular weights in two-dimensional films [1], and the preparation of mixed monolayers to be transferred onto suitable supports as Langmuir-Blodgett multilayers for a variety of applications [2].

In previous papers, we have shown that two-dimensional compatibility derives:

1) for non-ionizable substances, from interactions between the hydrophobic chains and, hence, it is conditioned by the almost parallel distribution of these chains at the interface [3];

2) if at least one of the components is ionizable, by interaction among the polar groups if these are sufficiently ionized, even if the hydrophobic chains of the two components are not distributed parallel to each other at the interface [4].

The aim of the present paper is to study the twodimensional compatibility of binary systems with the following features:

a) different interfacial distribution of the components

b) presence of an ionizable component, whose interfacial distribution and ionization may be affected by the pH of the subphase.

For this purpose, we chose as ionizable components two anphiphilic compounds, stearic acid (SA) and stearylamine (SAm), which have the same hydrophobic chains and differ only in their polar groups, and whose interfacial distribution has been widely studied [4, 5]. In both cases, it is vertical and its surface ionization occurs for quite different support pHs, as already observed [4, 6]. As non-ionizable substances having different surface distribution, we chose polymethylmethacrylate (PMMA), which has been studied previously [3] and valinomycin (VAL) (7). This substance is able to function in membranes as a ion carrier and in monolayers is able to bind to ions placed in the aqueous subphase [7] PMMA and VAL have different molecular conformations, but both are horizontally distributed on the surface [3, 7].

Finally, given the importance of VAL as an ion carrier and its capacity for building mimetic membranes, we also studied the VAL/methyl stearate (MS) system, bearing in mind the ability of VAL to fit into the hydrophobic membrane environment. In fact, although the MS hydrophobic chain is the same length as those of the two ionizable amphiphilis and the same interphasal distribution [8], it is non-ionizable and has a less hydrophilic polar group.

Experimental

The SA, SAm, and MS were supplied by Carlo Erba (Milan), PMMA with average molecular weight 90 000 was from Aldrich Chem. Co., and VAL from Sigma.

As spreading solvent for the isotherms pure chloroform of analytical grade (Merck) was used. The subphase was formed of aqueous solutions obtained with twice-distilled water purified with a Milli-Q apparatus, as previously [7c].

Analytical grade boric acid, borax, sodium chloride, and potassium chloride (all Merck) were used to obtain solutions with controlled pH and ionic strengths in the subphase.

The spreading isotherms were measured with a Lauda balance interfaced with a PDP 11/23 digital calculator through discontinuous compressions, again, as previously described [7c].

The temperature was 25 °C ± 0.1 for the SA/VAL and SAm/VAL systems and 15°, 25°, and 30 °C for the MS/VAL system regulated by means of a Haake thermostat with a probe.

For the transfer of the monolayers from liquid subphase to solid support (Germanium), and to obtain multilayers, a KSV 5000 instrument was used as previously [9]. The germanium support was cleaned with chloroform and placed to sonicate in chloroform; each support was used for only one measurement; purity was verified by FTIR-ATR spectrum.

The transfer ratio was 0.98 for the first layer and 0.40 for the second and subsequent ones; the extraction speed was 12 mN/min.

FT-IR spectra were obtained with a Fourier IFS HR instrument from Bruker.

Results and discussion

The interfacial properties of SA, SAm, PMMA and VAL in a two-dimensional state a the W/A interface have been widely discussed in previous studies [3, 4]. The binary bidimensional systems SA/PMMA and SAm/PMMA have also been studied [4, 6].

Figure 1 shows the trends of the areas at π = 10 mN/m, of the collapse pressures, of the surface compressional moduli in relation to the molar ratios for the SA/VAL system at 25 °C on a subphase of 0.1 M NaCl at pH = 5.6, and on a subphase buffered with borax/boric acid at pH = 7.35.

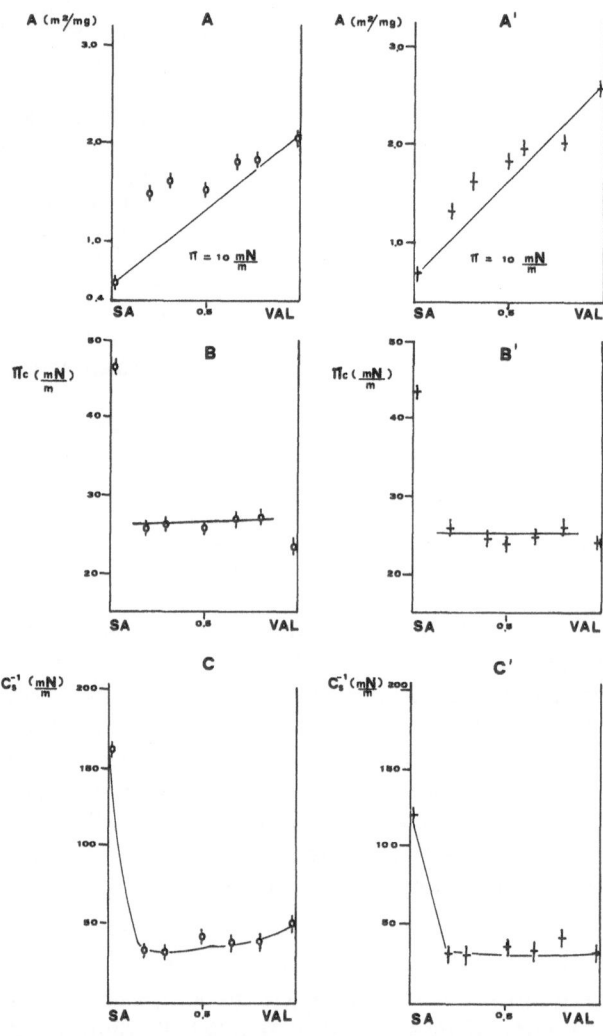

Fig. 1. Plot of the molar ratios for SA-VAL system as a function of: A) area (A) on subphase at pH 5.6; A') area (A) on subphase at pH 7.4; B) collapse pressure (π_c) on subphase at pH 5.6; B') collapse pressure (π_c) on subphase at pH 7.4; C) surface compressional modulus (Cs^{-1}) on subphase at pH 5.6; C') surface compressional modulus (Cs^{-1}) on subphase at pH 7.4

The surface compressional moduli are calculated on the basis of their definition ($Cs^{-1} = -A(\delta\pi/\delta A)$) for the monolyer in the condensed phase by means of analysis of the recorded findings, i.e., deriving the π-A curve in the whole range of areas considered and choosing the highest value of the product $A \cdot (\delta\pi/\delta A)$.

In Fig. 2 are the same trends for the SAm/VAL with the same conditions of temperature and subphase pH. These pH values were chosen because it had been demonstrated elsewhere [4] that each value guaranteed sufficient ionization of one of the choosen amphiphiles (i.e., pH 5.6 for SAm and pH 7.35 for SA) and, thus, guarantee the two-dimensional compatibility, even between components with hydrophobic chains not oriented in a mutually favorable way [4]. The ionic strength of the subphase solutions was in the same order of magnitude for all the subphases.

From Figs. 1 and 2 the following observations are possible.

For the SA-VAL mixture the trend of the areas as a function of the molar ratios is more or less additive for the support at pH = 7.4, while there are positive deviations from additivity at pH = 5.6, above all for the mixtures richest in SA.

Neither of the trends can be retained as being finally indicative of miscibility or immiscibility since, in the case of addivity, this might indicate immiscibility or ideal miscicibility [1, 10] and in the case of deviation, even non-additive trends may indicate an untrue miscibility [10—11]. The same trends in the areas are seen in the case of the SAm-VAL mixture, for which there are positive deviations from additivity, especially for the mixture on subphase at pH = 7.4 which is richer in the amine.

The above considerations, are, therefore, necessary criteria for deciding on the miscibility. This is undoubtedly provided by the trends of the collapse pressure (π_c) as a function of molar ratios shown in the same figures. In fact, the invariance of π_c for both the systems, by simple application of the phase rule for surfaces [12], shows the presence at the surface of two separate phases, i.e., immiscibility between the components.

The indication provided by the collapse pressure is particularly reliable, both because the π_c found for all the mixture is close to that of the component which collapses at the lowest pressure, i.e., valinomycin, but also because the collapse pressures were measured for all mixtures under identical compression conditions and, moreover,

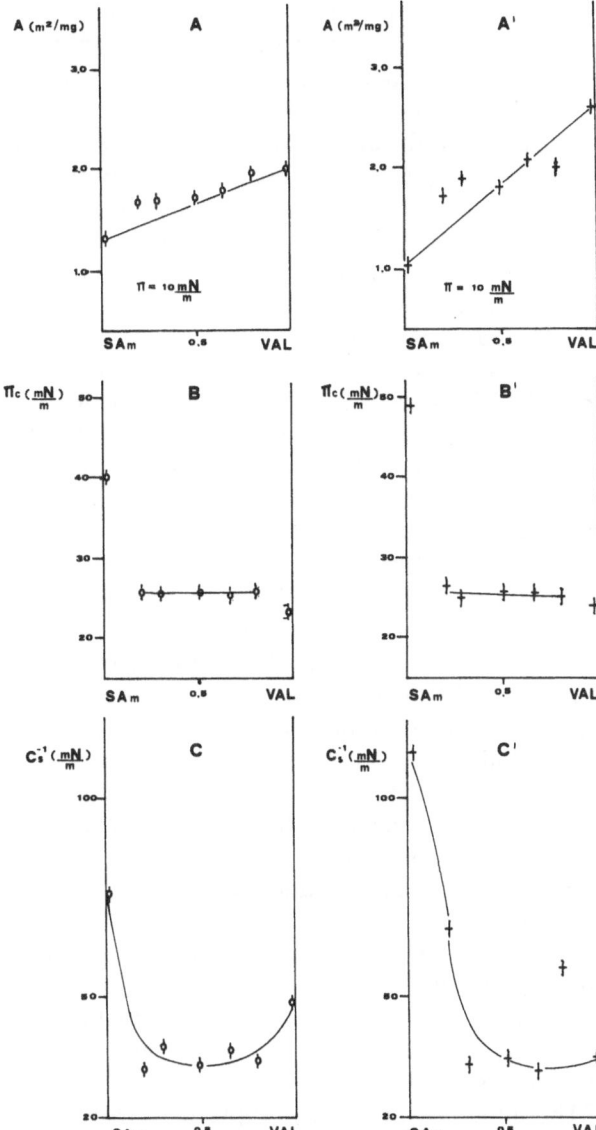

Fig. 2. Plot of the molar ratios for SAm-VAL system as a function of: A) area (A) on a subphase at pH 5.6; A') area (A) on subphase at pH 7.4; B) collapse pressure (π_c) on subphase at pH 5.6; B') collapse pressure (π_c) on subphase at pH 7.4; C) surface compressional modulus (Cs^{-1}) on subphase at pH 5.6; C') surface compressional modulus (Cs^{-1}) on subphase at pH 7.4

bearing in mind that in the π-A isotherm there is generally a further rise in the π after the collapse one which is ascribable to the uncollapsed component.

Having thus shown the immiscibility between the components, it seems probable that the positive deviations in the area additivity which are found may be ascribed, not to a real repulsive interactions between the components, but rather to an expansion of the surface phases due to the introduction of a component such as VAL, which has a more expanded phase and very different area as compared to the acid and the amine. This hypothesis seems to be supported by the Cs^{-1} behavior; this decreases as compared to the pure acid and amine, even for small amounts of VAL. Another validation is that the deviations occur in the subphase pH conditions for which the acid and the amine would give more condensed monolayers.

A different trend had previously been observed for SA-PMMA and SAm-PMMA mixture for the same pH values of the subphase (4,6), and the miscibility found in these cases had been mainly attributed to the interaction between polar groups of the two components.

This discrepancy may be attributed to the disadvantageous distribution of the ionized polar groups of the acid or of the base as compared to the VAL polar groups. This, in fact, fits into the hydrophobic part of the double lipid layer of the membrane and functions as a carrier, being made up of a hydrophilic cavity surrounded by a hydrophobic shield (10). This position of the VAL polar groups in membranes is maintained in VAL monolayers and is taken on by other antibioctics which are able to function as carriers, as demonstrated in a recent study [13].

A further confirmation of the above-mentioned distribution of VAL polar groups is given by the transfer of the K^+ ion from the subphase to monolayers of the VAL, as previously demonstrated by ESCA spectra [7c].

The ESCA results were confirmed with ATR-FT-IR spectra of multilayers of equimolar mixture of SA and VAL on a subphase of 0.1 M KCl transferred as L.B. films.

Figure 3 shows the ATR-FTIR spectra obtained after transfer of three monolayers of an equimolar SA/VAL mixture.

A shift in the C=O amidic and esteric bands may be seen in the passage from multilayers transferred from aqueous subphase to those on KCl; this shift may be attributed to the complexation of K^+ in the VAL cavity, as already described in the literature [7, 14].

The distribution of the polar groups which are able to combine with K^+ ions would, thus, seem to be responsible for the impossibility of setting up polar-type interactions with SA and SAm and, hence, for the incompatibility which was noted,

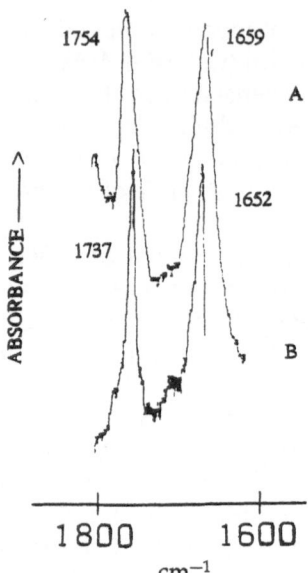

Fig. 3. ATR-FTIR spectra of three transferred monolyers of mixture SA-VAL: A) from aqueous subphase; B) from 0.1 M KCl subphase. For the sake of clarity this spectrum ends at 1665 cm^{-1}

also for different dissociations of the acid and amine, since this incompatibility was found for both pH subphase values tested.

MS/VAL system

Figure 4 shows the trends of the areas of collapse pressures and of compressibility moduli in relation to molar ratios at 25°C on a subphase of 0.01 M NaCl.

As it appears from Fig. 4, the behavior of the areas as a function of molar ratios shows positive deviations from additivity in the whole range of examined molar ratios.

Also, in this case this is not sufficient to choose between miscibility or immiscibility, but the sharp variation of π_c as a function of molar ratios indicates the miscibility between the two components.

The variation of π_c with the molar ratios is particularly evident if we consider the nonrelevant difference between the collapse pressure of the two components and the rigorously constant compression conditions for all the mixtures.

The area at 10 mN/m and the π_c of VAL are slightly lower than those reported for other systems: this can attributed to the different ionic strengths of the subphase.

The miscibility is prevalently characterized by repulsive-type forces as found in the case of other binary systems in which the components show different bidimensional phases [15], and as appears from the increase in the area occupied at constant pressure as compared to the ideal mixture, and by a decrease in the compressional modulus.

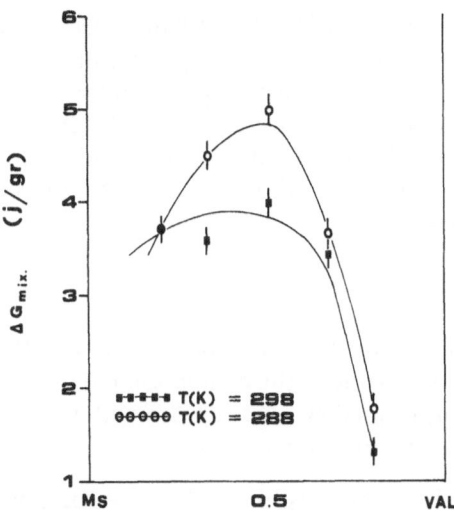

Fig. 5. Plot of ΔG mix as a function of molar ratios for the system MS-VAL

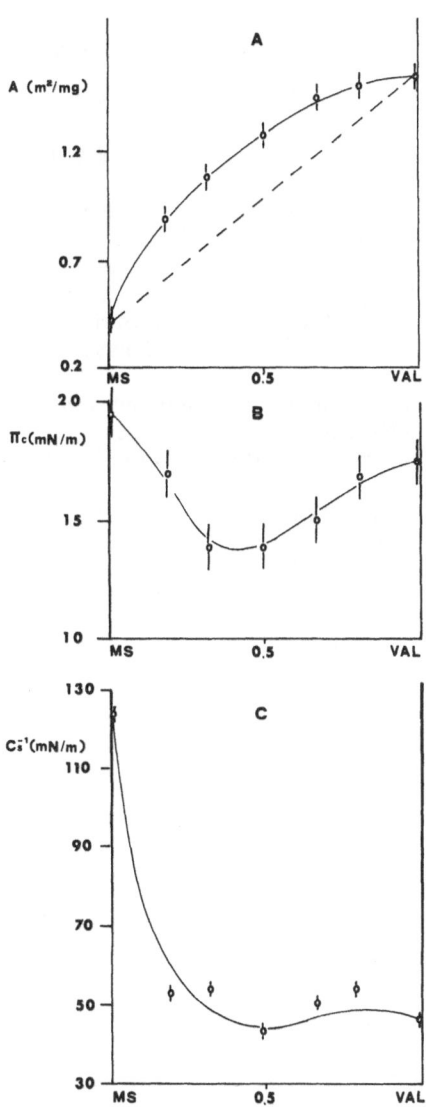

Fig. 4. Plot of the molar ratios for MS-VAL system at 25 °C on 0.01 NaCl as a functionof: A) surface area (A); B) collapse pressure π_c; C) surface compressional modulus Cs^{-1}

This compatibility between the VAL and MS, in contrast to SA and SAm, could be attributed to interactions, even of a repulsive type, between the methyl groups of MS and the hydrophobic part of VAL, which in a two-dimensional phase are positioned towards the periphery of the molecule, i.e., towards the other component of the binary system.

In view of the compatibility between MS and VAL, the ΔG_{mix} value calculated at two temperatures (288 K and 298 K) at different molar ratios was also determined as in previous studies [3] (Fig. 5).

As expected, the ΔG_{mix} value was positive, confirming interactions between components, but also confirming the lower stability of the mixture than for the pure components. This pattern was maintained at all the pressures examined. Furthermore, it was noted that the ΔG_{mix} has a pattern characterized by a maximum around the VAL-MS 1-1 molar ratio, in agreement with the minimum value of the collapse pressure at the same molar ratio, indicating a lower stability of this mixture at the surface.

The contribution of enthalpic (ΔH_{mix}) and entropic mixing ($T\Delta S_{mix}$) to the mixing ΔG at the two temperatures were also calculated. They were all found to be positive and, thus, the miscibility would seem to be favored by the entropic factor, as confirmed by the trends of the areas of the mixture and by the phase, which is more expanded as compared to one of the components, which also gives greater mobility to the components in the mixture.

For this system, too, the ATR-FT-IR spectrum of LB multilayers transferred onto a germanium support from a subphase containing K^+ and from an aqueous subphase is exactly equivalent to that obtained for the SA/VAL system, demonstrating that even if MS is present, VAL binds to the K^+ placed in the subphase.

Conclusions

1) Comparison of the trends of the areas of the collapse pressures and of the compressional moduli of the systems formed by SA and SAm with PMMA and VAL has shown that:

a) the two ionizable amphiphils become compatible with PMMA when the pH of the support allows sufficient ionization;

b) the two amphiphiles are, instead, incompatibile with VAL. This incompatibility may be attributed to the impossibility of interaction between the polar groups of the two components.

2) The examination of the ATR-FT-IR spectra on transferred multilayers of VAL confirms the transfer of K^+ ions from the subphase to the VAL molecular cavity in the monolayer and, hence, confirms the distribution of the antibiotic polar groups as being towards the inside of the cavity.

3) When the comparison is extended to the MS/VAL system, it is seen that there is the possibility of interactions (even though these are of a repulsive type) between the methyl groups of MS and the VAL hydrophobic periphery.

4) Studies of VAL binary system with amphiphiles having different hydrophobic groups are currently in progress to confirm point 3).

Acknowledgements

This work supported by CNR and MURST.

References

1. Gaines GL Jr (1966) "Insoluble Monolayers at liquid-gas Interfaces" Intersc. Publ. New York. — Adamson AW (1982) "Physical Chemistry of Surfaces" John Wiley p 139. — Birdi KS (1989) "Lipid and Biopolymer Monolayers at liquid Interfaces" Plenum Press, London, p 107
2. Roberts G (1990) "Langmuir-Boldgett Films" Plenum Press, New York. — Biesmans G, Van der Auweraer M, Schryver FC (1990) Langmuir 6:277. — Fukuda K, Sugi M (1989) Proc IV Conference on Langmuir Blodgett Films, Tsukuba Elsevier
3. Gabrielli G, Puggelli M, Baglioni P (1982) J Coll Interface Sci 86:485
4. Puggelli M, Gabrielli G (1985) Coll Polymer Sci 263:879
5. Harkins WD (1952) "Physical Chemistry of Surface Films" Reinhold, New York
6. Puggelli M, Gabrielli G (1987) Colloid Polym Sci 265:432
7a. Howar VA, Petty MC (1988) Thin Solid Films 160:483
7b. Howart VA, Petty MC, Davies GH, Yarwood J (1989) Langmuir 5:330
7c. Gabrielli G, Puggelli M, Prelazzi S (1991) Progr Coll Polymer Sci 84:232
8. Gabrielli G, Niccolai A, Dei L (1986) Colloid Polymer Sci 264:972
9. Bettarini S, Bonosi F, Gabrielli G, Martini G (1991) Langmuir 7:1082
10. Shah DO, Capps RW (1968) J Coll Interface Sci 27:320
11. Malcom BR (1973) in Progress in Surface and Membrane Science 7:223; Danielli JF, Rosenberg MD, Cadenhead DA (ed) Academic Press; Gabrielli G, Baglioni P, Fabbrini A (1981) Coll and Surface 3:147
12. Crisp DJ (1949) in "Surfaces Chemistry Suppl Research" London
13. Asmonay H, Hochapfel A, Hadj Sahraoni A, Jafrrain M, Peretti P (1991) Paper presented at V International Conference on LB films — Paris — Thin Solid Films (in press)
14. Gliozzi A, Rolandi R (1984) in "Membranes and Sensory Trasduction" Colombetti G, Leuci F (eds) Plenum Press, New York
15. Margheri E, Niccolai A, Gabrielli G, Ferroni E (1991) Coll and Surface 53:135

Authors' address:

G. Gabrielli
Department of Chemistry
University of Florence
v. Gino Capponi 9
50121 Florence, Italy

Progress in Colloid & Polymer Science   Progr Colloid Polym Sci 89:233—234 (1992)

Aggregation of cyanine dyes in Langmuir-Blodgett films

M. I. Sluch and A. G. Vitukhnovsky

Lebedev Physical Institute, Russian Academy of Science, Moscow, Russia

Abstract: We report fluorescence studies of aggregates in Langmuir-Blodgett (LB) films using the time-correlated single-photon counting technique and fluorescence spectroscopy. The kinetics and spectra of aggregates are presented. The results demonstrate that the aggregates in LB-films belong to J-type aggregates.

Key words: Aggregation; single-photon counting; fluorescence; Langmuir-Blodgett

Experimental evidence for the association of cyanine dyes molecules in Langmuir-Blodgett (LB) films has existed for many years [1], but, recently, dynamical properties of aggregates have been observed [2]. J aggregates are characterized by an intense narrow absorption and luminescence bands which are shifted to longer wavelength relatively to the monomer absorption. The J-aggregate can be described as a sum of N dipoles of dipole moment oscillating in phase where N is the number of dye molecules in the aggregate. The resulting dipole is more than an individual one, therefore, the fluorescence lifetime of a J aggregate is shorter than that of the monomer. We have deposited LB films on different substrates (GaAs, Si, quartz, Au). Measurements of fluorescence and absorption spectra and kinetics of fluorescence decay were made to determine the types of aggregates and the number of aggregate molecules.

Spectra fluorescence and absorption were measured with an optical multichannel analyzer OMA-2. Fluorescence decay curves were measured with an Edinburgh Instruments spectrofluorimeter model 199. The repetition rate of the flash-lamp is 20 kHz and the pulse duration is 1.2 ns (fwhm). Excitation wavelength was 405 nm. Fluorescence decay curves were analyzed by using a nonlinear, least-squares iterative convolution method based on the Marquardt algorithm. All experiments were made at room temperature. Curve fitting and control experiments were carried out with an LSI-11 computer.

Cyanine dye was obtained from the Institute of Organic Chemistry, Ukrainian Academy of Sciences (Fig. 1). The dye was spread from benzene solution (3.5×10^{-4} M) onto the water surface containing 2.5×10^{-4} M $KClO_4$, in the Langmuir through (Joyce-Loeble). Monolayers of the dye were transferred onto quartz, GaAs, Au, and Si substrates at 25 mNm^{-1}.

Figure 1 shows the absorption and fluorescence spectra of the dye in LB films and benzene solution. Absorption spectrum of LB films are characterized by a narrow absorption band (maximum 497 nm, fwhm $\approx$ 1251 cm^{-1}) which is shifted to longer wavelengths relative to the monomer absorption in benzene solution (maximum 463 nm, fwhm $\approx$ 2898 cm^{-1}). In addition, the fluorescence spectrum of LB films (maximum 520 nm, fwhm $\approx$ 1351 cm^{-1}) is narrower than that of the monomer of the dye in solution (maximum 497 nm, fwhm $\approx$ 1879 cm^{-1}). These features are characteristics of J-type aggregate formation. Knapp has proposed a formula for the narrow absorption band of linear J aggregate [3]:

$$\Delta\omega_J/\Delta\omega_m \sim 1/\sqrt{N} , \qquad (1)$$

where $\Delta\omega_J$ and $\Delta\omega_m$ are fwhm absorption bands of J aggregates and monomers, respectively and N is the number of coupled molecules in the aggregate. The formula allows to determine the size of the J aggregate (the number of coherently coupled molecules), which is roughly four molecules.

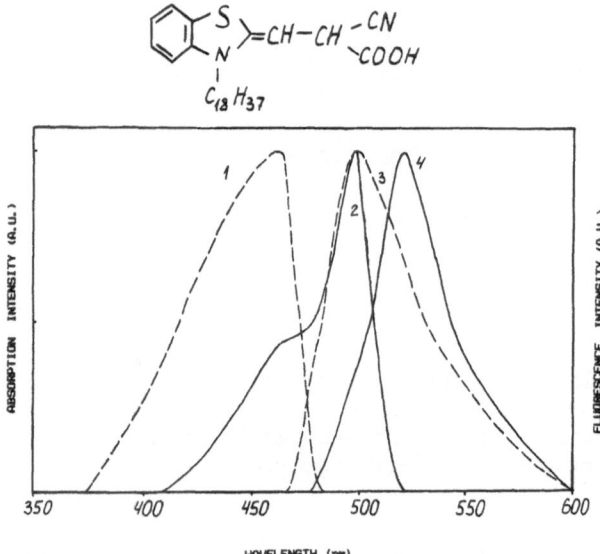

Fig. 1. Absorption spectra of cyanine dye in benzene solution (1), in LB Films (3), fluorescence spectra of cyanine dye in benzene solution (2), and in LB films (4) at room temperature

Table 1. The best-fit parameters for fluorescence decay of cyanine dye LB films

Substrate	τ (ns)	χ^2
GaAs	0.6	1.4
Quartz	0.7	1.2
Au	0.6	1.5
Si	0.6	1.3
Benzene solution 10^{-4} M	2.0	1.3

Acknowledgement

The authors thank Prof. M. Galanin for helpful discussions.

References

1. Kuhn H, Mobius D, Bucher H (1972) In: Techniques of Chemistry, Weissberger A, Rossiter BW (eds): Wiley, New York, vol 1, part 3B, pp 577—702
2. Boer SD, Wiersma DA (1990) Chem Phys Letters 165:45—53
3. Knapp EW (1984) Chem Phys 85:73—82
4. Mukamel S, Spano II (1989) J Chem Phys 91:683—700

Mukamel and Spano proposed a theory [4] for time dependence of fluorescence intensity of J aggregate:

$$\tau_J / \tau_m \sim 1/N \, , \qquad (2)$$

where τ_J and τ_m are the lifetimes of the J aggregate and monomer, respectively.

Table 1 shows the best-fit fluorescence decay parameters for LB films (em 520 nm) and the dye in benzene solution (em 497 nm). The lifetime of the cyanine dye in LB films is shorter than in benzene solution. These features are also characteristics of J-type aggregate formation. Unfortunately, the margin of error in determiing the lifetime is 50% (± 0.3 ns), because the lifetime of the J aggregate is shorter than that of the excited light pulse. The error in determining the fwhm absorption band is ± 1 cm^{-1}. Therefore, more precise data were obtained from absorption measurements.

Authors' address:

M. I. Sluch
Lebedev Physical Institute
Russian Academy of Science, Moscow 117924
Leninsky 53, Russia

Progress in Colloid & Polymer Science

Progr Colloid Polym Sci 89:235—238 (1992)

Spreading isotherms and electron spin resonance of nitronylnitroxide mono- and multilayers

F. Bonosi, A. Caneschi, and G. Martini

Department of Chemistry, University of Florence, Italy

Abstract: Nitronylnitroxides are promising materials for the preparation of unusual magnetic systems, particularly when coordinated with transition metal ions. In this work mono- and multilayers of the amphiphilic radical 2-tridecyl-4,4,5,5-tetramethyl-imidazolidine-1-oxyl-3-oxide radical were characterized from its spreading isotherms and by optical and ESR spectroscopy. The time stability of the films was also studied.

Key words: Monolayer; LB films; ESR; spreading isotherms

Introduction

Nitronylnitroxides, with general formula:

(2-R-4,4,5,5-tetramethyl-imidazolidine-1-oxyl-3-oxide, NIT-R) with two equivalent coordination centers have a wide versatility in coordinating metal ions [1], with the formation of chains characterized by low-dimensional magnetic properties [2]. In addition, when R is a sufficiently long hydrocarbon chain the nitronylnitroxide becomes an amphiphilic molecule for which surfactant properties are expected.

In this paper, we report the results obtained in a study on the tridecyl derivative ($R = C_{13}H_{27}$) of NIT-R, NIT-C_{13}, carried out by spreading isotherm determination, electron spin resonance (ESR) and electronic (UV-VIS) spectroscopy. Both mono- and Langmuir-Blodgett multilayers were investigated.

Experimental section

NIT-C_{13} was prepared according to the methods given in [3, 4] for NIT-R compounds, through the condensation of 2,3-dimethyl-2,3-bis(hydroxylamine)-butane with tetradecylaldehyde. The crude radical obtained after oxidation with NaIO$_4$ was purified twice by column chromatography on Act. III alumina with CH$_2$Cl$_2$ as the eluent. Analysis: calculated for $C_{20}H_{39}N_2O_2$: C, 70.79; H, 11.50; N, 8.25; found: C, 70.64; H, 11.51; N, 8.21.

In order to prevent NIT-C_{13} decomposition, the solid was stored at 4°C in the dark.

A chloroform (Merck, purity >99%) solution of NIT-C_{13} (1.57 mmol/l) was used for both isotherm determination and LB film deposition. Twice-distilled water, purified with the Millipore-Q Water System (Millipore) with a resistivity >18 MΩ · cm was used as the subphase. The spreading isotherms were obtained with the Lauda FW2 film balance at a continuous compression rate of 5 mm/min. LB films were prepared at 288 K with the KSV 5000 LB apparatus.

ESR spectra were recorded with the aid of the Bruker model 200D ESR spectrometer on quartz slides (40 × 8 × 0.5 mm) on which 37 layers of NIT-C_{13} were deposited with dipping speeds of 5 mm/min for the first seven layers and 10 mm/min for the subsequent ones.

UV-VIS spectra were recorded by using the Perkin-Elmer Lambda 5 spectrometer on quartz, slides (50 × 12.5 × 1 mm) on which a NIT-C_{13} monolayer was deposited at a dipping speed of 1 mm/min.

Other experimental conditions and instrumental settings were as reported in [5, 6].

Results and discussion

Figure 1 shows the room-temperature ESR spectrum of a 1.57 × 10^{-3} mol/l NIT-C_{13} solution in $CHCl_3$. There was the typical five-line absorption due to one unpaired electron interacting with two equivalent ^{14}N nuclei (A_N = 0.774 mT), both of them belonging to the imidazolidine ring. Each ^{14}N manifold was further split into a triplet due to the interaction with the two equivalent ^1H nuclei of the α-CH_2 in the hydrocarbon chain.

0.8 mT

Fig. 1. ESR spectrum of a 1.57 × 10^{-3} mol/l NIT-C_{13} solution in $CHCl_3$ at room temperature

Figure 2 shows the 293 K spreading isotherms of NIT-C_{13} monolayer on water. Only a small temperature dependence was observed in the range 288—313 K. NIT-C_{13} gave stable monolayers up to a surface pressure of 30 mN/m, where the monolayer collapse took place. The monolayer occurred as a liquid expanded phase at the water surface, with a maximum value of the compressibility modulus, C_s^{-1}, of 52 mN/m. The limiting area A_0 was 62 Å²/molecule. This value was greater than the value calculated for the polar head of NIT-C_{13}, which is about 40 Å²/molecule. The C_{13} hydrocarbon chain took on a randomly tilted configuration at the surface. The configuration was intermediate between those of the closed packed solid phase and of the gaseous phase whose molecules were nearly flat at the interface. The chain tilting thus caused an increase in the mean molecular area of NIT-C_{13}.

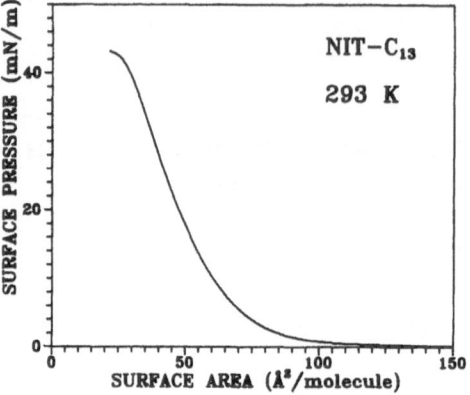

Fig. 2. Surface pressure vs surface area of NIT-C_{13} on Milli-Q water at 293 K

Information on the magnetic interactions in the multilayered systems was obtained from the ESR spectra. All of the investigated samples gave single-line absorptions. The resolution of the hyperfine splitting was never observed in pure NIT-C_{13} films. This was not unexpected because the value of A_N in NIT-C_{13} was about one-half of the A_N value in usual nitroxides, such as doxyl stearic acids whose LB films have been studied either as pure compounds or as mixtures with stearic acid [6]. For the sake of comparison, 5-doxyl stearic acid [2-(3-carboxypropyl)-4,4-dimethyl-2-tridecyl-3-oxazolidine-1-oxyl] only shows a clear resolution of the hyperfine structure when mixed with stearic acid in a 1:10 molar ratio [6]. Casting film, collapsed material, and quartz-deposited multilayer (37 layers) of pure NIT-C_{13} gave different linewidths (ΔB = 0.68, 1.36, and 3.1 mT, respectively) indicating different molecular packing in the three cases. All spectra were dominated by spin-spin effects, namely, the Heisenberg spin exchange, described by the Hamiltonian $\hat{H} = -2 \sum J S_i \cdot S_j$. This led to a homogeneous line broadening which was dependent upon the spin-exchange frequency w_{exch}. The width of the observed exchange-narrowed spectra was related to w_{exch} by: $\Delta B_{exch} = A_N^2 / w_{exch}$. Thus, the increased ΔB from casting to LB films indicated a weakened spin-exchange interaction due to a higher mean distance between unpaired electrons, i.e., a more open structure in the LB film, as expected from the more flexible structure of the multilayer film.

The film formation seemed to strongly influence the time stability of the radical. Nitronylnitroxides are indeed known to undergo decomposition [7]. A partial radical decomposition was, in fact, observed in the casting film, as was evident from ΔB which increased from 0.68 mT in the freshly prepared sample to 0.85 mT in the same sample 2 days after preparation. The decomposition rate was markedly increased in collapsed and LB films. For instance, the 2-day old collapsed film did not give any further ESR signal.

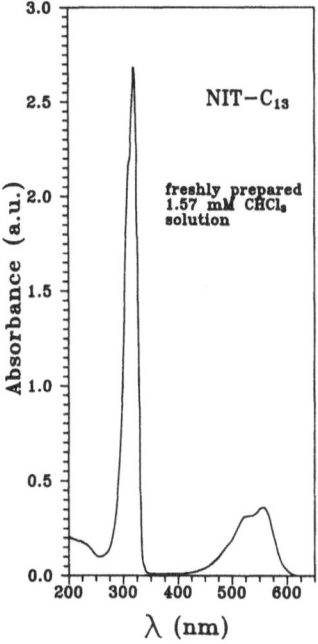

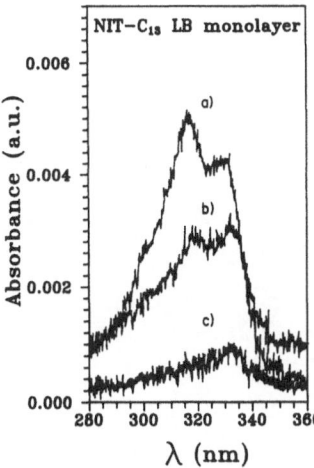

Fig. 3. Optical spectra at room temperature of NIT-C$_{13}$ in CHCl$_3$ solution (top) and as monolayer on a quartz plate (bottom) immediately after the preparation (a), after 2 days (b), and after 6 days (c)

The same effect was also observed with electronic spectroscopy, which gave additional interesting features. Figure 3 (top) shows the UV-VIS spectrum of NIT-C$_{13}$ in CHCl$_3$. This spectrum was stable for several hours. The two main absorptions at 557.2 nm (17,950 cm^{-1}) and at 318.6 nm (31,390 cm^{-1}) are typical of a nitronylnitroxide [4, 7] and are attributed to $n \rightarrow \pi^*$ and $\pi \rightarrow \pi^*$ transitions, respectively. Both transitions were resolved into doublets. The two peaks at 319.2 nm (31,330 cm^{-1}) and at 309 nm (32,360 cm^{-1}) are attributed to the $\pi \rightarrow \pi^*$ transition and the two peaks at 525 nm (19,050 cm^{-1}) and at 557 nm (17,450 cm^{-1}) are attributed to the $n \rightarrow \pi^*$ transition. Figure 3 (bottom) shows the UV-VIS spectra in the region 280—360 nm of a single layer of NIT-C$_{13}$, immediately after preparation and after 2 and 6 days of aging on a quartz plate oriented perpendicularly to the direction of the light. The absorption disappeared 1 week after the preparation. Practically the same spectrum was observed in the casting film (peaks at 306, 318.5, and 334 nm). The very low absorbances of these samples prevented an accurate analysis. However, a significant blue-shift of the $\pi \rightarrow \pi^*$ peaks (to 304 and 315 nm, respectively) was clearly seen, together with the occurrence of a new peak at 330 nm (30,300 cm^{-1}), whose origin is, at present, unknown. With time, the intensity of the 330 nm peak in the monolayer absorption increased with respect to the $\pi \rightarrow \pi^*$ doublet, probably because of a variation in the structural arrangement of the film and of partial decomposition of the radical.

More details on the orientation dependence of mono- and multilayers of NIT-C$_{13}$ and its mixture with stearic acid will be given in a forthcoming paper.

Conclusion

NIT-C$_{13}$ nitronylnitroxide has typical properties of an amphiphilic molecule and gives spreading monolayers which are easily transferred onto a solid support. Although both ESR and optical spectroscopies showed a relative stability of the film with time, this class of materials seems to be promising for thin films with peculiar magnetic properties, particularly when their metal coordination capacity is taken into account.

Acknowledgements

Thanks are due to the National Council of Research (CNR) and to Ministero della Universita' (MURST) for financial support. The authors are indebted to Prof. G. Gabrielli and D. Gatteschi for useful discussions.

References

1. Caneschi A, Gatteschi D, Rey P (1991) Progr Inorg Chem 39:331—929
2. Caneschi A, Gatteschi D, Sessoli R (1989) Acc Chem Res 22:392—398
3. Lamchen M, Wittag TW (1966) J Chem Soc 2300—2303
4. Ullman EF, Call L, Osiecky JH (1970) J Org Chem 35:3623—3631
5. Bonosi F, Gabrielli G, Martini G, Ottaviani MF (1989) Langmuir 5:1037—1043
6. Martini G, Bonosi F, Ottaviani MF, Gabrielli G (1989) Thin Solid Films 178:271—279
7. Ullman EF, Osiecky JH, Boocock DGB, Darcy R (1972) J Am Chem Soc 94:7049—7059

Authors' address:

Prof. Giacomo Martini
Department of Chemistry
University of Florence
Via G. Capponi 9
50121 Firenze, Italy

Progress in Colloid & Polymer Science   Progr Colloid Polym Sci 89:239—242 (1992)

Adsorption properties of soluble surface active stilbazium dyes at the air-water interface

K. Lunkenheimer, A. Laschewsky[1])

Max-Planck-Institut für Kolloid- und Grenzflächenforschung, Berlin Adlershof
[1]) Institut für Organische Chemie, Universität Mainz

Abstract: The adsorption behaviour of a homologous series of amphiphilic hemicyanines is studied at the air-water interface. The dyes exhibit an unusual, marked odd-even effect with respect to the standard free energy of adsorption and the minimum surface area demand. The reasons for such strong, hitherto unknown odd-even effect is not yet clear.

Key words: Surface-active dyes; hemicyanine; adsorption isotherm; surface tension; odd-even effect

Introduction

The application of dyes in dying processes, photography, etc., requires surfactants or other organized assemblies such as micelles, vesicles or layers. Amphiphilic dyes are of particular interest as the concepts of color and surface-active properties coexist within the same molecular framework [1].

In elucidating the interactions in the dye-surfactant system it is first necessary to understand the separate behavior of the two single components. Here, we report our investigations of the adsorption properties of some model amphiphilic hemicyamine dyes at the air/water interface.

Experimental

Materials

The hemicyamine dyes N-n-alkyl-4-(dimethylamino)-stilbazium bromides ($3 \leqslant n_c \leqslant 8$) were prepared by condensation of the corresponding N-alkylated 4-methylpyridines with 4-dimethylamino benzaldehyde [2, 3]. The products were purified by repeated recrystallization from ethanol.

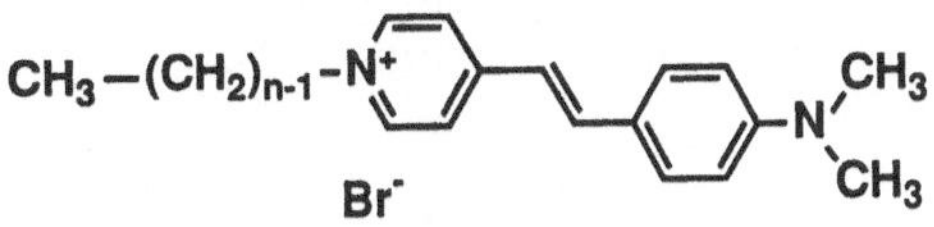

$$CH_3-(CH_2)_{n-1}-N^+ \cdots =N \begin{smallmatrix} CH_3 \\ CH_3 \end{smallmatrix}$$

Br⁻

n = 3, 4, 5, 6, 7, 8

Scheme 1. N-n-alkyl-4'-dimethylamino stilbazium bromides synthesized

Methods

Surface-tension measurements of aqueous solutions were performed by a modified [4, 5] ring tensiometer method at 295 K. To obtain reliable results on adsorption properties of surface-active dyes, stock solutions were repurified by an elaborate procedure [6] until adulterating surface-active trace contaminants had been completely removed. The equilibrium surface tension vs. concentration isotherms of the dyes were evaluated by applying Szyszkowski's (Eq. (1)) or Frumkin's (Eq. (2)) surface equations of state, which stand for ideal and regular surface tension behavior, respectively [7].

$$\sigma_0 - \sigma_e = RT\Gamma_\infty \ln(1 + c/a_L) \qquad (1)$$

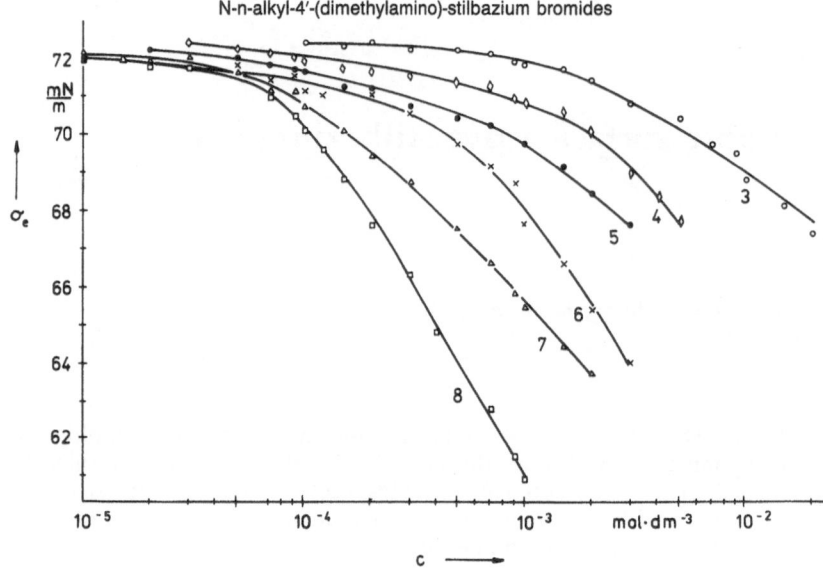

Fig. 1. Equilibrium surface tension (σ_e) vs. concentration (c) isotherms of aqueous solutions of hemicyanines at 295 K. The numbers denote the alkyl chain length n_c

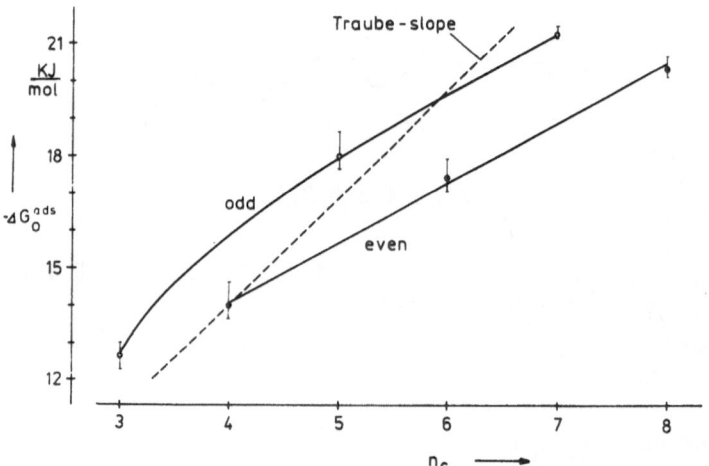

Fig. 2. Standard free energy of adsorption ΔG^{ads} of hemicyanines in dependence on number of carbon atoms n_c of n-alkyl chain

$$\sigma_0 - \sigma_e = RT\Gamma_\infty \cdot \ln(1 - \Gamma/\Gamma_{00})$$
$$+ H_s/RT \cdot (\Gamma/\Gamma_{00})^2 \qquad (2a)$$

$$c = a_L \cdot \Gamma/\Gamma_\infty (1 - \Gamma/\Gamma_{00})^{-1}$$
$$\times \exp(-2H_s/RT \cdot \Gamma/\Gamma_\infty) ; \qquad (2b)$$

σ_0 denotes the surface tension of pure water, σ_e is the equilibrium surface tension of the dye solution. Γ and Γ_{00} refer to the excess surface concentration at bulk concentration c and at surface saturation. H_s denotes the partial molar free energy of surface mixing of dye and solvent; a_L stands for the surface activity parameter of the dye which is related to its free energy of adsorption ΔG^{ads} by

$$a_L = \exp(\Delta G^{ads}/RT) . \qquad (3)$$

The standard deviations s of the best fit procedures were $0.06 < s < 0.21$ mN/m. The mean errors of the equation of state parameters are indicated by bars in the figures.

Results

The hemicyanines studied represent surface-active substances which are scarcely soluble in water. They do not form micelles, but they are surface active. The surface tension of pure water is decreased by 12 mN/m at maximum (Fig. 1). The standard free energy of adsorption ΔG^{ads} exhibits a marked odd-even effect (Fig. 2). For $n_c \geqslant 4$, ΔG^{ads} shows a linear dependence on the carbon number n_c of the n-alkyl chain for both for the even and the odd-numbered members each:

$$(\Delta G^{ads})_{even} = -1.58\, n_c - 7.79 \tag{4}$$

$$(\Delta G^{ads})_{odd} = -1.65\, n_c - 9.75 . \tag{5}$$

The two straight lines have practically an identical slope. However, it should be noted that these slopes are less teep than the slope predicted by Traube's rule. The difference:

$$\Delta\Delta G = (\Delta G_0)_{n+1} - (\Delta G_0)_n$$

is about -2.9 kJ/mol for typical surfactants according to Traube. But, $\Delta\Delta G \approx -1.6$ kJ/mol for the hemicyanines investigated ($4 \leqslant n_c \leqslant 8$). Most strikingly for these dyes, the surface activity of the following even-numbered chain homolog is still lower than that of the preceding odd-numbered homolog because of the enormous differences between the surface activity of the odd-numbered and even-numbered members.

$$|(\Delta G_0)_{2n+1}| > |(\Delta G_0)_{2n+2}| \quad \text{for} \quad n_c \geqslant 4 . \tag{6}$$

The minimum surface area demand per molecule adsorbed A_{min} is calculated from the saturation adsorption value Γ_{00} of (eq. 1 and 2) (Fig. 3), according to

$$A_{min} = (\Gamma_\infty \cdot N_L)^{-1} . \tag{7}$$

Apart from the shortest chain homolog with $n_c = 3$, for the even and the odd-numbered members each, A_{min} is roughly constant within the homologous series. In fact, the A_{min} values of the odd-numbered members are almost twice as large as those of the even-numbered members.

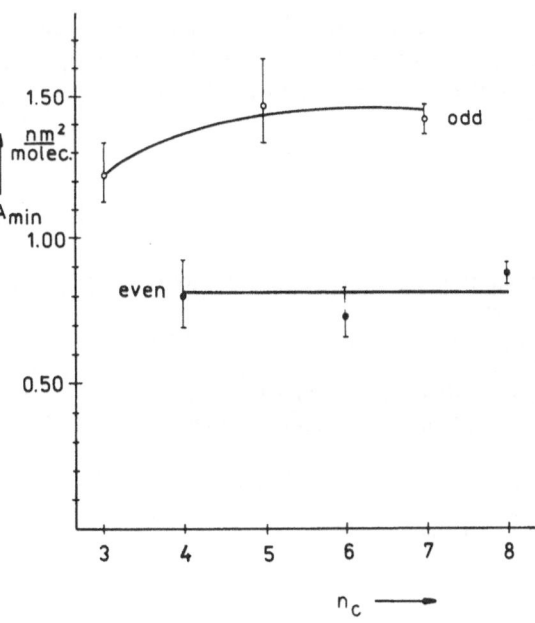

Fig. 3. Minimum surface area demand per hemicyanine molecule adsorbed in dependence on number of carbon atoms n_c of n-alkyl chain

Discussion

The hemicyanines studied reveal striking peculiarities in their adsorption behavior in comparison with ordinary surfactant systems: Distinct, unusual even-odd effects of the n-alkyl chain have been found concerning i) the standard free energy of adsorption ΔG^{ads}, and ii) the minimum surface area demand per molecule adsorbed A_{min}.

Even-odd effects of soluble surfactants are exceptional, and have been observed only recently, for cis 2-n-alkyl-5-hydroxy-1,3-dioxanes [8]. They have been ascribed to donor-acceptor interactions between the surfactants and water. Of particular interest, Shimomura et al. reported on a peculiar even-odd effect of redox potentials and absorption spectra of viologen polymers [9, 10] when acting as polymeric counterions of anionic bilayer membranes. They assume that the peculiar even-odd effect is due to the restricted conformation of the alkyl spacer between the viologen groups fixed on the regularly oriented charged surface of the bilayer assembly.

In the case of the hemicyanines studied, the cause of the distinct odd-even effects is not yet clear. Donor-acceptor interactions, as discussed in [8], may be considered, although the effects seem to strong to be accounted for. Alternatively, orientational effects or aggregation of the chromophores may be discussed. Further investigations are in progress.

References

1. Barni E, Savarino P, Viscardi G (1991) Acc Chem Res 24:98
2. Phillips AP (1949) J Org Chem 14:302
3. Lupo D, Prass W, Scheunemann U, Laschewsky A, Ringsdorf H (1988) J Opt Soc Am B 35:300
4. Lunkenheimer K, Wantke KD (1981) Colloid Polym Sci 259:354
5. Lunkenheimer K (1989) J Colloid Interface Sci 131:580
6. Lunkenheimer K, Pergande HJ, Krüger H (1987) Rev Sci Instrum 58:2313
7. Lucassen-Reynders EH (1976) Progr Surf Membr Sci 10:253
8. Lunkenheimer K, Burczyk B, Piasecki K, Hirte R (1991) Langmuir 7:1765
9. Shimomura M, Utsugi K, Okuyama K (1986) J Chem Soc Chem Commun 1805
10. Shimomura M, Utsugi K, Horikoshi J, Okuyama K, Hatozaki O, Oyama N (1991) Langmuir 7:760

Authors' address:

K. Lunkenheimer
Max-Planck-Instit. Kolloid- u. Grenfl. Chemie
Rudower Chaussee 5
D-O-1199 Berlin Adlershof, FRG

Progress in Colloid & Polymer Science   Progr Colloid Polym Sci 89:243—248 (1992)

Incorporation of membrane proteins into lipid surface monolayers: Characterization by fluorescence and electron microscopies

M. Schönhoff*), M. Lösche*), M. Meyer #), and C. Wilhelm #)

*) Institut für Physikalische Chemie
#) Institut für Allgemeine Botanik, Johannes-Gutenberg-Universität Mainz, FRG

Abstract: The preparation of oriented protein samples is an attractive goal, e.g., to gain more detailed information from spectroscopic experiments. Our approach towards this aim was to prepare monolayers of phospholipids at the air-water interface and to incorporate the proteins into these ordered structures. Subsequently, we used the Langmuir-Boldgett (LB) transfer technique to obtain samples of oriented proteins on solid supports. — Incorporation was achieved by spreading the proteins from a detergent solution onto a prespread lipid monolayer on the water surface. We characterized successful incorporation by in situ fluorescence microscopy and by electron microscopy, and investigated the topology of transferred monolayers. — Our results show that the choice of the lipid matrix is pivotal. We present an optimized lipid matrix system that facilitates protein incorporation, while enabling LB transfer at the same time. — Fluorescence microscopic investigations revealed that the lateral phase structure of the lipid films remains qualitatively unaffected by the reconstitution, with the proteins incorporated into the fluid phases. Quantitative evaluations of the protein content of the thin film samples indicated that ~15% of the proteins spread were incorporated, whereas evaluation of the additional area of incorporation (of protein and detergent) showed that a substantial amount of detergent was incorporated as well.

Key words: Membrane protein; incorporation; fluorescence microscopy; electron microscopy; lipid monolayers

Introduction

Spectroscopy on chromoproteins is a wide field for investigation of the features of the photosynthetic process. Much work has been done on reaction centers of photosynthetic bacteria [1] by time resolved absorption spectroscopy [2] and related techniques [3]. By Stark spectroscopy on an isotropic sample, the magnitude of the dipole moment change on excitation $|\Delta \vec{\mu}|$ was determined [4]. To obtain further information on the direction of $\Delta \vec{\mu}$, samples of oriented proteins may be valuable [5].

Although three-dimensional protein crystals are such ordered structures, they provide no macroscopic preferential orientation of the molecules and their extremely high optical density may complicate

spectroscopic experiments. Earlier attempts to perform spectroscopic experiments on preferentially oriented samples included air drying of detergent solutions on solid supports [6], stretching of protein-containing gelatin films [7], and photo-selection experiments [8].

A different approach has been made with a film of reaction centers on a water surface [9]. In continuation of earlier efforts in our lab [10], our strategy was to use phospholipid layers as a matrix system carrying oriented proteins. Then, by LB transfer of alternating (protein-containing and pure lipid) layers, samples with a defined protein orientation may conceivably be obtained. Here, we characterize the lipid/protein system on the water surface, and after transfer of single films onto solid supports as a first step for further spectroscopic investigations.

Materials and methods

Two different membrane proteins were investigated. One was the reaction center (RC) of the photosynthetic baterium *Rhodospirillum rubrum*. RCs were isolated by sucrose gradient centrifugation and column chromatography of detergent solubilized membranes, and stored in 0.1% LDAO solution.

As a second model protein, we used the light harvesting complex (LHC II) from the green alga *Chlorella fusca*. This protein is very similar to that described in [11]. LHC II was isolated from the thylakoid membrane by non-denaturating SDS-PAGE [12].

The lipids used were behenic acid (purity >99%, Fluka), dipalmitoyl-phosphatidic acid (DPPA, Na salt, >99%, Fluka), dioleoyl-phosphatidic acid (DOPA, Na salt, ~98%, Sigma). For the subphase $2.5 \cdot 10^{-4}$ M calcium acetate (Fluka, p.a.), dissolved in Millipore-filtered water and buffered at pH = 7.5 with Tris/HCl (Merck, p.a.) was used [13].

For incorporation, first the matrix lipid or lipid mixture was spread on a Langmuir film balance. A compression isotherm (continuous compression) for DPPA/DOPA (4:1) is shown in Fig. 1 (dot symbols in trace a). After expansion to a lateral pressure of π = 10 mN/m (cross symbols in trace a) a protein (here, LHC II) detergent solution was spread in small, single drops distributed over the whole trough area as homogeneously as possible. While spreading, an immediate increase of the surface pressure up to about 20 mN/m (highest pressure in trace b) was observed, indicating incorporation of protein and detergent molecules.

Subsequently, the film is expanded to π = 0 mN/m (trace b in Fig. 1) to achieve further insertion of protein from the subphase into the molecular layer. The final compression isotherm of the lipid layer containing protein, trace c in Fig. 1, shows an area increase of δA = 300 Å^2 per spread protein molecule at π = 10 mN/m as compared to trace a. Variations of the initial lipid pressure in the range between 5 and 15 mN/m did not lead to different results of the area increase by the protein, whereas it was found that the final pressure after the expansion of the protein/lipid film had to be zero in order to obtain maximum area increase.

Fluorescence microscopy was performed on a film balance with the microscope optics included in the bottom of the trough [14]. Excitation was achieved by a Xe lamp, and for the detection of the lateral

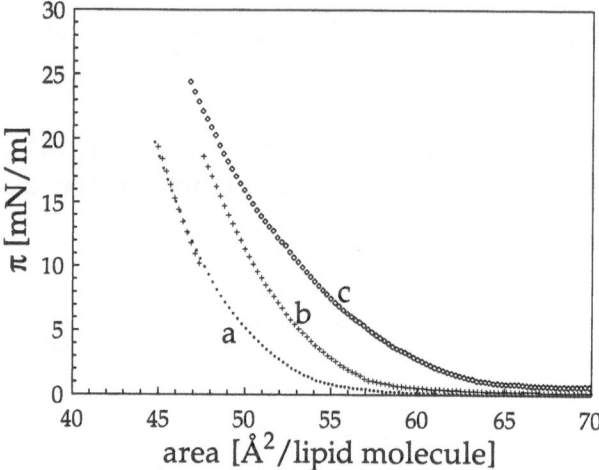

Fig. 1. Compression isotherms of DPPA/DOPA (4:1) at room temperature. The lipid monolayer was compressed to π = 20 mN/m after spreading (trace a, dot symbols) and expanded to 10 mN/m (trace a, cross symbols). Subsequently, LHC II/detergent solutions was spread (pressure rise to trace b at constant area). The protein/lipid monolayer was expanded to 0 mN/m (trace b). After 30 min equilibration, a final compression isotherm of the protein/lipid monolayer was measured and is displayed in trace c

distribution of proteins or lipid fluorophores within the monolayer a sensitive TV camera (Hamamatsu) was used. For electron microscopy, the samples were prepared as follows: A monolayer of the lipid or lipid/protein film was transferred to a hydrophilic glass support using the LB technique. The samples were shadowed with carbon/platinum under an angle of 15°. For stabilization, an additional deposit of carbon was added. Subsequently, the replica was removed from the glass substrate in dilute HF and transferred to electron microscope copper grids.

The lipid matrix problem

In search of a suitable lipid matrix, one has to take into account two opposing requirements: On the one hand, there should be lipid molecules with disordered chains available for the solubilization of the proteins. On the other hand, LB transfer of the film requires the structural stability of a highly ordered lipid film, e.g., in a solid-phase state.

Our first approach was to use single-component lipid films of behenic acid or phosphatidic acid

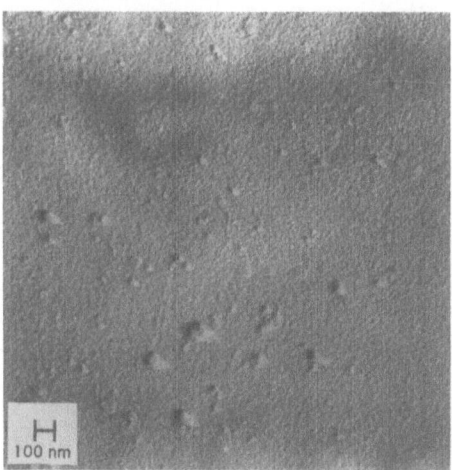

Fig. 2. Electron micrograph of a replica of RCs incorporated into a pure DPPA monolayer

which are easily deposited as LB multilayers. Figure 2 shows electron micrographs of DPPA with RCs incorporated. The film was transferred at π = 10 mN/m. Large structures with heights rising above the monolayer up to 350 Å (note shadwos due to platinum deposition) and diameters up to 750 Å were observed. Such structures were not detected on reference replicas of films when detergent solution (without protein) was spread onto the lipid layer. Hence, they are attributed to the protein. The large size of the structures compared to the protein dimensions (RC diameter: ~70 Å) leaves an oriented incorporation of proteins highly improbable. If the films were deposited at π = 20 mN/m, fewer large clusters were observed, as they seem to dissolve into the subphase at progressively higher π. The situation is similar for behenic acid, showing that both lipids are by far too rigid to facilitate protein incorporation such that ordered structures are created.

Our search for an optimized lipid matrix led us to use a mixture of DPPA and DOPA (4:1). Here, the unsaturated chains of DOPA are thought to act as a mediator between the hydrophobic part of the protein surface, which is presumably irregular in its topology, and the ordered chains of the majority component, DPPA, in a solid state [15]. Without protein, Figs. 3a and b show the phase structure of DPPA/DOPA, visualized by fluorescence microscopy with 1 mol% of NBD-DPPE in the lipid mixture. Two different regions occur, each containing two coexisting phases. Region 1 (Fig. 3a) con-

sists of dark domains in a fluid phase, whereas region 2 (Fig. 3b) contains brighter patches in a more rigid, net-like environment. Both regions occurred over the whole range of the isotherm and no conversion between them was observed, while about 60% of the whole area was occupied by region 1. Assuming negligible DOPA solubility in the phases of high chain order in both regions, an evaluation of the area fraction occupied by the different phases led to a DOPA amount of 35% and 54% in the fluid areas of regions 1 and 2, respectively. These values revealed no significant dependence on lateral pressure.

Results on protein incorporation

Figure 4a shows a fluorescence micrograph of DPPA/DOPA with LHC II incorporated. Here, the protein fluorescence indicates that the phase structure is similar to that in the pure lipid monolayer, Figs. 3a and b. As the protein is incoporated into the fluid phases, the phase structure is qualitatively preserved, while the area fractions of the fluid phase are expanded. The expansion is most pronounced with the bright patches of region 2, presumably, because this fluid phase contains the largest amount of DOPA and should therefore be the least ordered.

An electron micrograph of LHC II in DPPA/DOPA is shown in Fig. 4b. As opposed to behenic acid or DPPA, the protein structures are flat with heights rising above the layer less than 30 Å, which corresponds to the height of the hydrophilic tops of LHC II of 7 and 20 Å, respectively; however, the resolution was not high enough to distinguish between these two values. The diameters of the aggregates vary from 100 to 300 Å (LHC diameter: 27 Å) and indicate that not single proteins, but two-dimensional clusters are incorporated into the lipid matrix. This leads us to speculate that the molecules are arranged in a oriented way.

Figures 5a and b show the results for RCs. For the fluorescence measurements (Fig. 5a) the lipid layer contained 1 mol% NBD-DPPE for indirect visualization of changes in the phase structure by the protein. The results are similar to the LHC II case, and the replicas show similar flat structures with diameters varying from 100 to 500 Å, and heights from 20 to 35 Å, indicating oriented 2D clusters.

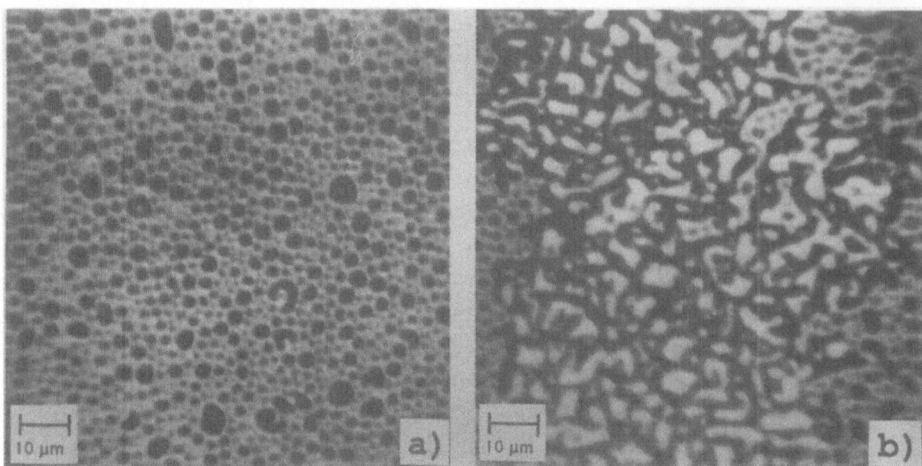

Fig. 3. Fluorescence micrographs of a DPPA/DOPA (4:1) monolayer incorporating 1 mol% of NDB-DPPE. Two characteristic morphologies coexisting on a macroscopic length scale within the lipid monolayer are shown in *a* and *b*

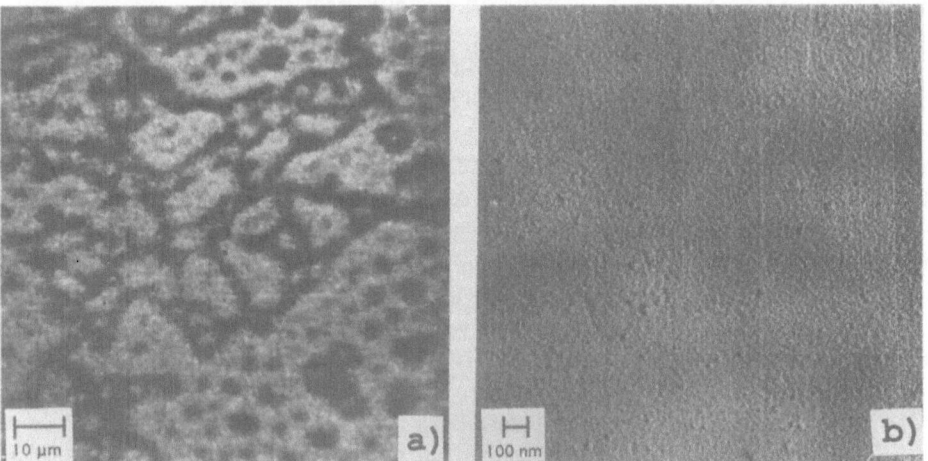

Fig. 4. LHC protein in DPPA/DOPA monolayers. a: fluorescence micrograph (intrinsic protein fluorescence), and b: electron micrograph of a replica after incorporation

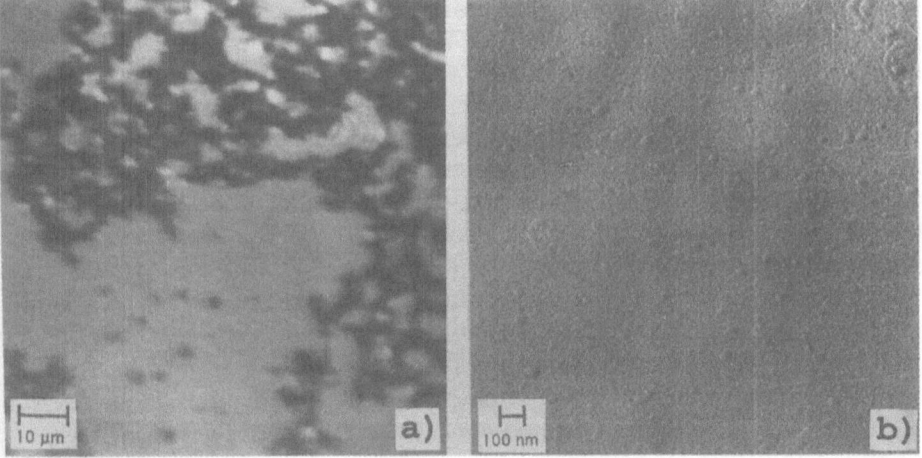

Fig. 5. RC protein in DPPA/DOPA monolayers. a: fluorescence micrograph (1 mol% NBD-DPPE as a trace), and b: electron micrograph of a replica after incorporation

Estimate of protein content

An approach for a quantitative evaluation of the protein content in the lipid films with different techniques has been made for the case of LHC II.

From fluorescence microscopy, information was obtained by comparison of the areas of the fluid phases in protein-reconstituted and in pure lipid films. Evaluation of 10 micrographs at π between 2 and 10 mN/m revealed an overall area increase of the fluid phase that amounted to 9% of the total film area*). This value poses a lower limit for the total area increase due to protein incorporation, as the amount of reconstituted material may be larger close to areas in the film, where it was spread, and where we observed a bright homogeneous fluorescence. The isotherms show an area increase of ~20% at 5 mN/m. These two techniques assess protein and detergent incorporation into the monolayer.

The most reliable value for the amount of incorporated protein was obtained from optical absorption spectra of transferred multilayers. They revealed that ~15% of the spread protein was incorporated, corresponding to a final area fraction of 4% for the proteins. In addition, the lateral density of the protein patches observed in electron microscopy was evaluated. Assuming an average diameter of 200 Å, this resulted in 2% of the area occupied by the protein.

Discussion

We obtained two sets of results for the area increase due to protein reconstitution: The values from fluorescence microscopy (FM) and from isotherms (>9% and 20%, respectively) are in fairly good agreement. The discrepancy can be accounted for if one assumes that a small amount of detergent may be incorporated into the solid lipid phases. Electron microscopy (EM) and absorption measurements result in a relative protein area of 2% and 4%, respectively. Here, the low value from EM observations may be due to structures (possibly as small as single proteins or oligomers) that were not resolved. The apparent mismatch between the FM and isotherm values on the one hand, and the EM

and optical absorption values on the other hand can be due to two alternative reasons:

i) In EM and absorption measurements only the proteins have an effect on the area evaluation, whereas in FM and isotherms the detergent from the protein solution will cause an additional area increase as it is incorporated into the monolayer.

ii) FM and isotherm measurements, both resulting in large values of area increase, were done on monolayers on the water surface. Possibly, the transferred layers contain less protein, because of protein dissolution into the subphase during the transfer.

Isotherms of pure detergent solution spread onto the lipid layer revealed that the detergent indeed incorporates into the layer, with a large area increase, thus favoring the first of the explanations for the discrepancy of results from different techniques.

Conclusions

We have demonstrated that membrane proteins can be reconstituted into lipid monolayers on the water surface by spreading the proteins from a detergent solution. DPPA/DOPA-mixtures were shown to be suitable matrices for two reasons: The unsaturated chains of the DOPA component provide a fluid environment for the protein and the DPPA condensed phase provides the structural stability necessary for the transfer to solid substrates.

The characterization of the layer revealed that ~15% of the proteins initially spread was incorporated into the monolayer. They form 2D aggregates with heights above the monolayer compatible with an assumption of oriented proteins. Incorporation takes place mainly into the fluid phases of the matrix, with area fractions of about 4% due to the protein and a larger area occupied by the detergent.

Acknowledgements

We thank M. Piepenstock for the isolation of reaction centers and H. Scheer for his advice. The work was financially supported by the Deutsche Forschungsgemeinschaft (Lo 352/3).

References

1. Clayton RC (1978) in: The Photosynthetic Bacteria, Clayton RC, Sistrom WR (eds) Plenum Press, New York

---

*) This results from an increase from 57% of 61% in region 1 (which occupied 60% of the total film area) and from 37% to 53% in region 2.

2. Vermeglio A, Clayton RC (1977) Biochem Biophys Acta 461:159—165. Martin J-L et al. (1986) Proc Natl Acad Sci USA 83:957—961. Klevanik AV, Ganago AO, Shkuropatov AY, Shuvalov VA (1988) 237:61—64
3. Johnson SG et al. (1990) J Phys Chem 94:5849—5855. Popovic ZD, Kovacs GJ, Vincett PS, Dutton PL (1985) Chem Phys Lett 116:405—410
4. Lösche M, Feher G, Okamura MY (1987) Proc Natl Acad Sci USA 84:7537—7541
5. Lösche M, Feher G, Okamura MY (1988) in: The Photosynthetic Bacterial Reaction Center, Breton J, Vermeglio A (1988) (eds) Plenum Press, New York
6. Vermeglio A, Clayton RK (1976) Biochem Biophys Acta 449:500—515
7. Rafferty CN, Clayton RK (1978) Biochem Biophys Acta 502:51—60
8. Vermeglio A, Breton J, Paillotin G, Cogdell R (1978) Biochem Biophys Acta 501:514—530
9. Erokhin VV, Kayaushina RL, Lvov YM, Feigin LA (1989) Stud Biophys 132:97-104
10. Heckl WM, Lösche M, Möhwald H (1985) Thin Solid Films 133:73—81
11. Kühlbrandt W, Wang DN (1991) Nature 350:130—134
12. Wild A, Urschel B (1980) Z Naturforsch 35c:627—637
13. Howarth VA, Petty MC, Ancelin H, Yarwood J (1990) Vibrat Spectr 1:29—33
14. Lösche M, Möhwald H (1984) Rev Sci Instrum 55:1972—1976
15. Albrecht O, Gruler H, Sackmann E (1978) J Physique (Paris) 39:301—313

Authors' address:

M. Schönhoff
Institut für Physikalische Chemie
Joh.-Gutenberg-Universität
Welder-Weg 11
6500 Mainz, FRG

Progress in Colloid & Polymer Science   Progr Colloid Polym Sci 89:249—252 (1992)

Interfacial properties of surfactant mixtures with alkyl polyglycosides

D. Nickel, C. Nitsch, P. Kurzendörfer, and W. von Rybinski

Laboratories of Henkel KGaA, Düsseldorf, FRG

Abstract: Alkyl monoglycosides and polyglycosides (APG) are characterized in terms of their interfacial properties which exhibit a dependence on alkyl chain length and degree of polymerization. A comparison is drawn to other nonionic and anionic surfactants. In hard water, clouding phenomena are observed for C8/10 and C12/14 APG. The critical micelle concentrations (CMC) of the alkylglycosides are similar to those of other nonionic surfactants and show a decrease with increasing alkyl chain length. The degree of polymerization has only a minor influence on the CMC. Above the CMC, alkylglycosides exhibit very low interfacial tensions against mineral oil compared to C12/14 fatty alcohol sulfate (C12/14 FAS). In contrast to C12/14 FAS and C12/14 fatty alcohol hexaglycol ether, respectively, C12/14 APG shows remarakbly high dynamic surface tensions even at a concentration far above the CMC. For mixtures of C12/14 APG and C12/14 FAS the static surface tensions and the CMC lie close to those of C12/14 APG, whereas the dynamic surface tension behavior is dominated by the anionic surfactant.

Key words: Alkyl polyglycoside; clouding behavior; static and dynamic surface tension; surfactant mixtures

1. Introduction

Alkyl polyglycosides (APG) are nonionic surfactants based on glycose and fatty alcohol. They can be extracted from renewable raw materials, and are non-toxic and readily biodegradable [1, 2]. Because their properties make them suitable for use as wetting agents, emulsifiers and hydrotropes, they are steadily gaining in importance for numerous technical appliations. The first systematic studies of their surface activity were made by Shinoda [3]. A more recent paper by Lüders and Balzer describes their synthesis and physicochemical properties [1]. Studies of the phase behavior of this class of surfactants are important with regard to their use in practical applications. The present paper aims to characterize these compounds and their mixtures with anionic surfactants. Specifically, we report on static and dynamic measurements of surface and interfacial tension. These properties, as exhibited by pure and technical grade surfactants, are described with particular reference to their dependence on alkyl chain length and degree of polymerization. A comparison is drawn with other nonionic and anionic surfactants.

2. Experimental

C8/10, C12, and C12/14 APG and C8, C10, and C12 monoglycosides were obtained by means of Fischer glycosidation with technical grade or pure fatty alcohols [4]. For the monoglycosides, distillation and fractional crystallization were also applied. In addition, C12/14 fatty alcohol sulfate (C12/14 FAS) and C12/14 fatty alcohol hexaglycol ether (C12/14 E6) were studied as technical grade surfactants. All of the substances were supplied by Henkel. The clouding behavior of APG was examined photometrically. Static surface tension was determined with an automatic tensiometer (Lauda). Dynamic surface tension was measured with a home-built appratus by means of the maximum bubble pressure method [6, 7]. Interfacial tension measurements were performed with a spinning-drop tensiometer (Krüss).

3. Results

3.1 Clouding behavior

The clouding behavior of C8/10 APG and C12/14 APG was studied as a function of concentration and temperature at a given water hardness. C12/14 APG behaves similarly to other nonionic surfactants in that it exhibits a concentration-dependent cloud point in distilled water. In contrast, C8/10 APG, which contains shorter chains, forms an isotropic liquid phase between 20° and 90°C in distilled water, even at high concentrations. However, when a mixture of Ca and Mg ions is added a narrow liquid/liquid coexistence region appears at low concentrations (Fig. 1).

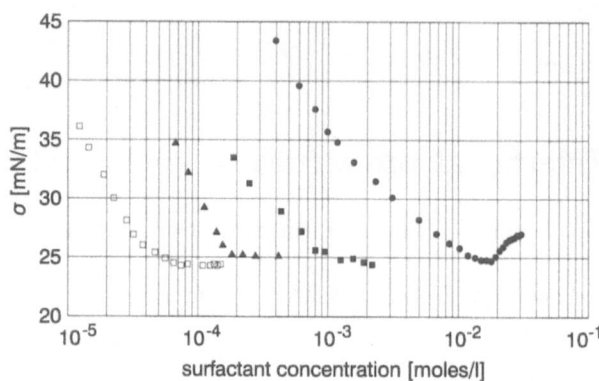

Fig. 2. Surface tension as a function of concentration for different alkylglycosides at 60°C in distilled water: (●) C8 monoglycoside; (■) C10 monoglycoside; (▲) C12 monoglycoside; (□) C12/14 APG

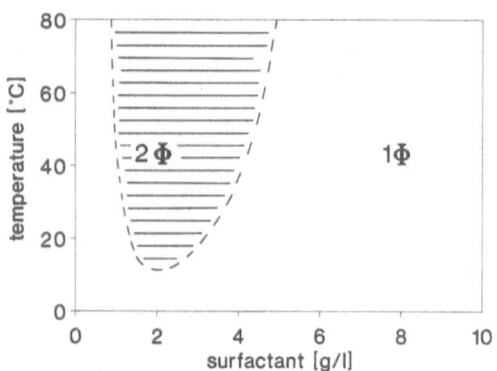

Fig. 1. Clouding behavior of C8/10 APG in water of 340 ppm hardness (Ca/Mg = 5/1)

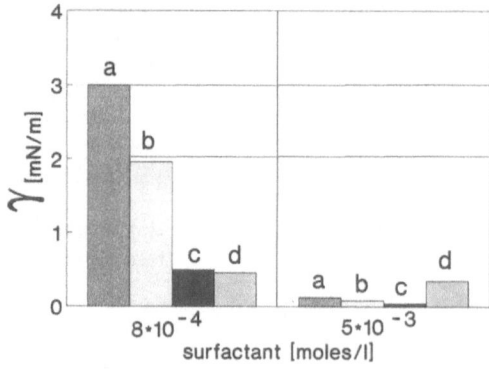

Fig. 3. Interfacial tension against mineral oil at 40°C for different concentrations: a) C8 monoglycoside; b) C10 monoglycoside; c) C12 monoglycoside; d) C12/14 FAS

3.2 Static surface tension

Figure 2 shows the dependence of surface tension on concentration of the pure and technical grade alkyl glycosides at 60°C. The critical micelle concentrations (CMC) are in the same region as those of conventional nonionic surfactants, and decrease markedly as the length of the alkyl chain increases. Compared with alkyl chain length, the number of glycoside groups of the APG exerts only a slight influence on the CMC.

3.3 Interfacial tension

Figure 3 shows the interfacial tensions against mineral oil of the three monoglycosides and C12/14

FAS obtained at two different concentrations and at 40°C. At the higher concentration (>CMC), low interfacial tensions (<1 mN/m) are measured in all cases. At the lower concentration the interfacial tension increases with decreasing alkyl chain length and increasing CMC (cf. Fig. 2). The alkyl glycosides therefore only exhibit marked interfacial activity at concentrations above or in the region of the CMC; with mineral oil, this activity is much higher than that exhibited by anionic surfactants.

3.4 Dynamic surface tension

Figure 4 exhibits the surface tension determined by the maximum bubble pressure method [5]. The

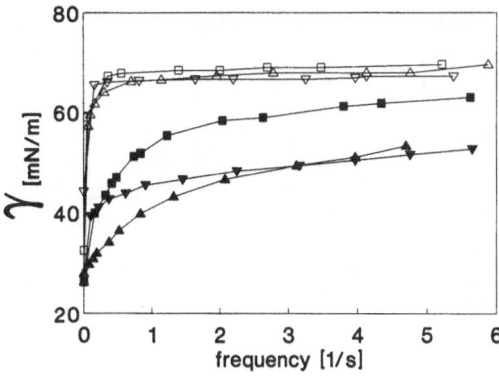

Fig. 4. Frequency dependence of dynamic surface tension of different surfactants at 40 °C for two concentrations: (□) $4 \cdot 10^{-5}$ moles/l C12/14 APG; (▽) $4 \cdot 10^{-5}$ moles/l C12/14 FAS; (△) $4 \cdot 10^{-5}$ moles/l C12/14 E6; (■) $8 \cdot 10^{-4}$ moles/l C12/14 APG; (▼) $8 \cdot 10^{-4}$ moles/l C12/14 FAS; (▲) $8 \cdot 10^{-4}$ moles/l C12/14 E6

experiments were performed as a function of bubble frequency. With increasing frequency, there is a reduction in the time available for the surfactant to diffuse to the surface, i.e., there is less time available for the surface tension to be reduced [6]. In Fig. 4 the dynamic surface tensions at 40 °C for C12/14 APG, C12/14 FAS and C12/14 E6 are compared for two different concentrations. At lower concentration, chosen to be lower than the respective CMC, all of the technical grade surfactants exhibit slow and almost identical surface adsorption rates. At higher concentration, identical to the CMC for C12/14 FAS and exceeding it for C12/14 E6, the surface tension is markedly smaller over the whole frequency range. By comparison, C12/14 APG exhibits relatively high surface tension, especially at higher frequencies, so that diffusion to the surface must be slower, even though the concentration is above the CMC.

A comparison to the frequency dependence of dynamic surface tension obtained with the shorter chained alkyl glycosides at a concentration below the CMC shows that, at low frequencies, C8/10 APG occupies a position between the C8 monoglycoside which adsorbs slowly, and the C10 monoglycoside which adsorbs more rapidly. At higher frequencies the mixture approaches the behavior of the C8 monoglycoside. This effect is also observed at concentrations above the CMC, i.e., at lower static surface tensions.

3.5 Interfacial properties of APG/anionic surfactant mixtures

The interfacial properties of APG/anionic surfactant mixtures are of special importance for practical applications. Figure 5 shows the dependence of surface tension on concentration for C12/14 APG, C12/14 FAS and two mixtures at 60 °C. The values characterizing the mixtures lie close to the curve for APG, even when the anionic surfactant content is high. It may be concluded that the mixed micelle formation facilitates favorable interaction betweeen the nonionic and the anionic surfactant [8].

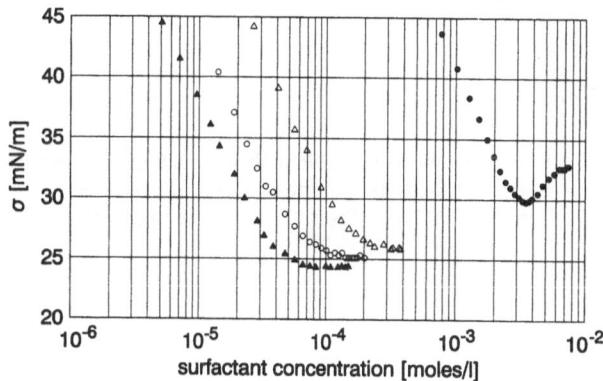

Fig. 5. Surface tension as a function of concentration for different mixtures of C12/14 APG and C12/14 FAS at 60 °C in distilled water: (▲) APG/FAS = 1/0; (○) APG/FAS = 4/1; (△) APG/FAS = 1/1; (●) APG/FAS = 0/1

Figure 6 shows the dynamic surface tension as a function of frequency measured at 40 °C by the bubble pressure method for the same series of mixtures as that in Fig. 5. In all cases, the concentration is above the CMC. In the rate-determining frequency range the behavior of the mixtures closely resembles that of C12/14 FAS, which adsorbs more quickly to the surface than that of C12/14 APG, which adsorbs more slowly. By contrast to the situation under static conditions, the change in surface tension of the surfactant mixture as a function of time seems to be dominated by the anionic surfactant. Both parameters, i.e., static and dynamic surface tension, indicate that mixtures of APG and anionic surfactants exhibit especially favorable interfacial properties.

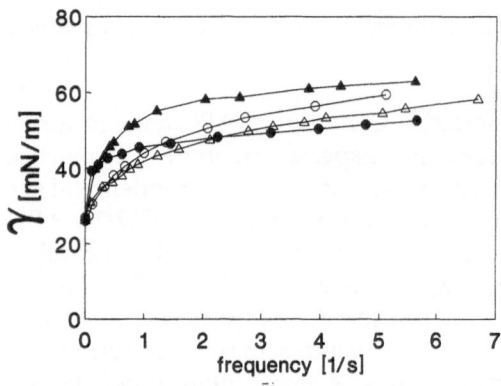

Fig. 6. Frequency dependence of dynamic surface tension of different mixtures of C12/14 APG and C12/14 FAS at 40°C: (▲) APG/FAS = 1/0; (o) APG/FAS = 4/1; (△) APG/FAS = 1/1; (●) APG/FAS = 0/1

References

1. Lüders H, Balzer D (1988) Proc 2nd CESIO International Surfactants Congress, Paris, pp 80—93
2. Schulz PT (1991) Proc Chemspec Europe 91 BACS Symposium, pp 33—37
3. Shinoda K, Yamaguchi T, Hari R (1961) Bull Chem Soc Japan 34:237—241
4. Fischer E (1893) Chem Ber. 26:2400—2412
5. Padday F (1969) In: Matijeric E (ed) Surface & Colloid Sci, Wiley Intersci, New York, pp 101—149
6. Mysels KJ (1990) Coll Surf 43:241—262
7. Engels T, Förster T, Mathis R, von Rybinski W (1991) Proc 7th Int Conference on Surface and Colloid Interface Sci 99:435—442
8. Zhu Y, Rosen MJ (1984) J Colloid Interface Sci 99:435—442

Authors' address:

Dr. W. von Rybinski
Laboratories of Henkel KGaA
W-4000 Düsseldorf, FRG

Progress in Colloid & Polymer Science   Progr Colloid Polym Sci 89:253—257 (1992)

Dynamical properties of lecithin-based microemulsions*)

F. Aliotta, M. E. Fontanella, S. Magazú, G. Maisano[1]), D. Majolino, and P. Migliardo[1])

Istituto di Tecniche Spettroscopiche del C.N.R. — Messina, Italy
[1]) Dipartimento di Fisica, Universitá — Messina, Italy

Abstract: Lecithin/isooctane/water gel is investigated by means of different experimental techniques, namely, depolarized light scattering, ultrasonic and dielectric measurements. Pure isooctane and lecithin/isooctane solution are also tested for comparison. A relaxation process, taking place in the 10^{-13} s time scale was revealed from the light-scattering experiment, both in the gel and in the solution. This is attributed to the reorientational relaxation of the solvent that keeps its bulk properties. The comparison between the activation enthalpies for the detected process in the gel and in the solution, together with the behavior of the relaxation times vs frequency showed the existence of some influence of the entangled network on the isooctane dynamics. Data from ultrasonic measurements show the existence of the same relaxation process, both in the gel and in the solution, with a relaxed value very near to the one obtained for pure isooctane. The observed dynamics were interpreted as due to the existence of the same elementary microspheres of lecithin in the bulk of the solvent in both systems. The relaxation phenomenon revealed by the dielectric data was analyzed in the frame of a Cole-Cole phenomenological law and associated to the role played by water molecules on lecithin reorientation. The temperature dependence of the observed phenomenon is in agreement with the existence of an entangled polymeric-network.

Key words: Organogels; reverse micelles; lecithin; gels; entangled networks

1. General consideration

In the last few years a number of papers [1, 2] have been devoted to the study of statical and dynamical properties of soybean lecithin gels. The appeal of this kind of system lies in some peculiarities that mark them as belonging to a new class of gels with respect to the more conventional gelatin or polymer gels. In particular, the observed differences forced researches to modify the existing models [3] or to create new ones in order to understand the microscopic interactions giving rise to the observed local topology and structural evolution. In any case, such systems have not yet been fully

elucidated, and a number of further experiments are planned in order to obtain a complete picture.

First of all, it must be stressed that lecithin is able to act as a surfactant when a ternary system organic solvent/lecithin/water is taken into account. The gelification process takes place when the small quantity of water necessary to create a conventional reverse micellar solution reaches a critical concentration value w_0; the situation could appear very close to that observed in a gelatin gel: In both cases, in fact, the gelification process is observed in a system that is originally a usual micellar solution. But, at the same time, the difference is quite evident: in our case no guest polymer is necessary to induce the sol-gel transition by means of the building up of an "infinite" gelatine cluster. Consequently, the basic process of the observed gelification has to be looked for in the growth process of

---
*) Work partially supported by C.N.R., progetto finalizzato Chimica Fine II. "Chimica e tecnologia dei polimeri".

the micelles triggered by the interaction between the water molecules and lecithin.

It is to be stressed that, when water is added to a solution of lecithin/organic solvent, reverse micelles are generated that grow linearly. Only a small quantity of water is needed to induce the existence of very long, flexible, cylindrical micelles. It was hypothesized [4] that, at the critical value of the water concentration at which the gelification process occurs, the cylinders reach a critical length at which the entanglement of the micelles becomes highly favorable and, as a consequence, a transient network is established. Following this picture, an additional difference between gelatin and lecithin gels should be pointed out: due to the very small quantity of water, no water pool can be hypothesized in the second case, and all the water molecules are to be considered as being firmly attached to the head groups of the lecithin molecules. This is confirmed by recent small-angle neutron scattering experiments [2] on the system soybean lecithin/isooctane/water, from which a gyration radius of ~ 15 Å was obtained to which a radius of ~ 20 Å for the micelles corresponds.

Due to this kind of topology, the situation could appear very similar to the one observed for a polymeric gel: the long micelles play the same role as the polymeric chains that generate the long-range correlation. But, in this case, a strong difference exists, too: it is not possible to define the analogous of the chemical cross-link interaction among the distinct polymeric units.

Furthermore, a number of rheological measurements were performed [1] in order to follow the gelification processes. In fact, the evolution of the micelle size (the analogous ● of the evolution of the molecular structure in a polymeric gel) has to originate pronounced effects on molecular mobility that can easily be monitored by the change in viscosity and elasticity. In particular, it was observed that the general trend of shear viscosity η_s, storage modulus $G'(\omega)$, and loss modulus $G''(\omega)$ vs frequency follows the characteristic power laws of a polymeric gel [4]. However, we have to point out that, while the scaling law for the elastic modulus is exactly the one expected for an entangled network, a deviation from the expectation for a polymeric-like model is observed in the relationship between η_s and concentration. The interpretation of the experimental results required some modifications of the polymeric-like model to match the experimental data [5]. The divergence can be at-

tributed to the fact that the micellar aggregation strongly depends on the lecithin volume fraction and, furthermore, to the fact that micelles are dynamical entities whose breaking and reforming mechanism give additional contributions to the stress relaxation. The application of the theory indicates that the scaling behavior of η_s is a strong function of the micelle kinetics. It is now evident that in such a system a hierarchy of structure exists, each one giving rise to different dynamics with its own relaxation phenomenon (i.e., the gel network and its structural relaxation process, the micelle and its breaking and reforming dynamics, the reptation modes of the local structure, the water dynamics inside the micelle, the diffusional modes of the bulk solvent, and so on).

The aim of this work is to open some windows in the complex dynamics of our system by means of different experimental techniques, namely, depolarized Rayleigh scattering, ultrasonic and dielectric measurements. From the obtained data, we are able to extract some information about the dynamic response of the system in the temporal range of 10^{-12} to 1 s.

2. Experimental procedure and results

High purity (97%) soybean lecithin was obtained by starting from a commercial product (Sigma Chemicals) and following the purification procedure described elsewhere [6] then testing the results by thin-layer chromatography. The purified lecithin was dissolved under stirring at room temperature in isooctane, reagent grade quality (Baker Chemicals) to obtain a ~ 100 mM solution (corresponding to a lecithin volume fraction $\Phi \simeq 0.072$). Then, the gel samples were obtained by adding water until the gelification of the sample was observed. The obtained stable and clear gels turned out to have a water-to-lecithin ratio $w_0 \simeq 2.5$.

The depolarized Rayleigh wing measurements were performed by means of a SPEX Ramalog 5 triple monochromator, in a 90° geometry, using the 4880 Å line of an Ar$^+$ laser. Besides the gel, pure isooctane was also tested for comparison. In all the runs, performed at different temperatures ($T = 10$, 15, 20, 30, 40, and 50°C), three different pass bands of 0.3, 0.6, and 1.0 cm^{-1} were used, in order to obtain a better compromise between intensity and resolution; they were then numerically matched. In Fig. 1, we report, as an example, the obtained spectrum at $T = 30$°C for the gel.

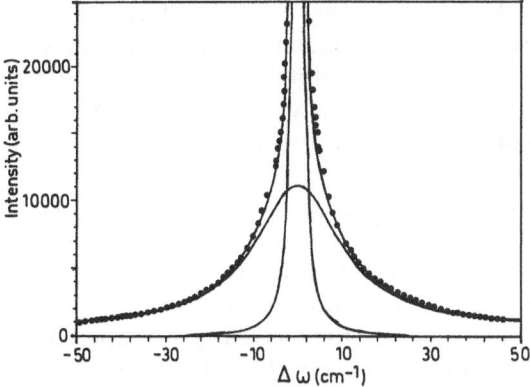

Fig. 1. DRS spectrum for the soybean/lecithin/water gel at $T = 30°C$. Continuous lines represent the two components revealed by the deconvolution procedure and the total fitting

The ultrasonic measurements were performed by a standard Matec pulse echo apparatus, in the same temperature range and in the 5 ÷ 45 MHz frequency range. Pure isooctane and the solution isooctane/lecithin were also tested. In Fig. 2 the obtained absorption data are reported.

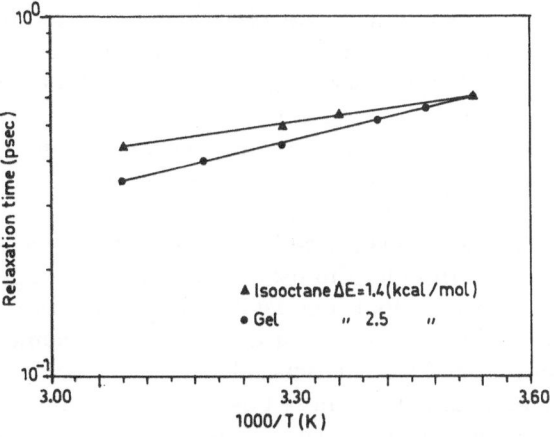

Fig. 2. Normalized ultrasonic absorption vs frequency for soybean lecithin/isooctane solution, lecithin/isooctane/water gel and pure isooctane at $T = 30°C$

For the dielectric measurements a temperature-controlled cylindrical cell was used. Data were taken in the frequency range $10 ÷ 10^6$ Hz, using an experimental set-up described elsewhere [7], on pure isooctane, isooctane/lecithin solution, and gel.

In order to obtain the higher frequency value of the dielectric function $\varepsilon_\infty = n^2$, additional refractive index data were measured by means of a conventional Abbé refractometer.

3. Discussion and conclusions

In order to extract information about the different mechanisms from the depolarized light scattering spectra contributing to the density-density correlation function the experimental data were analyzed according to the following expression:

$$I^{tot}_{anis}(\omega) = S(\omega) + \Sigma_i L_i(\omega) + I^{vibr}_{anis}(\omega) , \qquad (1)$$

where in the term $S(\omega)$, resolution enlarged, all the slow processes taking place in the gel have collapsed; the second term is a superposition of symmetric lines originated by fast mechanisms occurring on a time scale accesible to our experiment, and the last term is the cooperative contribution of the vibrational density of states that behaves like a constant background in the low-frequency region. In our spectra, besides the resolution line, only one enlarged contribution was detected.

As stressed above, in the slow term (HWHM $\simeq$ 0.15 cm^{-1}) all the dynamics inaccessible to our experiment are collapsed (translational diffusion and reorientational dynamics of micelles plus coupling with transverse spectral contributions). The fast term clearly gives evidence of some process taking place on a 10^{-13}-s time scale. In Fig. 1 the continuous lines represent the total fitting and the two mentioned contributions. From the HWHM value (Γ) of the symmetric Lorentzian line the relaxation time can be derived ($\tau = 1/2\pi c\Gamma$), whose behavior as a function of reverse temperature is reported in Fig. 3 in an Arrhenius plot, both for the gel and the pure solvent. The fact that the two processes take place on the same time-scale suggests that the observed motion is connected with the reorientational dynamics of bulk isooctane. Such a hypothesis is in close agreement with the general observation that in organo-gels the organic solvent is not directly involved in the "polymeric" network and preserves all its bulky behavior [1].

Indeed, an inspection of Fig. 3 shows some difference between the behavior vs temperature in the pure solvent and in the gel. In particular, the two straight lines are the result of the observed process of ~2.5 Kcal/mole in gel and 1.4 Kcal/mole in pure isooctane are obtained. The obtained data seem to indicate a more hindered character for the observed process in the gel, together with faster relaxation

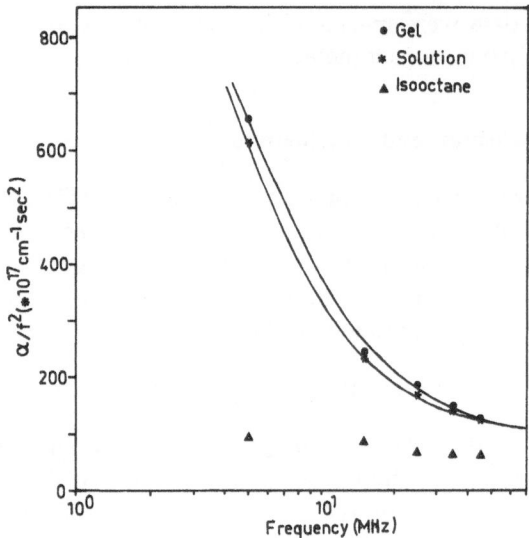

Fig. 3. Relaxation time for the fast process vs $1/T$. Continuous lines represent the Arrhenius law fits

dynamics, at least at higher temperatures. Such results could be rationalized by taking the existence of some interactions between the entangled network and the isooctane into account. The existence of a polymeric-like network could create more stable equilibrium positions for the solvent molecules trapped in the cages. Such a hypothesis could explain both the higher activation enthalpy observed in the gel and the faster relaxation process towards a new configurational equilibrium.

Obviously, we are observing only one of the aspects of an exceedingly complicated structural evolution. Unfortunately, as stressed above, all the other processes taking place on longer time scales are collapsed into the resolution enlarged central contribution and are not accessible to our technique.

From our ultrasonic results, the sound velocity turns out to be almost independent of the system, passing from pure isooctane through the isooctane/lecithin solution to the gel ($v_s \simeq 1050$ m/s). The most interesting results are observed when we look at the normalized ultrasonic absorption reported in Fig. 2. It is quite evident that some relaxation process is taking place, both in the isooctane/lecithin solution and in the gel. When we tried to fit the data with a single relaxation law

$$\frac{a}{f^2} = \frac{N}{1 + \left(\dfrac{f}{f_c}\right)} + M , \qquad (2)$$

where N, f_c and M are the strength of the relaxation, the relaxation frequency, and the relaxed value of the absorption, an almost temperature-independent relaxation frequency of ~ 5 MHz was obtained. From an inspection of Fig. 3, it turns out that the relaxed value for the observed process is nearly the same as the one obtained for pure isooctane. Our data seem to be consistent with a behavior of our system very close to that of a polymeric critical gel or of a nearly critical gel [8]. In such systems a self-similar shape of the relaxation time spectrum is observed that turns out to behave like $(\tau/\tau_0)^{-n}$, where τ_0 is the shortest relaxation time for the system under consideration. Furthermore, a cut-off at a maximum value of the relaxation times τ_{max} is to be taken into account when a nearly critical gel is considered, due to the fact that the clusterization process was stopped at a maximum correlation length. Our system could be very close to the latter situation since, owing to the hypothesis that micelles are dynamical entities, continuously breaking and reforming, the formation of a truly "infinite" network is hindered. Such a point of view suggests the existence of a distribution of the micelle size that is both temperature- and concentration-dependent.

In our experiment, we observe the tail of the relaxation process at the faster relaxation time. In other words, we observe the dynamical evolution of the minimal correlated entities existing in our system, namely, the microspheres of lecithin in isooctane representing the embryos of the micelles existing in the ternary system. On this basis, it is possible to rationalize both the similarity between the results from lecithin/isooctane solution and gel with respect to their temperature dependence.

From the dielectric measurements the obtained values of $\tan\delta$ show a typical behavior with a well-defined maximum, whose position is temperature-dependent in the case of the gel. A constant behavior is detected for the non-polar solvent, while the lecithin/isooctane solution shows the characteristic behavior of a polar liquid at the lowest frequencies.

Our data were well reproduced by a Cole-Cole relaxation model that assumes a reorientational motion of molecules in the system, which is very slow with respect to the diffusional translation of some mobile defects. In our system, the stronger dipoles with which the dielectric probe can interact are the lecithin molecules. They are tightly boned among themselves and "frozen" in the entangled network and, at the same time, are connected via hydrogen bonds to the water inside the micelles. From such a local topology, we can expect any reorientation of lecithin molecules to be triggered by the breaking of the H-bond towards the water molecule with subsequent diffusion of the latter. Such a process will obviously be orders of magnitude faster than the reorientation of the lecithin molecule. In Table 1 the values of the frequency of the maximum of $\tan\delta$, the width of the distribution of the relaxation times a, and the activation enthalpy ΔH as a function of T are reported. The dependence of the maximum position with T is in agreement with the hypothesized T-dependence of the width of the micellar size distribution and of the mean length of the gel "fibers". A confirmation of our interpretation comes from the obtained value of ~ 29.7 KJ/mole for the activation energy of the observed process, which is very close to the activation enthalpy for the water molecules engaged in hydration shells.

In conclusion, we have to state that the trend of a to increase with temperature is consistent with the hypothesis of a local topology and of dynamics very close to that of nearly critical gels. When temperature increases, the gel-fiber-size distribution is reduced, thus, the gel becomes more swollen and the cut-off in the relaxation times distribution shifts towards lower values.

References

1. Luisi PL, Scartazzini R, Haering G, Schurtenberger P (1990) Colloid Polym Sci 268:356—374 (and references therein)
2. Schurtenberger P, Scartazzini R, Magid LJ, Leser ME, Luisi PL (1990) J Phys Chem 94:3695—3701
3. De Gennes PG (1982) Ann Rev Phys Chem 33:49—61
4. Schurtenberger P, Scartazzini R, Luisi PL (1989) Rheol Acta 28:372—381
5. Candau SJ, Hirsh E, Zana R, Adam N (1988) J Colloid Interface Sci 122:430
6. Aliotta F, Fontanella ME, Magazu' S, Vasi C, Crupi V, Maisano G, Majolino D (1991) Molecular Cryst and Liquid Cryst (in press)
7. Aliotta F, Fontanella ME, Galli G, Lanza M, Migliardo P, Salvato G (1991) (submitted to J Phys Chem)
8. Winter HH (1991) Material Research Soc Bull 16:44—48 (and references therein)

Table 1. Temperature-dependence of the Cole-Cole distribution width a, of the frequency of the maximum loss tangent $\langle f_{max} \rangle$ and of the mean activation enthalpy ΔH for soybean lecithin/isooctane/water gel

T (K)	$\langle f_{max} \rangle$ (Hz)	a	ΔH (KJ/mol)
303	2870	0.37	29.7
307	3294	0.38	—
310	3899	0.41	—
313	4176	0.44	—

Authors' address:

Maria Elena Fontanella, Dr.
Istituto di Tecniche Spettroscopiche
Salita Sperone, Contrada Paperdo 31
98166 Vill. S. Agata, Messina, Italy

Progress in Colloid & Polymer Science   Progr Colloid Polym Sci 89:258—262 (1992)

Local hydration effects in reversed micellar aggregates

F. Aliotta, P. Migliardo[1]), D. I. Donato[2]), V. Turco-Liveri[2]), E. Bardez[3]), and B. Larrey[3])

Istituto die Tecniche Spettroscopiche del C.N.R.-Messina, Italy
[1]) Dipartimento di Fisica and INFM, Universitá-Messina, Italy
[2]) Dipartimento di Chimica-Fisica Universitá-Palermo, Italy
[3]) Laboratoire de Chemie Generale (CNRSURA 1103) Conservatoire National des Arts et Metiers, Paris, France

Abstract: FT-IR spectra in the hydroxyl region in reversed micelles where the sodium counterions of the surfactant Aerosol-OT have been replaced by Ca^{2+} and Zn^{2+} ions are presented. The IR spectra, taken as a function of the molar ratio R = [H_2O]/[surfactant], are analyzed in the framework of a twostate model for the water molecules in which the O—H vibration can be split into two contributions: one originated by the water "tightly-bonded" to the surfactant ions, and the other induced by the "bulk" water of the micellar water pool. We also evaluate the percentage $a(R)$ of the bond water in the micellar growth process. Furthermore, an explanation of the O—H stretching band features is suggested by comparison between our spectra and those simulated through an MD "experiment".

Key words: Reverse micelles; hydration; IR spectroscopy

1. General considerations

The investigation of the water properties near interfaces and in nanoscopic environments [1] represents one of the most challenging problems in chemical-physics. Very recently [2], it has been shown that the properties of water, i.e., of a system that organizes itself in a local tetrahedral structure [3] imposed by H-bond intermolecular interaction, change drastically near interfaces. In addition, the knowledge of the structural and dynamical properties of water confined in reverse micelles of amphiphilic molecules, a few tens of angstroms in size, is important in order to "simulate" the behavior of the water confined in biological membranes or tightly bonded to biopolymers, enzymes, and proteins [4, 5].

It is now well established [6] that, for the smaller sized droplets, water mainly exists as "interstitial" water bonded to the surfactant negative head groups (SO_3^- in the case of aerosol-OT) and to the corresponding counterions (Na^+ in the case of AOT). As the droplet's size increases, bulk water domains that are in "dynamical" equilibrium with the bonded ones can be postulated. Many ex-

perimental techniques such as DSC, ESR, NMR [2, 7] and also thermodynamical [6] and spectroscopic [8, 9] measurements have been employed in the past in order to determine (i) the amount of "perturbed" or bonded water, and ii) the structural and dynamical properties of this kind of water with respect to that of the "unperturbed" or bulk water.

We have unambiguously shown that the Raman O—H stretching band analysis on the n-dodecane/potassium oleate/hexanol/water [10] and on the reversed micelle n-heptane/sodium bis(2-ethylhexyl)sulfosuccinate(AOT)/water [11] can be fully understood if the O—H stretching band is split into two contributions associated to bonded and bulk water with the existence of an isosbestic point. This "two-state model" also explained the dynamical behavior of water in reversed micelles as probed by IR [9] and IQENS [12]. In their latest paper, the authors showed that the water solubilized inside the micelles has "two" different translational diffusion coefficients, one of these corresponding to that of pure bulk water. Furthermore, some authors [8, 9] claimed to have distinguished two distinct kinds of structures for the perturbed water: "interstitial" (i.e., water monomers trapped between the surfactant ion pairs) and bonded water (i.e., water polarized by the counterions). Such an event was not confirmed by our Raman results [10, 11].

In the present paper, we try to investigate the structure of the various water species present in reversed micelles of aerosol-OT, where the Na^+ counterions have been replaced by Ca^{2+} and Zn^{2+} at very low water content by FTIR spectroscopy. This technique seems to be more fruitful with respect to Raman spectroscopy because of its enhanced sensitivity in the O—H stretching region when concentration values of R (R being the H_2O/surfactant molar ratio) are of the order of unity. As we will show, such a preliminary experiment furnishes relevant information on the growth processes of reversed micelles as a function of the ionic strength of the surfactant salt, as well as on the different role of the anionic hydration on the O—H stretching band. The results will be compared with very recent results on the counterion reactivity [13] and on NMR proton relaxation [14], where a clear difference between the various counterions (Na^+, Ca^{2+}, Mg^{2+}, Zn^{2+}, Al^{3+}) on the polarization of the strictly bonded water has been reported. We anticipate that a two-state model for the O—H stretching vibration for the water molecules restricted in such nanoscopic structures can fully explain the O—H fundamental band at any R value in the investigated range. In addition, a new insight into the understanding of the tightly bonded water will be given through a comparison with molecular dynamic results [15, 16].

2. Experimental procedure and results

Calcium and zinc bis(2-ethylhexyl) sulfosuccinates have been prepared from the metathesis reactions of methanolic sodium salts (AOT) and aqueous calcium or zinc chloride. The precipitated salts were washed many times with water until no chlorine ions were detected in water, and in order to get rid of residual sodium salt. Thus, the molar percentages of sodium ions are lower than 0.5% for the Ca-salt and 0.4% for the Zn-salt (atomic absorption analysis). Special attention must be paid to reduce the residual sodium content as much as possible, because *the properties of the apolar solutions of surfactants are strongly affected by the presence of residual sodium*, especially the maximum amount of water that can be solubilized. The drying of the

samples was carried out in vacuum at room temperature for several days. The residual water conent is 0.7—0.75 mole of water per cation in each of the two surfactants (NMR determination).

AOT was purchased from Sigma and used without further purification, after having been checked to be free of residual acidity. Solvents were of spectroscopic grade.

For the IR measurements, we used a solution of calcium bis(2-ethylhexyl) sulfosuccinate (CaA_2) in cyclohexane at the molar concentration of $c = 4.32 \times 10^{-2}$ M and a molar ratio of water to metal ion R up to the maximum limiting value of 25. The CaA_2 surfactant has also been used in solution with heptane at the molar concentration of $c = 4.81 \times 10^{-2}$ M and R going from 0.78 to 25. Finally, the zinc bis(2-ethylhexyl) sulfosuccinate (ZnA_2) was dissolved in cyclohexane at the molar concentration of $c = 2.64 \times 10^{-2}$ M, and R lying between 0.7 and 12.

Infrared absorption spectra of all the samples were taken with a BOMEM DA8 FT-IR spectrometer purged with dry nitrogen at room temperature ($T = 20\,°C$). The FT-IR spectrometer was equipped with a Globar lamp source, a KBr beamsplitter, and a DTGS/KBr detector. Thirty-two repetitive scans for each sample, contained in a Perkin Elmer IR absorption cell equipped with CaF_2 windows, with a spectral resolution of 4 cm^{-1} in the hydroxyl stretching region (3000—3800 cm^{-1}) have automatically been added in order to obtain a good signal-to-noise ratio. We used a MicroVax II as an interface computer for the acquisition and the analysis of data. FT-IR measurements were taken by using light (H_2O) and heavy (D_2O) water in the reversed micelles for the same value of R in cyclohexane and in heptane as non polar solvents. In order to eliminate the disturbing C—H stretching bands, we used both deuterated sample spectra and pure solvent spectra as reference signals for the Fourier analysis. As expected, the C—H stretching bands at various water content have been found to remain unchanged in all the microemulsions investigated and, hence, were automatically subtracted with the above described procedure. Absorption spectra with a sufficient signal-to-noise ratio have been taken, measuring samples with a thickness of ~4 μm for pure water and ~140 μm for the other cases. The IR spectra in the hydroxyl region were properly normalized in order to take into account the effective number of absorbers (H_2O molecules) in the absorption volume, and then processed following the procedure described in some detail in [11].

3. Discussion and conclusions

In order to extract relevant information regarding the polarization effects on the O—H stretching induced by the various counterions and by the SO_3^- anion from our experimental IR spectra, we hypothesize that in a reversed micelle there exists a boundary layer of depth d of "bonded" water (that takes into account the "interstitial" or tightly bonded water trapped between each ion pair and the hydrated water "oriented" towards the center of the water pool and polarized by the cationic charges) whose dynamical properties dramatically differ from those of the bulk or unperturbed water.

With such an initial hypothesis, the O—H normalized IR band $A(v)$ could be split as the sum of two spatially and dynamically separated contributions.

$$A(v) = a(R) \cdot A^{bonded}(v) + [1 - a(R)] \cdot A^{bulk}(v) ,$$
(1)

where $A^{bonded}(v)$ and $A^{bulk}(v)$ represent the two normalized absorbance contributions at a given wavenumber v. Furthermore, we reasonably assume that the shape of the $A^{bonded}(v)$ corresponds to the IR absorbance of the "interstitial" water obtained at the minimum value of R, i.e., the spectrum at $R = 0.7$ for CaA_2 in cyclohexane, the spectrum at $R = 0.78$ for CaA_2 in heptane and $R = 0.7$ for ZnA_2 in cyclohexane; the "bulk" water spectrum $A^{bulk}(v)$ being represented by pure water IR absorbance. In Figs. 1, 2, and 3 the IR normalized absorbances of bulk and bonded water, as experimentally measured by our spectrometer, in the wavenumber region between 3000 cm^{-1} and 3750 cm^{-1} are represented as dotted lines in the sections (a) and (b) respectively, in the case of CaA_2/cyclohexane/water, CaA_2/heptane/water and ZnA_2/cyclohexane/water reversed micelles. In the same figures sections (c), (d), and (e) ((c) and (d) in the case of ZnA_2) show the experimental IR normalized absorbance (points) and the fitting results with Eq. (1) (solid lines). As can be seen, in spite of its simplicity, the model is able to fulfill the spectral features. The continuous line "mimics" the experimental spectrum at each R value very well. An interesting result that can be extracted from our

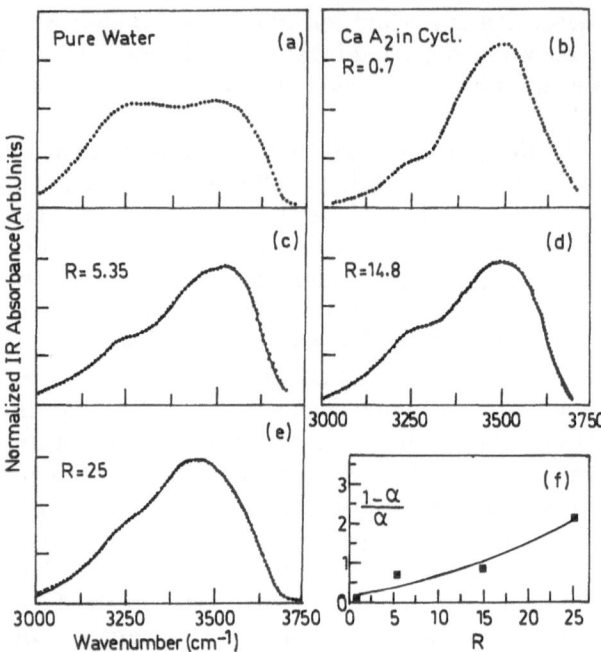

Fig. 1. Normalized O—H stretching IR absorbance spectra of water in the CaA_2/cyclohexane/water system as a function of $R = $ [water]/[CaA_2] molar ratio. In sections (a) and (b) the dotted curves refer to the experimental spectra of pure "bulk" water and "bonded" water, respectively. In sections (c), (d) and (e) the dots are the experimental data and the continuous lines represent the best fit with Eq. (1) (see text). In the section (f) the ratio between "bulk" and "bonded" water IR contributions vs. R is plotted

results in the IR data analysis is that the water bonded to the interface has the larger content of nontetrahedrically bonded water [17] (whose frequency center is at ~3300 cm^{-1}), thus, confirming that the polarization effects of the sulphonate polar heads of the micelle surface and the positive counterions partially destroy the structure of the so-called open-water (i.e., the water developing a full H-bond in a regular tetrahedral connectivity [18]). The water molecules "bonded" to the surfactant develop a more "compact" structure in which hydration effects play the main role. Another result that can be inferred is that both the IR and Raman responses [10, 11] are sensitive only to the first hydration shell around the ions. As shown in [19], the existence of an isosbestic point [18, 19] (as a direct consequence of the dynamical model used in this paper) centered at ~3375 cm^{-1} of the IR spectra in the CaA_2 system is a strong argument for the existence of a two-state equilibrium in our reversed micelles.

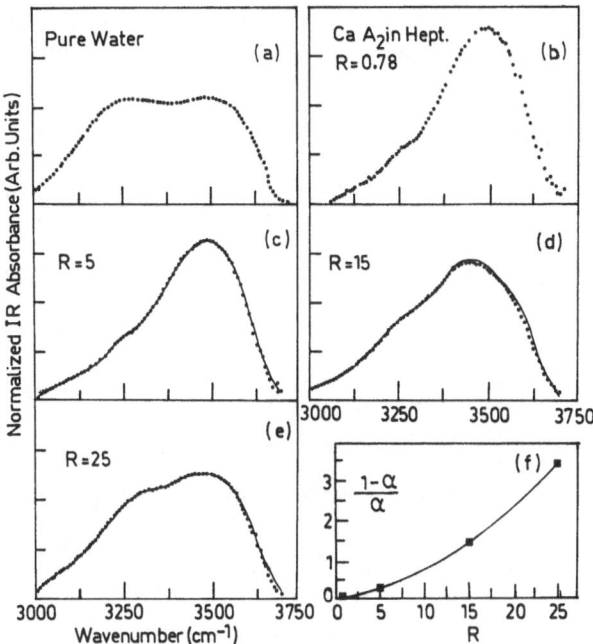

Fig. 2. Normalized O—H stretching IR absorbance spectra in CaA$_2$/heptane/water reversed micelles. Meaning of symbols as in Fig. 1

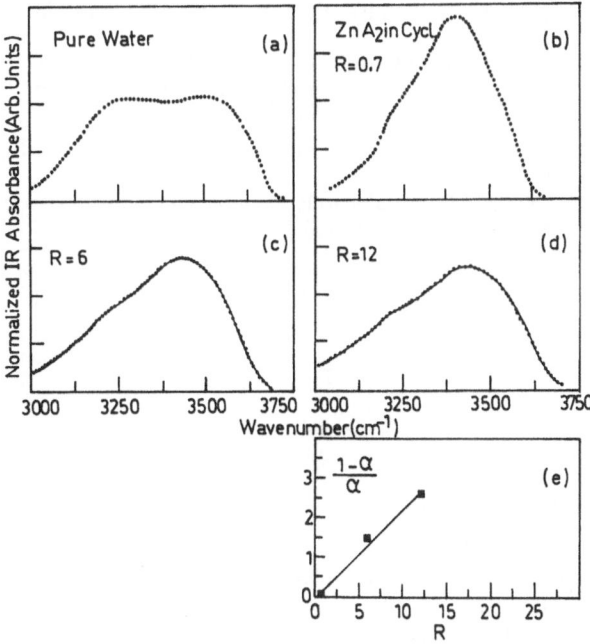

Fig. 3. Normalized O—H stretching IR absorbance spectra in ZnA$_2$/cyclohexane/water reversed micelles. Meaning of symbols as in Fig. 1

In order to quantify the concentration dependence of the two different water structures, the ratio between the number of absorbing H$_2$O molecules associated to the "bulk" water $(1 - a)$ and the number of absorbing H$_2$O molecules "bonded" to the surfactant polar groups (a) is reported as a function of R for the three investigated reversed micelles as section (f) (Fig. 1), (f) (Fig. 2), and (e) (Fig. 3).
$\frac{1-a}{a}$ vs. R represents the "dynamically" seen water-pool growth process, starting from the minimum value of the reversed micelle radius until some tens of angstroms that could correspond to $R \simeq 25$. The positive slope of this ratio indicates that, as R increases, the relative number of the "bulk" water molecules increases with a non-linear rate in the case of CaA$_2$ in cyclohexane and heptane, and almost linearly in the case of ZnA$_2$ in cyclohexane. The different behavior can be explained on the basis of a stronger acidity in the case of ZnA$_2$ reversed micelles [13], with respect to the one measured on CaA$_2$. On the other hand, from a chemical-physics point of view, in the zinc-based aqueous solutions the ion-ion interaction and, hence, the ion-pair population, tends to play a relevant role [20], consequently reducing the bonded water percentage in the growth process. Unfortunately, the lack of knowledge about the local structure and of the correct phase diagram in such a system does not allow any hypothesis about the growth process.

Finally, by turning out attention to the shape of the "bonded" water O—H spectrum (we will not discuss the spectral features of the "bulk" water, which have accurately been discussed in the past [17]), we find that at least two large contributions are clearly evident from an inspection of Figs. 1, 2, and 3, section (b). The two large bands are centered at about 3200 cm^{-1} and 3500 cm^{-1} of the wavenumber value. The physical origin of these absorption bands can be understood if one recalls that "bonded" water O—H vibrational features could be similar to those observed in highly concentrated ionic solutions. Recent molecular dynamic investigations [15, 16] have shown that the total-symmetric and antisymmetric O—H stretching vibrations of water hydrated by Ca^{2+} give rise to a band centered at wavenumber ~3200 cm^{-1}, whereas the symmetric and antisymmetric O—H bands in water locally coordinated by anions (Cl$^-$ in the case reported in [15, 16]) originate a large band centered at ~3500 cm^{-1}. The position of the peak maxima

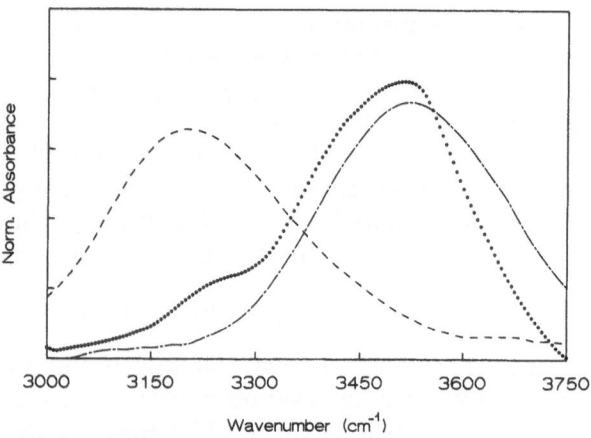

Fig. 4. Comparison among the experimental normalized IR absorbance in the hydroxyl region of "bonded" water spectra (dotted line) in the CaA₂/cyclohexane/water micelle at $R = 0.7$ and MD "simulated" spectral densities of hydration water of Ca^{2+} (dashed line) and of Cl^- (long-dash short-dash line) as reported in [15, 16]

of the "simulated" water is nearly coincident with that obtained by our studies, in spite of the difference between the cationic species. In Fig. 4, we compare the normalized IR absorbance of the bonded water, in the case of CaA₂ in cyclohexane (dotted line), with the normalized power spectra of Ca^{2+} hydrated water contribution (dashed line) and anionic (Cl^-) hydrated water contribution (long-dash/short-dash line). The comparison shows that the "dynamically seen" bonded water can be viewed as a distribution of water molecules trapped between the anionic polar heads and cationic counterions. Further investigation is in progress, in order to clarify the role of the different counterions in the growth processes of water-in-oil reversed micelles, by also performing a more quantitative spectroscopic comparison between the charged surface hindered water and the water present in "bulk" isotropic electrolyte solutions.

References

1. Angell CA (1991) In: Dore JC, Teixeira J (eds.) Hydrogen bonded liquids. Khever Acad Press, Netherlands, pp 73—75
2. Hanser H, Harring G, Pande A, Luisi PL (1989) J Phys Chem 93:7869—7876 (and references therein)
3. Dore JC (1986) In: Neilson GW and Enderby JE (eds) Water and aqueous solutions, A. Hilger Pub, Bristol, pp 89—98
4. Clifford J (1975) In: Franks F (ed) Water: a Comprehensive Treatise. Plenum Press, NY, Vol V, pp 75—132
5. Phillips MC (1975) In: Franks F (ed) Water: a Comprehensive Treatise. Plenum Press, NY, Vol V, pp 133—172
6. Luisi PL, Magid LJ (1986) CRC Crit Rev Biochem 20:409—429
7. Llor A, Rigny P (1986) J Am Chem Soc 108:7530—7541
8. Biocelli CA, Giomini M, Giuliani AM (1984) Appl Spectr 38:537—539
9. Jain TK, Varshney M, Maitra A (1989) J Phys Chem 93:7409—7416
10. Mallamace F, Migliardo P, Vasi C, Wanderlingh F (1981) Phys Chem Liq 11:47—58
11. D'Aprano A, Lizio A, Turco Liveri V, Aliotta F, Vasi C, Migliardo P (1988) J Phys Chem 92:4436—4439
12. Deniz U, Sumanan L, Goyal PS, Parvathanathan PS, Datta G (1992) Physica B (in press)
13. Bardez E, Larrey B, Zhu XX, Valeur B (1990) Chem Phys Lett 171:362—368
14. Zhu XX, Bardez E, Dallery L, Larrey B, Valeur B (1992) Langmuir (submitted)
15. Heinzinger K (1985) Physica B 131:196—216
16. Bopp P (1987) In: Bellisent-Funel MC, Neilson GW (eds) The Physics and Chemistry of aqueous ionic solutions. Reidel Pub Comp, London, pp 217—243 (and reference therein)
17. Walrafen GE (1972) In: Franks F (ed) Water: a Comprehensive Treatise. Plenum Press, New York, Vol I, pp 151—214
18. D'Arrigo G, Maisano G, mallamace F, Migliardo P, Wanderlingh F (1981) J Chem Phys 75:4264—4270
19. Verrall RE (1973) In: Franks F (ed) Water: a Comprehensive Treatise. Plenum Press, N.Y., Vol III, pp 211—264
20. Maisano G, Migliardo P, Fontana MP, Bellisent-Funel MC, Dianoux AJ (1985) J Phys C 18:115—1133

Authors' address:

Prof. P. Migliardo
Dipartimento di Fisica Universita' die Messina
P.O. Box 55
98166 Vill. S. Agata
Messina, Italy

Progress in Colloid & Polymer Science   Progr Colloid Polym Sci 89:263—267 (1992)

Similarities of aqueous and nonaqueous microemulsions

K.-V. Schubert, R. Strey, and M. Kahlweit

Max-Planck-Institut für biophysikalische Chemie, Göttingen, FRG

Abstract: The phase behavior of ternary mixtures of water, hydrocarbons, and nonionic amphiphiles is, by now, well known. In this paper the effect of replacing water by formamide is studied. This polar protic solvent reduces the hydrophobic effect and, thereby, the repulsive interactions with surfactant tails. The result is a higher mutual solubility of formamide and surfactant, a weaker adsorption at the internal interface and, hence, a lower solubilization capacity of the amphiphile. The reduced amphilphilic strength of surfactants in formamide corresponds to $C_{i-4}E_{j-1}$ in aqueous systems. As with aqueous systems the present mixtures form microstructures, namely oil-droplets in formamide and formamide-droplets in oil. For equal volumes of formamide and oil, small-angle neutron scattering (SANS) spectra indicate the existence of bicontinuous structures.

Key words: Nonaqueous microemulsion; phase behavior; microstructure; n-alkyl polyglycol ether; small-angle neutron scattering

Introduction

Microemulsions are isotropic, thermodynamically stable colloidal dispersions of polar and nonpolar solvents stabilized by amphiphiles. They usually contain salt, water, hydrocarbons, alcohols, and surfactants. The essential features of the phase behavior of such systems are already observed in more easily accessible ternary systems composed of water (A), alkanes (B), and n-alkyl polyglycol ethers (C) (denoted as C_iE_j) [1—5]. That aqueous microemulsions form microstructures has been confirmed by small-angle scattering, NMR-selfdiffusion measurements, and electron microscopy [6—11]. The properties of aqueous solutions of amphiphiles were interpreted by Tanford [12] to represent the consequence of competing forces, namely, the repulsive hydrophobic interaction between the hydrocarbon tails of the amphiphiles and water, and the attractive hydrophilic interaction between their headgroups and water. This raises a question about interactions between amphiphiles and solvents other than water or alkanes. The modification of the nonpolar or the polar solvent in microemulsion systems has been discussed [13—24]. Here, we study in detail the effect of form-

amide (denoted as FA) on aqueous microemulsions containing C_iE_j, as well as the phase behavior and microstructure of nonaqueous microemulsions [25—28].

Results

Effect of formamide on aqueous surfactant systems

We started from the binary system water-$C_{10}E_4$ [29, 26, 27] and replaced water sucessively by mixtures of water/formamide (the weight fraction of formamide in the water/formamide mixture is denoted as ψ, in wt%) and determined the upper miscibility gap with varying formamide content (Fig. 1, top). As noted earlier, water and formamide form a true pseudocomponent [25].

The lowest curve in the top panel represents the upper miscibility gap of $C_{10}E_4$ in water ($\psi = 0$). With increasing formamide content (ψ) the loop shrinks and is shifted to higher temperatures. For the nonaqueous system no miscibility gap was detected. In aqueous systems similar shifts are observed if one reduces the chain length of the amphiphile. The amphiphile becomes effectively less

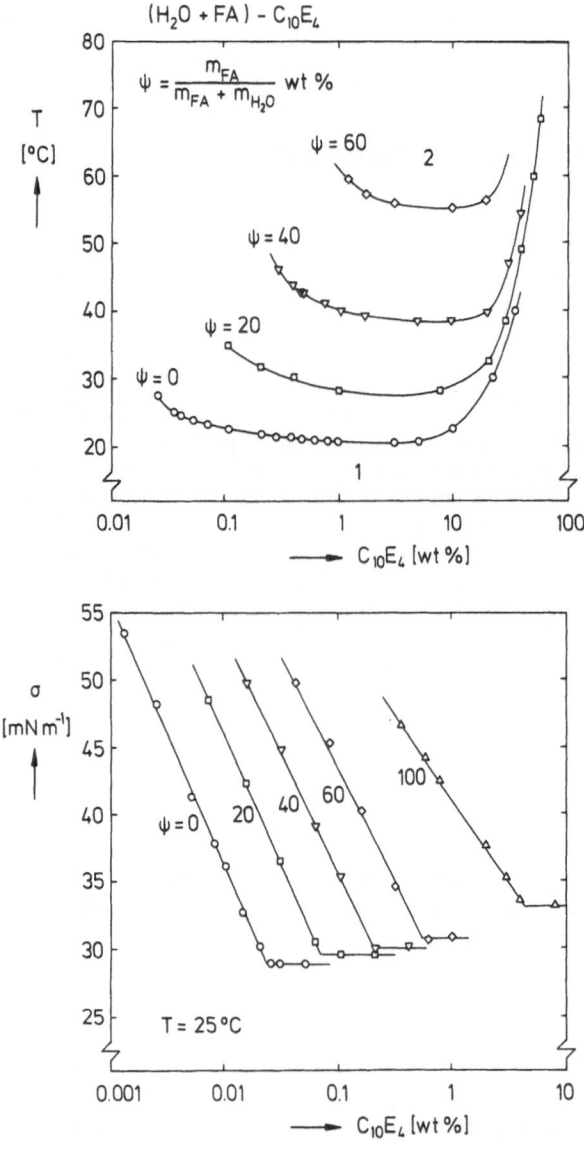

decreases with increasing ψ, indicating a weaker tendency of the surfactant molecules to adsorb at the liquid/air interface.

Nonaqueous microemulsions

In Fig. 2, vertical sections through phase prisms for a water — n-octane — $C_{12}E_4$ (bottom) and a formamide — n-octane — $C_{12}E_4$ (top) system are shown. For the formamide system the extension of the three-phase body far exceeds that seen in the aqueous counterpart. The minimal surfactant concentration necessary to solubilize equal amounts of polar solvent and oil is 3 wt% for the aqueous system, while we find 19 wt% for the formamide system. Thus, the surfactant is much less efficient in formamide than in water.

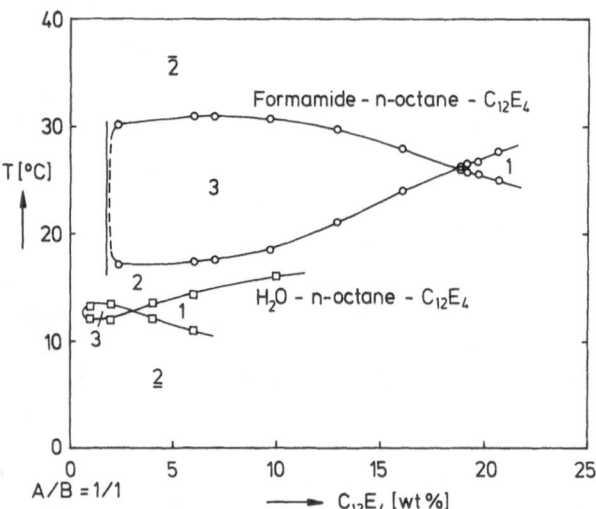

Fig. 1. (Top) Upper miscibility gap of the pseudobinary system H_2O/FA — $C_{10}E_4$ with increasing formamide content (ψ in wt%). (Bottom) Surface tension (σ) vs log of surfactant concentration for the H_2O/FA — $C_{10}E_4$ system with increasing formamide content (ψ)

Fig. 2. Vertical sections through phase prisms at equal volume fractions of polar and nonpolar components (A/B = 1/1). (For discussion see text.)

hydrophobic. The same trend is observed on the critical micelle concentration (CMC). The bottom panel of Fig. 1 shows the dependence of surface tension (σ) on surfactant concentration at $T = 25\,°C$ for different values of ψ. As usual, the breakpoint of each curve is taken as the CMC: this is seen to increase with increasing formamide content. A well-defined CMC is observed even in the pure nonaqueous system. The slope of the curve

In Fig. 2, we note another interesting property of the formamide system: the low-concentration end of the three-phase body in the nonaqueous system is located at higher surfactant concentrations that it is in the aqueous counterpart. This fact is related to the monomeric solubility of the surfactant and is an indication that a substantial amount of surfactant is dissolved in the excess phases. It is a special feature of the formamide systems reflecting the fact that both the CMC's in formamide and octane are of the same order of magnitude.

This is borne out by the data of Fig. 3, showing the corners of the Gibbs triangular phase diagram. This is divided into two parts: on the lefthand side, we show the formamide-rich region, measured at a temperature below the three-phase body; on the righthand side, we display the oil-rich region at a temperature above the three-phase body. Remarkably, each binodal forms a straight line. On the left, the intersection at the A—C-axis (oil-free mixture) is called CMC^a. The dotted line represents the CMC-tie line [5]. On the right, the same features are apparent for the oil-rich system; again the binodals form a straight line. The intersection point with the B—C-axis (formamide-free mixture) is called CMC^b. For aqueous systems, CMC^a is usually substantially lower than CMC^b [5].

The constancy of the slope of the binodals indicates a constant ratio of the surfactant-to-interior of the microemulsion droplets, formamide on the oil-rich side and oil on the formamide-rich side (see test tubes in Fig. 3). The fact that the microemulsions indeed form droplets is confirmed by SANS (Fig. 4).

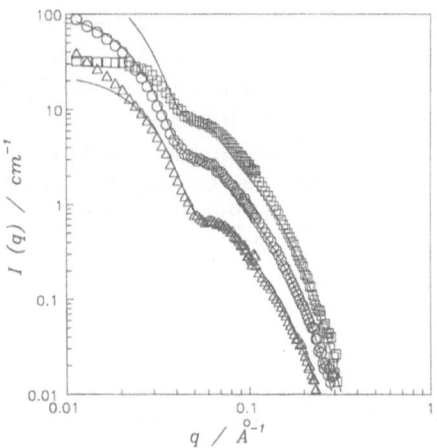

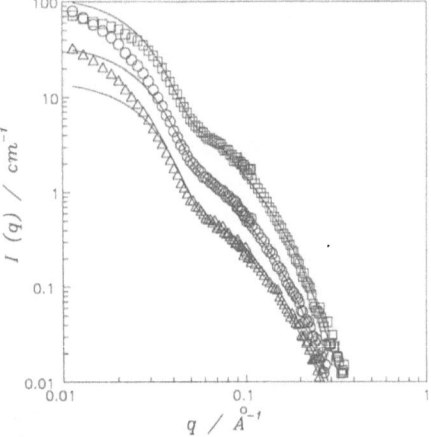

Fig. 4. SANS spectra on a log-log scale for film contrast samples with different volume fractions, for a formamide-rich microemulsion (top), and for an oil-rich microemulsion (bottom). The spectrum with the highest intensity (□) represents the sample with the highest volume fraction of surfactant (indicated as □ in Fig. 3). The spectra are plotted on an absolute scale. The full lines are fits for non-interacting polydisperse spherical shells (full details will be given elsewhere [32])

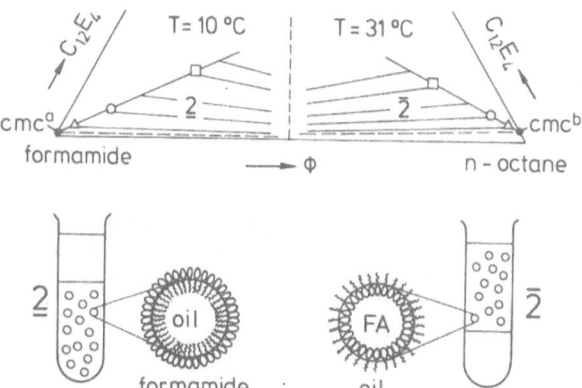

Fig. 3. Corners of a Gibbs triangular diagram for the system FA — n-octane — $C_{12}E_4$. The test tubes show schematically which type of droplets is formed

The film contrast scattering spectra show a characteristic dip which occurs for every curve (concentration) at the same position, indicating a constant radius of the shells [30] with increasing surfactant concentration. Oil and formamide droplets have a radius of 52 and 62 Å, respectively. The maximum of the contrast lies close to the polar-non-

polar interface, decorated by the surfactant molecules. The scattering spectra are well fitted by a model of polydisperse shells; deviations in the low-q-regime are interpreted to indicate attractive droplet interactions (for low droplet volume fractions) or repulsive droplet interactions (for high droplet volume fractions). Since the measurements had been extended to sufficiently high q, a background could be precisely determined. After substraction of the background the spectra show a steeper than q^{-2} decay in the high q range ($q > 0.2$ Å^{-1}); this is attributed to the diffuse nature of the interface [31]. The convolution of a model for polydisperse shells with a function representing a diffuse interface yields the fits in Fig. 4. A detailed derivation will be published elsewhere [32].

The phase behavior of formamide and aqueous systems may be matched by varying the chain length of surfactant (Fig. 5). Based on the extensions of the three-phase bodies, we conclude that these two systems exhibit comparable behavior.

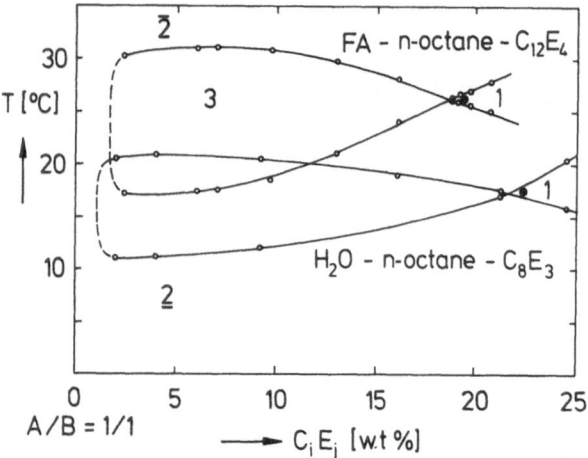

Fig. 5. Vertical sections through phase prims at equal volume fractions of polar and nonpolar components (A/B = 1/1). Note the same extensions of the respective three-phase bodies. Samples (full points) near the point where the three-phase region and one-phase region meet are studied by SANS (see Fig. 6)

In the homogeneous region the aqueous system forms a bicontinuous microstructure typical for symmetric microemulsions [9]. The scattering behavior of a symmetric aqueous microemulsion in bulk contrast and of the corresponding nonaqueous microemulsion is shown in Fig. 6; both the aqueous and the formamide microemulsion scatter similar.

At large q, both scattering spectra exhibit a q^{-4} decay (Porod limit), due to well-defined internal interfaces. The flattening of the curve at very large q is caused by the incoherent background. The scattering peak indicates a characteristic length, the quasi-periodic octane-octane repeat distance. For the nonaqueous system the peak position occurs at somewhat lower q due to the lower volume fraction of the surfactant, as compared to the aqueous

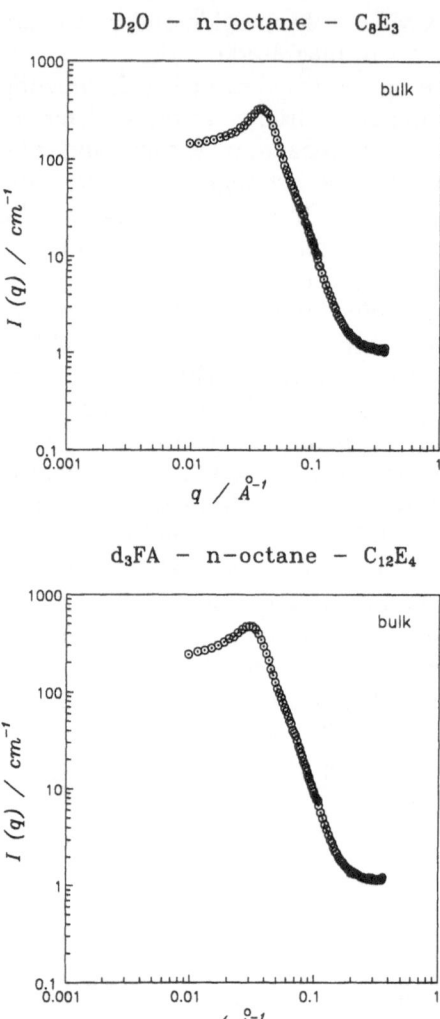

Fig. 6. Comparison of SANS spectra for bulk contrast samples, on a log-log scale of bicontinuous microemulsions; the corresponding compositions are indicated by full points in Fig. 5

system (compare the full points in Fig. 5). As for aqueous systems, the proof of the bicontinuity will have to rely on NMR-selfdiffusion [7] and electron microscopy [11]. More details will be published in a forthcoming paper [28].

The formation of microstructure seems to be a generic property of the ternary system and is not simply a consequence of the binary systems [1,4], because $H_2O—C_8E_3$ shows an upper miscibility gap, whereas the formamide—$C_{12}E_4$ system does not.

Conclusion

The same qualitative features in phase behavior of formamide systems as for aqueous systems are observed. Since formamide reduces the hydrophobic effect, the surfactants are less efficient in formamide. Roughly, C_iE_j in formamide corresponds to $C_{i-4}E_{j-1}$ in water. As for aqueous systems, microstructure also exists in formamide systems containing surfactants with correspondingly longer chain lengths.

References

1. Kahlweit M, Strey R (1985) Angew Chem Int Ed Engl 24:654
2. Kahlweit M, Strey R, Firman P (1986) J Phys Chem 90:671
3. Kahlweit M, Strey R, Firman P, Haase D, Jen J, Schomäcker R (1988) Langmuir 4:499
4. Kahlweit M, Strey R, Haase D, Firman P (1988) Langmuir 4:785
5. Kahlweit M, Strey R, Busse G (1990) J Phys Chem 94:3881
6. Lichterfeld F, Schmeling T, Strey R (1986) J Phys Chem 90:5762
7. Olsson U, Shinoda K, Lindman B (1986) J Phys Chem 90:4083
8. Carnali JO, Ceglie A, Lindman B, Shinoda K (1986) Langmuir 2:417
9. Kahlweit M, Strey R, Haase D, Kunieda H, Schmeling T, Faulhaber B, Borkovec M, Eicke H-F, Busse G, Eggers F, Funck Th, Richmann H, Magid L, Söderman O, Stilbs P, Winkler J, Dittrich A, Jahn W (1987) J Colloid Interface Sci 118:436
10. Chen S-H, Chang S-L, Strey R (1990) J Chem Phys 93:1907
11. Jahn W, Strey R (1988) J Phys Chem 92:2294
12. Tanford C (1980) In: The Hydrophobic Effect, 2. Ed, J. Wiley, New York
13. Wormuth KR, Kaler EW (1989) J Phys Chem 93:4855
14. Mc Donald C (1970) J Pharm Pharmac 22:774
15. Couper A, Gladden GP, Ingram BT (1975) Farad Disc 59:63
16. Ramadan MS, Evans DF, Lumry R (1983) J Phys Chem 87:4538
17. Rico I, Lattes A (1986) J Phys Chem 90:5870
18. Auvray X, Petipas C, Anthore R, Rico I, Lattes A, Ahmah-Zadeh Sammii A, de Savignac A (1987) Colloid Polym Sci 265:925
19. Belmajdoub A, Marchal JP, Canet D, Rico I, Lattes A (1987) New J Chem 11:415
20. Wärnheim T, Jönsson A (1988) J Colloid Interface Sci 125:627
21. Wärnheim T, Bokström J, Williams Y (1988) Colloid Polym Sci 266:562
22. Wärnheim T, Sjöberg M (1989) J Colloid Interface Sci 131:402
23. Martino A, Kaler EW (1990) J Phys Chem 94:1627
24. Jonströmer M, Sjöberg M, Wärnheim T (1990) J Phys Chem 94:7549
25. Schubert K-V, Strey R (1991) J Chem Phys 95:8532
26. Schubert K-V, Kahlweit M, Strey R In: Lindman B, Friberg SE (eds) Organized Solutions. Marcel Dekker, New York (in press)
27. Schubert K-V, Kahlweit M, Strey R In: Chen S-H, Huang JS, Tartaglia P (eds) Structure and Dynamics of Supramolecular Aggregates and Strongly Interacting Colloids. NATO ASI Series, Kluwer Academic Press, Dordrecht, Netherlands (in press)
28. Schubert K-V, Kahlweit M, Strey R (in preparation)
29. Lang JC, Morgan RD (1980) J Chem Phys 73:5849
30. Glatter O (1982) In: Glatter O, Kratky O (eds) Small Angle X-ray Scattering. Academic Press, New York
31. Strey R, Winkler J, Magid L (1991) J Phys Chem 95:7502
32. Strey R, Magid L (in preparation)

Author's address:

K.-V. Schubert (Abt. 040)
Max-Planck-Institut für Biophysikalische Chemie
Postfach 2841
3400 Göttingen, FRG

Progress in Colloid & Polymer Science

Progr Colloid Polym Sci 89:268—270 (1992)

Influence of stigmastanyl phosphorylcholine on the size, mass, and shape of taurocholate/lecithin/cholesterol mixed micelles

J. P. Caniparoli[1]), N. Gains[1]), and M. Zulauf[2])

Pharma Division, [1]) Preclinical Research and
[2]) New Technologies, F. Hoffmann-La Roche Ltd, Basel, Switzerland

Abstract: Stigmastanyl phosphorylcholine, an inhibitor of cholesterol absorption was shown by light scattering to rearrange the size, mass, and shape of mixed lecithin/taurocholate/cholesterol micelles.

Key words: Stigmastanyl phosphorylcholine; taurocholate/lecithin mixed micelles; light scattering; intermicellar concentration; cholesterol absorption

Introduction

Stigmastanyl phosphorylcholine (S-PC) has been previously shown [1] to decrease the absorption of cholesterol in the intestine, lower plasma cholesterol, and decrease atheroma formation [2] (Fig. 1). In order to determine its mode of action, we are investigating its in vitro interaction with micelles. Here, we report on the structural changes that S-PC induces in model biliary micelles composed of taurocholate, lecithin, and cholesterol.

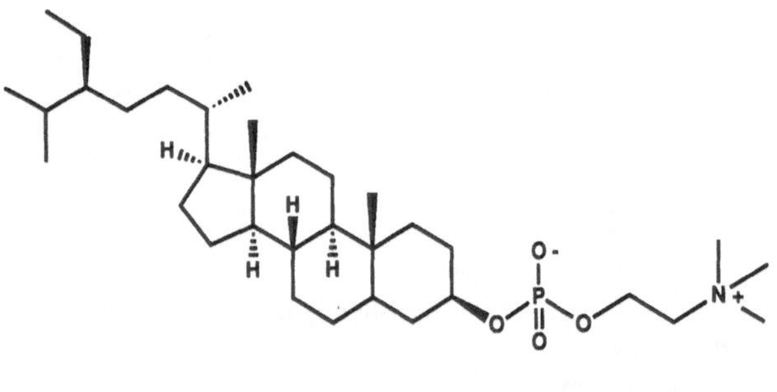

Fig. 1. Chemical structure of S-PC

Experimental

S-PC was dispersed in performed micelles that had been made by coprecipitation. The stock micellar solution contained 6.4 mM taurocholate (TC), 2 mM lecithin, and 0 or 200 μM cholesterol in 150 mM NaCl, 2 mM BES, plus NaOH to pH 7.2. These S-PC dispersions were monophasic, micellar solutions. At higher cholesterol (~ 400 μM) or S-PC ($\geqslant 10$ mM) concentrations, polyphasic dispersions were obtained. These are not considered in this report.

The intermicellar concentration of TC which is the concentration of monomeric TC was measured by centrifuging (500 g) the solutions through an Amicon polycarbonate membrane (MW cut-off 30 kD) to separate the TC molecules present as monomers from the micelles. The hydrodynamic radius and mean aggregate mass of the micelles were determined from dynamic and static light scattering using standard procedures [3].

Results

The intermicellar concentration of TC was found to decrease with increasing S-PC concentration (data not shown). From this, we calculated the mole fraction of S-PC in the micelles (Xm) which is needed for the determination of the micelle mass.

The measn hydrodynamic radius of the micelles containing 0 and 200 μM cholesterol initially decreases, then remains constant and finally increases as the mol fraction of S-PC is increased (Fig. 2). Cholesterol has little effect on it.

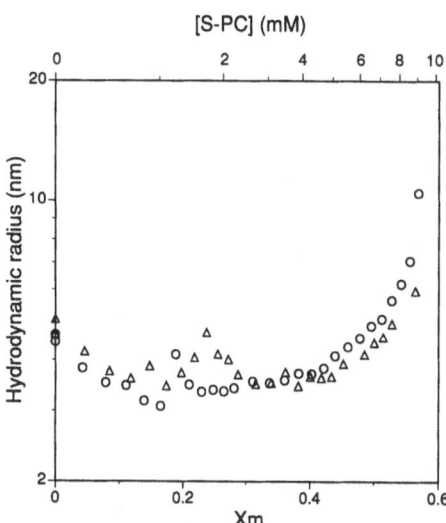

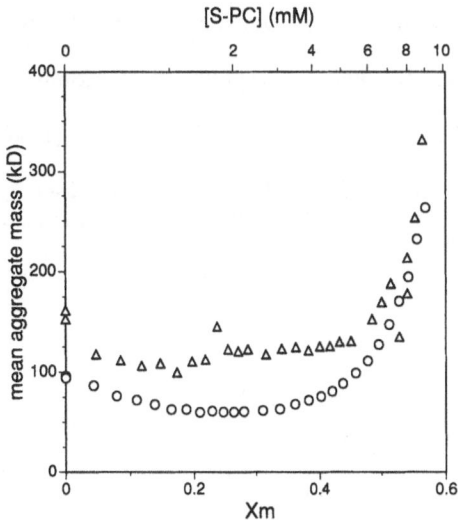

Fig. 2. Hydrodynamic radii of TC/lecithin/cholesterol dispersions with 0 (o) or 200 μM (Δ) cholesterol as a function of the mole fraction (Xm) of S-PC at 20°C. Xm is calculated from the measured intermicellar concentration of TC. The hydrodynamic radii are extracted from the data using the program CONTIN [4]

Fig. 3. Mean mass of aggregates in TC/lecithin/cholesterol dispersions, with 0 (o) or 200 μM (Δ) cholesterol as a function of the mol fraction (Xm) of S-PC. The masses are derived from the observed scattered intensities (Rayleigh ratio). The refractive index increments are 0.153 ml/g, 0.146 ml/g, and 0.167 ml/g for TC, lecithin and S-PC respectively. The contribution of cholesterol is neglected

The mean aggregate mass is shown in Fig. 3. As the S-PC mol fraction is increased to 0.5, the hydrodynamic radius and mean aggregate mass remain constant for the micelles containing 0 and 200 μM cholesterol. This is only feasible if the number of particles increases.

The particle shape can be inferred by fitting the data by models which assume the micelles are discs with a constant thickness, or rods of a constant diameter [5]. The radius of the rods derived using a free fit to the data is 1.8 nm, which is approximately the length of egg-lecithin and S-PC molecules (Fig. 4). In contrast, the growth of disk-like micelles with a thickness of 3.6 nm is expected to follow the broken curve. Rod-like growth is also supported by the observation that a higher S-PC concentration (Xm = 0.8) these micelles form a transparent gel. Rod-shaped aggregates have also been observed using neutron scattering in glycocholate/lecithin systems [6].

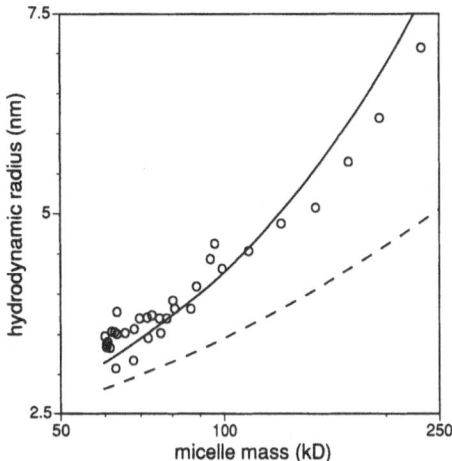

Fig. 4. Hydrodynamic radius as a function of the mass of TC/lecithin micelles containing no cholesterol, and S-PC at mol fractions (Xm) between 0 and 0.6. The solid line is the result of a fit to the experimental data, assuming rod-like growth; this gives a radius of 1.8 nm. The dashed line corresponds to disc-like groth with a disc thickness of 3.6 nm.

Conclusion

Addition of S-PC to TC/lecithin mixed micelles leads to a rearrangement of the micelle structure. Similar changes might be induced by the presence of S-PC in the intestine. This could result in a decrease in cholesterol uptake by the enterocytes and, thereby, explain the observed in vivo inhibition of cholesterol absorption.

References

1. Cassal JM (1987) US patent 4680290
2. Cassal JM, Gains N, Kuhn H, Lengsfeld H, Zulauf M (1988) in "Poster Sessions Abstract Book" of 8th International Symposium on Atherosclerosis, Rome, p 130

3. Berne BJ, Pecora R (1976) "Dynamic Light Scattering", Wiley
4. Provencher S (1982) Computat Phys Commun 27:229—242
5. Mazer NA, Benedek GB, Carey MC (1980) Biochemistry 19:601—605
6. Hjelm RP, Thiyagaragan P, Sivia DS, Lindner P, Alkan H, Schwahn D (1990) Progr Colloid Polym Sci 81:225—231

Author's address:

Dr. J. P. Caniparoli
PRPM 68/319
F. Hoffmann La Roche AG
CH-4002 Basel, Switzerland

Progress in Colloid & Polymer Science   Progr Colloid Polym Sci 89:271—273 (1992)

Polymorphism of phosphatidylcholines varied in the hydrophobic part

G. Förster, G. Brezesinski, and S. Wolgast

Martin-Luther-Universität Halle-Wittenberg, Institut für Physikalische Chemie, Halle, FRG

Abstract: Variations of the hydrophobic part of phosphatidylcholines (PCs) such as chain shortening, chain branching, and chain linkage lead to different homologous series in which, surprisingly, a uniform behavior of the phase transition parameters is observed, namely, the parameters pass through a minimum. The first member of the monobranched 1-ether-2-ester-lecithins with a methyl branching in the sn-2-chain was investigated in mixtures with excess water by DSC and x-ray diffraction. It shows a polymorphism, which more resembles that of the 1,2-di-ester-PC than that of an unbranched 1-ether-2-ester-PC. In the subgel phase a bilayer structure with tilted, orthorhombic packed chains were detected which are oppositely arranged. On heating, a gel phase with tilted, hexagonal packed chains appears which becomes fluid above the main transition at 313.8 K. The idea of frustration is applied to gel phases to understand the drastic structural variations in PCs, in which only small chemical modifications in the hydrophobic part were created.

Key words: Branched lecithins; polymorphism; chain linkage; frustration

Introduction

In recent years much information relevant to the lipid matrix of cell membranes has been derived from model studies of synthetic lipid dispersions. Here, we focus on the structure and properties of phosphatidylcholines (PCs), in which chain length, chain branching, and chain linkage have been systematically varied. These variations lead to different homolgous series in which, surprisingly, a uniform behavior of the phase transition parameters is observed, namely, the parameters pass through a minimum [1—3]. For PCs varying in the chain length (CL) the change in the main-transition temperature T_m is described quantitatively by the empirical Huang parameter $\Delta C/CL$, i.e., the normalized chain length difference ΔC between the sn-1 and sn-2 chains [4]. Recent structure investigations have shown that with changing the molecule shape from a double- to a single-chain PC, different interdigitated gel phase structures appear [1, 5] (partially, mixed, and fully interdigitated bilayers [4]). From the minimum in T_m it can be concluded

that, in dependence on the asymmetry of the molecules, the packing density within a homologous series should go through a minimum, too. Furthermore, because the two chains are not equivalent, both the thermodynamic parameters and the structural behavior differ in the homologous series, in which the sn-1- and sn-2-chain is shortened [1, 3].

The introduction and elongation of branches in one or both chains of the PCs result in triple- or quadruple-chain lecithins. In the new homologous series of branched PCs a thermodynamic and structural behavior was observed [2, 3] which is comparable to that of asymmetric lecithins.

A third variation of the hydrophobic part, alterations of the chain linkage, especially influences the gel-phase structures. This molecule shape modification is small compared to the chain shortening or branching, and is presently under investigation [6—9]. It is necessary to note that the mixed ester- and ether-linked lecithins are the first members in the homologous series of monobranched PCs, which have a branched acyl chain and an unbranched alkyl chain.

In the following, new experimental results are presented from a monobranched 1-ether-2-ester-PC/water mixture, and its polymorphism is compared to that of the unbranched 1,2-di-ester-PC and 1-ether-2-ester-PC, respectively.

Results

The lecithin 1-hexadecyl-2-(2-methyl-palmitoyl)-sn-glycerophosphocholine [abbreviation 1-H-2-($2C_1$-16:0)-PC] and the mentioned unbranched PCs were investigated in excess water (50 wt%) mixture. Experimental details for the standard methods DSC and x-ray diffraction are reported elsewhere [2]. In the DSC scan of the monobranched PC a very small and broad sub-transition peak around 295 K and the main-transition peak at 313.8 K were observed. In Table 1 the transition parameters are compared with those of the unbranched 1,2-di-ester-PC and 1-ether-2-ester-PC.

Table 1. Comparison of the temperature (K) and enthalpy (kJ/mole) values of the sub-, pre-, and main transitions of symmetric, mixed linkage, and monobranched lecithins

Lipid	T_{sub}	T_p	T_m	ΔH_{sub}	ΔH_p	ΔH_m
1,2-di(16:0)-PC	288.0	307.5	314.7	17.1	6.1	38.1
1-H-2-(16:0)-PC [9]	—	310	313.2	—	Σ 42.3	
1-H-2($2C_1$-16:0)-PC	295	—	313.8	1.5	—	43.4

On heating the 1-H-2-($2C_1$-16:0)-PC/water system, only lamellar phases were observed. In addition to a hysteresis, which is connected with the freezing of the free water and melting of ice, respectively [10], the repeat distances d_L show a linear temperature-dependence over all phases below T_m, and a remarkable thermal expansion coefficient (Fig. 1). Unusually high bilayer spacings were observed above T_m in the $L\alpha$ phase. To characterize the chain packing within the subgel phase the wide angle x-ray diffraction scans are shown in Fig. 2 together with those of the phase $L\beta'$ of 1,2-di(16:0)-PC (deformed hexagonal packing of tilted, oppositively arranged chains [11]) and the phase $L\rho_i$ of 1-($3C_1$-16:0)-2-H-PC (orthorhombic packing of interdigitated chains [10]). It is obvious from the peak positions and the intensity ratio of both peaks

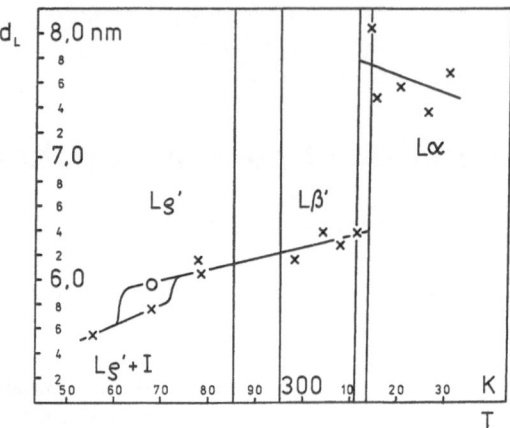

Fig. 1. Temperature dependence of the long spacings in a 1-H-2-($2C_1$-16:0)-PC/50 wt% water mixture. (Because of a weight averaging of all observed orders, an error of ±1 Å was estimated in gel phases.) ○ = sample with supercooled water

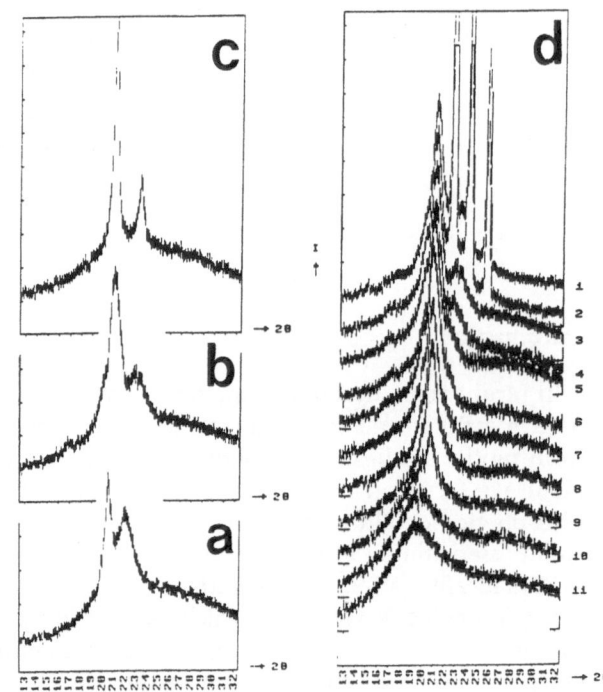

Fig. 2. Normalized x-ray intensity as a function of scattering angle 2Θ of:
a) 1,2-di(16:0)-PC/water (277.3 K, phase $L\beta'$, d_1 = 4.35 Å, d_2 = 4.04 Å, Σ = 19.84 Å²); b) 1-H-2-($2C_1$-16:0)-PC/water (268.2 K, phase $L\rho'$, d_1 = 4.23 Å, d_2 = 3.87 Å, Σ = 18.41 Å²); c) 1-($3C_1$-16:0)-2-H-PC/water (266.1 K, phase $L\rho_i$, d_1 = 4.21 Å, d_2 = 3.82 Å, Σ = 18.05 Å²); d) Temperature-dependence of the wide-angle x-ray diffraction scans in 1-H-2-($2C_1$-16:0)-PC/50 wt% water mixtures: 1) 255.6 K with ice; 2) 268.2 K with ice; 3) 268.2 K supercooled; 4) 278.2 K; 5) 298.2 K; 6) 304.4 K; 7) 308.0 K; 8) 311.4 K; 9) 313.6 K; 10) 315.2 K; 11) 330.4 K

that the character of packing tends to an orthorhombic subcell. From the pronounced half-width of the peaks it can been assumed that the chains are tilted. This subgel phase $L\rho'$ transforms first into a gel phase $L\beta'$ with a hexagonal packing of the tilted chains before melting (Fig. 2d). To date, we have found no evidence of a pre-transition in the system investigated.

Discussion

The introduction of one methyl branching in the sn-2 chain of a 1-ether-2-ester-PC gives rise to a polymorphism, which more resembles that of the 1,2-di-ester-PC than that of an unbranched 1-ether-2-ester-PC. The antiparallel packing of lecithin molecules, which appears after changing the chain linkage in the sn-1 chain of 1,2-di-ester-PC, is cancelled in the mixed-linkaged, monobranched PC. Again, a tilted packing of oppositely arranged molecules was detected, and a reduction of the hydration limit from a value of 23.6 mole water/mole lecithin (295 K [9]) to 19 is measured (17 as bonded water, 2 as trapped water, 273 K [10]). On the other hand, remarkable differences exist in the polymorphism of the PCs with oppositely arranged molecules. 1,2-di-ester-PC shows a pronounced stepwise melting process with nearly constant bilayer thicknesses in each phase, but clear changes at the phase transitions [13]. The monobranched 1-ether-2-ester-PC shows a "soft" melting process, characterized by a more continuous change of the structure parameter. The large lamellar spacings above the main transition temperature are the first experimental evidence of a symmetric lipid conformation, which was already theoretically predicted [15].

To understand the drastic structural changes in PCs, in which only small chemical modifications in the hydrophobic part were created, the idea of frustration [14] is applied to gel phases. Frustration arises at the interface between the hydrophilic and hydrophobic parts of the molecules and relates to the cross-sectional areas of the chains and the headgroup of the PC. In respect to this, local stress is created by the chemical variations in the hydrophobic parts, namely, chain shortening, chain branching, and chain linkage. Such modified molecules reach their minimum in free energy in mixtures with excess water in very different structures. Many physical parameters like hydration, tilting, interdigitation or molecule conformation must change in order to establish the thermodynamic equilibrium.

Acknowledgement

Thanks are due to the group of P. Nuhn (Pharmaceutical Faculty of the Halle University), especially to B. Dobner. This work was supported by the Deutsche Forschungsgemeinschaft (Mo 283/21-1) and the Fonds der Chemischen Industrie.

References

1. Shah J, Sripada PK, Shipley G (1990) Biochemistry 29:4254—4262
2. Nuhn P, Brezesinski G, Dobner B, Förster G, Gutheil M, Dörfler HD (1986) Chem Phys Lipids 39:221—236
3. Brezesinski G, Förster G, Rettig W, Kuschel F (1991) Ber Bunsenges Phys Chem 95:1507—1511
4. Huang C (1990) Klin Wochenschrift 68:149—165
5. Mattai J, Sripada PK, Shipley GG (1987) Biochemistry 26:3287—3297
6. Ruocco MJ, Siminovich DJ, Griffin RG (1985) Biochemistry 24:2406—2411 and 24:4844—4851
7. Kim JT, Mattai J, Shipley GG (1987) Biochemistry 26:6592—6598 and 26:6599—6603
8. Laggner P, Lohner K, Degovics G, Müller K, Schuster A (1987) Chem Phys Lipids 44:31—60 and 44:61—70
9. Haas NS, Sripada PK, Shipley GG (1990) Biophys J 57:117—124
10. Förster G, Brezesinski G (1989) Liq Crystals 5:1659—1668
11. Tardieu A, Luzzati V, Reman FC (1973) J Mol Biol 75:711—733
12. Ruocco MJ, Shipley GG (1982) Biochim Biophys Acta 684:59-66
13. Füldner HH (1981) Biochemistry 20:5707—5710
14. Sadoc JF, Charvolin J (1986) J Physique 47:683—691
15. Kreissler M, Bothorel P (1978) Chem Phys Lipids 22:261—277

Authors' address:

Doz. Dr. G. Brezesinski
Martin-Luther-Universität
Halle/Wittenberg
Institut für Physikalische Chemie
Mühlpforte 1
O-4020 Halle, FRG

Progress in Colloid & Polymer Science   Progr Colloid Polym Sci 89:274—277 (1992)

A sphere to flexible coil transition in lecithin reverse micellar solutions

P. Schurtenberger[a]), L. J. Magid[b]), P. Lindner[c]), and P. L. Luisi[a])

[a]) Institut für Polymere, ETH Zentrum, Zürich, Switzerland
[b]) Department of Chemistry, University of Tennessee, Knoxville, USA
[c]) Institut Max von Laue-Paul Langevin, Grenoble, France

Abstract: We report a dynamic (QLS) and static (SLS) light scattering and small-angle neutron scattering (SANS) investigation of lecithin reverse micellar solutions with novel polymer-like properties. We present evidence for a water-induced transition from small colloid-like to giant polymer-like cylindrical aggregates. Under suitable conditions, we can determine the overall dimension, the flexibility, and the local cylindrical structure of these reverse micelles using SLS and SANS. This permits us to directly confirm the existing structural model and verify the postulated analogy between the structural properties of polymer chains and lecithin reverse micelles.

Key words: Lecithin reverse micelles; flexible coils; equilibrium polymers; small-angle neutron scattering

Introduction

The formation of gel-like, viscoelastic, reverse micellar solutions in the system lecithin/organic solvent/water was recently discovered [1]. In particular, it was found that the viscosity of reverse micellar solutions of soybean lecithin in a number of different organic solvents increases dramatically upon the addition of very small quantities of water [2, 3]. This observation led us to propose that the addition of water to lecithin reverse micellar solutions induced one-dimensional micellar growth into long cylindrical reverse micelles [4]. A cross-over lecithin volume fraction Φ^* was predicted, above which these micelles were envisioned to entangle and to form a transient network similar to that found in semidilute polymer solutions. This structural model was subsequently tested for lecithin/isooctane solutions, and strong evidence was found for the presence of long and cylindrical reverse micellar aggregates and the formation of a transient network above Φ^* [2—5].

However, the hypothesis of a water-induced cylindrical growth for the lecithin molecules could only be supported indirectly by making analogies to classical polymer theory. Experimentally, it is difficult to quantitatively assess the water-induced micellar growth and to determine the concentration dependence of the micellar size distribution in isooctane. This is due to the very narrow range of molar ratios of water to lecithin w_0, where these phenomena can be observed. It is also not possible to work at concentrations well below Φ^* due to the existence of a phase boundary for liquid-liquid phase separation at high values of w_0 and low values of Φ. Consequently, single particle properties have so far not been determined in the original lecithin/isooctane system. We have now performed, and report here, scattering measurements on lecithin/cyclohexane solutions, for which much higher values of w_0 can be achieved even at low lecithin volume fraction without subsequent phase separation. A detailed account of the phase behavior of this system is given elsewhere [7].

Experimental

Soybean lecithin was obtained from Lucas Meyer (Epikuron 200) and used without further purification. Isooctane (2,2,4-trimethylpentane, spectroscopic grade) and cyclohexane (spectroscopic grade) were purchased from Fluka, and fully deuterated d_{12}-cyclohexane from Dr. Glaser AG (99.5% isotopic purity). Samples were prepared as described previously [3, 6].

SANS measurements were made using the D11 small-angle instrument at the Institute Laue-Langevin, Grenoble (details are given in [6]). Static (SLS) and dynamic light-scattering measurements were made with a Malvern 4700 PS/MW spectrometer. From the dynamic light-scattering experiments, a cooperative diffusion coefficient D_c was obtained by means of a cumulant analysis of the intensity autocorrelation function. From D_c an apparent hydrodynamic radius R_h was calculated [6]. Static light-scattering experiments were performed at 13 different angles ($30° \leqslant \theta \leqslant 150°$). The data was then corrected for background (cell and solvent) scattering and converted into absolute scattering intensities (i.e., "excess Rayleigh ratio" $\Delta R(\theta)$) [6, 8].

Results and discussion

For lecithin/cyclohexane solutions, much larger values of w_0 can be achieved without phase separation. In this system, the maximum of the viscosity is shifted to approximately $w_{0,\max} \approx 10$, which should be compared with the corresponding value for isooctane ($w_{0,\max} \approx 3.0$) [2, 3]. An example of the w_0-dependence of the zero shear viscosity for lecithin/cyclohexane solutions is given in Fig. 1A.

The addition of a small amount of water to lecithin/cyclohexane reverse micellar solutions induces strong micellar growth and results in the formation of very large micelles. We have performed static and dynamic light-scattering experiments at a surfactant volume fraction of $\Phi = 3.6 \cdot 10^{-3}$, i.e., well below Φ^* at all values of w_0 investigated [9]. The results from QLS and SLS are summarized in Fig. 1B, C.

At a low water content of $w_0 = 2.0$, the data are consistent with the formation of small and almost spherical reverse micelles with $R_h \approx 35$ Å and an apparent aggregation number of $N \approx 120$. At higher values of w_0, we observe a dramatic increase of the apparent hydrodynamic radius. For $w_0 \geqslant 4.0$, the particles are sufficiently large to produce a significant angular dependence of the scattering intensity $I(Q)$, where Q denotes the scattering vector. This enables us to derive additional information on the size and shape of the aggregates from the Q-dependence of the intensity. In the limit of small values of $Q \cdot R_g$, where R_g is the (z-average) radius of gyration of the individual particles, the scattered intensity can be expressed as [8]

$$\frac{1}{I(Q)} \approx \frac{1}{I(0)} (1 + Q^2 R_g^2/3) . \tag{1}$$

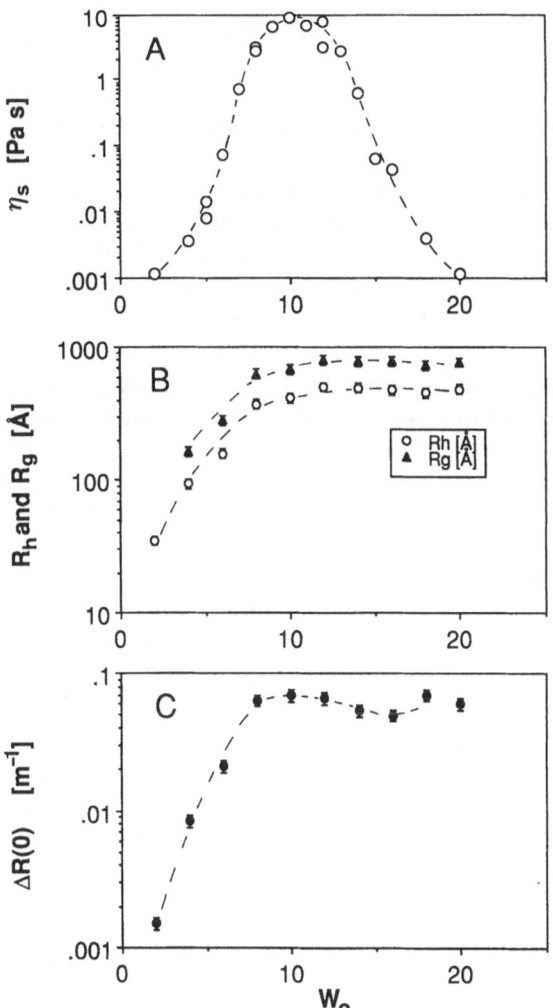

Fig. 1. A) Zero shear viscosity η_s versus added water to lecithin molar ratio w_0 for soybean lecithin in cyclohexane, $\Phi = 0.036$, at a temperature of 25.0 °C. (See [3] for details.) B) Dependence of the apparent hydrodynamic radius R_h (o) and the z-average radius of gyration R_g (▲) upon w_0 for soybean lecithin in cyclohexane at $\Phi = 0.0036$ and $T = 25.0$ °C. C) Dependence of the excess Rayleigh ratio $\Delta R(0)$ upon w_0 for soybean lecithin in cyclohexane at $\Phi = 0.0036$ and $T = 25.0$ °C

We can thus estimate the z-average radius of gyration and determine $I(0)$ (or $\Delta R(0)$) from the observed Q-dependence of the scattered intensity using Eq. (1). Figure 1 demonstrates the dramatic increase of R_h and R_g with increasing w_0.. The ratio of $R_g/R_h \approx 1.65 \pm 0.07$ obtained for $w_0 \geqslant 4.0$ is in good agreement with theoretical models for semiflexible (wormlike) chains, but clearly inconsistent with the classical picture of a droplet-like microemulsion. The results from SLS and QLS measurements at different values of w_0 and Φ are summarized in Fig. 2, together with model calculations for various geometrical shapes. The light-scattering data are in good agreement with the wormlike chain model by Yamakawa and Fuji [10, 8], although the data are not sufficiently accurate in order to quantitatively determine the persistence length l_p, a measure of the flexibility of the lecithin reverse micelles.

However, we are able to directly confirm this picture of a water-induced formation of flexible cylindrical micelles and verify the postulated analogy between the structural properties of polymer chains and lecithin reverse micelles using a combination of SLS and small-angle neutron scattering. We can combine the light and neutron scattering data in order to obtain $I(Q)$ over a wide range of Q values. This is illustrated in Fig. 3A, where an example of the scattering data from lecithin reverse micelles at $w_0 = 8.0$ and $\Phi \ll \Phi^*$ is shown. $I(Q)$ for these systems has several distinct regimes which permit a quantitative study of the different length scales characterizing overall dimension (R_g), flexibility (l_p), and local cylindrical cross-section (R). Here, we only focus on the determination of the persistence length. A description of the analysis of $I(Q)$ employed to determine local structure (R) and overall size (R_g) is given elsewhere [6].

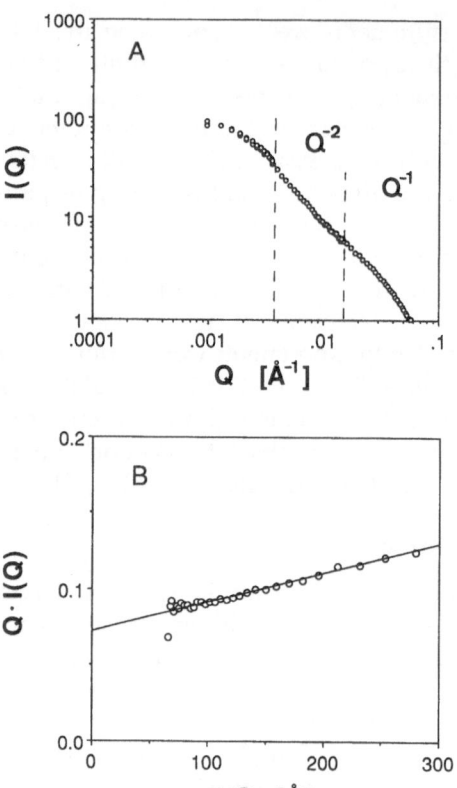

Fig. 3. A) Plot of $I(Q)$ versus Q for solutions of soybean lecithin in deuterated cyclohexane ($w_0 = 8.0$) at $\Phi = 0.0036$. Data shown here were obtained from light and neutron scattering experiments. Also indicated are the regimes where a different characteristic Q-dependence can be observed (see [6] for details). B) Plot of $Q \cdot I(Q)$ versus $1/Q$ for solutions of soybean lecithin in deuterated cyclohexane ($w_0 = 8.0$) at $\Phi = 0.0036$. Also shown are linear least-squares fit to the data used for the deduction of the persistence length l_p

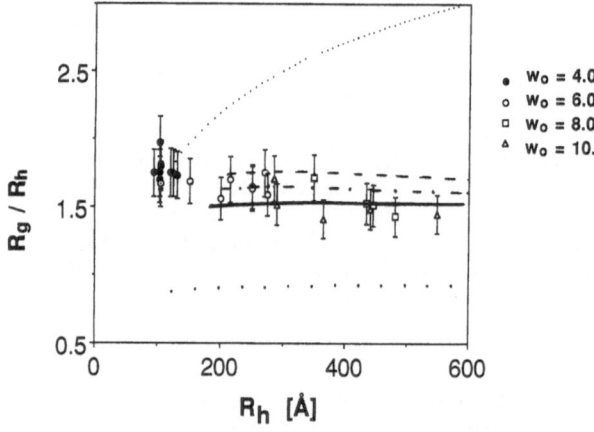

Fig. 2. R_g/R_h as a function of the hydrodynamic radius R_h for soybean lecithin in cyclohexane at $T = 25.0\,^{\circ}\text{C}$. The data were obtained with different values of w_0 (see figure legend) and for $\Phi \ll \Phi^*$. Also shown are theoretical model calculations for: disks, thickness $t = 50$ Å ($\cdot\,\cdot$), rigid cylinders, diameter $d = 50$ Å ($\cdots$), and semiflexible chains, $d = 50$ Å, $l_p = 110$ Å (—), $l_p = 200$ Å (—·—), $l_p = 300$ Å (——)

For very long and flexible cylinders, the asymptotic Q-dependence of the scattering intensity $I(Q)$ crosses over from $1/Q$ for rigid cylinders to $1/Q^2$ for Gaussian coils. If $L \gg l_p \gg R$, where L is the contour length of the cylinder, we can observe this crossover at low Q, since it is obscured neither on the high-Q side by the exponential decay of the scattered intensity caused by the finite radius of the cylinder, nor on the low-Q side by the effect of the finite overall size of the flexible coil. Such a crossover from $1/Q$ to $1/Q^2$ should occur at $Q \cdot l_p \approx 1.9$, and the following relation for $I(Q)$ should be valid in this Q-range [6, 11]:

$$Q \cdot I(Q) \sim \frac{\pi}{L} + \frac{2}{3\,l_p L Q} \ . \qquad (2)$$

We can quantitatively test whether such a Q^{-2} decay of $I(Q)$ exists, i.e., if there is a regime where the micelles appear as flexible coils on a length scale comparable to the characteristic probing distance of the scattering experiment. A plot of $Q \cdot I(Q)$ vs. Q^{-1} (Fig. 3B) reveals that the data for $w_0 = 8.0$ and $\Phi = 0.0036$ are indeed well represented by relation (2). From the intercept and slope we can estimate the persistence length l_p. We obtain an average value of $l_p = 110 \pm 30$ Å, which is in good agreement with the theoretical calculations for the wormlike chain model shown in Fig. 2.

Scattering experiments on lecithin/cyclohexane solutions have thus permitted the study of individual structural properties of the reverse micelles present in these solutions, and we are able to directly confirm the postulated model of flexible rodlike aggregates. However, Fig. 1 also illustrates the limits of a direct application of polymer theory to surfactant solutions. Due to the dynamic nature of the micellar aggregates, η_s exhibits a w_0-dependence which is far more complex than what would be expected on the basis of classical polymer models: the reptation model would simply predict a M^3-dependence [2]. A more detailed knowledge of the micellar kinetics and the thermodynamics of lecithin reverse micelle formation is clearly required for an understanding of the dynamic properties. However, based on the data we have presented, we believe that reverse micellar solutions of lecithin in cyclohexane may serve as an ideal model system for structural and dynamic studies of "equilibrium polymer" solutions [12].

Acknowledgements

This work has been supported in part by the Swiss National Science Foundation (NF-19) and the U.S. National Science Foundation (Magid, CHE-9008589). We gratefully acknowledge the help of Dr. R. Scartazzini and C. Cavaco.

References

1. Scartazzini R, Luisi PL (1988) J Phys Chem 92:829
2. Schurtenberger P, Scartazzini R, Luisi PL (1989) Rheol Acta 28:372
3. Schurtenberger P, Scartazzini R, Magid LJ, Leser ME, Luisi PL (1990) J Phys Chem 94:3695
4. Luisi PL, Scartazzini R, Haering G, Schurtenberger P (1990) Colloid Polym Sci 268:356
5. Ott A, Urbach W, Langevin D, Schurtenberger P, Scartazzini R, Luisi PL (1990) J Phys: Condens Matter 2:5907
6. Schurtenberger P, Magid LJ, King SM, Lindner P (1991) J Phys Chem 95:4173
7. Schurtenberger P, Peng Q, Leser M, Luisi PL, submitted to J Colloid Interface Sci
8. Schurtenberger P, Augusteyn RC (1991) Biopolymers 31:1229
9. Schurtenberger P, to be submitted to Colloid Polym Sci
10. Yamakawa H, Fujii M (1973) Macromolecules 6:407
11. Marignan J, Appell J, Bassereau P, Porte G, May RP (1989) J Phys France 50:3553
12. Cates ME, Candau SJ, J Phys: Condens Matter (1990) 2:6869, and references therein.

Authors' address:

PD Dr. Peter Schurtenberger
Institut für Polymere
ETH Zentrum
CH-8092 Zürich, Switzerland

Progress in Colloid & Polymer Science

Progr Colloid Polym Sci 89:278—280 (1992)

Thermodynamic properties of water in reverse micelles

D. Bertolini, M. Cassettari, G. Salvetti, E. Tombari, S. Veronesi, and G. Squadrito*)

Istituto di Fisica Atomica e Molecolare del C.N.R., Pisa, Italy
*) Istituto di Tecniche Spettroscopiche del C.N.R., Messina, Italy

Abstract: Microcalarimetry is a powerful tool in many research fields; it has been widely utilized to investigate the processes inducing change in the thermodynamic properties of the sample and, thus, it has played an important role in the study of the liquid state, especially in gaining more information on solute-solute interaction in liquid mixtures. — We report here recent measurements on AOT-n-heptane-water micelles performed by means of a new microcalorimeter for liquid samples, from which the apparent molar heat capacity of water confined in the reverse micelle core versus $n = [H_2O]/[AOT]$ has been investigated at 25°C.

Key words: Micelle; specific-heat

Introduction

In order to study thermodynamic properties of confined water, we investigated micellar systems which, in addition to other interesting features, permit to obtain the heat capacity of water confined in very small spherical droplets, the diameter of which can be measured with sufficient accuracy [1]. The diameter of the micellar water pool is usually smaller than 150 Å. The radius of the water pool can be easily varied by changing the molar ratio $n = [H_2O]/[AOT]$; as a consequence, we have different spatial constraints for water.

Our results show the dependence of apparent molar heat capacity vs. n at fixed micellar volume fraction Φ (defined as the sum of volume fractions of AOT and water) and at $T = 25$°C. The present data can be analyzed to obtain information on the structure of water in the presence of spatial constraint [2—5]. Such an analysis is not the aim of this short contribution.

Experimental details

Materials

We studied two sets of solutions at $\Phi = 0.03$ and $\Phi = 0.1$ prepared with: n-heptane by Carlo Erba (purity 99.5%; RPE grade); AOT (sodium bis 2-ethylhexylsulfosuccinate) by Fluka (purity 98%) has been dried under vacuum for 3 days and utilized without further purification, and bidistilled water. The solutions were made by dilution with heptane of a concentrated AOT-heptane solution and addition of water. After suitable mixing, these solutions were stored at room temperature ($T = 24 \pm 0.3$°C), avoiding exposure to light, for several days before their use.

Calorimetry

The differential calorimeter and the experimental technique have been previously described [6]. The reported results refer to measurements performed at 25°C (temperature stability ± 0.01°C). The reproducibility of heat capacity measures was $\pm 8 \times 10^{-4}$ J/°C, working with a sample volume of 1.567 ± 10^{-4} cm^3. The sample cell was filled by using a microsyringe kept at constant temperature ($T = 32 \pm 0.01$°C). The reference cell was filled with heptane following the same procedure; this procedure allows us to calculate the sample mass when its density is measured at the filling temperature. The mean value of the sample's temperature during the time necessary for taking a measurement (≈ 180 s) increases by about 0.1°C [6].

work. In the second part, we give information about our experimental procedure. In the third part, we use the pair interaction parameters that we determined from CMC measurements, for a detailed study of critical concentrations and compositions of TTAB/OP(EO)$_8$/$n\Phi Cm$ micelles.

Ternary mixed micelle formation

The development of a pseudo phase separation model for the formation of nonideal ternary mixed micelles [7, 8] shows that the CMC of an anionic/cationic/nonionic surfactant system, the mole fractions x_a of the anionic surfactant, and x_c of the cationic surfactant in the mixed micelle obey the equations

$$a_a C_{acn} = x_a C_a \exp[\beta_{ac} x_c^2 + \beta_{an}(1 - x_a - x_c)^2$$
$$+ (\beta_{ac} + \beta_{an} - \beta_{cn})x_c(1 - x_a - x_c)] \quad (1)$$

$$a_c C_{acn} = x_c C_c \exp[\beta_{ac} x_a^2 + \beta_{cn}(1 - x_a - x_c)^2$$
$$+ (\beta_{ac} - \beta_{an} + \beta_{cn})x_a(1 - x_a - x_c)] \quad (2)$$

$$(1 - a_a - a_c)C_{acn} = (1 - x_a - x_c)C_n$$
$$\cdot \exp[\beta_{an} x_a^2 + \beta_{cn} x_c^2$$
$$+ (-\beta_{ac} + \beta_{an} + \beta_{cn})x_a x_c] \cdot$$
$$(3)$$

In these expressions:

— a_a and a_c, the mole fractions of the anionic surfactant and of the cationic one in the total mixed solution, are known.
— C_a, C_c, C_n, the CMC of each pure surfactant, can be measured.
— β_{ac}, β_{cn}, and β_{an}, the pair interaction parameters, can be deduced from CMC measurements of the binary mixtures [9].

Thus, the three unknowns C_{acn}, x_a, x_c, (so $x_n = 1 - x_a - x_c$) can be calculated from Eqs. (1), (2), (3). As C_{acn} can also be directly measured, it is possible to compare model with experiment.

Experimental procedure

All the CMC have been measured in water at 25°C.

The tetradecyltrimethylammonium bromide was provided by Aldrich-Chimie; its composition in quaternary ammonium is higher than 99%. The polyoxyethylene octylphenol was synthesized at the Institut de Recherche de Chimie Appliquée (IR-CHA); its purity grade was given as higher than 95%, and a gas chromatography analysis showed that it was really monodistributed. The alkylbenzene sulfonates are branched. Branching position and chain length are provided as a function of the carbon atoms of the alkyl chain: n indicates the position of the branching of the polar head on the alkyl chain, m is the length of the alkyl chain. These surfactants were synthesized at the University of Texas [10]; their purity grade was higher than 99%. Water was deionized and redistilled.

The CMC were mainly determined from surface tension versus concentration data, measured according to the Wilhelmy plate method, with a Krüss digital tensiometer. When surface tension measurements were difficult to interpret, alternative methods were used. Thus, CMC of ionic surfactants or of some mixtures including principally an ionic surfactant were measured by conductivity with a Wayne and Kerr bridge, Model B 331. CMC of nonionic surfactants or CMC of mixtures including principally a nonionic surfactant were measured by studying the ultraviolet absorption of the surfactant solutions as a function of the surfactant concentration with a Beckman spectrophotometer, Model 34.

Results and discussion

The CMC of the pure surfactants are reported in Table 1. For symmetrically branched anionic surfactants (5ΦC10, 6ΦC12, 7ΦC14), we found that the CMC decreases when the alkyl chain length increases. For anionic surfactants which have the same alkyl chain length (3ΦC10, 5ΦC10), the CMC was found to decrease when the branching was shifted toward the end of the alkyl chain. The CMC of the nonionic surfactant is about 10 times lower than the lowest of the anionic surfactant CMC.

We deduced the pair interaction parameters of the studied surfactants from CMC measurements performed on the binary mixtures; they are reported

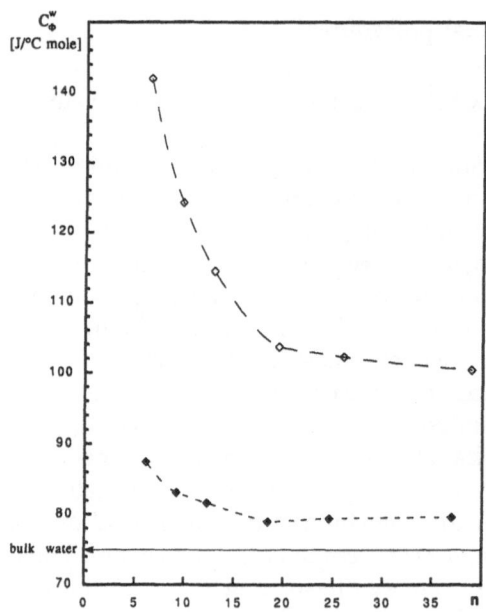

Fig. 4. Apparent molar heat capacity for the solutions at $\Phi = 0.03$ (◇) and $\Phi = 0.1$ (◆)

water pool became increasingly greater. As a consequence, the difference between C_ϕ^w and the heat capacity of pure water for $n > 20$ is related not only to "structured" water, but also to the change of $C^{Hep + AOT}$. To obtain a deeper insight into the behaviour of C_ϕ^w, very accurate measurements at constant AOT molality and temperature as a function of n are necessary. (This study is in progress.)

References

1. Zulauf M, Heike HF (1979) J Phys Chem 83:480—486
2. Boicelli CA, Conti F, Giomini M, Giuliani AM (1985) In: Conti F, Blumberg WE, de Gier J, Pocchiari F (eds) Physical Methods on Biological Membranes and Their Model Sytems, New York, pp 141—162
3. Boicelli CA, Giomini M, Guiliani AM (1984) App Spectr 38:537—539
4. D'Aprano A, Lizzio A, Turco Livieri V (1987) J Phys Chem 91:4749—4751
5. D'Aprano A, Lizzio A, Turco Livieri V, Aliotta F, Vasi C, Migliardo P (1988) J Phys Chem 92:4436—4439
6. Bertolini D, Cassettari M, Salvetti G, Tombari E, Veronesi S (1990) Rev Sci Instrum 61:2416—2419
7. Morel JP, Morel-Desrosiers N, Lhermet C (1984) J Chimie Physique 81:109—112

observed increase of C_ϕ^w for $n < 20$, with respect to the value of bulk water (see Fig. 4) can be related to the increase of the "structured"/bulk water ratio, which at $n > 20$ assumes an about constant value, as suggested by the observed plateau. For a quantitative analysis it is necessary to utilize other experimental techniques such as Raman spectroscopy, from which the "structured" water fraction can be evaluated [5].

The plateaus observed for $n > 20$ are higher than the C_p value of bulk water for both the micellar concentration values. This unexpected result requires some explanation.

Equation (1) is derived on the assumption that interactions between the solute (H_2O) and the solvent (hept. + AOT) are negligible (additivity of the heat capacity). In our system, we can neglect interaction among micelles, but we cannot neglect water/solvent interaction which is expected to change the solvent specific heat, mainly owing to structural deformation of the AOT shell when the

Authors' address:

Dr. Giuseppe Salvetti
I.F.A.M. (CNR)
via del Giardino 7
56100 Pisa, Italy

Time-dependent heat capacity of aqueous solutions of biomolecules

D. Bertolini, M. Cassettari, G. Salvetti, E. Tombari, S. Veronesi, and G. Squadrito*)

Istituto di Fisica Atomica e Molecolare del C.N.R., Pisa, Italy
*) Istituto di Tecniche Spettroscopiche del C.N.R., Messina, Italy

Abstract: We present heat capacity (C_p) and thermal conductivity measurements on water-lysozyme solution in the water-rich region versus both time and protein concentration. The aim is to prove the occurrence of structural rearrangements vs. time in water-lysozyme solutions already observed in the high-concentration region (10 + 15 wt%) using different techniques. — The observed increase of heat capacity in time occurs independently of the lysozyme concentration and the pH value of the solvent. The characteristic time to reach equilibrium can change from a few minutes to hours, and it seems to be independent of pH values of the solution. Evolution of a spatial order of lysozyme in water and/or of water molecules interacting with the protein is considered as a possible explanation for the observed phenomenon, which is absent when the solute is an amino acid (alanine, glycine, etc.). — The thermal conductivity of these solutions does not show the time dependence observed for the heat capacity.

Key words: Lysozyme; heat-capacity

Introduction

Much experimental evidence exist that proteins in aqueous solutions are involved in processes in which the measured macroscopical properties of the system evolve with time. This behaviour has been attributed to the development of an order similar to that in thyxotropic structures [1].

In order to study these properties according to their thermodynamic aspects, we planned a systematic research, focusing our efforts on water lysozyme solutions, lysozyme being one of the most available and elucidated proteins.

Previous calorimetric experiment on these systems were designed to shed light on the unfolding process of the enzyme [2], and on the onset of the biological activity [3], but no evidence of time evolution has been reported.

Our results show a clear heat-capacity time evolution, with characteristic times ranging from minutes to hours. The occurrence of the phenomenon is pH-independent and the equilibrium reached can be easily destroyed by gentle mechanical shaking. After this shaking the system again undergoes a time evolution toward the previous equilibrium state. The absence of the phenomenon in water-amino acid solution illustrates the role of the size of the solute molecule, and of the correlated long-range interaction which a protein is able to set up in an aqueous environment.

Experimental details

Materials

We studied solutions prepared both with bidistilled water (pH = 7) and with different buffer solutions in order to observe possible effects of the pH. Lysozyme from chicken albumen (3× crystallized, dialyzed, containing 95% of protein and primary buffer salt, from Sigma Chemical Co.) was stored following the prescriptions and used without any further purification. We have used commercial buffer solutions (0.1 M) for analysis (Normex line of Farmitalia Carlo Erba). Alanine by Sigma Chem. Co. (purity 99%) was utilized without further purification. Solutions are carefully prepared to avoid contamination and mixed for about 10 min, using a mechanical stirrer or sonication, at room temperature (T = 24 ± 0.3°C). At the same temperature the pH value of the solutions was measured.

Calorimetry

The differential calorimeter and the experimental technique have been previously described [4]; in addition to the features reported in [4] the calorimeter was programmed to perform automatic measurements in time. The heat capacity of the sample was measured at constant pressure, starting 5—60 min after the filling of the calorimetric cell, at fixed time intervals of 10—60 min. The reported results refer to measurements performed at 25°C (temperature stability ± 0.01°C) and at various lysozyme concentrations and pH values. The repeatability of heat capacity measurements was ± 8 × 10^{-4} J/°C, working with a sample volume of 1.567 ± 10^{-4} cm^3. To obtain this precision, we have utilized a microsyringe, kept at constant temperature (± 0.01°C), in a suitable mechanical set up. By means of the same procedure, we filled the sample cell and the reference one with the same volume of solution and bidistilled water, respectively. This procedure allows to calculate the sample mass when its density is measured at the filling temperature. The mean value of the sample's temperature during the time necessary for taking a measurement (≈ 180 s) increases by about 0.1°C [4]. Specific heat values were not calculated, the density of the sample not being available at that time.

Results

In Fig. 1 are shown experimental curves referring to water-lysozyme (Fig. 1b) and to water-alanine solution (Fig. 1a). The heat capacity of the water-lysozyme solution shows an exponential approach to an equilibrium value, whereas the water-alanine solution remains practically constant, the small fluctuation present being also observed in the stability test performed with a sample of pure water. This fluctuation is due to slowly varying temperature gradients in the calorimeter. On the contrary, the thermal conductivity of the solutions does not show any time dependence.

The as-obtained data in Fig. 1b for heat capacity were fitted to an exponential equation of the type

$$C_L(t) = C_L(0) + \Delta C(1 - e^{-t/\tau}) \,, \tag{1}$$

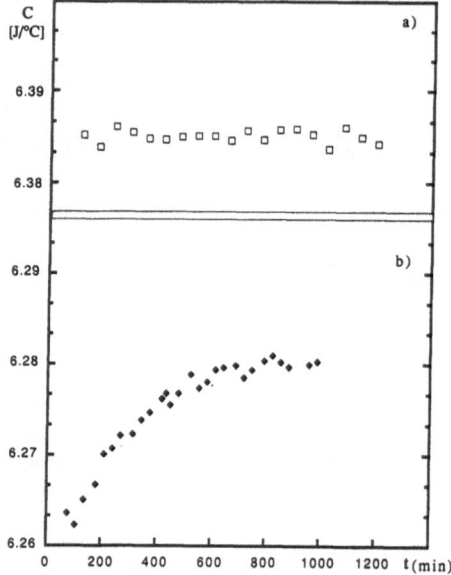

Fig. 1. a) Typical time test performed with the reference cell filled with water and sample cell filled with a water-alanine solution. (T = 25°C). b) The heat capacity C_L vs. Time of a sample prepared with lysozyme (w.f. 11%) in the buffer solution (0.1 M; pH = 6.18) using as reference the same solution at T = 25°C. The curve is fitted by $C_L(t) = 6.2525 + 0.2576 [1 — \exp(-t/275)]$

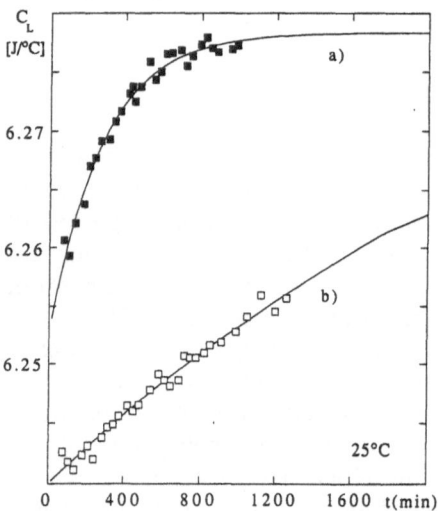

Fig. 2. C_L vs. time for the same sample (Lysozyme w.f. 11% in the buffer solution, pH = 6.18). The curve a) refers to the sample just prepared; the curve b) to the same sample, gently shaken and then re-introduced in the cell

where τ is the characteristic time for the process. From Eq. (1) and the plot in Fig. 1b, it is possible to calculate the equilibrium value of heat capacity attained with a time constant τ = 247 min. We have

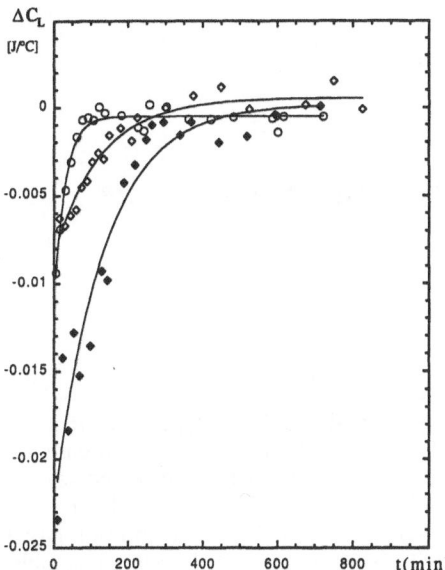

Fig. 3. The amplitude of the effect ΔC_L vs. time at 25°C, wf 1% lysozyme concentration at three different pH values: (o) pH = 4; (♦) pH = 6; (◇) pH = 10

observed a wide range of values for τ and no evidence of a correlation with the sample's pH, lysozyme's weight fraction, or sample preparation procedure has been found yet. The results obtained at 25°C with a sample of 11 wt% lysozyme in the buffer solution are shown in Fig. 2. The upper curve (a) refers to the sample just prepared, and the lower curve (b) to the same sample which was taken out from the cell, gently shaken, and then re-introduced in the calorimetric cell. The difference between the curves (a) and (b) suggests that the equilibrium state attained by the lysozyme solution is destroyed upon mechanical agitation and an approach towards a new equilibrium condition begins, but with a different rate and initial value. In Fig. 3 are the results obtained at 25°C with 1 wt% lysozyme solutions at three pH values.

Discussion

The results reported here undoubtedly show that the heat capacity of the lysozyme-water solution increases in time and an equilibrium value is attained in times ranging from a few tens of minutes to many hours at 25°C. The relative variation of the heat capacity is in the range of 0.2—2%, a value which is mucher greater than the experimental error. The process which determines the time-dependence of $C_L(t)$ is fundamental for understanding biological solu-

tions, and is intimately related to protein dynamics and protein-water interaction which, up to now are not completely known [5]. Owing to the preliminary character of these results and the complexity of the involved processes, only a schematic and qualitative analysis of the results can be given here.

The information now available from calorimetry shows that entropy fluctuations in the sample increase with time [6]. These fluctuations can be related to the appearance of ordered regions, the statistical weight of which increases until a dynamic equilibrium is attained. These regions are not necessarily "aggregates" of lysozyme and the possibility of an order induced in the solvent by protein-water interactions cannot be disregarded.

Indeed, water-amino acid solution shows a time-independent heat capacity also when the solute-solute interaction is widely changed (by varying concentration, pH, and the amino acid in the solution). Thus, the different behavior of lysozyme solution can be traced back to the size and/or the peculiar interaction between lysozyme and water. The role of the size of the molecule and the properties reported here and in [1] suggest to study the role of the electrostatic and van der Waals forces in the frame of the classical experiments on colloidal systems. With this aim, we have planned to investigate the properties of water-protein solution as functions of temperature, pH, dielectric constant, and ionic strength of the solvent [7].

References

1. Farsaci F, Fontanella ME, Salvato G, Wanderlingh F, Giordano R, Wanderlingh U (1989) Phys Chem Liq 20:205
2. Privalov PL, Khechinashvili NN (1974) J Mol Biol 86:665
3. Rupley AJ, Pang-Hsiong Yang (1979) Biochemistry 18:2654
4. Bertolini D, Cassettari M, Salvetti G, Tombari E, Veronesi S (1990) Rev Sci Instrum 61:2416—2419
5. Richards FM (1991) Scientific American 264(1):34—41
6. Landau LD, Lifschitz EM (1969) Statistical Physics, 2nd ed. (Pergamon, London), p. 351.
7. Prost J, Rondelez F (1991) Nature supplement to Vol. 350:11—23

Authors' address:

Dr. Giuseppe Salvetti
I.F.A.M. (CNR)
via del Giardino 7
56100 Pisa, Italy

Progress in Colloid & Polymer Science Progr Colloid Polym Sci 89:284—287 (1992)

Is the AOT/water/oil system really simple?
Conductivity measurements in ionic and nonionic microemulsions

W. Sager, W. Sun, and H.-F. Eicke

Institut für Physikalische Chemie, Universität Basel, Switzerland

Abstract: Conductivities of the AOT and nonionic $C_{12}E_5$ systems are reported. A peculiarity of the AOT system is the possibility to prepare w/o microemulsions with more than 50 wt% of water, which show, however, a low conductivity. In contrast, a nonionic system with comparable amounts of water and oil displays a conductivity several orders of magnitude larger. Upon addition of salt the conductivity pattern of the AOT system resembles that of a nonionic with opposite temperature dependence.

Key words: Water/oil microemulsions; AOT; $C_{10}E_5$; conductivity

Introduction

Microemulsions are thermodynamically stable mixtures of water, oil, and a surfactant, which consist of oil and water domains separated by a surfactant monolayer. The structure of the water and oil domains depends crucially on the water-to-oil ratio. Microemulsions containing a comparable amount of oil and water usually form a bicontinuous structure in a certain range of temperature, whereas in the oil or water concerns of the Gibbs triangle nanometer sized droplets of water in oil or oil in water will be formed. The AOT (sodium di-2-ethylhexylsulfosuccinate)/water/alkane system, however, allows preparation of oil-continuous microemulsions with more than 50 wt% of water. In order to show the peculiarity of the AOT system a comparison of conductivity measurements with the nonionic system $C_{10}E_5$ (pentaethyleneglycol mono-decylether)/water/hexane has been made. Hexane has been used as oil because it provides the largest solubilization capacity for water in these systems.

Starting from the oil corner of the Gibbs triangle and proceeding towards the water-rich side, the conductivity rises sharply by three to four orders of magnitude with increasing water content due to the percolation of water droplets [1]. The percolation transition is governed by power laws characterized by critical exponents. For a given droplet radius, i.e., constant water to surfactant weight ratio (r_w),

the critical volume fraction of the percolation transition depends strongly on temperature. Due to the reverse phase behavior of ionic and nonionic surfactants with respect to temperature, percolation is induced by increasing the temperature in ionic surfactant systems and occurs with decreasing temperature in systems of nonionic surfactants [2].

Experimental section

The microemulsions were prepared by mixing appropriate weight fractions of water or brine solution, hydrocarbon and surfactant. AOT (Fluka) was purified with activated charcoal [1], whereas $C_{10}E_5$ (Nikkol) was used without further purification. A dilution series was obtained by fixing the r_w-value ($r_w = w/s$) and varying c_w ($c_w = (w + s)/(w + s + 0)$), where w, s and o correspond to masses of water, surfactant and oil, respectively. For the determination of phase diagrams samples were equilibrated by using a magnetic stirrer in a thermostated water bath. Phase separation took place within a few minutes up to several hours. Liquid crystalline phases were observed with crossed polarizers. The position of the phase boundaries was reproducible for both raising and lowering the temperature. The conductivity was measured in a thermostated glass cell with two platinum electrodes and an autobalancing Wheatstone bridge KONDUX 1 (Kamphausen) operating at 3 kHz or 380 Hz.

Results and discussion

The Gibbs triangle of the AOT/H$_2$O/hexane system at 25 °C (Fig. 1) shows an isolated two-phase region [3], where a w/o microemulsion is in

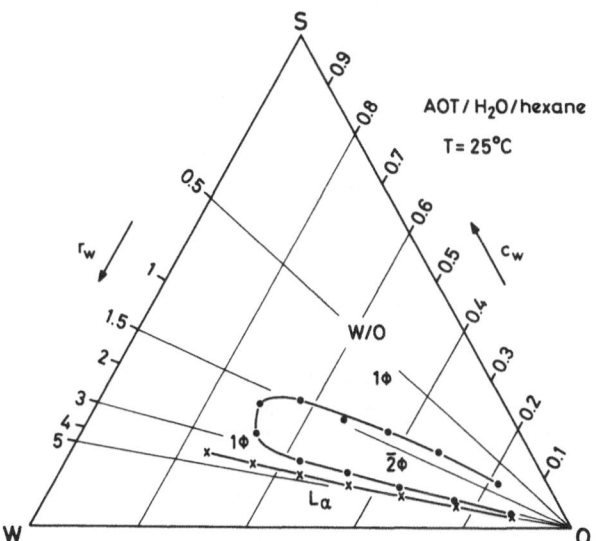

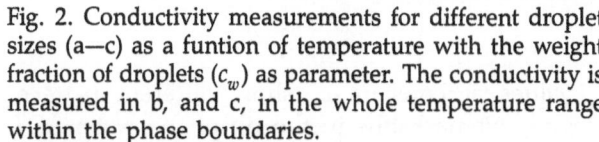

Fig. 1. Gibbs phase triangle of the AOT/H$_2$O/hexane system at 25 °C. r_w denotes the weight ratio of water to surfactant ($r_w = w/s$), whereas c_w is the weight fraction of water plus surfactant ($c_w = (w + s)/(w + s + o)$). The phase diagram shows a broad one-phase region (1Φ) at small r_w-values, an isolated two-phase region ($\overline{2}\Phi$) where a w/o microemulsion is in equilibrium with an excess water phase, a small one-phase channel at a r_w-value of about 4, and a lamellar phase (L_a) at higher r_w-values

equilibrium with an excess water phase, and a small one-phase channel at a nearly constant r_w-value of about 4. With increasing temperature the two phase region shrinks toward the oil corner and a lamellar phase expands into the triangle. The conductivity was measured as a function of temperature and c_w-values for different droplet sizes (Fig. 2). For the

Fig. 2. Conductivity measurements for different droplet sizes (a—c) as a funtion of temperature with the weight fraction of droplets (c_w) as parameter. The conductivity is measured in b, and c, in the whole temperature range within the phase boundaries.
a) AOT/H$_2$O/dodecane: $r_w = 0.5$;
b) AOT/H$_2$O/hexane: $r_w = 2.5$;
c) AOT/H$_2$O/hexane: $r_w = 4.5$

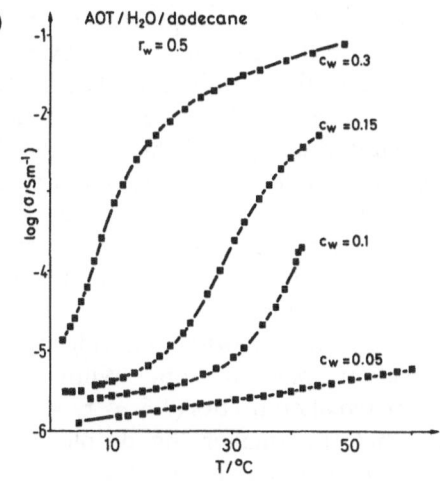

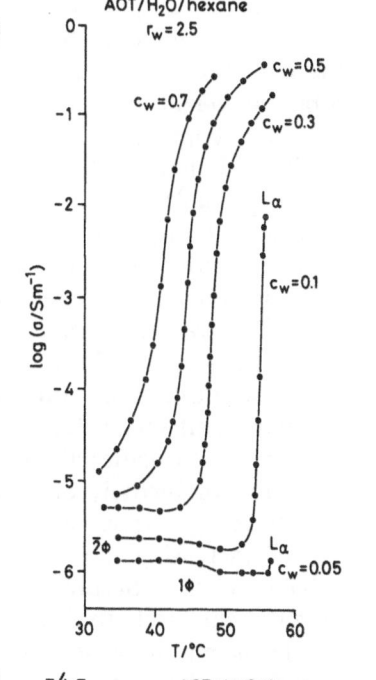

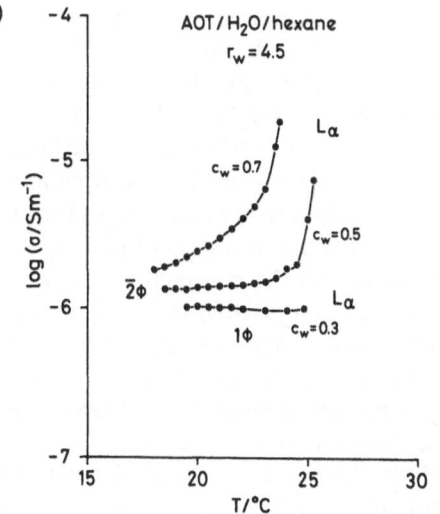

AOT/H$_2$O/hexane system the dependence of the radius of gyration on the r_w-value is given by $r_{g,z}$ = (0.82 + 3.5 r_w) nm [4]. In order to measure the conductivity in a convenient temperature region, dodecane was chosen for r_w = 0.5, because it shifts the one-phase region to lower temperature. At small droplet sizes (r_w = 0.5) the conductivity increases continuously as a function of temperature within the whole one-phase region at not too small c_w-values. For larger droplets (r_w = 2.5) a sharp increase of the conductivity is observed, indicating a typical percolation phenomenon. Within the regime where percolation occurs (1.5 < r_w < 3.5), we can state that the smaller the droplets, the stronger is the dependence of the percolation onset on the weight fraction. For very large droplets (r_w = 4.5) the temperature-dependent conductivity remains almost constant in the one-phase region and is only little affected by the weight fraction (c_w) of the droplets. Microemulsions with a r_w-value of 4.5 are found in the one-phase channel shown in the Gibbs triangle, where it is possible to get low conductive w/o microemulsions with up to 70 wt % of water. This appears quite surprising since we are in the region of geometric percolation. The three plots corresponding to different r_w-values seem to indicate different mechanisms governed by the interplay of aggregation and permeation phenomena.

In systems with nonionic surfactants [2] (e.g., C$_{10}$E$_5$ (pentaethylenglycol monodecylether)/0.1 wt% NaCl/hexane), however, the conductivity of small droplets (r_w = 0.1—0.2) is not influenced by temperature, whereas for increasing amounts of water the conductivity rises sharply with temperature, emphasizing the fact that larger amounts of water lead to bicontinuous structures (Fig. 3). Contrary to nonionic systems, the ionic strength of the water pools changes in the AOT system with the droplet size. Increasing the r_w-values cause the ionic strength to drop and, thus, increases the spontaneous radius of the surfactant layer. Adding salt reduces this effect and balances the hydrophilic and lipophilic properties of the surfactant due to a screening of the polar head group. The influence of salt on the conductivity is shown for small droplet sizes (r_w = 0.5) in Fig. 4. Increasing salt concentration decreases the conductivity. Due to the nearly zero spontaneous curvature of the surfactant layer, bicontinuous structures are formed at comparable amounts of water and oil [5].

Isoconductivity lines in the tongue-shaped one-phase domain as a function of temperature and

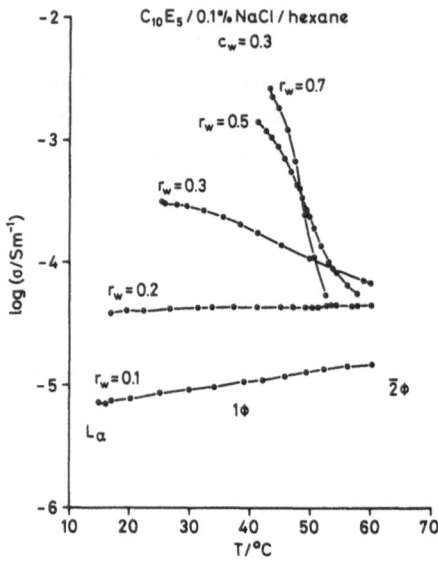

Fig. 3. Conductivity as a function of temperature for different water-to-surfactant ratios (r_w) of the nonionic system: C$_{10}$E$_5$/0.1 wt% NaCl/hexane at a constant c_w-value (c_w = 0.3)

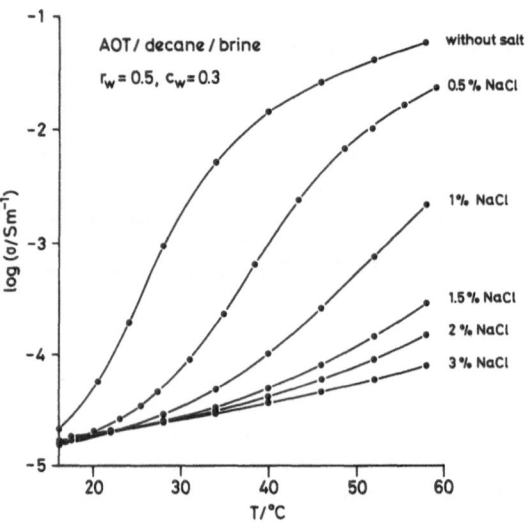

Fig. 4. Conductivity of the AOT/brine/dodecane system as a function of temperature at const. r_w- and c_w-value (r_w = 0.5, c_w = 0.3) for different salt concentrations (wt%) of the aqueous phase

r_w-values at constant c_w show different patterns (Fig. 5). Remarkably, in the tip of the water/AOT/hexane tongue, only low conductivity is observed. In the AOT-brine system the one phase-region becomes symmetric and the domain of high con-

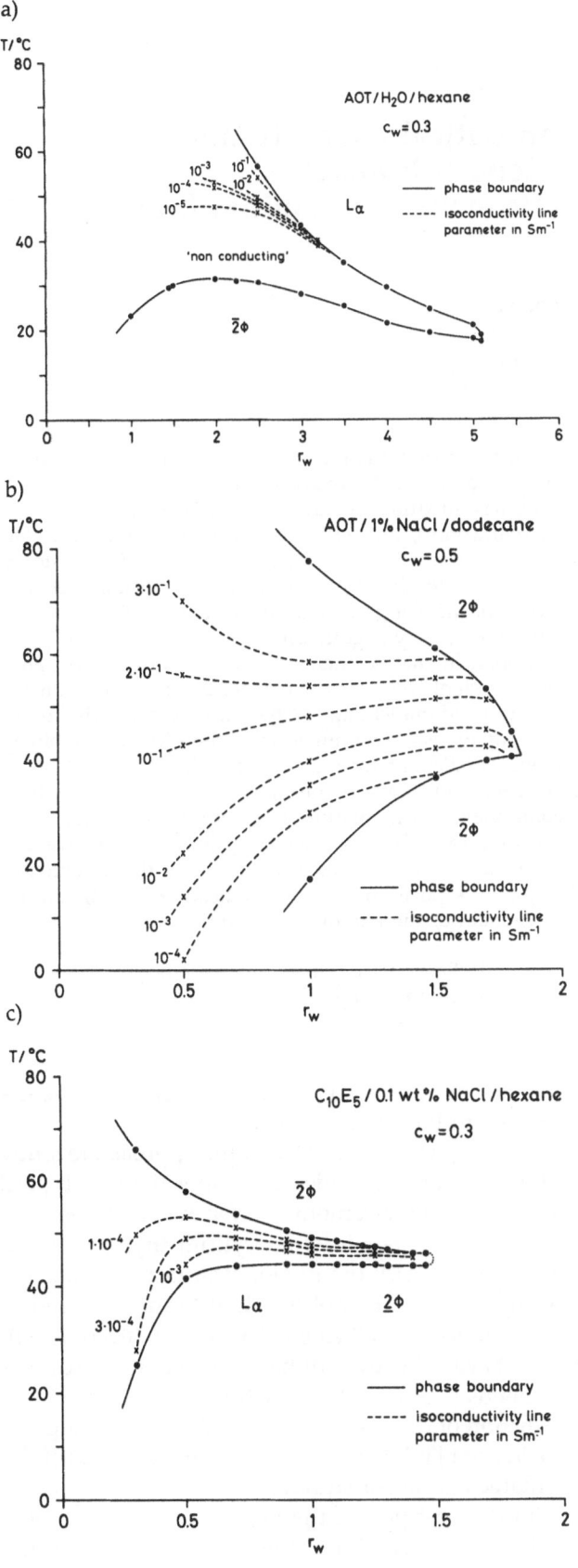

ductivity extends into the tip of the tongue, and with respect to temperature shows, thus, a reverse but analogous pattern when compared to the nonionic system.

Acknowledgements

We are grateful to the Swiss National Science Foundation and to the Ciba-Stiftung for financial suppport. We thank Y. Hauger for carrying out the measurements.

References

1. Borkovec M, Eicke H-F, Hammerich H, and Das Gupta B (1988) J Phys Chem 92:206
2. Kahlweit M, Strey R, Haase D, Kunieda H, Schmeling T, Faulhaber B, Borkovec M, Eicke H-F, Busse G, Eggers F, Funck Th, Richmann H, Magid L, Söderman O, Stilbs P, Winkler J, Dittrich A, Jahn W (1987) J Colloid Interface Sci 118:436
3. Kunieda H and Shinoda K (1979) J Colloid Interface Sci 70:577
4. Hilfiker R, Eicke H-F, Sager W, Steeb C, Hofmeier U, Gerke R (1990) Ber Bunsenges Phys Chem 94:677
5. Chen SH, Chang SL, Strey R (1990) J Chem Phys 93:1907

Authors' address:

Prof. Dr. H.-F. Eicke
Institut für Physikalische Chemie
Klingelbergstr. 80
CH-4056 Basel, Switzerland

Fig. 5. A cut through the Gibbs phase triangle at const. c_w-value. The tongue-shaped one-phase region is shown as a function of temperature and r_w-values. The solid lines present the phase boundaries, whereas the dotted lines are isoconductivity lines in S/m. The solubilization limit curve corresponds to the lower phase bondary in the ionic system and to the upper in the nonionic system.
a) AOT/H_2O/hexane: $c_w = 0.3$;
b) AOT/1 wt% NaCl/hexane: $c_w = 0.5$;
c) $C_{10}E_5$/0.1 wt% NaCl/hexane: $c_w = 0.3$

Progress in Colloid & Polymer Science

Progr Colloid Polym Sci 89:288—292 (1992)

Effect of anionic surfactant structure on critical concentrations and compositions of sodium alkylbenzene sulfonate/ polyoxyethylene octylphenol/tetradecyltrimethylammonium bromide mixed micelles

A. Graciaa, M. Ben Ghoulam, G. Marion, and J. Lachaise

L.T.E.M.P.M., Centre Universitaire de Recherche Scientifique, Pau, France

Abstract: The model for multicomponent nonideal mixed micelles used via the regular solution approximation allows accounting for the critical micelle (CMC) of binary or ternary mixtures obtained with $n\Phi Cm$ (anionic surfactant), TTAB (cationic surfactant), and $OP(EO)_8$ (nonionic surfactant). The anionic/cationic interaction is much higher than the ionic/nonionic interactions; this is probably due to the high electrostatic attraction between molecules which have opposite charges. We have found that anionic/cationic interaction appears to depend very slightly on the anionic surfactant structure, while anionic/nonionic interaction increases with the length of the alkyl chain of the anionic surfactant, but does not depend significantly on the position of the polar and head branching. — The introduction in the model of the pair interaction parameters, determined from the CMC of the binary mixtures, gives valuable predictions for the CMC and for the micelle compositions of the ternary mixtures. CMC of the ternary mixture are lower than the CMC of each component of the mixture. Contrary to that which is expected for an ideal mixing, the mixed micelles are principally composed of ionic surfactant molecules. Our results seem to show that these behaviors could depend slightly on the position of the polar head branching and on the length of the alkyl chain of the anionic surfactant.

Key words: CMC; mixed micelles; anionic surfactant; cationic surfactant; nonionic surfactant; regular solution theory

Introduction

Models for binary mixtures have been developed on the basis of pseudo phase separations. Generally, they assume ideal mixing of the surfactants in the micelle for nonionic *or* ionic surfactants, particularly when the surfactants have the same hydrophilic group [1—4]. But for nonionic *and* ionic surfactants, or surfactants with different hydrophilic groups, it is necessary to assume nonideal mixing in the micelle which, in most cases, can be treated via a regular solution approximation [5, 6].

More recently, the above treatment has been extended to multicomponent nonideal mixed micelles [7, 8]. A good agreement has been found between theory and experiment for ternary nonideal mixed surfactant systems using pair interaction measurements on binary mixtures [7, 9].

In this paper, we use the same approach to study the CMC and the micelle compositions of typical cationic/nonionic/anionic surfactant mixtures. The cationic surfactant is tetradecyltrimethylammonium bromide (TTAB), the nonionic surfactant is polyoxyethylene octylphenol ($OP(EO)_8$), and the anionic surfactants are alkylbenzene sulfonates ($n\Phi Cm$). We have already studied similar mixtures for synergism of micelle formation, for instance, as a function of the oxyethylene number of the nonionic surfactant [10]. Here, we study the influences of the anionic surfactant structure.

In the first part of the paper, we recall the main relations required for the understanding of the work. In the second part, we give information about our experimental procedure. In the third part, we use the pair interaction parameters that we determined from CMC measurements, for a detailed study of critical concentrations and compositions of TTAB/OP(EO)$_8$/$n\Phi Cm$ micelles.

Ternary mixed micelle formation

The development of a pseudo phase separation model for the formation of nonideal ternary mixed micelles [7, 8] shows that the CMC of an anionic/cationic/nonionic surfactant system, the mole fractions x_a of the anionic surfactant, and x_c of the cationic surfactant in the mixed micelle obey the equations

$$a_a C_{acn} = x_a C_a \exp[\beta_{ac}x_c^2 + \beta_{an}(1 - x_a - x_c)^2$$
$$+ (\beta_{ac} + \beta_{an} - \beta_{cn})x_c(1 - x_a - x_c)] \quad (1)$$

$$a_c C_{acn} = x_c C_c \exp[\beta_{ac}x_a^2 + \beta_{cn}(1 - x_a - x_c)^2$$
$$+ (\beta_{ac} - \beta_{an} + \beta_{cn})x_a(1 - x_a - x_c)] \quad (2)$$

$$(1 - a_a - a_c)C_{acn} = (1 - x_a - x_c)C_n$$
$$\cdot \exp[\beta_{an}x_a^2 + \beta_{cn}x_c^2$$
$$+ (-\beta_{ac} + \beta_{an} + \beta_{cn})x_a x_c] . \quad (3)$$

In these expressions:

— a_a and a_c, the mole fractions of the anionic surfactant and of the cationic one in the total mixed solution, are known.
— C_a, C_c, C_n, the CMC of each pure surfactant, can be measured.
— β_{ac}, β_{cn}, and β_{an}, the pair interaction parameters, can be deduced from CMC measurements of the binary mixtures [9].

Thus, the three unknowns C_{acn}, x_a, x_c, (so $x_n = 1 - x_a - x_c$) can be calculated from Eqs. (1), (2), (3).
As C_{acn} can also be directly measured, it is possible to compare model with experiment.

Experimental procedure

All the CMC have been measured in water at 25 °C.

The tetradecyltrimethylammonium bromide was provided by Aldrich-Chimie; its composition in quaternary ammonium is higher than 99%. The polyoxyethylene octylphenol was synthesized at the Institut de Recherche de Chimie Appliquée (IRCHA); its purity grade was given as higher than 95%, and a gas chromatography analysis showed that it was really monodistributed. The alkylbenzene sulfonates are branched. Branching position and chain length are provided as a function of the carbon atoms of the alkyl chain: n indicates the position of the branching of the polar head on the alkyl chain, m is the length of the alkyl chain. These surfactants were synthesized at the University of Texas [10]; their purity grade was higher than 99%. Water was deionized and redistilled.

The CMC were mainly determined from surface tension versus concentration data, measured according to the Wilhelmy plate method, with a Krüss digital tensiometer. When surface tension measurements were difficult to interpret, alternative methods were used. Thus, CMC of ionic surfactants or of some mixtures including principally an ionic surfactant were measured by conductivity with a Wayne and Kerr bridge, Model B 331. CMC of nonionic surfactants or CMC of mixtures including principally a nonionic surfactant were measured by studying the ultraviolet absorption of the surfactant solutions as a function of the surfactant concentration with a Beckman spectrophotometer, Model 34.

Results and discussion

The CMC of the pure surfactants are reported in Table 1. For symmetrically branched anionic surfactants (5ΦC10, 6ΦC12, 7ΦC14), we found that the CMC decreases when the alkyl chain length increases. For anionic surfactants which have the same alkyl chain length (3ΦC10, 5ΦC10), the CMC was found to decrease when the branching was shifted toward the end of the alkyl chain. The CMC of the nonionic surfactant is about 10 times lower than the lowest of the anionic surfactant CMC.

We deduced the pair interaction parameters of the studied surfactants from CMC measurements performed on the binary mixtures; they are reported

Table 1. CMC of pure surfactants

Surfactant	CMC (mol/l)
3ΦC10	5.0×10^{-3}
5ΦC10	8.0×10^{-3}
6ΦC12	2.4×10^{-3}
7ΦC14	1.0×10^{-3}
TTAB	3.3×10^{-3}
OP(EO)$_8$	1.7×10^{-4}

in Table 2. We observed that the $n\Phi Cm$/TTAB interaction was much higher than $n\Phi Cm$/OP(EO)$_8$ and TTAB/OP(EO)$_8$ interactions have already been found in other systems of the same type [5, 9]. This large difference is due to the high electrostatic attraction between ionic molecules whose polar heads have opposite charges. Alkyl chain length and branching position have a low influence on $n\Phi Cm$/TTAB interaction: the variations of the pair interaction parameter do not exceed 20%. But the $n\Phi Cm$/OP(EO)$_8$ interaction increases singificantly with the alkyl chain length.

Table 2. Pair-interaction parameters

Mixture	β_{ac}	β_{an}	β_{cn}
3ΦC10/TTAB	−24.05		
5ΦC10/TTAB	−22.30		
6ΦC12/TTAB	−24.10		
7ΦC14/TTAB	−20.70		
3ΦC10/OP(EO)$_8$		−4.2	
5ΦC10/OP(EO)$_8$		−3.9	
6ΦC12/OP(EO)$_8$		−5.0	
7ΦC14/OP(EO)$_8$		−8.2	
TTAB/OP(EO)$_8$			−4.0

Below, we give results on critical concentrations and compositions of TTAB/$n\Phi Cm$/OP(EO)$_8$ ternary mixed micelles issued from equimolar $n\Phi Cm$/TTAB blends to which were added known quantities of OP(EO)$_8$. For each ternary surfactant system, the variation of the interfacial tension as a function of the surfactant concentration presents a single discontinuity, so the CMC is determined without ambiguity.

For binary surfactants mixtures, a negative pair interaction coefficients is sufficient to be sure that a single type of micelle is created [12]. This condition

extended to the three pair interaction coefficients is not sufficient for ternary surfactant mixtures. However, the observation of a single discontinuity in the variation of the interfacial tension as a function of the surfactant concentration allows us to suppose that a single type of micelles is really created in these systems, although the three pair interaction coefficients are negative.

CMC of the $n\Phi Cm$/TTAB/OP(EO)$_8$ mixtures

The measured CMC appear as increasing functions of the OP(EO)$_8$ added quantities (Figs. 1, 2). They are always lower than the CMC predicted for an ideal mixing of the three surfactants. Being maximal for the initial $n\Phi Cm$/TTAB blend, the difference decreases as the OP(EO)$_8$ proportion increases in the ternary mixture. These differences increase the higher the $n\Phi Cm$/TTAB interaction.

A relatively good agreement is obtained when the experimental measurements are compared with the CMC values calculated from Eqs. (1), (2), (3), in which are included the CMC of the three pure surfactants, the pair interaction parameters, and the a_a, a_c experimental values (Figs. 1, 2). The use of the CMC of the three pure surfactants and of the indicated values of a_a and a_c means that, as a first approximation, we have ignored the counterion effect. We think this approximation is correct, because we suppose the electrostatic attraction between a $n\Phi Cm$ molecule and a TTAB one to be sufficiently high to obtain a pair of molecules having a nonionic character; this pair of molecules can then participate in the formation of the ternary mixed micelles without other significant interaction with the actual OP(EO)$_8$ surfactant molecules.

Compositions of the $n\Phi Cm$/TTAB/OP(EO)$_8$ mixed micelles

The observed good agreement between experiment and theory for CMC allows us to calculate the compositions of the ternary mixed micelles from Eqs. (1), (2), (3). The results of these calculations are reported in Figs. 3 and 4. We see that for low OP(EO)$_8$ concentration in the mixture, the mixed micelle is principally composed of $n\Phi Cm$ and TTAB. The mole fractions of $n\Phi Cm$ and TTAB are then close to 0.5. The $n\Phi Cm$ mole fraction seems to

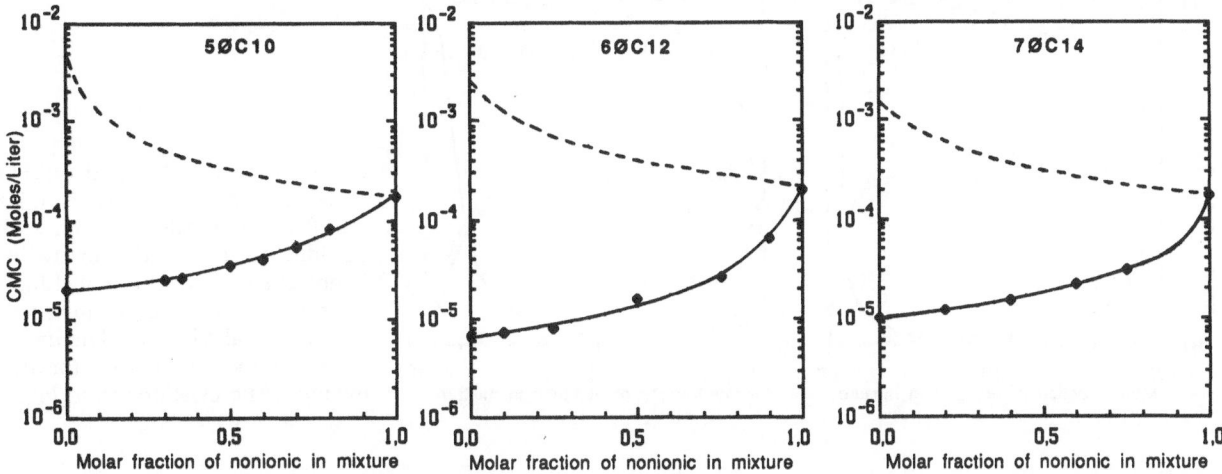

Fig. 1. CMC of $n\Phi Cm$/TTAB/OP(EO)$_8$ mixtures for different lengths of the alkyl chain of the anionic surfactant. The plotted points are experimental data, and the solid lines are the predictions of the nonideal mixed micelle model. The upper dashed lines are the predictions for ideal mixing

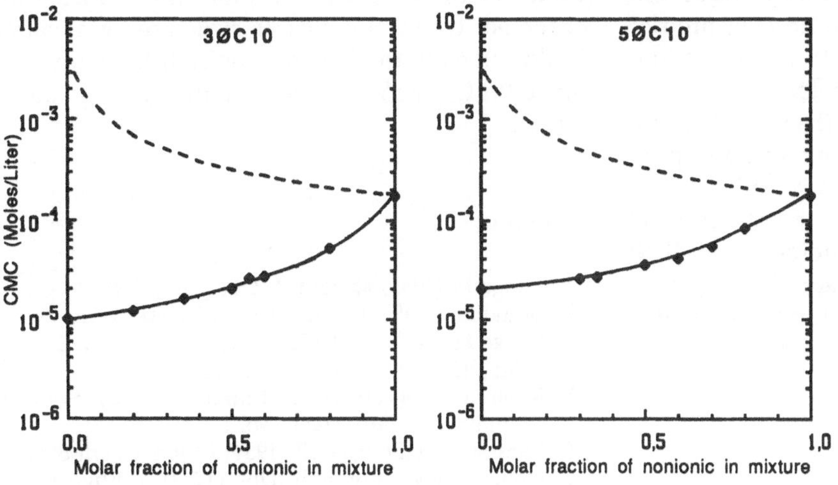

Fig. 2. CMC of $n\Phi Cm$/TTAB/ OP(EO)$_8$ mixtures for two different positions of the branching of the polar head of the anionic surfactant on the alkyl chain. The plotted points are experimental data, and the solid lines are the predictions of the nonideal mixed micelle model. The upper dashed lines are the predictions for ideal mixing

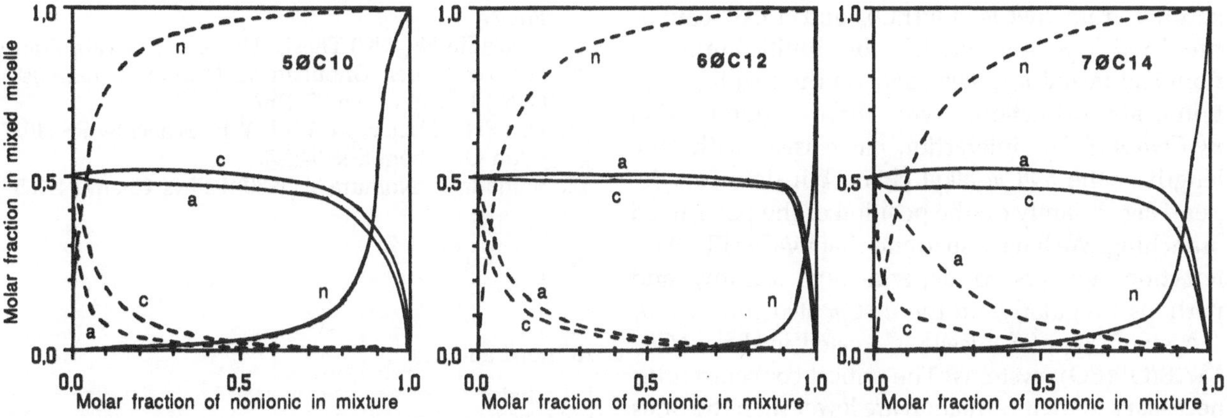

Fig. 3. $n\Phi Cm$, TTAB, and OP(EO)$_8$ mole fractions in the ternary mixed micelles calculated from the nonideal micelle model for different lengths of the alkyl chain of the anionic surfactant. The dashed lines are the predictions for ideal mixing. (Abbreviations: a = anionic; c = cationic; n = nonionic.)

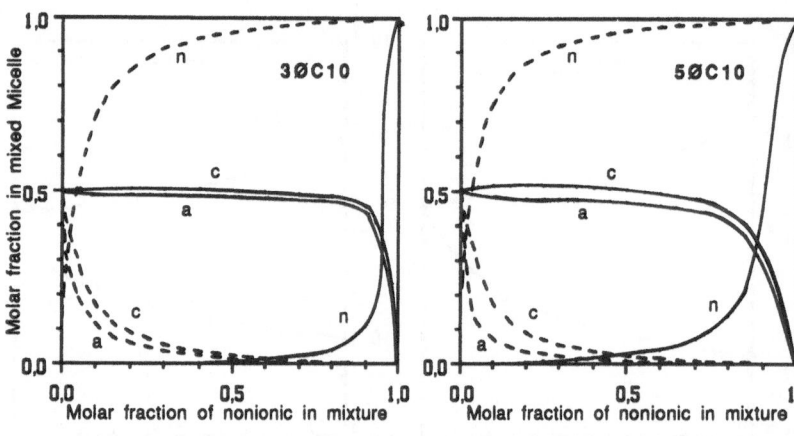

Fig. 4. $n\Phi Cm$, TTAB, and OP(EO)$_8$ mole fractions in the ternary mixed micelles calculated from the nonideal micelle model for two different positions of the branching of the polar head of the anionic surfactant on the alkyl chain. The dashed lines are the predictions for ideal mixing. (Abbreviations as in Fig. 3.)

be slightly higher than the TTAB one for long alkyl chains; it seems to be slightly lower for short alkyl chains or branching position shifted towards one end of the alkyl chain. The mixed micelles are rich in OP(EO)$_8$ only for high OP(EO)$_8$ concentrations in the ternary mixture. This behavior is exactly opposite to that of an ideal mixing (Figs. 3, 4). Thus, the domination of $n\Phi Cm$/TTAB interactions principally favors the presence of $n\Phi Cm$ and TTAB in the ternary mixed micelles, with the structure of $n\Phi Cm$ introducing only low differences in their mole fractions. Finally, it appears that OP(EO)$_8$ could act as a diluent for the salt formed by $n\Phi Cm$ and TTAB in the ternary micelles.

Conclusion

Micellization of binary and ternary systems constituted from $n\Phi Cm$, OP(EO)$_8$ and TTAB can be predicted by the model for multicomponent nonideal mixed micelles used via the regular solution approximation. We have found that $n\Phi Cm$/OP(EO)$_8$ interaction increases with the length of the $n\Phi Cm$ alkyl chain, but does not depend signficiantly on the position of the polar head branching. We have also found that $n\Phi Cm$/TTAB interaction appears to depend very slightly, and perhaps irregularly, on the $n\Phi Cm$ structure.

A single type of micelle is created in the $n\Phi Cm$/TTAB/OP(EO)$_8$ systems. The critical concentrations necessary to their formation are lower than the ones predicted for an ideal mixing; contrary to that which is expected from an ideal mixing, they are principally composed of $n\Phi Cm$ and TTAB, with

OP(EO)$_8$ acting only as a diluent for the salt formed by the two ionic surfactant molecules. The high values of the $n\Phi Cm$/TTAB interaction which were observed when the $n\Phi Cm$ structure was varied could intensify the lowering of both the CMC and the OP(EO)$_8$ concentration in the ternary mixed micelles.

References

1. Lange H (1953) Kolloid Z Z Polym 131:96
2. Shinoda K (1954) J Phys Chem 58:541
3. Lange H, Beck KH (1973) Kolloid Z Z Polym 251:424
4. Clint J (1975) J Chem Soc 71:1327
5. Rubingh DN (1979) in Mittal KL (ed) Solution Chemistry of Surfactants Vol I, 337
6. Kamrath RF, Franses EI (1984) J Phys Chem 88:1642
7. Holland PM, Rubingh DN (1983) J Phys Chem 87:1984
8. Ben Ghoulam M (1984) Thesis, University of Pau, France
9. Moatadid N (1987) Thesis, University of Pau, France
10. Graciaa A, Ben Ghoulam M, Marion G, Lachaise J (1989) J Phys Chem 93:4167
11. Doe PH, El Emary M, Wade WH, Schechter RS (1977) J Am Oil Chem Soc 54:570
12. Shinoda K, Nomura T (1980) J Phys Chem 84:365

Authors' address:

J. Lachaise
L.T.E.M.P.M.
Centre Universitaire de Recherche Scientifique
Avenue de l'Université
64000 Pau, France

Progress in Colloid & Polymer Science   Progr Colloid Polym Sci 89:293—296 (1992)

An infrared study of micelle formation in AOT-H$_2$O-CCl$_4$ solutions

G. Onori and A. Santucci

Dipartimento di Fisica, Università di Perugia, Italy

Abstract: The concentration dependence of the infrared spectra of CCl$_4$ solutions of AOT [bis(2-ethylhexyl)sodium sulfosuccinate] has been determined in a concentration range (0—1.6 mM) encompassing the critical micelle concentration (CMC $\cong$ 0.2 mM). — Effects detected on the molar extinction coefficients of the bands due to C—H stretching vibrations can be used as indicators of micelle formation. — The same spectroscopic technique has also been used to investigate the structure of water in oil microemulsion. The interaction of water with reversed micelles of AOT has been studied as a function of the ratio W = [H$_2$O]/[AOT] in the range 0—12 by using the absorption band due to O—H stretching mode, in the 3800—3000 cm^{-1} range, and the absorption band in the range 1700—1500 cm^{-1}, arising from the H—O—H bending vibration. Our results indicate that at low W values (W < ~4) the water in micellar core is structurally and motionally different from unperturbed bulk water. On increasing W, water displays more marked "bulk" properties, and the absorption bands approach that of bulk water.

Key words: Surfactants; bis(2-ethylhexyl)sodium sulfosuccinate; infrared spectroscopy; micellization

1. Introduction

Aerosol OT [bis(2-ethylhexyl)sodium sulfosuccinate] molecules aggregate in non polar solvents and form reversed micelles in which large amounts of water can be dissolved. Several features of these systems remain to be solved. Some of these pertain to the very debated question of water structure close to the interface, and to whether a well-defined critical concentration (CMC) exists in these micellar systems [1—3]. The present investigation deals with both these problems by using infrared spectroscopy. Although most of the properties of water in reversed micelles have been closely investigated, little attention has been paid to their vibrational spectra. Infrared (IR) spectroscopy is a non-invasive technique, which is functional-group selective, sensitive to the chemical environment, and is used extensively in fields in which a direct evaluation of molecular order and organization is of primary interest.

In this paper, the interaction of water with reversed micelles of AOT in CCl$_4$ has been investigated by following the changes in the C—H, O—H and H—O—H vibrational bands as a function of AOT and H$_2$O concentration. In CCl$_4$ almost all the water is solubilized in the reversed micelle and, therefore, the data analysis is more direct.

2. Experimental section

AOT 99% (Alfa Product) purified by recrystallization from methanol and drying in vacuo was stored in vacuo over P$_2$O$_5$. Bidistilled water and CCl$_4$ (purity 99.5%) were used without additional purification. The AOT/H$_2$O/CCl$_4$ mixtures were prepared by weight. Some residual water molecules remain bound to the AOT molecules after the drying process of the surfactant. Analysis of the water content of purified AOT and CCl$_4$ mixtures with a Karl Fisher titrator revealed, in several samples, the presence of 0.15 mole of water residual per mole of AOT. Such a small residual of water was considered as a part of the total water in the mixture under study.

IR spectra were recorded by means of a Shimdazu Mod. 470 infrared spectrophotometer equipped with a variable path-length cell and CaF_2 windows. Typical path-lengths employed were 50 µm for $AOT/H_2O/CCl_4$ mixtures. Pure water spectra were taken with path-lengths ranging from 12.0 to 2.0 µm.

The molar-extinction coefficients were calculated by using the expression $\varepsilon = A(c \cdot d)$, where A is the absorbance, c the concentration of absorbing specie in moles/l, and d the cell depth in centimeters.

3. Results and discussion

Figure 1 shows the IR spectra relative to $AOT/H_2O/CCl_4$ system at two values of molar ratio of $W = [H_2O]/[AOT]$ ($W = 0.15$; $W = 5.2$). The spectra were recorded in the 4000—1300 cm^{-1} range. The AOT concentration was the same in both solutions.

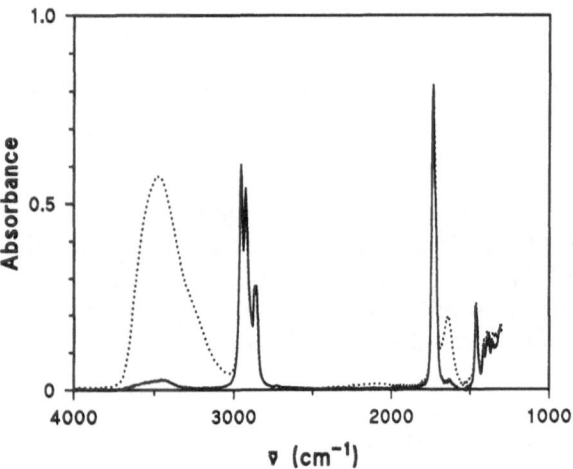

Fig. 1. IR spectra of $AOT/H_2O/CCl_4$ mixtures. [AOT] = 0.149 mol $\cdot$ l^{-1}. (—): $[H_2O]/[AOT]$ = 0.15; [$\cdots$]: $[H_2O]/[AOT]$ = 5.2

The contribution to the spectrum due to the vibrational modes of water and AOT are well characterized, because they appear in distinct spectral ranges. In the 4000—3000 cm^{-1} range appears the absorption due to symmetric and antisymmetric stretching modes of water. The $-CH_3$ and $-CH_2$ stretchings vibrations show absorptions in the 3000—2700 cm^{-1} range. At wavenumbers lower than 1700 cm^{-1} and higher than 1500 cm^{-1} absorp-

tion-bands originate from the $-C=O$ stretching modes of AOT and the H—O—H bending mode of water. At wavenumbers lower than 1500 cm^{-1}, we observe only AOT absorptions. No perturbative effects on the peaks shapes of $-CH_3$ and $-CH_2$ stretching modes due to the presence of water are observed. On the other hand, very small contributions attributable to the interaction of the water with sulphonic and ester groups are observed in the wavenumber region lower than 1500 cm^{-1}.

The concentration dependence of the infrared spectra of CCl_4 solutions of AOT ($W = 0.15$) has been determined in concentration ranges encompassing the CMC. Effects detected on the molar extinction coefficient of the bands due to C—H stretchings vibrations (Fig. 2) can be used as indicators of micelle formation. From the data, on increasing AOT concentration, a noticeable lowering in the molar extinction coefficient measured at the maximum of the antisymmetric and symmetric $-CH_3$ and $-CH_2$ stretching bands peaked at 2965, 2935, 2865, and 2878 cm^{-1} is observed. The ε values largely decrease at very low concentrations of surfactant and level off if one increases the AOT concentration (see Fig. 3). The concentration corresponding to the break in the ε vs [AOT] curve (Fig. 3) can be interpreted as CMC; its value ($\cong 0.2$ mM) coincides very satisfactorily with comparable findings obtained by other techniques [4]. Furthermore, the initial decrease of ε may indicate a possible premicellar structure.

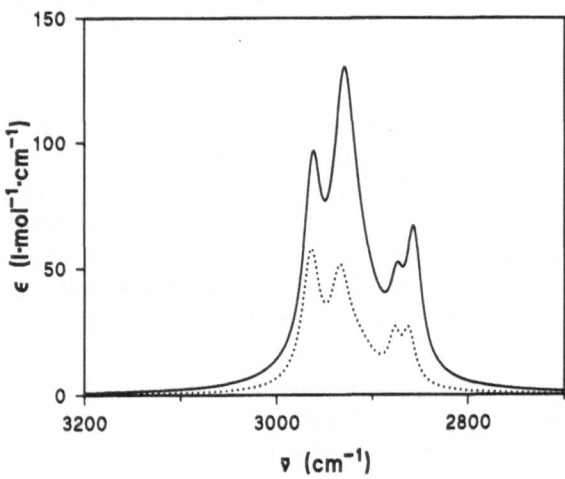

Fig. 2. Molar extinction coefficients vs. wavenumbers for $-CH_3$ and $-CH_2$ stretchings in AOT/CCl_4 mixtures. (—): [AOT] = 0.1 mol $\cdot$ l^{-1}; [$\cdots$]: [AOT] = 1.2 mol $\cdot$ l^{-1}

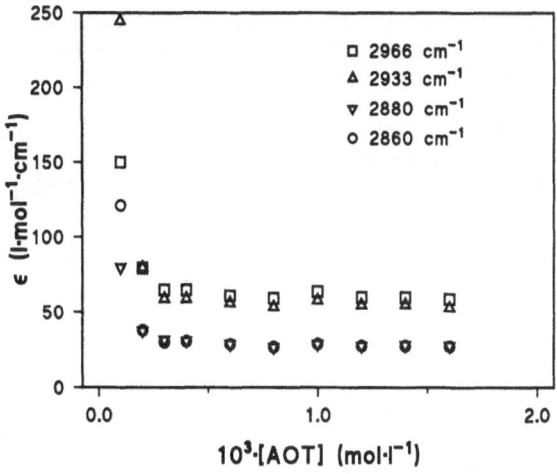

Fig. 3. Molar extinction coefficients measured at peak maximum for —CH$_3$ and —CH$_2$ stretchings as a function of [AOT] in AOT/CCl$_4$ mixtures

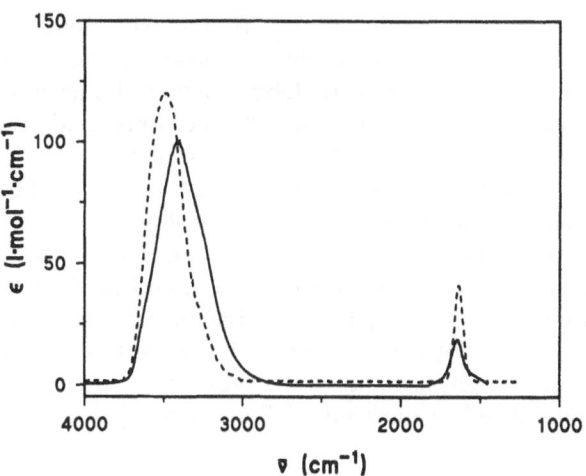

Fig. 4. Molar extinction coefficients vs. wavenumbers for water (—) and water in micellar core at [H$_2$O]/[AOT] = 0.7 (---)

Various spectroscopic parameters of water in reversed micelles have indicated that micelle solubilized water is perturbed by the highly charged interior surface of micelle [5]. "Perturbed" and "bulk" water can be distinguished by their characteristic vibrational band. According to our results, none of the peak frequencies or peak shapes due to the —CH$_3$ and —CH$_2$ stretching bands change on increasing W (see Fig. 1). Thus, it is possible to obtain the contribution of the vibrational modes of the water to the IR-spectrum simply by subtracting the IR spectra of the AOT/H$_2$O/CCl$_4$ and the AOT/CCl$_4$-system.

As an example, we show in Fig. 4 the stretchings and bending vibration bands of water in micellar core at W = 0.7. For comparison, in the same figure the ε of pure water is also reported; there is a significant difference in the spectra. A shift towards higher wavenumbers for the —OH stretching band is observed, while no appreciable frequency shift is found in the H—O—H bending absorption. A remarkable increase of the molar extinction coefficients is observed for both vibrational modes of the water. These differences indicate that water inside the micellar core is not hydrogen-bonded in the same manner as bulk water. At low water content all

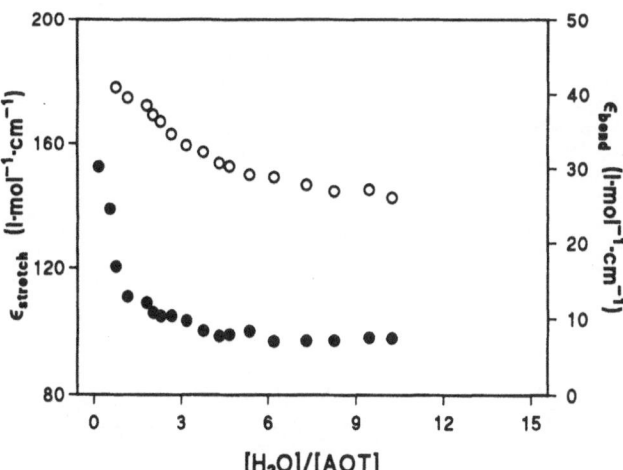

Fig. 5. Molar extinction coefficients measured at peak maximum as a function of [H$_2$O]/[AOT] for bending (○○○) and stretchings (●●●) of water in micellar core

the water molecules are expected to exist mainly as H_2O-bounded molecules whose static and dynamic properties are determined by the local interactions with the Na^+ counter-ions and the strong dipole of the AOT polar groups.

On increasing W, water displays more marked "bulk" properties and the absorption bands approach that of bulk water (Fig. 5). The ε values change significantly up to $W \cong 4$ and level off at higher water content. This behavior is qualitatively in agreement with data in the literature referring to several physico-chemical properties of water solubilized in reversed AOT micelles [5].

References

1. Eicke HF (1980) Top Current 87:86—145 (and references therein)
2. O'Connor CJ, Lomax TD, Ramage RE (1984) Adv Colloid Interface Sci 20:21—97 (and references therein)
3. Chevalier Y, Zemb T (1990) Rep Prog Phys 53:279—371 (and references therein)
4. Chen SH (1986) Ann Rev Phys Chem 37:351—399 (and references therein)
5. Hauser H, Haering G, Pande A, Luisi PL (1989) J Phys Chem 93:7869—7876

Authors' address:

G. Onori
Dipartimento di Fisica
Universita' di Perugia
V. A. Pascoli
I-06100 Perugia, Italy

Progress in Colloid & Polymer Science   Progr Colloid Polym Sci 89:297—301 (1992)

Effect of 1-alcohols on micelle formation and hydrophobic interactions

G. Onori and A. Santucci

Dipartimento di Fisica, Universita' di Perugia, Italy

Abstract: The effect of methanol, ethanol, and 1-propanol on the critical micelle concentration (CMC) of several surfactant aqueous solutions has been studied in the 10—40 °C temperature range by surface tension and conductivity measurements. Experimental parameters such as the standard free energy, enthalpy and entropy of micellization were also determined. — The results show that on increasing alcohol concentration the CMC first reaches a minimum at an intermediate cosolvent mole fraction $x_2 = x_2^*$ typical for each alcohol, and then increases with increasing x_2. The value of x_2^* is close to that at which structural changes in the mixtures occur as inferred from compressibility, UV, IR absorption spectra, and neutron and x-ray scattering measurements. — The present experiments support the assumption that the dominant mechanism by which 1-alcohols affect the micellization process is through their effect on the structure of solvent. It is suggested that the same conclusion could be valid for a number of other systems of biological and chemical interest.

Key words: Surfactants; water-alcohol mixtures; hydrophobic interactions; micellization; ribonucleic acid

1. Introduction

Recent results from our laboratories [1—3] show that the effects due to alcohols on two very different systems and processes, the thermal denaturation of transfer-ribonucleic acid molecules and the micellization of several surfactant molecules, are strikingly similar and are closely parallel in simpler properties of alcohol-water mixtures themselves. Accordingly, these results support the hypothesis that the dominant mechanism by which an alcohol affects these processes is through its effect on the structure of water.

The main results of these investigations are reviewed and discussed in this paper.

2. Results and discussion

A variety of properties of alcohol-water mixtures [4—6] strongly suggest the formation, at low alcohol concentration, of low-entropy structures or "cages" of fairly regular and longer-lived H bonds around hydrophobic groups. On increasing alcohol concentration a progressive interference among these structures are expected to cause a progressive loss of the low-entropy, high connectivity character of cages. Quite often, the mixtures' properties show an anomalous behaviour that occurs in the water-rich region and, for each alcohol, single out the critical cosolvent mole fraction, x_2^*, that corresponds to the maximum structuring in the solvent caused by the presence of alcohol. Further evidence for the water-methanol, water-ethanol, and water-1-propanol systems has been recently collected from compressibility [7, 8], UV [9] and NIR [10] absorption spectra, and neutron and x-ray [11] measurements done in our laboratories.

As an example of such a behavior, we report in Fig. 1 the apparent molar compressibility of methanol, ethanol, and 1-propanol in water as a function of alcohol mole fraction x_2. This quantity is nearly constant at low alcohol concentration ($x_2 < x_2^*$), but increases steeply as more alcohol is added to the solutions. The transition observed at x_2^* can be ascribed to structural changes that take

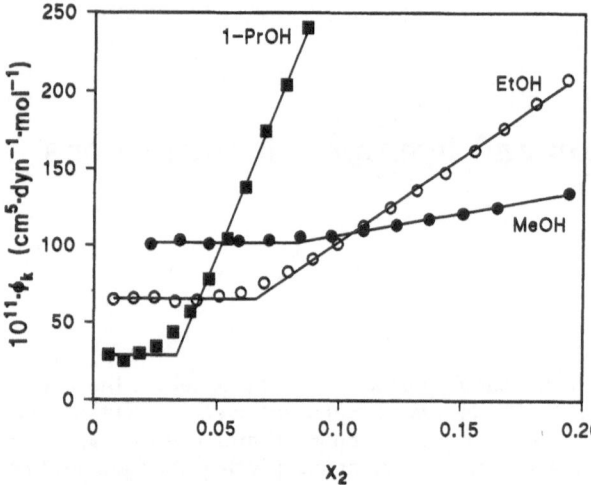

Fig. 1. Apparent molar isentropic compressibility Φ_k of methanol, ethanol, and 1-propanol in water at 20°C as a function of alcohol mole fraction x_2

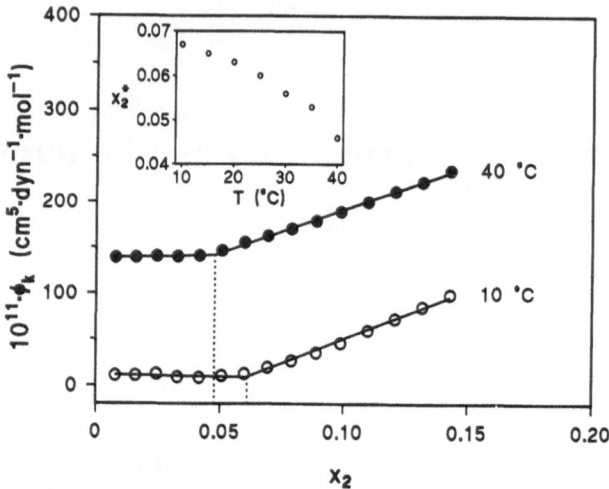

Fig. 2. Apparent molar isentropic compressibility Φ_k of ethanol in water at 10° and 40°C vs. mole fraction of ethanol. A discontinuity in the slope is observed at a value $x_2 = x_2^*$ that shifts towards lower values of x_2 with increasing temperature. (See inset.)

place in the hydrate sphere of alcohol. The x_2^* value decreases on going from methyl alcohol to propyl alcohol and divides the water-rich region into two subregions, in each of which distinct structural effects predominate. x_2^* is slightly dependent on temperature, shifting towards lower alcohol concentrations with increasing temperature (Fig. 2).

These results are of interest not only in themselves, but also for their relevance to the effects of solvent perturbation on the conformation of biological macromolecules [12], on the chemical reaction rates and equilibria [13], and on the micellization process [14]. In this regard, the effect of alcohol, often considered merely a means for lowering the dielectric constant of solvent and to affect the interactions among groups charged, is found to be more complex [1—3]. Figures 3 and 4 clearly show that alcohols affect the micellization process. Addition of low quantities of alcohol reduces the CMC, but high concentrations tend to increase it. This behavior depends on the carbon atoms number of the surfactant and the nature of the counterion and head group [2]. Nevertheless, in all the cases it can be observed that the CMC values decrease on increasing the mole fraction of alcohol until a minimum value is reached at $x_2 = x_{2m}$; thereafter, they increase. Similar results can be found in the literature, both for ionic and nonionic surfactants [14]. The x_{2m} concentration shifts to lower values with increasing alkyl group size

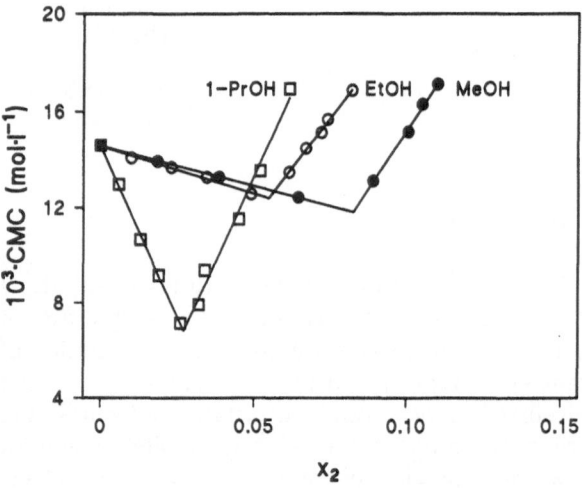

Fig. 3. The CMC values of dodecyl-trimethyl-ammonium-bromide (DTABr) as a function of mole fraction of alcohol at 20°C

(Fig. 3) and with increasing temperature (Fig. 4), assuming values surprisingly close to those at which a transition in the compressibility of plain water-alcohol mixtures has been observed (Figs. 1 and 2). Thus, it appears that the stabilization of micellar structure in the water-alcohol mixtures is closely linked to the properties and anomalous behavior of the solvent system.

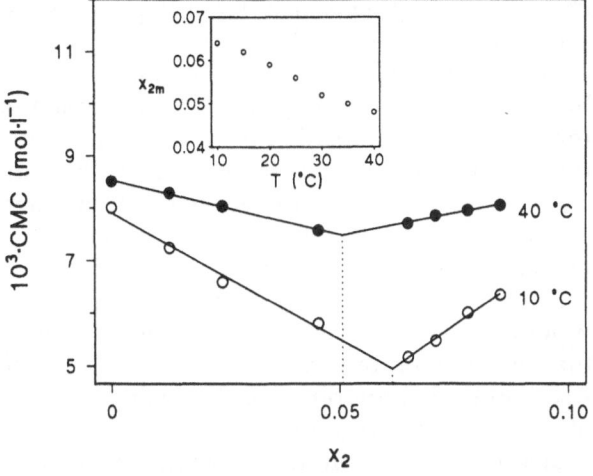

Fig. 4. The CMC values of sodium dodecyl-sulfate (SDS) as a function of mole fraction of ethanol at 10° and 40°C. The concentration x_{2m} corresponding to a minimum in the CMC shifts towards lower values with increasing temperature (see inset)

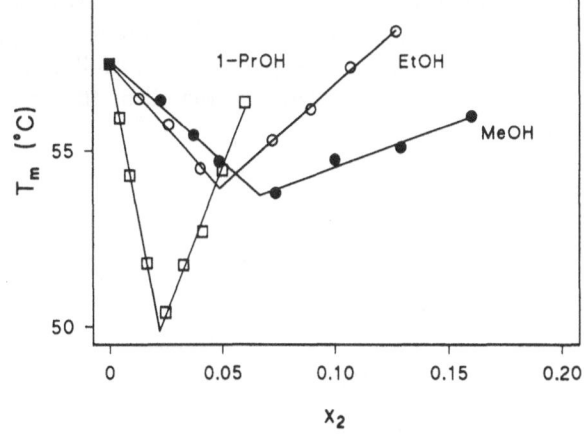

Fig. 5. Variation of the melting temperature T_m of an aqueous solution of transfer ribonucleic acid molecules containing various amount of methanol, ethanol, and 1-propanol as a function of the alcohol mole fraction

It is interesting to note that, in the same concentration range at which we found a minimum in the CMC, we also found an inversion of trend in several processes of chemical and biochemical interest studied in the water-alcohol solvent; this, for example, can be observed on studying the effect of alcohols upon the stability of protein and nucleic acid conformations [1, 3, 12], or the equilibria and rates of chemical reactions [13]. As an example, the melting temperature (T_m) of transfer ribonucleic acid (tRNA) solutions containing various amounts of methanol, ethanol, and 1-propanol are plotted in Fig. 5 as a function of the alcohol mole fraction x_2 [3]. Similar to that found for the CMC (Fig. 1) the observed effects depend on the size and concentration of the alcohol added. The T_m of tRNA at low values of x_2 decreases gradually on increasing the alcohol concentration, and the effect is more pronounced on increasing hydrocarbon content of the alcohol. The observed decrement of T_m is roughly linear in x_2 and continues up to a minimum at an intermediate composition typical for each alcohol; thereafter, an inverted trend is observed, i.e., the T_m increases with increasing x_2. The x_2 values at which the melting temperatures of tRNA change their trend are close to those at which we observed a minimum in the CMC (Fig. 1). The close similarity between the behavior of these two very different systems and processes is striking. It is relevant that

the only common feature is the same solvent system.

These considerations support the contention that changes in the solvent structure due to addition of alcohol are an important factor in the formation of micellar aggregates and in the conformational properties of biomolecules. The structure of the solvent is strictly related to the hydrophobic interactions. The primary contribution to the strength of these interactions derives from changes in the structure of water when non-polar groups interact with one another. Hydrophobic interactions are believed to be closely related to the micellization process and to contribute to the maintenance of ordered macromolecular structures. Such interactions arise from the unique three-dimensional structure of water and should be changed considerably by variations in the solvent structure due to addition of alcohol.

According to Benzinger [15, 16], and as developed by Frank and Lumry [17, 18], the structuring of water around non-polar groups is largely a compensated process, so that the entropy and enthalpy associated with such a structuring nearly compensate each other and, thus, make only a small contribution to the free energy. As a consequence, one would expect that enthalpy and entropy of micellization rather than free energy are affected by alcohol addition.

We use the temperature dependence of CMC for sodium dodecyl sulfate (SDS) solutions (Fig. 4) to find enthalpy and entropy of micellization as a function of ethanol mole fraction [19]. In particular, the standard state value of the free energy ΔG_m^0, enthalpy ΔH_m^0, and entropy ΔS_m^0 of micellization per mole of surfactant were calculated by the well-known relations

$$\Delta G_m^0 = RT \ln(\text{CMC})$$

$$\Delta H_m^0 = T^2 \frac{\partial}{\partial T}\left(\frac{\Delta G_m^0}{T}\right) \tag{1}$$

$$\Delta S_m^0 = \frac{\Delta H_m^0 - \Delta G_m^0}{T},$$

where the critical micelle concentration was taken in mole fraction scale. The results at 35°C were plotted in Fig. 6 as a function of x_2. Large but compensating changes in the enthalpy and entropy appear from the data. These thermodynamic quantities go to a minimum near the concentration x_2, where a maximum structuring in the water-ethanol mixtures is expected [8].

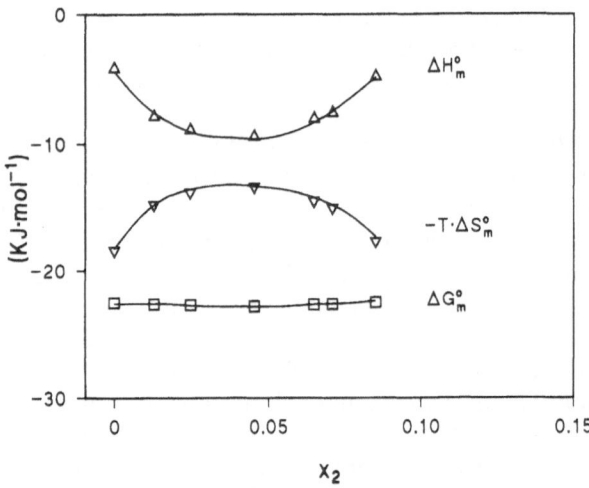

Fig. 6. Micellization of SDS in water-ethanol mixtures at 35°C. Standard state values of the free energy, enthalpy, and entropy vs. mole fraction of ethanol

The overall enthalpy and entropy of micellization is probably due to many different effects. An important contribution, ΔH^{hyd} and ΔS^{hyd}, to these thermodynamic quantities is usually attributed to the release of structured water around the isolated chains on formation of micellar aggregates (hydrophobic effect). It is usually accepted that this process is endothermic ($\Delta H^{\text{hyd}} > 0$) and accompanied by a large entropy change ($\Delta S^{\text{hyd}} > 0$). Based on the discussion of evolution of water-alcohol structure given above, one would expect that ΔH^{hyd} and ΔS^{hyd} are modulated by the alcohol concentration. As a matter of fact, the overall enthalpy and entropy of micellization change with x_2 (Fig. 6) in a way consistent with the suggested evolution in the water-ethanol structure and with the hypothesis that the addition of ethanol affects essentially the ΔH^{hyd} and ΔS^{hyd} contributions. The initial decrease of ΔH_m^0 and ΔS_m^0 (Fig. 6) on increasing x_2 can be related to a decrease in the local enthalpy and entropy due to cage formation by ethanol molecules; these thermodynamic quantities go to a minimum for $x_2 = x_2^*$ where the change in the degree of "structure" induced in the water by the addition of new hydrophobic groups is presumably zero and then increases for $x_2 > x_2^*$ where a progressive distortion of cages is expected. Consistent with the above suggestions, the observed changes in the enthalpy and entropy (Fig. 6) nearly compensate each other and, thus, give a net free energy which is almost independent of alcohol concentration.

In conclusion, the present experiments support the assumption that the dominant mechanism by which short chain alcohols affect the micellization process is through their effect on the structure of the solvent. Available literature data [1, 3, 12] provide further support that the same conclusion could be valid for a number of other systems of biological and chemical interest.

References

1. Beneventi S, Onori G (1986) Biophys Chem 25:181—190
2. Cipiciani A, Onori G, Savelli G (1988) Chem Phys Letters 143:505—509
3. Onori G, Passeri S, Cipiciani A (1989) J Phys Chem 93:4306—4310
4. Franks F (1974) In: Franks F (ed) Water: A Comprehensive Treatise. Plenum, New York, Vol. 2: pp 1—54
5. Franks F. Reiol DS (1974) In: Franks F (ed) Water: A Comprehensive Treatise. Plenum, New York, Vol. 2: pp 323—380
6. Franks F (1975) In: Franks F (ed) Water: A Comprehensive Treatise. Plenum, New York, Vol. 4: pp 1—94
7. Onori G (1987) J Chem Phys 87:1251—1255

8. Onori G (1988) J Chem Phys 89:4325—4332
9. Onori G (1987) Il Nuovo Cimento 90:507—515
10. Onori G (1989) Chem Phys Letters 154:212—216
11. Petrillo C, Onori, G, Sacchetti F (1989) Mol Phys 67:697—705
12. Eagland D (1975) In: Franks F (ed) Water: A Comprehensive Treatise. Plenum, New York, Vol 4 pp 305—518
13. Blandamer MJ (1977) In: Gold V (ed) Advances in Physical Organic Chemistry, Academic Press, New York, Vol 14, pp 203—351
14. Kresheck GC (1975) In: Franks F (ed) Water: A Comprehensive Treatise. Plenum, New York, Vol 4 pp 95—167
15. Benzinger TH (1969) In: Held F (ed) Thermodynamics of Life Growth, Appleton-Century-Crofts, New York, chapter 14
16. Benzinger TH (1971) Nature (London) 229:100
17. Lumry R, Frank H (1978) proc Int Biophys Congr 6th VII-30, 554
18. Lumry R (1980) In: Braibanti A (ed) Bioenergetics and Thermodynamics: Model Systems, Reidel, Dordrecht, pp 405—423
19. Onori G, Santucci A (1992) Chem Phys Letters 189:598—602

Authors' address:

Prof. G. Onori
Dipartimento di Fisica — Universita' di Perugia
V. A. Pascoli
I-06100 Perugia, Italy

Progress in Colloid & Polymer Science   Progr Colloid Polym Sci 89:302—306 (1992)

Molecular theory of rod-shaped micelles

A. Heindl, H.-H. Kohler, and J. Strnad

Institute of Physical and Macromolecular Chemistry, University of Regensburg, FRG

Abstract: Rod formation is primarily governed by the characteristic Gibbs free energy of rod formation, ΔG_{rod}. Geometrically, a rod-shaped micelle consists of a central cylindrical part of hydrocarbon radius r_c and of more or less hemispherical parts at the ends of hydrocarbon radius r_s. To smooth out the transition region between the spherical part of an endcap and the cylindrical part a catenoid is introduced. The model includes hydrophobic, electrostatic, and steric interactions, and allows direct adsorption of counterions to the micelle surface. It is used to calculate the optimal radius of the cylindrical part and to determine the according equilibrium configuration of the endcaps. The theoretical results are compared with empirical data for cetylpyridinium salts. Direct counterion adsorption is found to amount to about 50%. A strong steric influence on counterion adsorption is given.

Key words: Association colloid; rod-shaped micelle; molecular theory; thermodynamics; small systems

1. Introduction

A thermodynamic model is presented that describes the formation of rod-shaped micelles from ionic surfactants.

Rod formation is primarily governed by the characteristic Gibbs free energy of rod formation, ΔG_{rod}. This free energy change is obtained on forming the endcaps from surfactant ions originating from the cylindrical part of hydrocarbon radius r_c and of more or less hemispherical parts at the ends of hydrocarbon radius r_s. To smooth out the transition region between the spherical part of an endcap and the cylindrical part a catenoid is introduced (see Fig. 1).

On a molecular level, the model includes hydrophobic, electrostatic, and steric interactions, and allows direct adsorption of counterions to the micelle surface. The core of the micelle is assumed to consist of a liquid hydrocarbon phase which is allowed to include water molecules. The model is used to calculate the optimal radius of the cylindrical part and to determine the according equilibrium configuration of the endcaps. On this basis, the value of the Gibbs free energy ΔG_{rod} is calculated at 298 K for cetylpyridiniumnitrate,

CpNO$_3$, and cetylpyridiniumchloride, CpCl. The theoretical results are compared with empirical data.

2. Geometry

If r_s, the radius of the spherical part of the endcap, is larger than the radius of the cylindrical part, r_c, a catenoid (which by definition has curvature zero) is used to connect the hemispherical and the cylindrical part of the micelle.

The radius of the endcap is allowed to become larger than the critical length of a hydrocarbon chain, l_{crit}, by permitting water penetration into the center of the hemispherical part of the micelle. The water is assumed to form a droplet of radius r_{H_2O}.

A more detailed picture of the micellar surface is given in Fig. 2. We assume the last methylene group of the C-16 tail to be hydrated, thus forming a layer of thickness d_m. The charge of the pyridinium ring is largely fixed to the nitrogen. So, we assume a Stern layer on top of the methylene groups of thickness d_{St} given by the distance of closest approach of the hydrated counterion. At

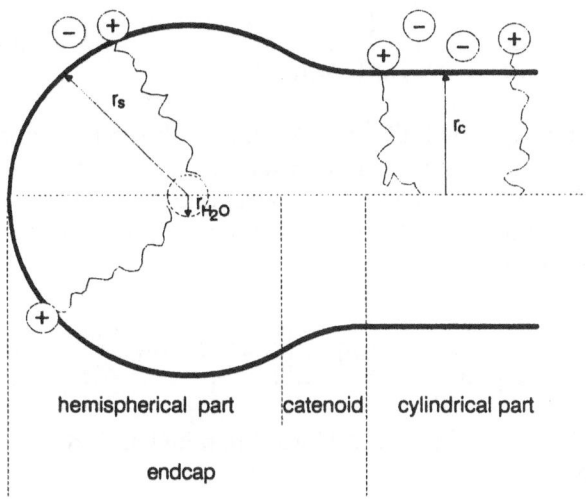

Fig. 1. Schematic drawing of a rod-like micelle

the outer Helmholtz surface (OHS) counterions may undergo direct adsorption which is controlled by the free head group area a_f (total head group area per single surfactant ion minus the projection area of the head itself).

3. Thermodynamics

Let g_c, g_s, g_{ca} denote the Gibbs free energies of a single surfactant ion in the cylindrical, spherical,

and catenoidal part of the micelles and a_c, a_s, a_{ca} the respective head group areas taken in the hydrocarbon surface (HCS). ΔG_{rod} is given by:

$$\Delta G_{rod} = N_{ca}g_{ca} + N_s g_s - (N_{ca} + N_s)g_c , \qquad (1)$$

where the N's are the number of surfactant ions forming the respective part of the micelle. The minimal value of ΔG_{rod} is determined from the following conditions:

$$\frac{\partial g_c}{\partial r_c} = 0 \qquad \frac{\partial \Delta G_{rod}}{\partial r_s} = 0 \qquad \frac{\partial \Delta G_{rod}}{\partial (a_s/a_{ca})} = 0 .$$

To calculate g (which may be g_c g_s or g_{ca}) the integrated form of the Gibbs fundamental equation for a charged and curved surface is used. (For simplicity, we assume that the surfactant ion is cationic and that all ionic species of the system have valency ± 1):

$$g - g^* = a^* \int_0^\tau \frac{\partial \tilde{\sigma}}{\partial \tilde{\tau}} \bigg|_{a^*,q=0} d\tilde{\tau} + \int_{a^*}^a \tilde{\sigma} \, d\tilde{a}$$

$$+ \int_0^{yq_0} \tilde{\varphi}_0 d\tilde{q} + kT \ln y + \varphi_0 q_0 \Theta ; \qquad (2)$$

σ is surface tension, τ curvature ($\tau = 2/r$ for a sphere, $\tau = 1/r$ for a cylinder, $\tau = 0$ for a catenoid),

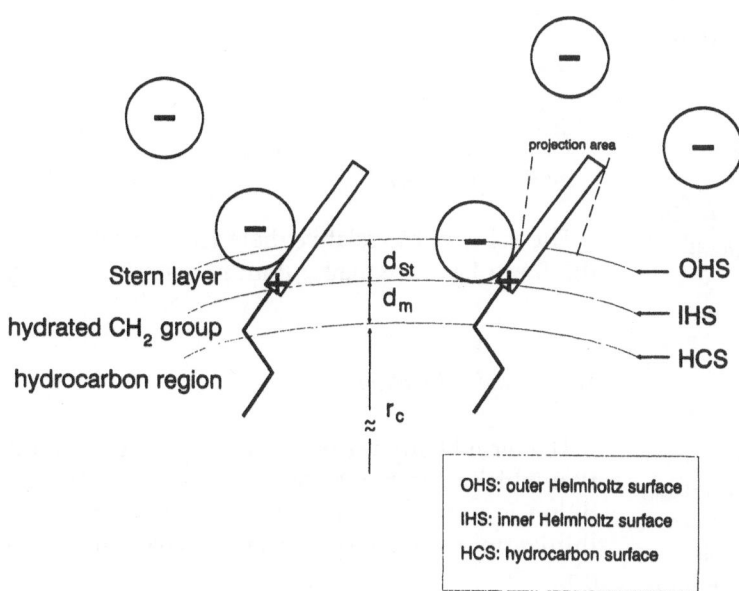

Fig. 2. Surface region of a micelle

φ_0 surface potential (IHS), q_0 elementary charge. Θ is the degree of direct counterion adsorption, and $y = 1 - \Theta$ the degree of counterion dissociation. The symbol * refers to a reference state in a plane surface with electrostatical interaction energy zero and head group area a^*.

Read from left to right, Eq. (2) implies that the plane surface is first curved up to curvature τ, then the head group area is brought to its final value a, and, subsequently, the curved surface is charged up to the final value φ_0 of the surface potential. The following two terms describe direct adsorption of the counterion, the last term ensures that the surface potential remains unaffected. The single terms now will be considered in more detail:

First term:

The relation between surface tension and curvature is assumed to be given by:

$$\sigma = \sigma_0 + \frac{\delta}{2}\,\tau^2\,, \qquad (3)$$

where σ_0 is the surface tension of a plane oil/water interface (50 mN/m) and δ a free parameter.

Second term:

The dependency of σ on the head group area a is due to steric repulsion of the head groups which is described by the two-dimensional Volmer equation:

$$\sigma^\tau - \sigma = \frac{kT}{a - \hat{a}}\,;. \qquad (4)$$

σ^τ is the surface tension obtained from equation (3), $\hat{a}$ is the excluded head group area (available from surface tension measurements).

Third term:

Neglecting, for the present, the Stern layer, adsorption of counterions coming from the immediate neighborhood of the surface is assumed to follow a Langmuir isotherm. Using Boltzmann's relation, the degree of counterion dissociation, y, is given by

$$y = \frac{K}{K + c \cdot \exp(\varphi_0 q_0/(kT))}\,, \qquad (5)$$

where c is the bulk concentration of the counterions and K the dissociation constant. K can be written as $K_0 \cdot (nm^2/a_f)$, where K_0 is independent of a_f.

According to Evans and Ninham [1], φ_0 is implicitly given by:

$$x = \frac{q}{a\sqrt{8RT\varepsilon c}} = \sinh\frac{\varphi_0 q_0}{2kT} + \frac{\tau}{\kappa}\tanh\frac{\varphi_0 q_0}{4kT}, \qquad (6)$$

where κ is the Debye-Hückel parameter. By partial integration, we get

$$\int_{q=0}^{q=yq_0} \tilde{\varphi}_0 d\tilde{q} = kTy\left\{\frac{\varphi_0 q_0}{kT} - \frac{2}{xy}\left(\cosh\frac{\varphi_0 q_0}{2kT} - 1\right)\right.$$
$$\left. - \frac{4\tau}{\kappa xy}\ln\left(\cosh\frac{\varphi_0 q_0}{4kT}\right)\right\}$$
$$+ \frac{\varepsilon}{2}\int E^2 dV\,, \qquad (7)$$

where the last term is due to the additonal electrostatic energy stored in the Stern layer and amounts to

$$\frac{q^2}{2a\varepsilon}\cdot d_{St} \qquad \text{for a plane,}$$

$$\frac{q^2 r_c}{2a\varepsilon}\cdot\ln\left(1 + \frac{d_{St}}{r_c}\right) \qquad \text{for a cylinder, and}$$

$$\frac{q^2 r_s}{2a\varepsilon}\cdot\frac{d_{St}}{r_s + d_{St}} \qquad \text{for a sphere.}$$

The value of the relative dielectric constant inside the Stern layer is smaller than in water (about 30).

Fourth and fifth terms:

The term $kT\ln y$ is due to pure Langmuir adsorption which causes φ_0 to remain constant. This condition is met by the fifth term (together with the third term, each head group area is given the charge q_0).

4. Results

The values for the molecular parameters are:

$$\hat{a} = 0.375 \ nm^2 \qquad l_{crit} = 2.05 \ nm$$

$$d_{St}(NO_3^-) = 0.110 \ nm \quad d_{St}(Cl^-) = 0.121 \ nm \ ,$$

while δ and K_0 are used as fitting parameters. Table 1 summarizes the results obtained from fitting the experimental values of ΔG_{rod} available from viscosity and surface tension measurements [2].

The theoretical results obtained from fitting the experimental ones are more completely represented in Fig. 3.

Table 2 shows details of calculation for $CpNO_3$ with $K_0 = 0.6$ mol/l (corresponding to $K \sim (3-6)$ mol/l) and $\delta = 1.39 \cdot 10^{-20}$ J (corresponding to $\sigma^\tau = 7$ mJ/m² for $\tau = 1/nm$).

5. Conclusions

In summary, we conclude from our investigations on rod-shaped Cp^+-micelles that:

— The head group areas in the hemispherical and in the cylindrical part of the micelle are nearly

Tab. 1. Rod formation: theory and experiment

	CpNO₃			CpCl		
$c/(mol \ l^{-1})$	0.05	0.1	0.2	0.05	0.1	0.2
	Experimental					
Rod formation	yes	yes	yes	no	no	no
$\Delta G_{rod}/(kT)$	11.0	12.4	13.8	—	—	—
	Theoretical					
Rod formation*	yes	yes	yes	no	no	no
$\Delta G_{rod}/(kT)$	10.3	12.3	13.9	5.1	6.9	8.6

*: results from aggregation model in [2].

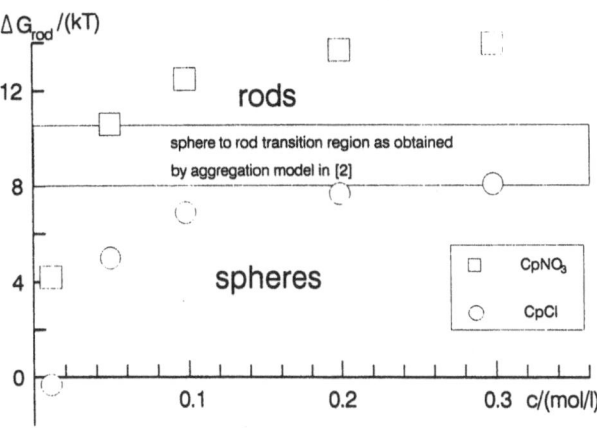

Fig. 3. Theoretical results of rod formation for $CpNO_3$ and CpCl

Table 2. Micelle parameters from curve fitting

$c/(mol \ l^{-1})$	0.05	0.1	0.2	$c/(mol \ l^{-1})$	0.05	0.1	0.2
r_c/nm	1.62	1.63	1.63	N_s	144	144	143
r_s/nm	2.05	2.05	2.05	N_{ca}	26	26	26
$a_c/nm^{2\dagger}$	0.532	0.530	0.529	φ_c/mV^{II}	116	99	82
$a_{c,f}/nm^{2\ddagger}$	0.121	0.120	0.118	φ_s/mV^{II}	106	89	73
$a_s/nm^{2\dagger}$	0.534	0.534	0.534	φ_{ca}/mV^{II}	110	93	76
$a_{s,f}/nm^{2\ddagger}$	0.175	0.175	0.175	$\Theta_c/\%$	47.3	47.7	48.4
$a_{ca}/nm^{2\dagger}$	0.676	0.676	0.675	$\Theta_s/\%$	47.6	48.5	49.8
$a_{ca,f}/nm^{2\ddagger}$	0.186	0.186	0.186	$\Theta_{ca}/\%$	53.2	53.6	54.2

†: HCS; ‡: OHS; II: IHS.

the same at all counterion concentrations, while the areas in the catenoidal part are greater by about 0.15 nm^2.

— Water penetration into the core of the micelle is negligible.

— The relatively weak dependence of ΔG_{rod} on c, found experimentally, requires direct counterion adsorption amounting to about 50%. Adsorption must occur preferentially in the endcap region of the micelle. This suggests a strong steric influence of the free surface on counterion adsorption.

— Rod formation by ionic surfactants requires positive values of the curvature parameter δ. Otherwise, due to electrostatic repulsion, the cylindrical part would be too unfavorable, energetically, to compete with the spherical part.

— Rod formation is favored by counterions with a small distance of closest approach (low energy of the Stern layer).

References

1. Evans DF, Ninham BW (1983) J Phys Chem 87:5025
2. Kohler H-H, Strnad J (1990) J Phys Chem 94:7628

Authors' address:

A. Heindl
Institute of Physical and Macromolecular Chemistry
University of Regensburg
Universitätsstr. 31
8400 Regensburg, FRG

Progress in Colloid & Polymer Science

Progr Colloid Polym Sci 89:307—314 (1992)

Behavior and properties of lyotropic-nematic and lyotropic-cholesteric phases

G. Bartusch[1]), H.-D. Dörfler[1]), and H. Hoffmann[2])

[1]) Lehrstuhl für Kolloidchemie, TU Dresden, FRG
[2]) Physikalische Chemie I, Universität Bayreuth, FRG

Abstract: The formation of lyotropic mesophases in systems of tetra-decyldimethylaminoxide/water/aliphatic alcohol was examined in dependence on alcohol concentration and alcohol chain length (heptanol, octanol, nonanol, decanol). The ternary phase diagrams were established. The following lyotropic phases were detected by microscopic texture observation: hexagonal phase (H), lamellar phase (L_a), rodlike nematic phase (N_C), and disklike nematic phase (N_D). The positions of the corresponding regions in the phase diagrams were determined. In particular, the N_C- and N_D-phase were of special interest. — Lyotropic-cholesteric phases were induced in the lyotropic-nematic phase region by adding optically active components such as cholesterol, tartaric acid, lithocholic acid. When subjected to the influence of a magnetic field, these cholesteric phases show the so-called spaghetti-like texture. The textures have been used for determining the pitch length. A linear relation between pitch length and concentration of the optically active component was established.

Key words: Lyotropic mesophases; nematics; cholesterics; microscopic textures; alignment in magnetic field; calorimetry

1. Introduction

Two goals were pursued in this paper. Hoffmann et al. [1] had investigated the system n-alkyl-dimethylaminoxide/water, where they found a rodlike nematic N_C-phase in the binary phase diagram. Based on these results, we analyzed which kinds of structural changes occurred in the system n-tetradecyldimethylaminoxide/water upon the addition of homologous, unbranched aliphatic alcohols such as heptanol, octanol, nonanol, and decanol. With the knowledge of these ternary phase diagrams of tetradecyldimethylaminoxide/water/n-alcohols, we intended, in a second step, to examine the behavior of the so-called "induced lyotropic-cholesteric phases". We started with the nematic phase of the ternary system tetradecyldimethylaminoxide/water/decanol and added optically active compounds, such as lithocholic acid, cholesterol, and tararctic acid. The aim of these

studies was to give a preliminary characterization of the properties of lyotropic-cholesteric textures under the influence of an external magnetic field and to describe alterations in the pitch length in the cholesteric phase in dependence on the concentration of the aminoxide and of added alcohols.

Cholesteric phases have previously been described in the literature [2—16]. Radley and Saupe [2] were successful in changing lyotropic-nematic phases into cholesteric ones by adding optically active components. These authors examined the following multicomponent systems: Cs-decylsulphate/water/decanol/tartaric acid; NH_4-decylsulphate/water/decanol/NH_2SO_4/brucine sulphate and NH_4-decylchloride/water/cholesterol. The cholesteric phases and their pitch lengths were determined by polarization microscopy.

Optically active surfactants have been applied to the formation of lyotropic-nematic structures for the first time by Acimis and Reeves [3]. In their paper, they distinguish between different types of cholesteric phases. In the system a-alanine hydrochloric decyl ester/water/Na$_2$SO$_4$ a dislike cholesteric phase has been detected. The optical axes of its disks can be oriented perpendicularly to the magnetic field.

Alcantara et al. [4—7] studied the systems K-laurate/H$_2$O/KCl/decanol/cholesterol; K-laurate/H$_2$O/KOH/undecanol/cholesterol; Na-decylsulphate/Na$_2$SO$_4$/H$_2$O/glycin/cholesterol and K-D, L-lauroyl serinate/KCl/KOH/decanol/water/cholesterol. They performed texture observations and ^{2}H-NMR measurements on deuterized K-laurate. The results of NMR-measurements demonstrated that the quadrupole splitting in the cholesteric phase was always larger than in the corresponding nematic phase of the same system. They also described a lyotropic-cholesteric phase whose helical axes were oriented in the direction of the magnetic field.

Neto et al. [8] tested similar systems, where K-laurate and Na-decyl sulphate served as surfactant and brucinesulphate as optically active component. For supporting the orientation process in the magnetic field, ferrofluoride was added. The authors examined the systems by x-ray diffraction measurements. They found a biaxial cholesteric phase and determined the pitch lengths. In this way, they detected that the pitch lengths of disklike and biaxial cholesteric phases which originated from the same mixture had identical numerical values.

Lee and Labes [9] induced lyotropic-cholesteric phases by applying optically active amino acids and dipeptides. Their nematic basic system was disodium chromoglycate/water. The aim of their studies was the evaluation of the so-called helical twisting power, also designated as HTP-value. To obtain this HTP-value the reciprocal pitch length was plotted against the concentration of the optically active amino acid or of the peptide component. The authors applied amino acids with hydrophobic lateral chains and obtained low HTP-values. This suggested a low pitch length for the lyotropic-cholesteric phases. Amino acids with hydrophilic lateral chains, however, showed the opposite behavior.

New results with interesting aspects are presented in a recent paper of Alcantara et al. [10]. The authors added various optically active monosaccharides and disaccharides to the nematic phase of active monosaccharides and disaccharides to the

nematic phase of the system K-laurate/H$_2$O/KCl/decanol. These sugars induced disklike cholesteric phases, but they had only a very low "twisting power". Therefore, the pitch lengths were extremely large. At sugar concentrations above $c = 2$ wt% the pitches could no longer be determined by microscopy. In addition to their texture observations and the lead deterination, the authors measured the rotary polarization angles in dependence on sugar concentration. In this way quantitative statements of the optical activity in connection with the properties of the cholesteric phase could be made.

2. Substrates and sample preparation

n-Tetradecyldimethylaminoxide (C$_{14}$DMAO) was a product of Hoechst AG, Gendorf. For purification, this substance was subjected to a threefold recrystallization with acetone (p.a.). The melting point ($F_p = 131.5$°C) served as purity criterion.

Heptanol, octanol, nonanol, and decanol (p.a.) were from Fluka, Switzerland, and used without further purification. Cholesterol (p.a.) was from Merck, Darmstadt, and lithocholic acid (p.a.) from SERVA. Both were also applied without further purification.

Required amounts of each component (C$_{14}$DMAO, alcohols, water) of the multicomponent system were weighed in 2-ml flasks and then homogenized by means of a powerful hot-air apparatus (type PHG 600 E; Bosch, FRG) and an intensive shaker (type VF 2; Jahnke & Kunkel, FRG). After being homogenized the samples were kept at room temperature, or at $T \approx 300$ K, respectively, for 1 or 2 days. Then, these samples were used for texture observations and calorimeteric measurements.

An overview of the mesophases existing in the multicomponent-systems at the temperature chosen was obtained by texture observations of contact samples [17 18]. Then, the textures of singular concentrations of the mixtures were examined. On the average, up to 16 singular mixtures per system were tested. The C$_{14}$DMAO concentration in the mixtures was varied in steps of 2 wt%. If necessary, smaller steps were applied. The alcohol concentrations were increased in steps of 0.2 wt%.

3. Methods

The lyotropic mesophases — nematic (N_C or N_D), cholesteric (Ch), hexagonal (H), lamellar (L_a), crystalline (C), isotropic (S) — in the phase diagram were classified by referring to texture observations carried out with a polarizing microscope (type Jenapol; Carl Zeiss, FRG). The temperature-dependent behavior of the samples was observed by applying a heating stage (type FP 82; Mettler, FRG), with a heating rate of 0.5 K min^{-1}.

For texture observation, the samples were put into microslides (Camlab, England) of a layer thickness of 0.3 mm. Some of the samples were oriented in the magnetic field of an electromagnet (type BE 20; Brucker, FRG), with a magnetic field strenght of 2.1 T. For orienting the samples in the magnetic field and for observing them by polarizing microscopy, the same microslides were used. During the magnetic field alignment the samples were kept at constant temperature.

The pitch length p of the lyotropic-cholesteric phase was determined by means of polarization microscopy. For this purpose, we used a micrometer eyepiece with an engraved line micrometer. Our test object was a 2-mm-thick sample that had been aligned in the magnetic field and showed the regular spaghetti-like texture, as illustrated by Fig. 5. Under these conditions the helical axis is parallel to the magnetic field, according to the drawing in Fig. 1. The texture, observed by polarizing microscopy, coincides with the vertical direction in the photograph and, in agreement with Fig. 1, it represents a section made across in the direction of the helical axis. The pitch length p is determined with the aid of the micrometer of the microscope.

Fig. 1. The principle of the determination of the pitch length p with the aid of spaghetti textures observed by a polarizing microscope (compare also Fig. 5). Symbols: B = direction of the magnetic field, H_x = direction of the helical axis

The temperatures and enthalpies of the phase transition $N_C \rightarrow S$ in the sytem C$_{14}$DMAO/water/decanol were obtained by calorimetric measurements. We applied a microcalorimeter (Setaram, France) with heating rates of 0.1—0.3 K and a sensitivity of 0.25 mW.

4. Results

4.1. Phase Diagrams of ternary systems tetradecyldimethylaminoxide/water/aliphatic alcohols

In determining the phase diagrams of these ternary systems, we refer to the phase diagrams of the system C$_{14}$DMAO/water already published in [1]. According to the phase diagram of this binary system, isotropic micellar solutions are formed at concentrations between 0—32 wt% of aminoxide. Between 32—35 wt%, we find the realtively narrow region of the rodlike lyotropic-nematic phase (N_C), which will be of particular interest in the following experiments. Then, between 36—66 wt%, we observe a large region of a hexagonal phase (H) and, at higher concentrations, a cubic phase (Q) which — in turn — turns into a lamellar phase L_a at 69 wt%. Finally, at concentrations above 72 wt% no mesophases are observed. The samples were of crystalline modification (C). All concentrations given refer to a temperature of T = 298 K. On increasing the temperature, we observed that the N_C-phase region shifts toward higher C$_{14}$DMAO concentrations and the region of the Q phase becomes broader.

It was now of interest to study the alterations in this binary phase diagram which were brought upon by the addition of decanol and other homologous alcohols. The results are illustrated in the ternary phase diagram of the system tetradecyldimethylaminoxide/water/decanol presented in Fig. 2. Comparing the ternary phase diagram of Fig. 2 with the corresponding binary one [1], we recognize that the single-phase regions are enlarged and an additional phase, the N_D-phase, appears. This lyotropic phase is characterized by a disklike nematic structure.

The N_C-phase extends up to decanol concentrations of 3.6 wt%. The heterogeneous two phase region is succeeded by the L_1-phase which exists between 3.8—4.4 wt% of decanol. It is an isotropic phase that becomes birefringent under the influence of shear forces. A further increase of the decanol concentration to 5.2—8.8 wt% leads to the formation of the N_D-phase. This demonstrates that decanol addition causes an enlargement or a splitting of the nematic phase regions (N_D and N_C).

In contrast to the two nematic phases N_D and N_C, which exist only within a small concentration range, the lamellar phase L_a extends over the

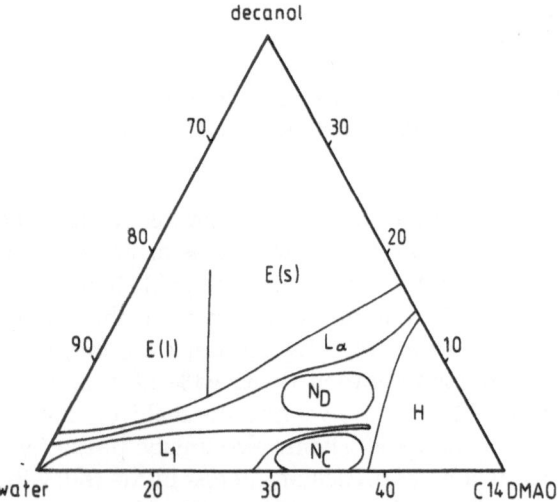

Fig. 2. Phase diagram of the ternary system n-tetradecyl-dimethylaminoxide/water/decanol at T = 298 K. Abbreviations for the phase regions: L = lamellar phase, H = hexagonal phase, N_C = rodlike nematic phase, N_D = disklike nematic phase, L_1 = isotropic phase that becomes birefringent under the influence of shear forces, E_l, aE_s = liquid (index 1) and solid (index s) emulsion (E)

whole concentration range of the quasi-binary system water/C_{14}DMAO. The L_a-phase region is also remarkably expanded by decanol addition, a fact which is also valid for the relatively large region of the hexagonal phase (H). Consequently, we can say that decanol is evidently incorporated into these two phases by forming mixed systems which exist

over considerably larger concentration ranges than the corresponding phases in the binary system [1].

When the decanol concentration exceeds the upper concentration limit of the L_a-phase region, liquid (index 1) and solid (index s) emulsions (E) are formed. These emulsions are of milky appearance and their viscosities vary extremely in dependence of the C_{14}DMAO concentration.

Next, we were interested in how the phase behavior of such ternary systems was affected by aliphatic alcohols of other chain lengths as heptanol, octanol, and nonanol. The results are summarized in Fig. 3. A comparison between the phase diagrams of these four ternary systems demonstrates that the phase behavior is identical in all these systems. Even positions and extensions of the phase regions in the ternary phase diagrams differ only slightly.

4.2. Formation of induced lyotropic-cholesteric phases from lyotropic nematics

Radley and Saupe [2] were the first to discover that lyotropic nematics can be transformed into lyotropic-cholesteric phases upon addition of optically active substances which are not surface active. Lyotropic-cholesteric phases obtained in this manner are designated as "induced cholesteric phases". Some years later, Acimis and Reeves [3] observed that lyotropic-cholesteric phases are also formed in water by optically active surfactants.

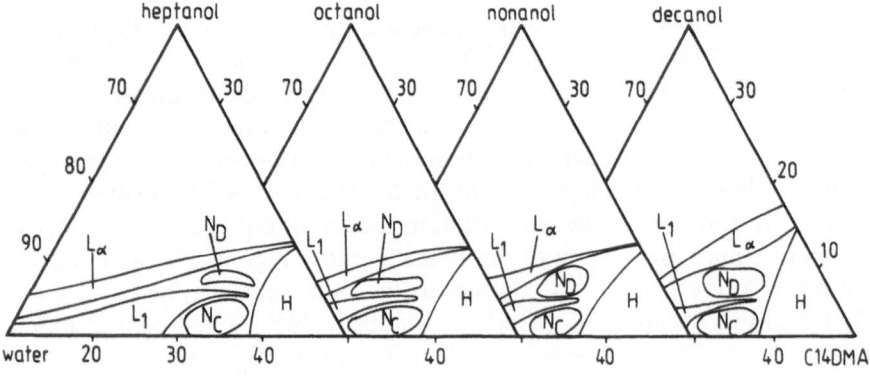

Fig. 3. Phase diagrams of ternary system tetradecyl-dimethyl-aminoxide/water/aliphatic alcohols in dependence on the alkyl chin length of the alcohol used (heptanol, octanol, nonanol, decanol) at T = 298 K. (or abbreviations of the phase region, see legend of Fig. 2)

The goal of the following experiments was therefore to transform nematic phases of ternary systems into "induced cholesteric phases" by adding optically active substances such as cholesterol, lithocholic acid, and tartaric acid. The phenomena which were caused by the addition were studied by texture observation in polarized light, with and without application of a magnetic field. The experiments were performed the same way as already described for the lyotropic-nematic phases.

The existence of a cholesteric phase can be established by the texture of a sample which is enclosed in the microslide and placed under the microscope for several hours. The cholesteric phase will be partially oriented under the influcnece of the glass walls. Then the typical fingerprint texture appears (see Fig. 4). It is generally known that the lyotropic-cholesteric phase is a twisted lyotropic-nematic structure. In analogy to the nematic phases the corresponding cholesteric phases can also be aligned by a magnetic phases. In the rodlike cholesteric phases the helical structure unwinds and the application of an external magnetic field leads to a nematic "Schlieren"-texture.

Fig. 4: Fingerprint texture of the non-aligned lyotropic-cholesteric phase in the system n-tetradecyldimethyl-aminoxide/water/decanol/cholesterol at $T = 298$ K. The phase consists of disklike micelles. Magnification: 100×

In the disklike cholesteric phase, however, the formation of cholesteric structures can easily be observed by means of polarizing microscopy. As shown in Fig. 4, only some domains of the fingerprint texture have an ordered morphology. These domains can be aligned by an external magnetic

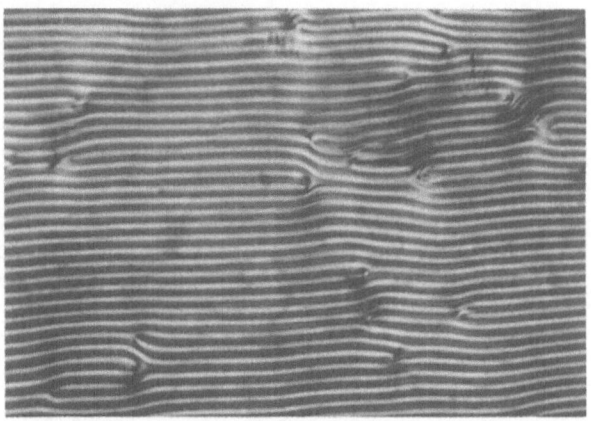

Fig. 5. Spaghetti-like texture of an oriented lyotropic-cholesteric phase in the system tetradecyldimethylaminoxide/water/decanol/cholesterol at $T = 298$ K. The phase consists of disklike micelles. It has been subjected to a magnetic field for 40 h. Magnification: 100×

field. Then, the typical spaghetti-like texture is observed, as shown in Fig. 5.

A point of special interest in texture observation is the change in pitch length effected by optically active substances such as tartaric acid, lithocholic acid, and cholesterol. Figure 6 illustrates the results. At low concentrations of optically active substance the reciprocal pitch length increases linearly. The

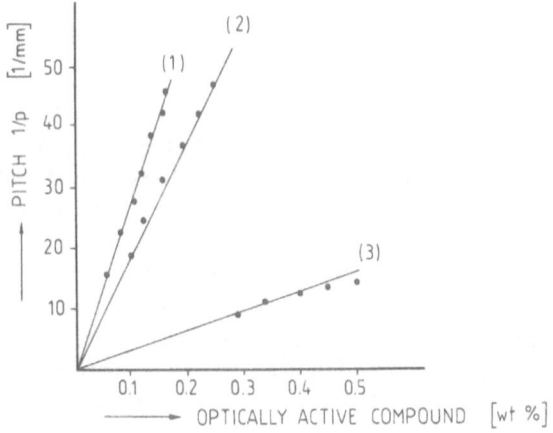

Fig. 6. Pitch lengths obtained by evaluating the spaghetti-like texture according to Fig. 1 in dependence on the concentration of the optically active component at $T = 298$ K. System: tetradecyldimethylaminoxide/water/decanol/optically active compound. 1) lithocholic acid; 2) cholesterol; 3) tartaric acid

slope of the lines (curves 1, 2, 3) is dependent on the chemical structure of the optically active compound.

The time required for forming a uniformly and completely aligned spaghetti-like texture by a magnetic field increases with increasing concentrations of tartaric acid, lithocholic acid, and cholesterol. This indicates that cholesteric phases with helical structures of small pitches take more time for sample orientation. This time-dependence is illustrated in Fig. 7 for the system $C_{14}DMAO/$ water/decanol/cholesterol.

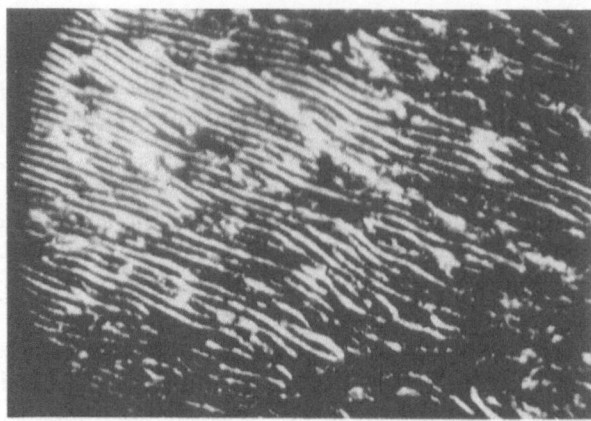

Fig. 8. Photograph of the phase transition from the cholesteric to the pseudoisotropic structure at $T = 309$ K. Crossed polarizers; magnification 100 fold; thickness of the samples $d = 0.2$ mm

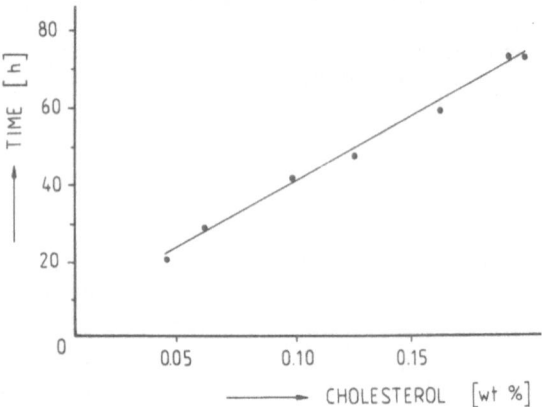

Fig. 7. Time required for complete alignment of phase by a magnetic field as a function of the cholesterol concentration at $T = 298$ K. The curves were obtained by interpreting cholesteric textures

The transformation of the cholesteric phase into the nematic phase was investigated. On applying low heating rates (0.5 K min^{-1}), we have succeeded in observing the disappearance of the typical cholesteric fingerprint texture at $T \approx 310$ K (compare Fig. 8). The texture melts gradually. With rising temperature the twisted cholesteric structures are probably loosened and afterward disappear. This process may be followed by the transition into a nematic phase. In principle, the arrangement of molecules and aggregates is not changed. For this reason, the transition enthalpy required is perhaps too small to be detected by calorimetric measurements. But these preliminary results should be supplemented by further calorimetric experiments performed with higher sensitivity and, if necessary, by another type of calorimeter.

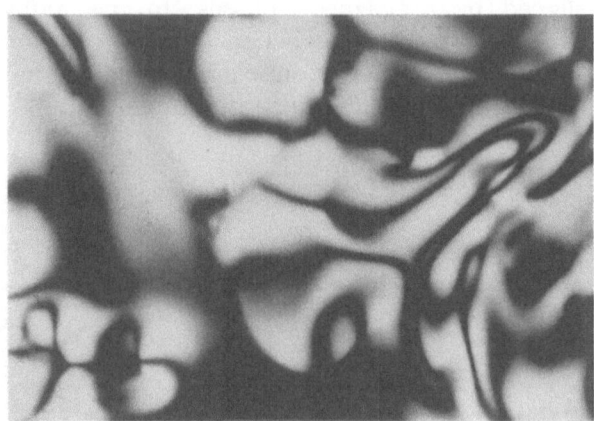

Fig. 9. "Schlieren" texture of the unaligned disklike nematic phase (N_D) of the system tetradecyldimethyl-aminoxide/water/decanol at $T = 298$ K (compare the phase diagram in Fig. 2). Magnification: $100 \times$

4.3. Orientation effects in lyotropic-nematic phases

A typical nematic "Schlieren"-texture of the N_D-phase consisting of disklike micelles is shown in Fig. 9. As described in detail by Hertel and Hoffmann [18], the glass walls of the microslides will be able to effect an orientation of the nematic phase. In the N_D-phase the disklike micelles are oriented parallel to the glass walls, so that their optical axes will be perpendicular to the glass surface. This orientation is indicated by a gradual disappearance of the "Schlieren"-texture. In contrast to the N_p-phase, an N_C-phase remains birefringent. This orientation process in nematic phases may be accelerated by an external magnetic field. Orienting the rods and disks solely by the influence exerted by the glass walls will take up to several days.

It is well known that the sign of the anisotropy of the diamagnetic susceptibility can be determined with the aid of orientation experiments in a magnetic field [17, 18]. For the systems under discussion, the magnetic susceptibility of the N_C-phase has a positive sign, and that of the N_D-phase a negative one. This means that the N_C-phase is oriented parallel to the magnetic field.

4.4. Calorimetric measurements of the phase transition $N_C \to S$

The nematic and the cholesteric phases (discussed in 4.2.) differ in their behavior with respect to temperature alteration. In the disklike nematic phase N_D, neither polarization microscopy nor calorimetric measurements indicated further phase transitions at temperatures between $T = 297—355$ K. But careful investigations of the rodlike nematic phase N_C performed within this temperature range proved the occurrence of transitions into the isotropic phase which depend on alcohol concentration. There is a trend to lower phase transition temperatures $N_C \to S$ with increasing decanol concentration. Figure 10 presents some of these DSC-curves recorded at two different alcohol concentrations and for various C_{14}DMAO concentrations (curves 1—8). The transition enthalpies H_U of the mixtures are of the order of $3.2 \cdot 10^{-2}$ J $\cdot$ g^{-1}, i.e., they are very small. The variation of the enthalpies ($\pm 1.5 \cdot 10^{-2}$ J $\cdot$ g^{-1}) is considerable.

In Fig. 11 calorimetric transition temperatures for rodlike nematics $\to$ isotropic phase of the ternary system C_{14}DMAO/water/decanol are plotted against the C_{14}DMAO concentration and for two different decanol concentrations. The curves obtained demonstrate that the temperature of the transition $N_C \to S$ is a linear function of the aminoxide concentration. Furthermore, this illustrates the influence of the decanol on the phase transition temperatures.

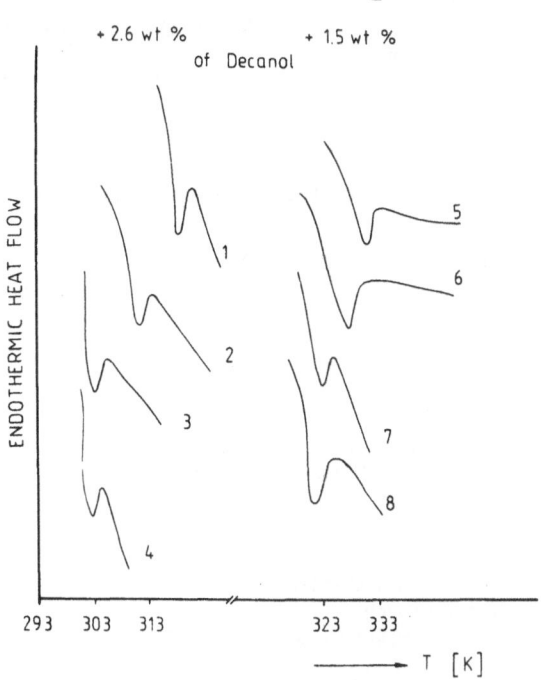

Fig. 10. Selected DSC-curves of the phase transition rodlike nematics $\to$ isotropic phase ($N_C \to S$) of the system tetradecyldimethylaminoxide/water/decanol in dependence on the concentration of tetradecyl dimethylaminoxide (C_{14}DMAO) at 2.6 and 1.5 wt% of decanol. Heating rate: 0.1 K min^{-1}. Surfactant concentrations are: curve 1: 35 wt% C_{14}DMAO; curve 2: 34 wt% C_{14}DMAO; curve 3: 33 wt% C_{14}DMAO; curve 4: 32 wt% C_{14}DMAO; curve 5: 34 wt% C_{14}DMAO; curve 6: 33 wt% C_{14}DMAO; curve 7: 32 wt% C_{14}DMAO; curve 8: 31 wt% C_{14}DMAO

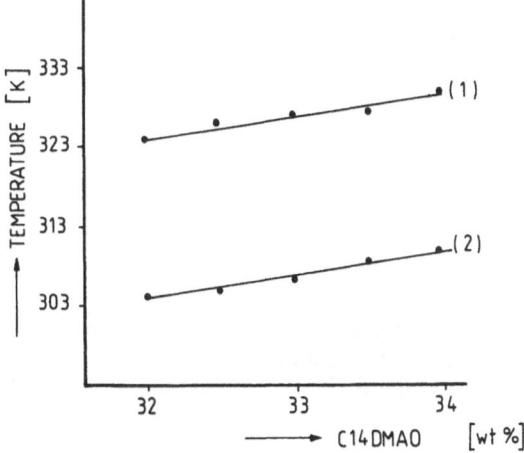

Fig. 11. Comparison between the temperatures of the phase transitions: rodlike nematics $\to$ isotropic phase ($N_C \to S$) in dependence on tetradecyldimethylaminoxide (C_{14}DMAO) concentration at two different alcohol concentrations. 1) 1.5 wt% decanol; 2) 2.6 wt% decanol. The phase transition temperatures were extrapolated from the DSC curves (see Fig. 10)

5. Conclusions

The results of our studies demonstrated that, in lyotropic systems, tetradecyldimethylaminoxide/water/aliphatic alcohols nematic phases can be transformed into lyotropic-cholesteric phases by adding optically active components. The aminoxide $C_{14}DMAO$ serves as nonionic surfactant. Hitherto, nonionics have not been utilized for forming cholesteric phases in lyotropic multicomponent systems. In the literature only ionic surfactants are described [2—16]. From our experiments and the results in the literature, it follows that the charge and chemical structure of the surfactants seems to be obviously insignificant for the transformation of nematic phases into cholesteric phases. Attention must only be paid to the prerequisite that the process has to start from a nematic phase of the multicomponent system. But the twisting mechanism occurring in the transformation process from the nematic phase into the cholesteric structure is not yet completely elucidated.

A comprehensive examination of the influence exerted by the optically active component on helix formation will be necessary. Therefore, a variety of optically active components must be tested. Furthermore, structural investigations, especially by small angle x-ray scattering will be required to obtain structural data of higher accuracy, which will give further information on the transition process nematic → cholesteric phase.

Acknowledgements

This work was financially supported by the SFB213 of the DFG in Bayreuth. One of us (G.B.) would like to thank the DFG for financial assistance.

References

1. Hertel G, Hoffmann H, Oetter G, Schwandner B (1987) Progr Coll Polym Sci 73:95—106
2. Radley K, Saupe A (1978) Mol Phys 35:1405—1412
3. Acimis M, Reeves LW (1980) Can J Chem 58:1533—1541
4. Alcantara MR, De Melo MVMC, Paoli VR, Vanin JA (1983) J Coll Interf Sci 93:560—561
5. Alcantara MR, De Melo MVMC, Paoli VR (1983) Mol Cryst Liq Cryst 95:299—307
6. Alcantara MR, Varnin JA (1984) Mol Cryst Liq Cryst 107:333—340
7. Alcantara MR, Varnin JA (1984) Mol Cryst Liq Cryst (Lett) 102:7—12
8. Figueiredo Neto AM, Galerne Y, Liebert L (1985) J Phys Chem 89:3939—3941
9. Lee H, Labes MM (1984) Mol Cryst Liq Cryst 108:125—132
10. Do Aido TMH, Alcantara MR, Felippe O, Jr, Pereira AMG, Vanin JA (1990) Mol Cryst Liq Cryst 185:61—66
11. Lawson KD, Flautt TJJ (1967) J Am Chem Soc 89:5489—5491
12. Yu LJ, Saupe A (1980) J Am Chem Soc 102:4879—4883
13. Figueredo Neto AM, Marcondes Helene ME (1987) J Phys Chem 91:1466—1469
14. Alcantara MR, De Melo MVMC, Paoli VR, Vanin JA (1983) Mol Cryst Liq Cryst 90:335—347
15. Tracey AS, Radley K (1984) J Phys Chem 88:6044—6048
16. Forrest BJ, Reeves LW, Vist M (1984) Mol Cryst Liq Cryst 113:37—58
17. Tracey AS, Radley K (1985) Mol Cryst Liq Cryst 122:77—87
18. Hertel G, Thesis (1989) University of Bayreuth, Phys Chem I
19. Hertel G, Hoffmann H (1989) Liq Cryst 5:1883—1898

Authors' address:

Prof. Dr. Heinz Hoffmann
Universität Bayreuth, Physikalische Chemie I,
Universitätsstraße 30, Postfach 101251
W-8580 Bayreuth, FRG

Prof. Dr. Hans-Dieter Dörfler
DLC Gerlind Bartusch
TU Dresden, Lehrstuhl für Kolloidchemie
Mommsenstr. 13
O-8027 Dresden, FRG

Progress in Colloid & Polymer Science   Progr Colloid Polym Sci 89:315—318 (1992)

Bilayer thickness from lipid-chain dynamics: Influence of cholesterol and electric charges. A ^{2}H-NMR approach

A. Léonard, J. C. Maillet, J. Dufourcq, and E. J. Dufourc

Centre de Recherche Paul Pascal, CNRS, Pessac, France

Abstract: The bilayer hydrophobic thickness l is estimated from the ^{2}H-NMR spectra of acyl chain perdeuteriated membrane phospholipids. Values obtained from individual quadrupolar splittings and those from the first moment of a powder spectrum are in close agreement. In the fluid (L_a) phase, the thermal variation of dimyristoylphosphatidylcholine (DMPC), dimyristoylphosphatidic acid (DMPA) at pH 4.2 and 8.2, and DMPC: cholesterol (7:3) bilayer thicknesses are compared; 30% cholesterol increases the thickness of DMPC by approximately 4—5 Å, whatever the temperature. DMPC and DMPA (whatever the pH) have the same hydrophobic thickness near their respective T_c, the gel-to-fluid phase transition temperature, and display a very similar l decrease when increasing the temperature. However, at high temperatures (60°C) DMPC is thinner than DMPA by approximately 1 Å. Steric, electrostatic, and temperature constraints are therefore seen to modulate the biomembrane hydrophobic thickness, as easily estimated by ^{2}H-NMR.

Key words: Bilayer thickness; ^{2}H-NMR; DMPC; DMPA; cholesterol

Introduction

Membrane dynamics have been extensively studied on model systems and their ordering properties are now well established. The modulation of the bilayer orientational order parameter under physical, chemical or biological constraints can now be estimated by a wide variety of techniques. ESR and fluorescence spectroscopy measure the ordering of an extraneous probe embedded in the membrane [1, 2], which is in turn utilized to estimate changes in the membrane order parameter. Alternatively, phosphorus (^{31}P) [3] and deuterium (^{2}H) [4, 5] NMR afford a direct measure of the local ordering. ^{2}H-NMR of chain perdeuteriated synthetic phospholipids is particularly suited to measure the order parameter of each of the chain segments. Subtle changes in bilayer core ordering can thus be proben.

It has been found that this minute information about methylene and methyl ordering may be correlated to the average length of a given segment. This information can be formally translated in terms of bilayer thickness within the frame of appropriate models [6, 7].

We present here a reliable method to determine membrane hydrophobic thickness l via ^{2}H-NMR of sn-2 chain perdeuteriated lipids. At first, two methods to determine l are compared:

i) measurement of the C-^{2}H bond order parameter $S_{C^{-2}H}$ of each of the dimyristoylphosphatidylcholine (DMPC) chain segments from "de-Paked" spectra, and

ii) calculation of the first spectral moment from powder spectra.

Secondly, we will report on bilayer thickness changes as promoted by a molecule (cholesterol) which goes into the hydrophobic core of DMPC, or alternatively, by modifications of the electric charge at the surface of dimyristoylphosphatidic acid (DMPA) bilayers.

Materials and methods

DMPC and DMPA perdeuteriated in the sn-2 chain were, respectively, synthesized according to procedures already published [8] and purchased from Avanti polar lipids (USA). ^{2}H-depleted water was provided from Aldrich (France). Cholesterol was obtained from Sigma (France).

Multilamellar dispersions were obtained by hydrating lipids with a buffered solution (pH 7.0 for DMPC and pH 4.2 and 8.2 for DMPA) on a vortex mixer. Several freeze-thaw cycles were performed to ensure sample homogeneity and pH stability.

NMR signals were acquired on a Bruker MSL-200 spectrometer, with quadrature detection and by means of quadrupolar echo pulse sequence [4]. Samples were allowed to equilibrate for at least 30 min before the NMR signal was acquired and the temperature was regulated to ±1°C. Data treatment was accomplished on a VAX/VMS 8600 computer.

Theoretical background

S_{C-^2H} can be directly estimated from measurement of the residual quadrupolar splitting $\Delta \nu_Q$ between the two most intense peaks (the so-called 90° orientations) of an individual deuterium powder spectrum [9]:

$$|S_{C-^2H}| = \frac{4}{3A_Q} \Delta \nu_Q , \qquad (1)$$

where $A_Q = 170$ kHz is the static quadrupolar coupling constant. When using perdeuteriated chains, the individual S_{C-^2H} are most easily measured on "dePaked" spectra [4] (Fig. 1B).

Seelig and coworkers [6, 7] proposed a method to determine the hydrophobic bilayer thickness l in model membrane fluid (L_a) phases from the measurement of the acyl chain orientational order parameter:

$$l = l_1(a\langle |S_{c-^2H}| \rangle + \beta) , \qquad (2)$$

where l_1 is the maximum value of the hydrophobic bilayer thickness (chain segments in all-trans conformation), $l_1 = 34.4$ Å for C14 acyl chains. $a = 1$ and $\beta = 1/2$ are constants calculated from the model mentioned earlier. $\langle |S_{C-^2H}| \rangle$ is the acyl chain orientational order parameter which can be obtained from the order parameter of the ith labeled carbon, $|S^i_{C-^2H}|$:

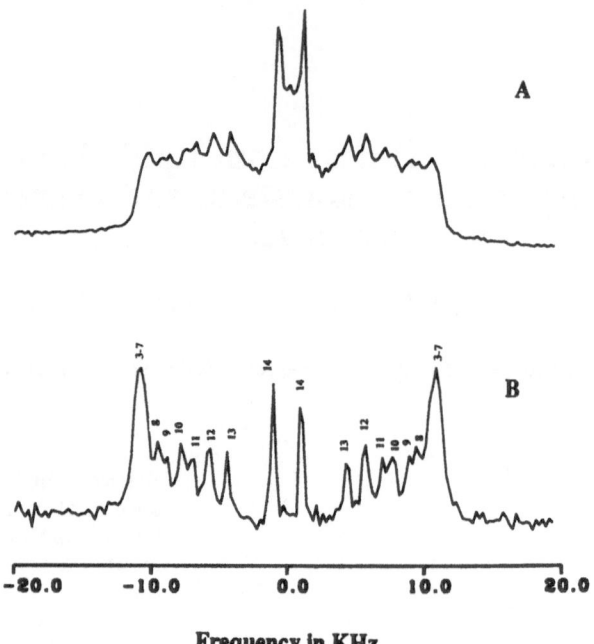

Fig. 1. A) ^{2}H-NMR powder spectrum of [sn-2-^{2}H$_{27}$]-DMPC at $T = 60$°C. Experimental parameters: 90° pulse length, 6.5 µs; pulse spacing in the echo sequence, 20 µs; recycling time, 2 s; spectral width, 500 kHz; 7000 acquisitions; quadrature detection and full-phase cycling of receiver and transmiter. B) Corresponding "dePaked" [11] spectrum. Numbers on spectrum represent the assignment of labeled carbon positions

$$\langle |S_{C-^2H}| \rangle = \frac{1}{N-1} \sum_{i=2}^{N} |S^i_{C-^2H}| . \qquad (3)$$

N stands for the number of labeled carbons ($N = 14$ for myristoyl chains).

The chain order parameter can also be obtained from the calculation of the first moment, M_1, of a powder spectrum originating from a fully labeled chain (Fig. 1A) [9]:

$$\langle |S_{C-^2H}| \rangle = \frac{\sqrt{3}}{\pi A_Q} M_1 . \qquad (4)$$

Results and discussion

Figure 1A displays a typical ^{2}H-NMR powder spectrum of [sn-2-^{2}H$_{27}$]DMPC multilamellar dispersions. Such a spectrum can be "dePaked" according to the algorithm initially proposed by Bloom et al. [10] and further developed by Sternin et al. [11], leading to Fig. 1B. The individual quadrupolar splittings Δv_Q^i of almost all labeled carbon positions can thus be easily measured from the frequency separation of each of the doublets. Numbers in Fig. 1B indicate the assignment to labeled carbon, according to previous work [4, 5]. Using these data, together with Eqs. (1) and (3), $\langle |S_{C^{-2}H}| \rangle$ can be calculated. It must be mentioned here that information from the C2 labeled position has not been used, since it is well established that the two deuterons are not equivalent [5]. The first moment of the powder spectrum (Fig. 1A) has also been calculated [4] and has led, by making use of Eq. (4), to a second estimate of $\langle |S_{C^{-2}H}| \rangle$. Both chain order parameter measurements were utilized to calculate (Eq. (2)) the bilayer hydrophobic thickness. These thermal variations of l are reported in Fig. 2. One notices that l is slightly lower when estimated from M_1 than from Δv_Q^i. This is due to the fact that Eqs. (3) and (4) are not strictly equivalent: i) there are three deuterons at the very disordered methyl terminal, instead of two for each of the methylene segments and, ii) as already mentioned, deuterons at C2 were not used in Eq. (3). Hence, $\langle |S_{C^{-2}H}| \rangle$ as estimated from Eq. (3) is systematically approximately 2% larger than when calculated from Eq. (4). However, both curves can be considered quasi-identical, within the experimental error. As a consequence, determination of l from measurement of M_1 should be preferred since it is simpler and does not require either "de-Paking" or attribution of Δv_Q^i to labeled positions. As a drawback, the calculation of moments requires high-quality spectra. Higher temperatures lead to a decrease in l, as shown in Fig. 2. For DMPC this shrinkage of the hydrophobic part of the bilayer is about 2 Å in a 35 °C temperature range.

Figure 3 reports the thermal variation of DMPC, DMPC:cholesterol (7:3). A 30% cholesterol concentration increases the bilayer thickness by 5—4 Å, which remains constant throughout the temperature range. This is to be related to the well documented ordering action of cholesterol on fluid phase lipids [12, 13]. The steroid molecule is indeed known to penetrate deeply into the bilayer core thereby reducing the amplitude of fluctuation of methyl and methylene segments. The proportion of trans conformers is thus increased, which is well correlated by the increase in l.

It is also interesting to compare the thermal dependence of DMPC and DMPA hydrophobic thickness (Fig. 3). DMPC is zwitterionic at pH = 7 and possesses a bulky phosphocholine head group,

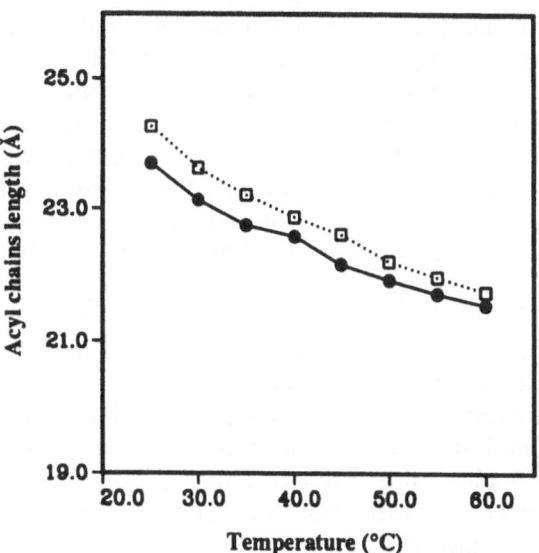

Fig. 2. Temperature dependence of the DMPC hydrophobic thickness l in the fluid (L_a) phase. (●) calculated from the first moment of ^{2}H-NMR powder spectra; (□) calculated from individual quadrupolar splittings measured on "de-Paked" spectra (see text). Experimental error is about 3%

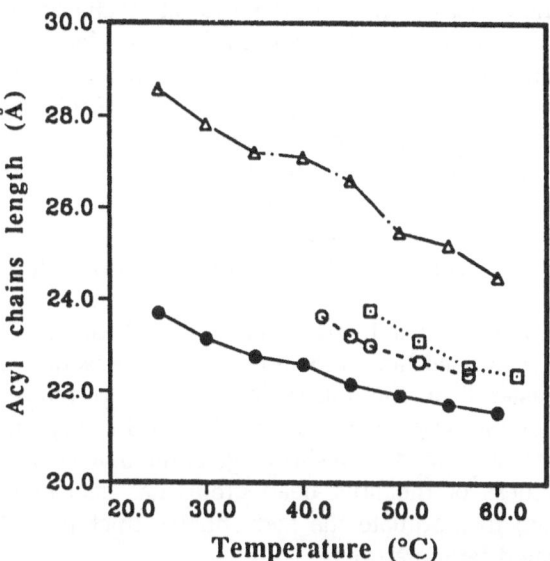

Fig. 3. Temperature dependence of the hydrophobic bilayer thickness, l, of: (●) DMPC; (○) DMPA pH 8.2; (□) DMPA pH 4.2 and (△) DMPC:cholesterol (7:3) as calculated from the first moment of ^{2}H-NMR powder spectra. Experimental error is about 3%

whereas DMPA bears, on its small phosphate group, one negative charge at pH 4.2 and two at pH 8.2. The curves in Fig. 3 do not have the same origin since the critical temperature T_C at which each of the systems goes into the fluid (L_a) phase is different. DMPC has a transition temperature independent of pH, $T_C = 20\,°C$, whereas DMPA has $T_C = 39\,°C$, and $42\,°C$ at pH 8.2 and 4.2, respectively. Reported T_C values are for lipids with the sn-2 chain perdeuteriated. This is known to induce a small temperature shift towards lower values [4]. DMPC and DMPA have almost the same bilayer thickness near T_C, whereas DMPA is thicker than DMPC by about 1 Å at elevated temperatures ($60\,°C$). It thus appears that, near the transition temperature, the presence of one or two negative charges on the head group promotes the same packing properties as does the bulky choline group. This indicates that both steric and electrostatic constraints play, under these conditions, the same role in ordering properties. Interestingly, the three systems have a very similar temperature dependence, i.e., l decreases by approximatively 1.5—2 Å on going to high temperatures. However, at $60\,°C$, DMPC is 1 Å thinner than DMPA. We believe that this directly reflects the presence of the bulky phosphocholine group which promotes more disorder than the small phosphate group.

The effect of one additional negative charge on DMPA induces a decrease in T_C, but does not greatly modify the overall temperature behavior. Interestingly, at high temperatures (55—60 °C) the DMPA hydrophobic thickness is the same whatever the pH. This indicates that the thermal energy increases the chain dynamics such that it overwhelms the electrostatic constraints.

As a conclusion, it appears that deuterium NMR of perdeuteriated lipid chains can be utilized to probe changes in the hydrophobic bilayer thickness. It has been used herein to show that the presence of cholesterol in the bilayer core leads to bilayer thickening and that, conversely, an increase in temperature leads to a shrinkage of the bilayer. The bulkiness of the lipid head group has also been shown to modulate the hydrophobic thickness at elevated temperatures.

References

1. Shimshick EJ, McConnell HM (1973) Biochem Biophys Res Commun 53:446—451
2. Deinum G, van Langen H, van Ginkel G, Levine Y (1988) Biochemistry 27:842—860
3. Dufourc EJ, Mayer C, Stohrer J, Althoff G, Kothe G (1992) Biophys J 61:42—57
4. Davis JH (1983) Biochim Biophys Acta 737:117—171
5. Seelig J (1977) Q Rev Biophys 10:353—418
6. Seelig A, Seelig J (1974) Biochemistry 13:4839—4845
7. Schindler H, Seelig J (1975) Biochemistry 14:2283—2287
9. Ipsen JH, Mouritsen OG, Bloom M (1990) Biophys J 52:405—412
8. Perly B, Dufourc EJ, Jarrell HC (1984) J Labeled Compd Radiopharm 21:1—13
10. Bloom M, Davis JH, MacKay AL (1981) Chem Phys Lett 80:198—202
11. Sternin E, Bloom M, MacKay AL (1983) J Magn Reson 55:274—282
12. Dufourc EJ, Parish EJ, Chitrakorn S, Smith ICP (1984) Biochemistry 23:6062—6071
13. Vist MR, Davis JH (1990) Biochemistry 29:451—464

Authors' address:

E. J. Dufourc
Centre de Recherche Paul Pascal
CNRS
Av. A. Schweitzer
F-33600 Pessac, France

Progress in Colloid & Polymer Science   Progr Colloid Polym Sci 89:319—323 (1992)

Structural alteration of lymphocyte membrane induced by gangliosides. A conductometric study

C. Cametti, F. De Luca, A. D'Ilario, B. Maraviglia, R. Misasi[1]), M. Sorice[2]), F. Bordi[3]), and M. A. Macrì[4])

Dipartimento di Fisica, Università degli Studi "La Sapienza", Rome, Italy
[1]) Dipartimento di Medicina Sperimentale, Università degli Studi "La Sapienza", Rome, Italy
[2]) Istituto di Malattie Tropicali, Università degli Studi "La Sapienza", Rome, Italy
[3]) Dipartimento di Medicina Interna, Università degli Studi "Tor Vergata", Rome, Italy
[4]) Istituto di Fisica Medica, Università di Chieti, Chieti, Italy

Abstract: The passive electrical properties of human lymphocyte membranes incubated with different gangliosides have been measured by means of radiowave dielectric spectroscopy in the frequency range where the Maxwell-Wagner effect occurs. — The alterations of the membrane structure and functionality reflect in an increase of the membrane conductivity σ_s and membrane permittivity ε_s, depending on the chemical composition of the ganglioside employed. The possible relation to the structural changes of the cytoplasmatic membrane are briefly discussed.

Key words: Lymphocyte membrane; gangliosides; electrical conductivity

Introduction

Gangliosides, natural anionic amphiphile molecules occurring in a variety of plasma membranes of mammalian cells [1], consist of a double-chain hydrophobic part (ceramide) and a hydrophilic head of approximately equal volume, containing a large number of hydrogen bonding groups which confer to them the possibility of performing more than one type of function in nature.

These molecules display strong amphiphilic properties and are implicated in a variety of cell surface phenomena [2], including signal recognition and signal transduction at the cell surface.

Gangliosides are inserted into the cytoplasmatic cell membrane, with their hydrophobic part anchored to the outer leaflet of the membrane and the oligosaccharide residue facing the extracellular interface, thus modifying, in principle, both the surface electrical properties and the molecular packing of the lipids in the membrane molecular organization.

The amounts and patterns of these lipids differ among cells from different lineages and between normal and malignant cells.

Because of the wide variety of glycolipid structure, these molecules perform different functions and differently influence cell surface phenomena such as fluidity [3], thermotropic properties [4], receptorial functions and cell-cell interactions [2].

Since gangliosides are easily incorporated into biological cell membranes over the normal (unaffected) concentration, it is possible to investigate whether they alter them and what sort of properties they confer on membranes, in particular, at the interface of the aqueous phase.

We have focused our attention on two distinct gangliosides, GM1 and GM3, which carry one residue of sialic acid, all with the same hydrocarbon portion, but with different oligosaccharide chains, and we have studied their influence and the possible alterations they produce on the cytoplasmatic membrane when added, in appropriate concentration, to human lymphocyte suspensions.

In the case of human lymphocytes, the gangliosides normally present in the plasmatic membrane consist essentially of GM3 (up to 70%) and GD1a (disialoganglioside), whereas GM1 is not present at detectable concentrations [5].

Since GM1 and GM3 differ in the hydrophilic part of the molecule, keeping the hydrophobic volume and the hydrocarbon length fixed, an increase of the GM3 concentration or the insertion of GM1 molecules in the lymphocyte membrane would produce different interactions with different parts of the membrane, thus providing useful information on the mechanisms, at the molecular level, that occur in ion transport processes.

Moreover, despite their importance because of their immunological functions, the study of the passive electrical properties and the response of the plasma lymphocyte membrane to electrical fields has not received much attention in the past, and very few studies have appeared on this subject [6, 7].

The aim of the present experiment was to determine the possible alteration of the passive electrical properties of human lymphocyte cells upon treatment with two different gangliosides. Measurements were performed by means of radiowave dielectric spectroscopy, by using conventional impedance measurements [8].

The technique provides a unique tool to determine the electrical behavior of the cytoplasmatic membrane in cell suspensions and has gained considerable attention in recent years in the area of physiological research, and among cellular biologists and biophysical chemists.

Theory

The method, which has been applied to a variety of heterogeneous systems of biological relevance [9], is based on the Maxwell-Wagner effect [10], consisting of a frequency dependence of the dielectric parameters of the system (permittivity ε and electrical conductivity σ) orginated by an accumulation of charge at the interface disjoining media with different dielectric properties.

In the present case, owing to the presence of the low-conductivity cytoplasmatic membrane which separates the cytosol from the external aqueous phase, for cell dimensions of the order of 10 µm, the Maxwell-Wagner dispersion occurs at radiowave frequencies with relaxation times of the order of 10^{-7} s, and can be easily measured, for example, through the overall electrical conductivity.

In fact, the conductivity spectrum of a cell suspension contains quantitative information on the electrical behvior of the cell membrane, char-

acterized from an electrical point of view, by the membrane conductivity σ_s and the membrane permittivity σ_s, and these parameters can be extracted by regression analysis of the data according to suitable cell suspension models.

The most common dielectric model is based on the so-called "single shell model" that takes into account the effects induced by the cytoplasmatic cell membrane, neglecting those attributable to the structure of the cytosol (nuclear membrane, proteins, small ions, organelles, etc.).

Within this model, biological cells are represented by a conducting sphere (the cytosol) of radius a and complex conductivity

$$\sigma_p^* = \sigma_p + i\omega\varepsilon_0\varepsilon_p$$

covered with a less-conductive shell (the plasma membrane) of thickness d and complex conductivity

$$\sigma_s^* = \sigma_s + i\omega\varepsilon_0\varepsilon_s ,$$

where ω is the angular frequency of the applied field and ε_0 is the permittivity of free space.

A suspension of shelled spheres, uniformly distributed in a continuous medium of complex conductivity

$$\sigma_m^* = \sigma_m + i\omega\varepsilon_0\varepsilon_m$$

displays an overall conductivity σ^* given by

$$\frac{\sigma^* - \sigma_m^*}{\sigma^* + 2\sigma_m^*} = \phi \, \frac{\sigma_{eq}^* - \sigma_m^*}{\sigma_{eq}^* + 2\sigma_m^*} ,$$

where ϕ is the fractional volume of the dispersed phase and

$$\sigma_{eq}^* = \sigma_s^* \, \frac{2\sigma_s^* + \sigma_p^* - 2v(\sigma_s^* - \sigma_p^*)}{2\sigma_s^* + \sigma_p^* + v(\sigma_s^* - \sigma_p^*)}$$

is the equivalent homogeneous complex conductivity of the shelled sphere, $v = [d/(d + a)]^3$ takes into account the cell shape and the membrane thickness.

This model, although highly idealized, has been widely used and applied to different biological systems, including cultured lymphoma cells [11], yeast cells [12], *Escherichia coli* [13], human erythrocytes in normal and pathological conditions [11, 12], and, finally, human lymphocytes [16].

Experimental

a) Sample preparation

Peripheral blood lymphocytes from healthy donors were prepared from gradient centrifugation with Ficoll-Hypaque (Lymphocyte Separation Medium-Sigma Chemical Co.). An analysis performed by a cell analyzer Technicon H1 System revealed a mean purity of lymphocyte preparation of 86%. Cells were incubated with GM1 or GM3 (Sigma) 500 µg/ml in the absence of serum at 37°C for 1 h.

The cells were then washed with PBS pH = 7.4 with serum 5%.

b) Conductivity measurements

The electrical conductivity of the cell suspension has been measured in the frequency range from 10 kHz to 100 MHz by means of two Impedance Analyzer HP models 4192A (10 kHz—10 MHz) and 4193A (0.5 MHz—100 MHz) using a conventional two-electrode conductivity cell. The temperature was fixed to 37°C within 0.1°C, and the overall accuracy over the entire frequency spectrum (based on a calibration procedure of the measuring set-up using liquids of known conductivity and dielectric constant) was within 0.5%.

Results and discussion

Figure 1 shows a typical conductivity spectrum over the frequency range from 10^4 to 10^8 Hz at a temperature of 37°C for a suspension of human lymphocytes incubated with GM1 and GM3 gangliosides, compared to that of unaffected control suspension.

The cell concentration was kept constant in the different samples and cell dimensions before and after the treatment with gangliosides were checked through accurate measurements performed by means of optical microscopy to make the comparison meaningful.

The gangliosides concentration was 500 µg per ml of suspension (corresponding to $2.7 \cdot 10^{-4}$ M for GM1 and $3 \cdot 10^{-4}$ M for GM3), whereas the cell concentration was maintained to $2 \cdot 10^8$ lymphocytes per ml of suspension.

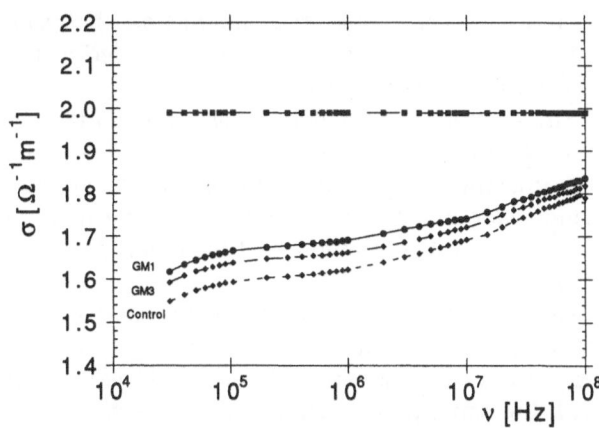

Fig. 1. The electrical conductivit σ of lymphocyte suspensions as a function of frequency at the temperature of 37°C. (●) lymphocyte suspension ($2.4 \cdot 10^8$ cells/ml) treated with GM1 (500 mg/ml); (♦) lymphocyte suspension ($2.4 \cdot 10^8$ cells/ml) treated with GM3 (500 mg/ml); (♦) control (lymphocyte concentration $2.4 \cdot 10^8$ cells/ml). The conductivity of physiological solution (upper curve) is also shown for comparison

Owing to their strong amphiphilic character, gangliosides in aqueous solution associate in micelles with aggregation number (of the order of 300) depending on the hydrophobic-hydrophilic balance.

Since the ganglioside critical micelle concentration (CMC) is in the range 10^{-6} to 10^{-4} M [17], well below the ganglioside concentration we have employed, the interaction with the plasma membrane would occur preferably with micelle associated gangliosides.

Moreover, these aggregating structures might be of different geometrical shape, since the packing parameter defined as

$$P = \frac{V}{Sl},$$

where V is the volume, l the maximum length of the hydrocarbon chain, and S the area per head group, varies from $P = 0.42$ for GM1 to $P \geqslant 0.5$ for GM3, for the sequence of gangliosides we have investigated.

This means that different aggregates might be built, i.e., non-spherical micelles for P lower than 0.5 and unilamellar vesicles for P larger than 0.5.

At this concentration, the incubation would increase the ganglioside concentration on the cell membranes or favor the interaction between micelles and the glycocalyx skeleton covering the plasma membrane surface.

Since before conductivity measurements the excess gangliosides have been removed from the supernatant by repeated washings, the membrane-ganglioside interactions would result in permanent alteration of the membrane functionality and/or structure.

Figure 2 shows the changes of the passive electrical parameters of the lymphocyte membrane upon the treatment with GM1 and GM3, compared with the same parameters in control cells.

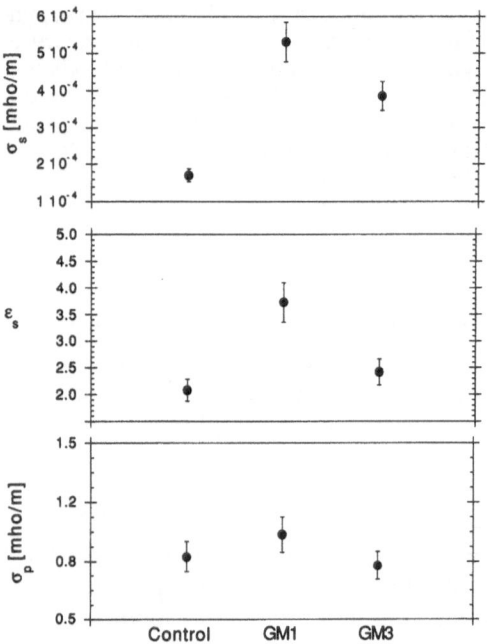

Fig. 2. The passive electrical properties of lymphocyte membrane derived from the single-shell model. From top to bottom: the membrane conductivity σ_s, the membrane permittivity ε_s and the conductivity of the cytosal σ_p

As can be seen, the membrane conductivity increases from $1.5 \cdot 10^{-4} \ \Omega^{-1} \ m^{-1}$ in control cells to $5.2 \ 10^{-4} \ \Omega^{-1} \ m^{-1}$ in GM1- and $3.8 \ 10^{-4} \ \Omega^{-1} \ m^{-1}$ in GM3-treated sample.

A similar behavior is observed in the membrane permittivity which passes from a value of about 2 in control cells to 3.7 in GM1- and to 2.4 in GM3-treated sample. As usually, no significative difference is observed in the cytosol, whose conductivity remains unchanged to the value of about $0.8 \ \Omega^{-1} \ m^{-1}$.

Within the single-shell model it is difficult to ascertain whether the membrane passive electrical properties have been modified by an effective alteration of the membrane structure or by a change of the surface aqueous environment, following ganglioside incubation.

Moreover, the possible molecular actions at the membrane level of ganglioside micellar aggregates are largely unknown.

Nevertheless, the changes observed both in the membrane conductivity and the membrane permittivity indicate an increase of the ionic permeation and an enhanced electrical polarizability of the overall membrane structure.

If gangliosides absorb only into the outer face of the bilayer membrane, preserving its overall structure, their accumulation could provoke a partial disruption of the lipid part of the membrane with a concurrent increase of the ionic permeation, observed through an increase of the membrane conductivity.

If the ionic transport across the membrane is simply due to the existence of conducting pathways embedded in the less-conductive hydrocarbon layer, the membrane conductivity can be written, to a first approximation, as

$$\sigma_s \cong (N_{pore}) \sigma_{pore} S_{pore} ,$$

where σ_{pore} is the conductivity of the conducting pores of area S_{pore}, and N_{pore} is their number per unit surface.

In this case, the observed increase of the membrane conductivity would result in an increase of the activated trans-membrane pores by a factor of 3 or 2 for GM1 and GM3 gangliosides, respectively.

It must be noted that these indications are in qualitative agreement with the effective ganglioside concentration present on the cell surface after the incubation phase. For example, we have found that incubation produces an increase of the GM3 density in the lymphocyte membrane from $3.6 \cdot 10^{12}$ molecules/cm^2 in control cells to $14 \cdot 10^{12}$ molecules/cm^2 in treated cells.

The membrane permittivity depends crucially on the distribution of the charged or polar membrane residues throughout the interphase between the membrane and the bulk solution, and the properties of the polar head group of the absorbed gangliosides can regulate the establishment of different types of surface cell organization. The increase of the permittivity in presence of GM1 and

GM3 can be associated with an increase of the surface polarizability due to the polar components of the gangliosides.

In fact, GM3, with one sugar less than GM1 and, consequently, a smaller head group area, induces a less marked increase of the membrane permittivity.

Further experiments are in progress to relate the observed induced alteration in the electrical properties of human lymphocyte membrane to the biological functions of these cells.

References

1. Steck TL, Dawson G (1974) J Biol Chem 249:2135. Ledeen RW (1978) J Supramol Struct 8:1
2. Fishman PH, Brady RO (1976) Science 194:407
3. Bertoli E, Masserini M, Sonnino S, Guidoni R, Cestaro B, Tettamanti G (1981) Biochim Biophys Acta 467:192
4. Masserini M, Freire E (1986) Biochemistry 25:1043
5. Kiguchi K, Henning CB, Chubb S, Huberman E (1990) J Biochem 107:8
6. Surowiec A, Stuchly SS, Izaguirre C (1986) Phys Med Biol 31:43
7. Jaroszynki W, Terlecki J, Sulocki J (1985) Studia Biophys 107:117
8. Foster KR, Schwan HP (1986) "C.R.C. Handbook of biological effects of electromagnetic fields" C. Polk and E. Postow (eds.), (C.R.C. Press Inc. Boca Raton)
9. Pethig R, Kell DB (1987) Phys Med Biol 32:933
10. Grant EH, Sheppard RJ, South GP Dielectric behaviour of biological molecules in solution (Clarendon Press, Oxford)
11. Irimajiri A, Doida Y, Hanai T, Inouye A (1978) J Memb Biol 38:209
12. Asami K, Hanai T, Koizumi N (1976) J Memb Biol 28:169
13. Asami K, Hanai T, Koizumi N (1986) Biophys J 31:215
14. Ballario C, Bonincontro A, Cametti C, Rosi A, Sportelli L (1984) Z Naturforsch 39c:160
15. Ballario C, Bonincontro A, Cametti C, Rosi A, Sportelli L (1984) Z Naturforsh 39c:1163
16. Bordi F, Cametti C, Di Biasio A (1990) Biochim Biophys Acta 1028:201
17. Formisano S, Johnson MI, Lee G, Aloj SM, Edelnoch H (1979) Biochemistry 18:1119

Authors' address:

C. Cametti
Dipartimento di Fisica
Università degli Studi "La Sapienza"
00185 Roma, Italy

Progress in Colloid & Polymer Science   Progr Colloid Polym Sci 89:324—327 (1992)

Dielectric dispersion in the ripple phase of DPL mixtures in water

C. Cametti, F. De Luca, A. D'Ilario, G. Briganti, M. A. Macrì*), and B. Maraviglia

Dipartimento di Fisica, Università degli Studi "La Sapienza", Roma, Italy
*) Istituto di Fisica Medica, Università di Chieti, Italy

Abstract: We have investigated, by means of radiowave dielectric spectroscopy measurements, the ripple phase of two different phospholipid — water mixtures (DPPC and DMPC), each of them containing different amounts of DPPE or DMPE, respectively, up to 7% wt/wt of total lipid. — Addition of DPPE or DMPE to the phospholipid aqueous systems should reduce the effective head group area and, hence, the orientational head group dipole correlation. The results, showing a marked damping of the dielectric increment at the pretransition in relation to the DPPE or DMPE concentration, provide further evidence that the anomalous polarization observed in these systems at the pretransition temperature must be attributed to the bilayer ripple modulation.

Key words: Phospholoipids; dielectric relaxation; ripple phase

Introduction

The phase behavior of lipid bilayer systems, like liposomes, vesicles or lamellae in excess water are extensively studied as simplified model systems in order to understand the structural and dynamical properties of biological membranes [1, 2].

Highly hydrated phospholipid bilayers display a large variety of structures, the most common of which is a repeated bilayer-water phase consisting of bimolecular leaflets of lipids separated by regions of water. These bilayers undergo many structural thermal transitions [3] which appear crucial to the understanding of the transport properties, the immunological activity, as well as the cell growth of biomembranes. At appropriate water concentration, lamellar phases exhibit two thermotropic phase transitions, the temperatures of which depend on the phospholipid chemical composition. The main transition, which takes place at high temperature, is mainly associated with the reduction of some of the degrees of freedom within the hydrocarbon chain, and the pre-transition, or ripple phase, which occurs at temperatures slightly below the main transition [4], is characterized by in-phase modulation of the layers.

It is generally accepted that the ripple phase is originated by a thermally-activated ripple distortion of the lipid lattice, as indicated, at various extents, by different experimental techniques [5—8]. However, the origin of the intermolecular interactions responsible for the bilayer corrugation is still unclear, especially the role of the chemical composition of the zwitterionic head groups [9—12].

Recently, on the basis of radiowave dielectric measurements, we have associated [11—13] the ripple phase to a thermally fluctuating rippled deformation of the bilayer, which may be decomposed in spatial modes called "ripplons" which are driven by the viscoelastic properties of the lipid-water interface.

To further investigate the structure and the properties of the pretransition phase, we have undertaken the present study by means of dielectric spectroscopy in the radiowave range on two phospholipid-water mixtures, dipalmitoylphosphatidiyl-choline (DPPC) and dimyristoylphosphatidylcholine (DMPC), to which different amount of dipalmitoylphosphatidiylethanolamine (DPPE) and dimyristoylphosphatidylelhanolamine (DMPE) has been added, respectively, up to 7% wt/wt of total lipid. The DPPC and DMPC molecules have the same polar group (phosphatidylcholine) with the hydrophobic chain composed of 16 and 14 atoms, respectively, while DMPE and DPPE have a different head group (phosphatidyletholonamine) with respect to DMPC and DPPC and similar hydrophobic tails.

The aim of this procedure, towards more complexity in membrane model systems, is to prove whether the addition of DPPE to DPPC- and DMPE to DMPC-water mixtures, respectively, could modify the rotational motion of the neighboring head groups and, hence, the dipole correlation [11—14].

In this case, the dielectric polarization should be damped at lower frequencies, since DPPE or DMPE molecules should reduce the correlation length over which the dipoles are spatially correlated.

Experimental and analysis

The sample preparation and the dielectric and calorimetric measurements have been described in details elsewhere [12].

In Fig. 1 is shown the heat capacity curve of DPPC-water system with different amounts of DPPE in the temperature interval from 30° to 55°C, obtained by differential scanning calorimetry. The calorimetric measurements allow us to correlate the dielectric anomalies observed at the pretransition temperature with the existence of the ripple phase. All samples contain a lipid concentration of 25% wt/wt of the total weight. The pretransition peak (that, in pure DPPC-water mixture, appears at about 36°C) shifts toward the main transition temperature of DPPC (about 45°C) as the concentration of DPPE is increased up to 7% wt/wt. The DMPC + DMPE-water mixtures show a similar behavior.

In Figs. 2 and 3, the permittivity of DMPC + DMPE- and DPPC + DPPE-water mixtures as a function of temperature at a typical frequency (3 kHz) are shown.

For the two systems investigated, the correlation between the dielectric "hump" in the pretransition region and the calorimetric measurements has been confirmed. For DPPC it is evident from the data of Figs. 1 and 3.

The frequency-dependent behavior of the dielectric hump $\Delta\varepsilon$ for the DMPC + DMPE-water mixtures is shown in Fig. 4 at 21°C, within the pretransition region. A similar behavior is also shown by the DPPC + DPPE-water mixtures.

For both the systems investigated, the dielectric increment $\Delta\varepsilon$, associated to the dipole correlation,

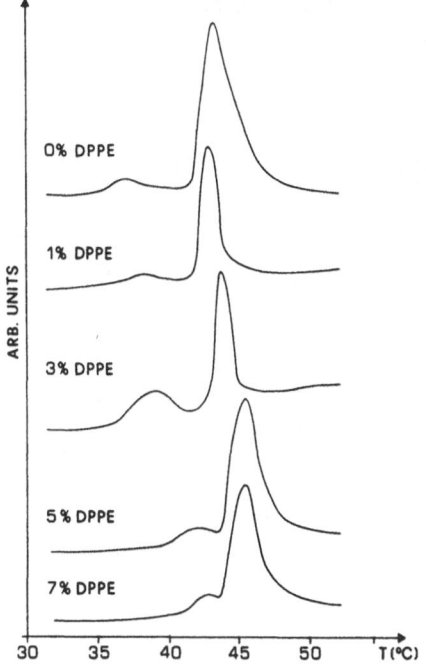

Fig. 1. The excess heat capacity of DPPC-water mixtures with varying concentrations of DPPE (up to 7% wt/wt) as a function of temperature. The scan rate was 8°C/min. The total lipid concentration is 25% wt/wt

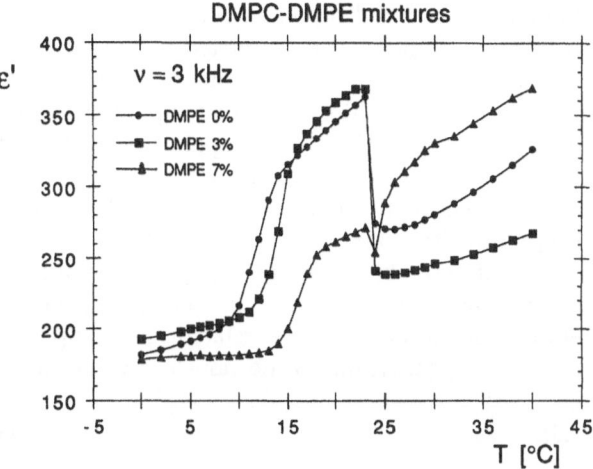

Fig. 2. The permittivity ε' of DMPC-water mixtures with varying concentrations of DMPE (up to 7% wt/wt) as a function of temperature at the frequency of 3 kHz. The total lipid concentration is 25% wt/wt

decreases with the increase of the DPPE or DMPE concentration, and at about 7% wt/wt the dielectric anomaly in the pretransition temperature interval disappears.

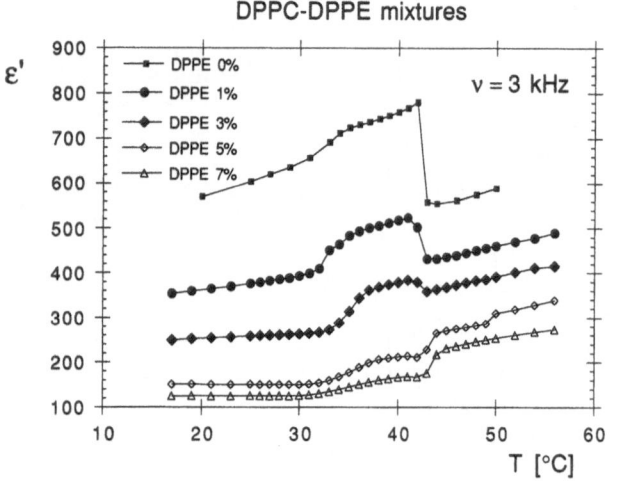

Fig. 3. The permittivity ε' of DPPC-water mixtures with varying concentration of DPPE (up to 7% wt/wt) as a function of temperature at the frequency of 3 kHz. The total lipid concentration is 25% wt/wt

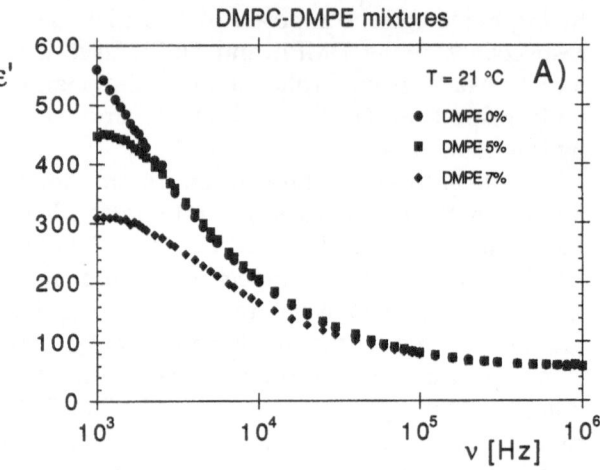

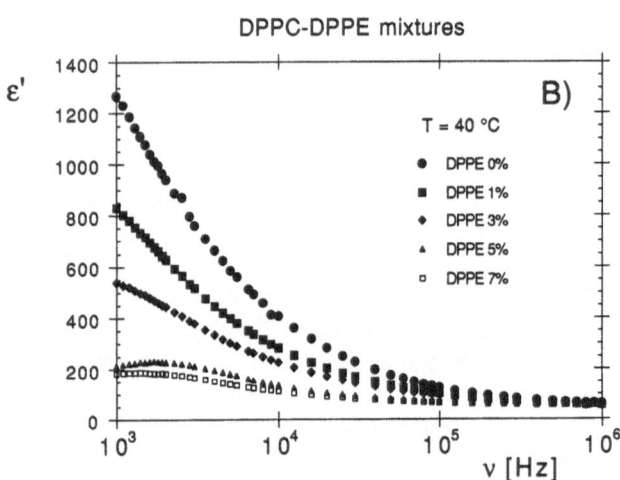

Fig. 4. The dielectric spectrum of a) DMPC-water mixtures with varying concentration of DMPE (up to 7% wt/wt) and b) DPPC-water mixtures with varying concentration of DPPE (up to 7% wt/wt), at selected temperatures within the pretransition — main transition temperature interval. The total lipid concentration is 25% wt/wt

The appearance of the dielectric anomaly in the pretransition region has been attributed to a molecular mechanism involving the bilayer head groups which give rise to the formation of extended domains of ordered dipoles [15]. The existence of these large domains can be related to the capillary waves which act as an orientational coherence transfer on the zwitterionic head dipoles [12]. As previously reported [14], the spatial extension of these in-phase dipole domains increases roughly by a factor 2—3 in the temperature interval between the pretransition and the main transition.

For symmetric bilayers separating identical fluids, the permittivity hump $\Delta\varepsilon$ in the pretransition temperature interval can be related to the ripplon frequency ν_r according to the power law [12, 16]

$$\Delta\varepsilon = A\nu_r^{-2/3}, \tag{1}$$

where the prefactor A includes numerical factors depending on temperature and lipid concentration.

Equation (1) gives a very good qualitative agreement with the experimental data in the case of pure DMPC- and DPPC-water mixtures [12, 14]. In the present case, for DMPC + DMPE- and DPPC + DPPE-water systems, Eq. (1) cannot be fully applied since the correlation factor that takes into account the dipole domain extension depends on the concentration of the DPPE or DMPE molecules.

In fact, while the dispersion relation of the ripplons is not appreciably modified at the low concentration of the dopant employed, the dipole correlation may occur on a spatial scale determined by the average distance between DMPE or DPPE molecules, respectively, in the bilayer lattice of the DMPC- or DPPC-water mixtures.

This means that the constant A (Eq. (1)) depends on the concentration of the dopant lipids and assumes values lower than those for the DMPC- and DPPC-water mixtures containing only one kind of lipid.

Moreover, since the average separation between dopant molecules is inversely proportional to its concentration [17], the dipole correlation extensions, which are mainly affected by the "impurities," are those related to the greater wavelength and, hence, to the lower frequency part of the dielectric spectrum. This behavior is consistent with the experimental findings shown in Fig. 4.

Conclusions

The model described here, despite its siplification, accounts for many of the observed characteristics of the DPPC + DPPE- and DMPC + DMPE-water systems. In the pure systems, without adding DPPE or DMPE molecules, the increment in the permittivity in the pretransition temperature region is directly associated with a change in the dipolar orientation due to the ripple deformation. Namely, the in-phase dipole domains are spatially extended on lengths of the order of the ripplon wavelength, whose full spectrum modulates the dielectric response. In this infinite plane bilayer approximate model, no boundary condition was imposed on the ripplon spectrum, because of the complexity of the effective lipids aggregation (lamellae of finite extension, multilayers). However, the introduction of adequate boundary conditions is expected to modify the low-frequency part (<1 kHz) of the dielectric spectrum [12].

The addition of DPPE or DMPE molecules to the DMPC or DPPC bilayers, on the assumption that the doping does not pose any inhibition of the ripplon propagation, neither appreciably modify the viscoelastic properties, nor that of the lipid aggregation, but does reduce the spatial extension of the dipole correlation, with the consequence that, as the dopant concentration is increased, the permittivity spectrum tends to be damped from low to high frequencies.

This means that, owing to the uniform distribution of the DPPE or DMPE molecules over the DMPC and DPPC bilayer lattice without the formation of clusters [17], the amplitude of the power law in Eq. (1) should be modulated by an appropriate filter function. Despite the difficulty in determining the dependence of the prefactor on all the effective parameters, crude fits to the data shown in Fig. 4 reveal that the basic characteristics of the DMPC- and DPPC-water bilayers at the pretransition can be qualitatively described in the framework of the ripplon model. It should be further developed to include more information about the molecular structure of lamellae distorted by periodic ripples.

Acknowledgement

We wish to acknowledge Prof. G. D'Ilario of the Chemistry Department at the University of Rome "La Sapienza" for his courtesy in performing the calorimetric measurements.

References

1. Nagle JF (1980) Rev Phys Chem 31:157
2. Nagle JF, Acott HL (1978) Physics Today 31:38
3. Cevc G, Marsh D (1978) Phospholipid Bilayers. Physical Principle and Models; Wiley, London
4. Malliaris A (1988) Progr Colloid Polym Sci 76:176
5. Atrenk LM, Westermann RW, Vaz NP, Doane JW (1985) Biophysical J 48:355
6. Enders A, Nimtz G (1984) Ber Bunsenges Phys Chem 88:512
7. Zasadzinski JAN, Achneider MB (1987) J Phys 48:200
8. McEihaney RN (1982) Chem Phys Lipids 30:229
9. Doniach S (1979) J Chem Phys 70:4587
10. Carlson JM, Aethna JP (1987) Phys Rev 36:3359
11. Cametti C, De Luca F, Macrì MA, Briganti G, Maraviglia B, Sorio P (1988) Liquid Cryst 3:839
12. Cametti C, De Luca F, D'Ilario A, Macrì MA, Maraviglia B,Sorio P (1990) Liquid Cryst 7:571
13. Cametti C, De Luca F, Maraviglia B, Sorio P (1985) Chem Phys Lett 118:626
14. Cametti C, De Luca F, D'Ilario A, Macrì MA, Briganti G, Maraviglia B (1991) Progr Colloid Polym Sci 84:465
15. Kaatze U, Gopel K, Pottel R (1985) J Chem Phys 89:2565
16. Crilly JF, Earnshaw JC (1983) Biophys J 41:197
17. Blume A, Ackermann T (1974) FEBA Lett 43:71

Authors' address:

C. Cametti
Dipartimento di Fisica
Università degli Studi "La Sapienza"
00187 Roma, Italy

Progress in Colloid & Polymer Science

Progr Colloid Polym Sci 89:328—331 (1992)

Structural change in AOT reverse micelles induced by changing the counterions

C. Petit[1,2]), P. Lixon[2]), and M. P. Pileni[1,2])

[1]) Université P. et M. Curie, Laboratoire S.R.S.I. Paris, France
[2]) C.E.N. Saclay, DRECAM.-S.C.M, Gif sur Yvette, France

Abstract: Divalent (2-ethyl hexyl) sulfosuccinate surfactant is used to form water in oil aggregates. By increasing the water content pronounced structural changes are observed. At low water content spherical droplets are formed. By increasing the water content, cylindrical aggregates are formed. The maximum amount of water solubilized using divalent surfactants is relatively low (the limit value of the ratio of water over AOT concentration w is less than 7, using divalent surfactants, and is equal to 60 using sodium AOT). At higher water content a phase transition is observed with two isotropic phases, with the lower phase containing mainly the surfactant.

Key words: AOT; microemulsion; structure; SAXS

Introduction

Sodium di(2-ethyl hexyl) sulfosuccinate is a very suitable surfactant to form spherical and monodisperse reverse micelles [1]. The micellar size exhibits a pronounced dependence on the water-content w = $[H_2O]/[AOT]$. Constant collision of the micelles allows the exchange of solute located in the water-pool. These properties lead to the development of chemical reaction in reverse micelles [1]. While the structure and properties of Na(AOT) reverse micelles are well-known, only a few papers recognize the influence of the AOT counterions on the physicochemical properties of the reverse micelles [2—4].

In the present paper, we report the change in the structure of divalent di(2-ethyl hexyl) sulfosuccinate-water-isooctane aggregates induced by varying the counterions. The surfactant used are cobalt {Co(AOT)$_2$}, copper {Cu(AOT)$_2$}, and cadmium {Cd(AOT)$_2$} di(2-ethyl hexyl) sulfosuccinate, respectively.

Results and discussion

The preparation of the $X(AOT)_2$ is carried out by ion exchange as described previously [5]. At low

water content an optically clear phase is obtained by solubilizing water in divalent AOT-isooctane solution. By increasing the amount of water in the system a phase separation is achieved; the two phases are isotropic and not birefringent. The behavior observed by solubilizing one of the three surfactants (Co(AOT)$_2$, Cu(AOT)$_2$, Cd(AOT)$_2$) in isooctane and by adding water is similar. As an example, we present the results obtained with Co(AOT)$_2$.

A) One phase solution:

In divalent AOT-water-isooctane solution the maximum amount of water still giving a homogeneous phase is such lower than the one observed for the sodium AOT (w = 60): The maximum water solubility is equal to 6.5, 7.5, and 5.5 for Co(AOT)$_2$, Cu(AOT)$_2$ and Cd(AOT)$_2$, respectively. Above this threshold value, two different optically clear and isotropic phases are observed. Photoabsorption spectra show that the cobalt, in contrast to the copper and cadmium ions, is not totally hydrated when the phase transition takes place. Below w = 5, the conductivity is unchanged and increases drastically at w up to 5. This indicates macroscopic changes in the structure of the aggregates. The scattered x-ray intensity at low angles was measured from 0.025 to 0.5 $Å^{-1}$ (which corresponds to a resolution in real space from 10 to 250 Å), using Cu $K\alpha$ radiation (1.54 Å). The experimental set-up has been described previously [6]. A log-log plot of the scattered intensity ($Ln(I(q))$ vs $Ln(q)$) shows a characteristic dependence of $I(q)$ on q^{-4} at large values of q. This is characteristic of a sharp interface and allows the determination of the total interface Σ in the system. Hence, by using Porod's equation [7], we can estimate the surface per AOT head σ, since the AOT concentration is known. No significant changes in the value of σ are observed in comparison to that obtained with sodium AOT (at $w = 5$, $\sigma = 24.5 \pm 0.5$ Å). This shows that the phase transition observed with divalent surfactants cannot be related to a change in the spontaneous curvature of the surfactant film, but could be explained by an increase of the interactions between the micelles.

The SAXS experiments described two structures varying with the water content:

i) w = 2.5: The behavior of the intensity $I(q)$, as observed with $Co(AOT)_2$-isooctane-water solution, is characteristic for a spherical structure (Fig. 1A): the radius of the microaggregate determined from the slope of the Guinier plot $Ln(I(q))$ versus q^2 is similar to that obtained from the minimum of the Porod plot $I(q) \cdot q^4$ versus q ($q \cdot R_{min} = 4.5$ for a sphere), and from the characteristic volume determined from the $I(q = 0)$ extrapolation. A good agreement between the experimental and simulated curves of the sattering intensity is obtained, especially at low q value. as is usual such case, the absence of the minima in the experimental curves is due to some polydispersity in form or in size [8]. These data confirm the formation of spherical reverse micelles at low water content ($w = 2.5$). Similar results have been observed with the copper and the cadmium sulfosuccinate surfactant. Table 1 gives the value of the water pool radius deduced for the three surfactants with the various counterions. There is no significant difference from pure AOT reverse micelles at very low water-content [6].

ii) w = 5 (near the phase transition): The behavior of the scattering spectrum, at very small wave vectors, no longer indicates a spherical structure. Especially in the $Ln(I)$ vs $Ln(q)$ plot, it can be noticed that the scattered intensity decreases with q^{-1}. This is a

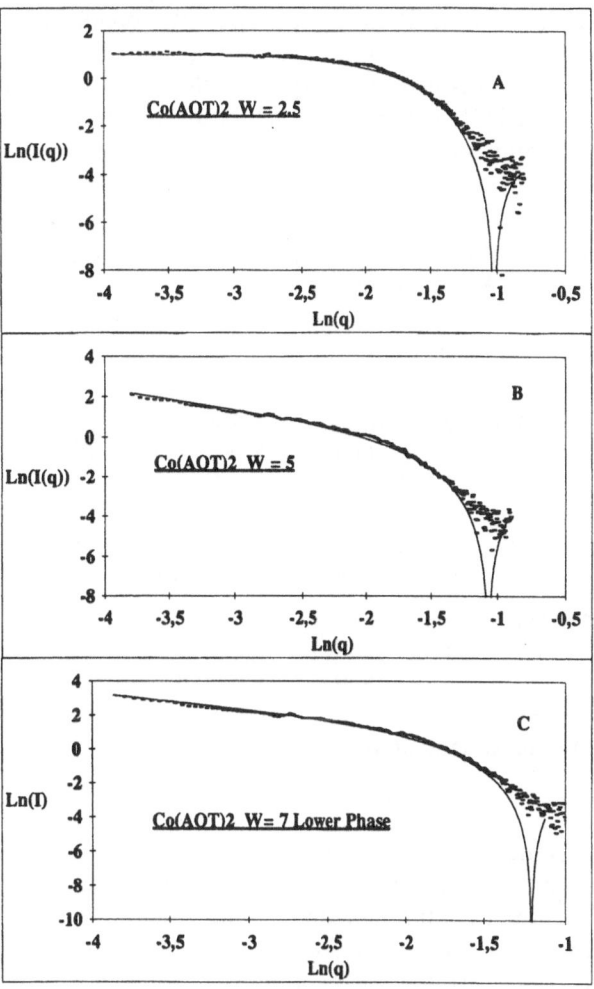

Fig. 1. Variation of the scattering intensity with q, $[Co(AOT)_2] = 0.05$ M A: $w = 2.5$; B: $w = 5$; C: $w = 7.5$ in the lower phase; (— theory; ∞ experiment)

characteristic of either a cylindrical structure or of a strong interaction between droplets [7] (Fig. 1B). Similar behavior is observed after dilution of the surfactant concentration by a factor 4. This confirms that the intesnsity $I(q)$ at low q values is not due to attractive interactions between the aggregates, but rather to the cylindrical shape of the aggregates. With the radius obtained by the minimum of the Porod's plot ($qR_{min} = 3.9$ for cylinders), we have simulated the x-ray scattering of cylinders and here, again, a good agreement at the absolute scale can be seen between the simulated curve and the experimental data. The radius determined from the slope of the linear relationship by plotting $Ln(q \cdot I(q))$ versus q^2 is also in good agreement with that deduced from Porod's plot. From the break in the linearity of this plot, the average length of the cylinder h is estimated [7] ($q = \pi/h$) and is given in the Table 1, which summarizes the results for the different samples.

Table 1. Change in the structural parameter of the different $X(AOT)_2$ reverse micelles

X-$(AOT)_2$	W	Structure	Radius (Å)	Length (Å)
$Co(AOT)_2$	2.5	Sphere	12.5 +/− 1	—
	5	Cylinder	7 +/− 0.5	≈30
	7.5	Cylinder*)	9 +/− 1	≈35
$Cu(AOT)_2$	2.5	Sphere	10 +/− 1	—
	5	Cylinder	7.5 +/− 1	≈30
	8.5	Cylinder*)	8.5 +/− 1	≈35
$Cd(AOT)_2$	2.5	Sphere	12 +/− 0.5	—
	5	Cylinder	7.5 +/− 0.5	≈30
	7.5	Cylinder*)	8 +/− 1	≈35

*) Studies done on the lower phase.

Hence, the structure of the micellar system with divalent counterions can be changed by increasing the water content when no changes are observed in the same conditions for the pure Na-AOT revers micelles. This effect should be a result of the presence of the divalent counterions and to the change in the geometry of the micelles induced with the change of AOT counterions.

B) Phase transition

By increasing the amount of water above 6.5, the system undergoes a transition and two isotropic and no birefringent liquid phases appear. Close to the phase transition the volume of the two phases is equal. By increasing the amount of water, the volume of the lower phase decreases. Above $w = 15$, the lower phase becomes turbid. This behavior is very different from what is observed in pure AOT reverse micelles: the transition is obtained at very high water-content ($w \approx 60$) and the lower phase is a turbid lamellar phase [1]. Similar phenomena are observed using copper and cadmium AOT. The phase transition is then observed for a w value equal to 7.5 and 5.5, respectively. The turbidity of the lower phase appears for a w value close to 12. These differences in the amount of water required to reach the phase transitions could be due to the differences in the hydration or to the relation between the ratio of the charge over the ionic radius [2].

The conductivity of the lower phase is much higher as compared to the homogeneous phase and changes with the kind of counterions (at $w = 8$, $\Lambda = 13\,000$ nS, 133\,000 nS, and 50\,000 nS for $Co(AOT)_2$, $Cu(AOT)_2$ $Cd(AOT)_2$, respectively). It increases with the water content and with the surfactant concentration in the lower phase. These strong differences in the conductivity obtained for the various surfactants can be partially attributed to the differences in the surfactant concentration in the lower phase. However, the physical properties of the ions have to be taken into account and, particularly, the degree of hydration of the counter-ions.

In the upper phase, with cobalt and copper ions, the AOT concentration is too low to form aggregates. From this point of view, such a transition is different from a simple liquid-gas transition.

In the lower phase, the variation of the scattered intensity with the wave vector q is always characteristic for cylinders (Fig. 1C). By plotting $Ln(q \cdot I(q))$ versus q^2, the average radius and the persistence length of the cylinders are deduced [7]. Table 1 summarizes the values obtained with the three surfactant systems. Because of the high conductivity of the lower phase, it seems reasonable to conclude that these cylinders are interconnected by a rigid section similar to that obtained just before the phase transition. Hence, the transition appears as a condensation of the cylindrical reverse micelles. The mechanism of such condensation is not clear. This could be explained in terms of charge distribution inside the aggregate: the formation of cylinder implies changes in the curvature of the interface. This could induces changes in the charge distribution with appearance of dipole moments in the water pool. This could induce strong enough interactions between droplets to reach a phase transition. However, no evident conclusions can actually be reasonalby claimed. The formation of different chelates by the divalent ions cannot be excluded [3].

Conclusion

Changes in the nature of the counterions of the AOT reverse micelles strongly perturbs the phase diagram of the system. Using divalent derivatives, we have shown that the limit in water solubilization drastically decreases in comparison with that of Na(AOT), depending on the counterion, and never reaches a value up to w 8. The shape of the aggregate changes with the water content. At low water content droplets are formed. The increase in the water content induces the formation of cylinders which are able to be connected at higher water content. On increasing the water content of the solution, a two-phase system is formed that consists of two optically clear and isotropic phases.

References

1. Pileni MP (ed) Structure and reactivity in reverse micelles (1989) Elsevier
2. Bardez E, Lorrey B, Zhu XX, Valeur B (1991) Chem Phys Lett (in press)
3. Eicke HF, Kivita P (1984) In: Luisi PL, Straub BE (eds) Reverse Micelles, Plenum Press, New York
4. Kitahara A, Ohashi O, Kijiro K (1969) J Coll Int Sci 29:48 and ibid (1974) 49:108
5. Petit C, Lixon P, Pileni MP (1990) J Phys Chem 94:1598
6. Pileni MP, Zemb T, Petit C (1985) Chem Phys Lett 118:414
7. Small Angles Scatterings of X-rays (1955) In: Guinier A, Fournet G (eds) Wileys and Sons: New York
8. Hilficker R, Eicke HF, Sager W, Steeb C, Hofmeier U, Gehrke R (1990) Ber Bunsenges Phys Chem 94:677

Authors' address:

Dr. C. Petit
Laboratoire S.R.S.I.
Université P. et M. Curie
Bât F
4 Place Jussieu
75005 Paris, France

Progress in Colloid & Polymer Science   Progr Colloid Polym Sci 89:332 (1992)

Dis-interdigitation of phospholipid bilayers by low amounts of cholesterol

P. Laggner[1]), R. Koynova[2]), and B. Tenchov[2])

[1]) Institute of Biophysics and X-Ray Structure Research, Austrian Academy of Sciences, Graz, Austria
[2]) Central Laboratory of Biophysics, Bulgarian Academy of Sciences, Sofia, Bulgaria

Interdigitation of hydrocarbon chains in phospholipid bilayers is generally considered to be a consequence of the geometric shapes of the constituent molecules [1]. Thus, DHPC (dihexadecylphosphatidylcholine) with its large mismatch between the areas required by the polar headgroups and the hydrocarbon chains, respectively, spontaneously forms interdigitated lamellar crystalline and gel phases [2]. Cholesterol at high concentrations (above 20 mol-%) is known to eliminate interdigitation. For a better understanding of the nature of this interesting effect and its possible physiological implications it is necessary to know the minimum amount of cholesterol sufficient to perturb or completely remove hydrocarbon chain interdigitation.

In the present work, we have studied by means of small-angle x-ray diffraction and differential scanning calorimetry the phase diagram of the fully hydrated DHPC/cholesterol binary mixture. An important result of this study is that cholesterol substantially perturbs the DHPC interdigitation already at concentrations as low as 0.1 mol-%, where a coexistence of interdigitated and non-interdigitated lamellar gel phases is observed. The coexistence range extends up to 5 mol-% of cholesterol. This concentration is sufficient to completely eliminate the interdigitated phase of DHPC.

Since these data cannot be sufficienctly explained within the framework of the geometric shape concept, we suggest that an important role in the mechanism of cholesterol action is played by the unfavorable line-boundaries between the interdigitated and non-interdigitated lipid domains.

References

1. Siminovitch DJ, Ruocco MJ, Makriyannis A, Griffin RG (1987) Biochim Biophys Acta 901:191—200
2. Laggner P, Lohner K, Degovics G, Müller K, Schuster A (1987) Chem Phys Lipids 44:31—60

Authors' address:

P. Laggner
Institute of Biophysics and x-ray structure research
Austrian Academy of Sciences
Steyrergasse 17
A-18010 Graz, Austria

Progress in Colloid & Polymer Science   Progr Colloid Polym Sci 89:333 (1992)

X-Ray studies on the near-equilibrium pathways of the pretransition in DPPC multilayer liposomes

A. Bóta, K. Lohner, and P. Laggner

Institute of Biophysics and X-Ray Structure Research, Austrian Academy of Sciences, Graz, Austria

Previous studies by millisecond time-resolved x-ray diffraction have demonstrated the existence of short-lived ordered layer structures as intermediates in the pretransition of saturated phosphatidylcholines (DPPC and DMPC) [1]. This result has been obtained under non-linear nonequilibrium conditions (IR-laser T-jump). The present work is aimed to charaterize the structural pathways of this transition under conditions close to equilibrium by small- and wide-angle x-ray diffraction.

In order to separate the effects of time and temperature in the thermal history of the hydrated lipid samples, a special incubation technique with a multi-temperature thermostat block has been used. In this technique, rather than sequentially exposing one sample to different temperatures, a series of samples is simultaneously held at different temperatures for controlled periods of time, up to 4 weeks. This has proven to be essential in studying transitions with complex slow kinetics.

It has been found that the relaxation kinetics depend on the temperature difference between quenching temperature and T_m in the sense of the linear-response concept. Near T_m, relaxation times of up to 5 days were observed. Neither during nor after equilibration could we observe intermediate ordered layer lattices, as seen previously by T-jump studies. To the contrary, lamellar lattice order is strongly reduced in the range of the pretransition, even after full relaxation. This indicates that the occurrence of ordered intermediates is not an equilibrium-thermodynamic, but rather a kinetic effect.

A pronounced memory-phenomenon has been observed in cycling through the pretransition: Samples that had been incubated for 8 days at temperatures around the pretransition, cooled into the $L_{\beta'}$ gel-phase (28 °C, 3 days) and then heated to 38 °C ($P_{\beta'}$), resulted in different degrees of lattice-order, despite the fact that in the gel-phase they all showed the normal, sharp diffraction pattern. Samples that had originally been incubated at the pretransition and hence, been disordered, did not recover the full appearance of the ripple phase. It is, thus, problematic to decide between frustrated and equilibrium nature in the discussion of the phases connected by the pretransition.

References

1. Laggner P, Kriechbaum M (1991) Chem Phys Lipids 57:121—145

Authors' address:

A. Bóta
Department of Physical Chemistry,
Technical University Budapest,
Egry J. u. 20—22, 1111 Budapest, Hungary

Progress in Colloid & Polymer Science   Progr Colloid Polym Sci 89:334 (1992)

Low-dose effects of melittin on phospholipid structure. Differences between diester and diether phospholipids

A. Colotto, K. Lohner, and P. Laggner

Institute of Biophysics and X-Ray Structure Research, Austrian Academy of Sciences, Graz, Austria

Metlittin, the main peptide component of bee venom, interacts with natural and artifical membranes and causes long-range effects, already at very low peptide-to-lipid ratios [1]. To further elucidate these effects and to determine specificities in the interaction with different lipid species, we have studied by x-ray diffraction and calorimetry the complexes formed by melittin and DPPC and DHPC (dipalmitoyl- and dihexadecylphosphatidylcholine, respectively) under different temperatures and hydrations.

The two lipid systems differ from each other merely by the linkage of the hydrocarbon chains to the glycerophosphorylcholine polar headgroups. While the liquid-crystalline L_a-phase is structurally similar in both systems, the gel-phase of DHPC differs in that it shows interdigitated hydrocarbon-chain packing [2], whereas DPPC has normal bilayer structure with tilted hydrocarbon chains arranged in a quasi-hexagonal subcell-lattice.

The results show significant differences in the interaction with melittin between the two systems, which are most pronounced in the L_a-phase.

For DPPC, the data show a rearrangement of the hydrocarbon chains into an untilted structure. This takes place already at peptide-to-lipid molar ratios (R) of less than 10^{-3} and is complete at $R = 5 \times 10^{-3}$. Above the main transition temperature, the L_a-structure persits up to this ratio.

For DHPC, the interaction with melittin leads to a transition from interdigitated to non-interdigitated structure in the gel-phase, again at R below 10^{-3}. Parallel to this effect, the multilamellar lattice loses coherence. Above the main transition temperature, non-lamellar (cubic or hexagonal) phase structures are observed. Their appearance starts already at $R = 10^{-4}$, depending on the degree of hydration in the range of 70—90% water.

References

1. Posch M, Rakusch U, Mollay C, Laggner P (1983) J Biol Chem 258:1761—1766
2. Laggner P, Lohner K, Degovics G, Müller K, Schuster A (1987) Chem Phys Lipids 44:31—60

Authors' address:

A. Colotto
Institute of Biophysics and
X-Rray Structure Research
Austrian Academy of Sciences
Steyrergasse 17
A-8010 Graz, Austria

Progress in Colloid & Polymer Science   Progr Colloid Polym Sci 89:335 (1992)

Fractal dimensions and BET-surfaces in wet and dry portland cement paste

G. Degovics[1]), P. Laggner[1]), and J. Tritthart[2])

[1]) Institut für Biophysik und Röntgenstrukturforschung, Österreichische Akademie der Wissenschaften, Graz, Austria
[2]) Institut für Werkstoffkunde, Festigkeitslehre und Materialprüfung, Technische Universität Graz, Austria

Small-angle x-ray scattering has proven to be a suitable, nondestructive method to analyze the submicroscopic pore structure of Portland cement paste [1]. Thus, the process of age-hardening in this material has been shown to be associated with an asymptotic increase in the mass fractal dimension D_m from 1.9 (first day after preparation) to 2.8 (28 days). A particular advantage over other techniques, such as vapor adsorption porosimetry, lies in the fact that samples can be investigated in the wet state. The present investigation, by systematic scanning of the time development of the x-ray small-angle scattering behavior of samples prepared under controlled conditions, was aimed at answering the following questions:

— does exhaustive drying, as used for BET measurements, by itself cause structural changes in cement microstructure?
— what are the effects of NaCl and $CaCl_2$ on the inner structure of cement paste?
— can a relationship between BET-specific inner surfaces and the fractal dimension be established?

The results have shown that all samples, wet or dry, can be described as mass fractals in the range of linear dimensions between 20 and 1000 Å. Generally, the wet samples showed lower D_m-values, indicating that drying leads to the formation of coarse particles from the gel. This would correlate with our observation that the speed of drying has pronounced effects on the measured BET-surface areas. The difference between wet and dry states is also significant with respect to the effects of NaCl and $CaCl_2$ (in the concentration range between 0.5 to 1.5%). There, the wet samples showed a small decrease in D_m with increasing salt concentration, indicating the stabilization of a more ramified structure. In the dry state, however, both salts (NaCl more strongly than $CaCl_2$) lead to an increase in D_m.

Concerning the comparability of fractal dimensions and BET-surface areas, as measured by small-angle x-ray scattering and gas-sorption, respectively, it must be concluded that a qualitative agreement in trends (D_m vs. W/Z, and area vs. W/Z) is apparent in dry samples; in wet samples, however, differing D_m-values and trends are observed, thus, rendering a direct comparison of wet and dry samples impossible.

References

1. Kriechbaum M, Degovics G, Tritthart J, Laggner P (1989) Progr Colloid Polym Sci 79:101—105

Authors' address:

P. Laggner
Institute of Biophysics and x-ray structure research
Austrian Academy of Sciences
Steyrergasse 17
A-8010 Graz, Austria

Author Index

Subject Index